Mobile Learning Applications in Early Childhood Education

Stamatios Papadakis
University of Crete, Greece

Michail Kalogiannakis
University of Crete, Greece

A volume in the Advances in Educational Technologies and Instructional Design (AETID) Book Series

IGI Global
DISSEMINATOR OF KNOWLEDGE

Published in the United States of America by
IGI Global
Information Science Reference (an imprint of IGI Global)
701 E. Chocolate Avenue
Hershey PA, USA 17033
Tel: 717-533-8845
Fax: 717-533-8661
E-mail: cust@igi-global.com
Web site: http://www.igi-global.com

Library of Congress Cataloging-in-Publication Data

Names: Papadakis, Stamatios, 1974- editor. | Kalogiannakis, Michail, 1969- editor.
Title: Mobile learning applications in early childhood education / Stamatios Papadakis and Michail Kalogiannakis, editors.
Description: Hershey, PA : Information Science Reference, 2020. | Includes bibliographical references and index. | Summary: "This book explores the methods and applications of mobile learning techniques and strategies within diversified teaching settings"-- Provided by publisher.
Identifiers: LCCN 2019030349 (print) | LCCN 2019030350 (ebook) | ISBN 9781799814863 (hardcover) | ISBN 9781799814870 (paperback) | ISBN 9781799814887 (ebook)
Subjects: LCSH: Early childhood education--Computer-assisted instruction. | Mobile communication systems in education. | Computers and children. | Mobile apps.
Classification: LCC LB1139.35.C64 M64 2020 (print) | LCC LB1139.35.C64 (ebook) | DDC 372.210285--dc23
LC record available at https://lccn.loc.gov/2019030349
LC ebook record available at https://lccn.loc.gov/2019030350

This book is published in the IGI Global book series Advances in Educational Technologies and Instructional Design (AE-TID) (ISSN: 2326-8905; eISSN: 2326-8913)

British Cataloguing in Publication Data
A Cataloguing in Publication record for this book is available from the British Library.

All work contributed to this book is new, previously-unpublished material. The views expressed in this book are those of the authors, but not necessarily of the publisher.

For electronic access to this publication, please contact: eresources@igi-global.com.

Advances in Educational Technologies and Instructional Design (AETID) Book Series

Lawrence A. Tomei
Robert Morris University, USA

ISSN:2326-8905
EISSN:2326-8913

Mission

Education has undergone, and continues to undergo, immense changes in the way it is enacted and distributed to both child and adult learners. In modern education, the traditional classroom learning experience has evolved to include technological resources and to provide online classroom opportunities to students of all ages regardless of their geographical locations. From distance education, Massive-Open-Online-Courses (MOOCs), and electronic tablets in the classroom, technology is now an integral part of learning and is also affecting the way educators communicate information to students.

The **Advances in Educational Technologies & Instructional Design (AETID) Book Series** explores new research and theories for facilitating learning and improving educational performance utilizing technological processes and resources. The series examines technologies that can be integrated into K-12 classrooms to improve skills and learning abilities in all subjects including STEM education and language learning. Additionally, it studies the emergence of fully online classrooms for young and adult learners alike, and the communication and accountability challenges that can arise. Trending topics that are covered include adaptive learning, game-based learning, virtual school environments, and social media effects. School administrators, educators, academicians, researchers, and students will find this series to be an excellent resource for the effective design and implementation of learning technologies in their classes.

Coverage

- Web 2.0 and Education
- Social Media Effects on Education
- Virtual School Environments
- Digital Divide in Education
- Educational Telecommunications
- Bring-Your-Own-Device
- Higher Education Technologies
- K-12 Educational Technologies
- Classroom Response Systems
- Instructional Design Models

The Advances in Educational Technologies and Instructional Design (AETID) Book Series (ISSN 2326-8905) is published by IGI Global, 701 E. Chocolate Avenue, Hershey, PA 17033-1240, USA, www.igi-global.com. This series is composed of titles available for purchase individually; each title is edited to be contextually exclusive from any other title within the series. For pricing and ordering information please visit http://www.igi-global.com/book-series/advances-educational-technologies-instructional-design/73678. Postmaster: Send all address changes to above address.

Titles in this Series

For a list of additional titles in this series, please visit:
https://www.igi-global.com/book-series/advances-educational-technologies-instructional-design/73678

Next Generation Digital Tools and Applications for Teaching and Learning Enhancement
Mary T. Grassetti (Framingham State University, USA) and Julie Zoino-Jeannetti (Framingham State University, USA)
Information Science Reference • © 2020 • 300pp • H/C (ISBN: 9781799817703) • US $195.00 (our price)

Handbook of Research on Digital Learning
Matthew Montebello (University of Malta, Malta)
Information Science Reference • © 2020 • 394pp • H/C (ISBN: 9781522593041) • US $245.00 (our price)

Handbook of Research on Fostering Student Engagement With Instructional Technology in Higher Education
Emtinan Alqurashi (Temple University, USA)
Information Science Reference • © 2020 • 567pp • H/C (ISBN: 9781799801191) • US $285.00 (our price)

Emerging Techniques and Applications for Blended Learning in K-20 Classrooms
Lydia Kyei-Blankson (Illinois State University, USA) Esther Ntuli (Idaho State University, USA) and Mohamed A. Nur-Awaleh (Illinois State University, USA)
Information Science Reference • © 2020 • 310pp • H/C (ISBN: 9781799802426) • US $195.00 (our price)

Profiling Target Learners for the Development of Effective Learning Strategies Emerging Research and Opportunities
Shalin Hai-Jew (Kansas State University, USA)
Information Science Reference • © 2020 • 183pp • H/C (ISBN: 9781799815730) • US $145.00 (our price)

Cases on Models and Methods for STEAM Education
Judith Ann Bazler (Monmouth University, USA) and Meta Lee Van Sickle (College of Charleston, USA)
Information Science Reference • © 2020 • 379pp • H/C (ISBN: 9781522596318) • US $195.00 (our price)

Examining Multiple Intelligences and Digital Technologies for Enhanced Learning Opportunities
Robert Z. Zheng (University of Utah, USA)
Information Science Reference • © 2020 • 371pp • H/C (ISBN: 9781799802495) • US $195.00 (our price)

Postgraduate Research Engagement in Low Resource Settings
Dennis Zami Atibuni (University of Johannesburg, South Africa)
Information Science Reference • © 2020 • 353pp • H/C (ISBN: 9781799802648) • US $185.00 (our price)

IGI Global
DISSEMINATOR OF KNOWLEDGE

701 East Chocolate Avenue, Hershey, PA 17033, USA
Tel: 717-533-8845 x100 • Fax: 717-533-8661
E-Mail: cust@igi-global.com • www.igi-global.com

Table of Contents

Detailed Table of Contents

Substantive research investigates the effects and impacts of tablets, in particular iPads, on children's education, but few papers discuss support for teachers in deciding which iPad applications can be integrated into the classroom. Even fewer articles are directed towards application developers. This chapter explores two standards for choosing apps for children - the four-pillar model of Hirsh-Pasek et al. (2015) and the rubric for the evaluations of educational apps for preschool children (REVEAC) by Papadakis, Kalogiannakis, and Zaranis (2017). This chapter draws from two standards for choosing iPad applications for young children in the classroom and through analysis of two applications will propose the REVEAC for educators and developers while also suggesting specific features that developers could consider when targeting children under 5 years old in educational contexts.

The chapter presents the results of a project in which tablets were used for teaching natural sciences to kindergarten students. The classification of animals depending on certain characteristics was the subject matter. Forty-five students participated, divided into three groups. The first used printed material, the second used computers and webpages, and the third used tablets and AR applications. Bybee's 5Es provided the teaching framework for all groups. Data were collected using evaluation sheets and structured interviews. The students in the tablets' group performed better in all the evaluation sheets compared to the ones who were taught using printed material, but there were no statistically significant differences compared to the computers' group. A positive impact on motivation and enjoyment was noted in the tablets' group. Thus, it can be concluded that tablets are an interesting alternative teaching tool for very young students. Implications for research and practice are also discussed.

Children's use of touch-screen tablets is increasing as these mobile devices become increasingly available. The interactive, tactile touch-screen interface and easily downloadable applications make tablets especially popular with pre-schoolers. This chapter's literature review provides an overview of recent research into tablets and emergent literacy development at home and in the pre-school setting. A multi-faceted synthesis of research on children's use of tablets and its impact in emergent literacy skills was conducted. The evidence indicated that mobile learning via touch-screen tablets has the potential to broadly enhance emergent writing and may facilitate the development of letter name/sound knowledge, print awareness, letter writing and name writing skills, and phonological awareness. The relation between the use of tablets and emergent literacy development seems to be complex as it is mediated by factors such as the type of multisensory experiences through literacy applications and the type of scaffolding used by adults. Recommendations and directions for future research are discussed.

Studies suggest that the exposure to STEM learning opportunities early in life is important because the development of STEM skills can further students' interest and educational attainment in STEM, expanding their career choices later in life. Smart mobile devices have become ubiquitous in schools and have been transforming educational practices at all ages and levels and almost all over the world. At the same time, there is evidence that teacher education departments lack the knowledge and skill to teach pre-service teachers about using these devices in their daily teaching practice. The findings of this chapter underline the need to develop teaching and learning processes that go beyond a mere transmission of the technical knowledge required to use mobile technologies with educational purposes, focusing instead on raising students' awareness about the educational benefits that the integration of mobile technologies can bring to formal education.

A decade ago, Computational Thinking (CT) and coding were typically considered part of the secondary education programs, as the focus was on programming and algorithm development. The early childhood classroom was not exactly the area expected to find students-developed coding skills. But as has been the case lately, CT and coding have been characterized as fundamental skills of the 21st century, not only for computer scientists but for all citizens. Yet, through the application of developmentally appropriate technologies, the development of coding skills is increasingly possible, and the result may be the advancement of CT fluency or at least familiarity in young age children. Given the enormous success of smart mobile devices and accompanying mobile apps the rationale for this chapter is to investigate if there are apps that provide the children of preschool and pre-primary school age with opportunities to cultivate their foundational coding and CT skills.

This chapter discusses the use of Duolingo App to enhance children's linguistic intelligence. Linguistic intelligence is expertise in applying vocabularies effectively and efficiently. Duolingo is an application based on the Android platform which helps children mastering other languages that they can practice

This chapter determines the factors that significantly influence pre-service teachers' acceptance to use mobile devices as resources for learning in a university context. Based on the methodological framework of the UTAUT, a modified contextualized model of evaluation was created. A data collection instrument was designed, validated contextually, and optimized for mobile learning in higher education. A total of 320 Greek university students from a Faculty of Education participated in the study. Results demonstrated that the instrument constructed showed a notable internal consistency, with a high validity for data collection in 8 of its 9 factors. Results indicated Performance Expectancy, Effort Expectancy, Social Influence, Facilitating Conditions, Hedonic Motivation, and Empowerment in Learning as factors which affected participants' Behavioral Intention to use mobile technology for learning. Behavioral Intention, Social Influence and Empowerment in Learning affected university students' behavior to use mobile devices for learning.

This chapter investigated young children's use of tablets at home and parents' views on tablet benefits and their concerns. A questionnaire was completed by the parents of 100 children aged 4-6. Young children engage in a range of activities such as playing games (76%), watching cartoons (75%), listening to music (65%), watching videos (60%), and using educational apps (54%). Fewer children look at pictures/photos/books or take pictures, while visiting websites and using email are never carried out by most of the children. Gender and age had an occasional isolated impact on children's tablet activities. 4-5 year olds tend to do tablet activities with an adult, while 5.5-6.5 year olds with siblings or alone. Most parents "agree and strongly agree" that tablets "teach basic technology skills" (85%), "teach foreign languages", and "can make learning fun". The parents' main concerns included dependence, reduction of communication, and inappropriate content. Implications regarding links between home and kindergarten are discussed.

This chapter investigates primary school students' perceptions regarding the use of games on mobile devices (i.e., smartphone, tablet) in classroom environment for teaching purposes. Data was collected from 10,381 students (Grades 4, 5, and 6) using a survey questionnaire. A series of semi-structured interviews were used to more deeply understand student perceptions. Those perceptions tended to range from neutral to positive regarding the use of games and their content. Students prefer games that enhance their knowledge and develop their thinking skills. They also believe lessons will be more enjoyable and interesting through the use of games, and learning will be easier and more effective. The results showed differences in students' perceptions in terms of their gender and grade level. Implications of this study and further research are discussed.

Foreword

It is often said that digital media have not only expanded the access to knowledge but have also changed the nature of knowledge. The advent of the Internet and Internet-equipped media devices have affected the ways in which we create knowledge and thus the definitions of what it means to know and understand. The Internet has been central to a vision of democratic knowledge societies that value information-sharing and multimedia communication. Apps are an embodiment of this vision. Apps are built to simplify and accelerate access to personal and social information and as such, they expand and constrain our meaning-making. With over two million apps available to download from the Apple and Google Play stores, there is no doubt that apps have had a vast influence on the meaning-making of millions of young and old across the globe. Apps that are designed for mobile touchscreens have grown exponentially in number and kind since the global availability of smartphones and tablets. The percentage of touchscreens' ownership has gone up in all countries and among all sections of the population, including young children. In the USA, for example, 53% of children below the age of 11 own a smartphone (Common Sense Media, 2019). The ubiquity and pervasiveness of apps in daily modern life raise the question of future developments in global app-production and app use. As more and more service providers, transport companies and communication channels squeeze their presence into apps, will soon all schools, teachers and individual students have their own app?

In their basic definition, apps are mobile applications but in their broader, historically grounded definition, apps are specially designed software programs that perform distinct functions. An app, thus, can take the form of a Khan Academic learning suite with video lectures on various learning topics, but it can also be an electronic calculator, performing simple mathematical computations. The different functions and functionalities of apps provide what Giddens (2003) described as knowledge 'access points' and what Säljö (2016) adopted in his theorizing of the value of apps for learning. Säljö (2016) highlights that apps provide access to knowledge that amplifies but not necessarily deepens our understanding of the world. He points out that while most users don't understand the formula or algorithms behind a step-counting app, for example, such apps can still augment an individual's knowledge about their physical health and daily regime. If we apply this understanding to education and children's learning of new facts and relationships, we can better understand why apps, on their own, cannot transform education.

Apps do not deepen knowledge in a linear sense, as Säljö (2016) writes, apps 'transform *what it means to know*, learn and remember in society. And schooling has to take this into account.' (p. 12, emphasis is mine). As part of the wider technological landscape, apps are redefining what it means to learn and to know, but they are not changing knowledge. Failure to remember this important point can lead to misunderstandings and the mistaken assumption that providing access to apps equals learning. Apps are contemporary, popular *access points* to knowledge, but their value to advance students' learning is continuously up for debate.

In comparison to other technological "access points" to knowledge, apps are unique because they are mobile: the devices on which we access apps are mobile and at the same time, the information is globally transferred. This mobility grants apps the status of *knowledge-scaling tools*. Apps can scale out and scale up knowledge, that is they can be used to simultaneously deliver information to all children in a classroom (scaling out) and they can also amplify or increase the information individual children receive (scaling up). Personalised feedback is an example of scaling up of information with apps. In classrooms with a high teacher-pupil ratio, personalized feedback and tailored content is a major asset, and the key reason why apps have been popular teaching tools in developing countries, large public schools and areas with teacher shortages (Outhwaite, Faulder, Gulliford, & Pitchford, 2019). Indeed, the vision of mobile technologies to bridge learning environments, socio-economic differences and geographical limits, has become reality with Internet-equipped apps. But to what extent is the vision of app-mediated knowledge merely about 'access', paying lip service to learning sciences?

As several contributors to this volume would concur, app-mediated learning is a rhetoric in many public schools and kindergartens. Teachers and educators are often immersed in a pro-technology discourse, dominated by enthusiastic promises of the transformational learning power of apps (and other technologies). Several authors in this book illustrate that many of these promises have not been fulfilled and caution against an uncritical adoption of mobile apps in public education. At the same time, the authors portray educational scenarios in which apps can widen the learning options offered to students in pre- and primary schools. These scenarios play out when apps are chosen with a specific learning goal and clear intention of use, which correspond to a curriculum purpose and individual children's learning goals. How to make that choice is an open question that requires strong pedagogical and theoretical frameworks.

The focus in this book is on pedagogical scenarios, particularly those scenarios that support app-mediated learning in early years. Whether it is the educational context of literacy, science or coding, there are some apps and some features of apps that can be particularly beneficial for reaching pre-defined learning goals and supporting specific skills. So that teachers can leverage the added learning value of these app features, they need to be secure with their technology skills, they need to know where and how to choose apps and how to combine them with their lesson plans. This requires planning time and the need for professional development in technology more broadly and apps more specifically. Unfortunately, time and technology training are sorely missing in national public-school systems, with increasing demands on teachers pressured to teach to narrow, testable, curriculum objectives (Richmond, Burn, Dougill, Raleigh, & Traves, 2017). Without adequate provision and secure knowledge base, it is not surprising that teachers' motivation to use apps is often low.

In contrast, when children are given the choice between apps and analogue resources, they choose apps for specific learning tasks such as retrieving information about weather (Eisen & Lillard, 2016). Some students, especially those who are reluctant to read, are more motivated to read stories on the screen in the form of apps, rather than paper-based stories (Picton & Clark, 2015). Whether this motivation translates into actual learning gains is very much dependent on the quality of instruction and teachers' ability to integrate app-mediated learning with curricular objectives. The authors in this book show that the educators who recognize the opportunity to connect home and school learning via apps and local and global knowledge-exchange, can offer unprecedented learning opportunities to their students.

These learning opportunities are more likely to occur if apps are of high educational quality. Most app providers/developers promise children magical and unforgettable experiences with their products. However, the results of quality analyses of most popular educational apps fly in the face of such promises: the majority of apps were found inadequate, with design based on outdated concepts, promoting mechanistic and behaviouristic learning. As the authors in the opening chapters note, the quality of some apps labelled as "educational" is a far cry from the actual definition of education in learning sciences. While commercially commissioned reports on individual apps claim educational benefits, researchers' observations show that children's actual engagement with many of these apps is less stellar. The reasons are multiple but a clear pattern that runs through the authors' critique in this book is that of technology determinism: the designers' assumptions that an app (or a suite of apps) can replace competent face-to-face teaching and address the complex task of learning. Such technology determinism has been embedded in the educational technology for decades. Instead of positioning teachers as co-designers of technologies, many apps are designed to replace teachers. And yet, as the authors of this book rightly point out, an app's learning influence is fully contingent upon the context of its use and pedagogy. A technocentric view on apps falls short of addressing the web of relationships that govern the learning process occurring in classrooms. How to communicate this complexity to the teaching profession? Researchers have developed several frameworks and models to facilitate the difficult task of integrating technology with existing pedagogies and current learning theories. But so that teachers not only recognise but also desire and capture the complexity of learning with educational technology, we need to have more radical theories.

As more advances in technology design are made, more the human and material become entangled and more the artificial and human intelligences merge. This blending of nature- and man-made knowledge is effortless and yet multifaceted with apps. With aesthetically pleasing, personalized and cleverly designed apps, one might easily succumb to a technology hype. One such hype is the aforementioned technology determinism, the belief that technology can be developed to accommodate all human feelings and thoughts. Technology determinism is an important motivational belief for inventing and re-inventing tools. It has contributed to the creation of all kinds of technologies, beginning with basic tools such as the wheel in 3500 BCE or the Internet of Things in the 21st century. Technology determinism is important for the progress of humankind, but it is also a dangerous belief because it can begin to challenge our relationship with nature and with other human beings. If technology can extend all human thoughts and feelings, then it could, some might believe, replace the human all together. This extreme version of technology determinism becomes visible with apps that replace the teacher and software programs designed to manipulate behaviour through artificially intelligent algorithms. Here comes a point when we need to look critically at how we make sense of technology influence and how we theorise it for learning.

In an effort to build a comprehensive theory of educational technology, Professor Rupert Wegeriff from Cambridge University suggests considering the philosophy of technology. He interprets the philosophy of Simondon (1958, 1964, cited in Dumouchel, 1992) by drawing attention to two aspects of Simondon's account: the first aspect relates to individuation, which could be understood as an active back-and-forth dynamic exchange between technology properties and technology users. Users are individuals who have their own properties and characteristics and who infuse these into the technologies they interact with. Simondon calls this process individuation because in using technologies, whether for a personal or pedagogic purpose, we individualise the properties these technologies offer us. The second aspect of Simondon's theory is that technology mediates - and does not control – the relationship between humankind and nature. Technologies are part of a natural cycle of development where both man and nature are necessary and dependent on each other. This implies that technologies can be deeply personal and public at the same time, their functionalities can have simultaneously a subjective and objective value. Wegeriff (2019, online) writes: 'For Simondon technical objects always have a subjective as well as an objective side - they have a phenomenology'.

These are not lofty scholarly ideas – these ideas are essential to understand the role of apps in learning and in our everyday lives. If we conceive of apps as phenomenological tools that we use to experience knowledge, individually and collectively, we can begin to understand the concept of technology as a network, with both material (technological) and human (social) "actants" (Latour, 1996). From this perspective, connectivity is the principal property of technology and consequently, of app-mediated learning. Indeed, connectivity has been described as the key resource for human-machine communication (van Dijck, 2013). Through the lens of connectivity, individual apps are techno- and socio-cultural platforms that offer opportunities for agency and participation.

In my own work with apps over the past two decades (e.g., Kucirkova, 2014; Kucirkova, Messer & Sheehy, 2014), I have seen how apps became usable and useful through two principles of connectivity: agency and participation. Well-designed apps brought together researchers, designers and teachers in a collaborative effort to amplify and deepen knowledge and their own agency. These interactions, however, were positive only in the contexts of socio-technical entanglements, where all involved stakeholders (adults and children in their various roles of learners, teachers, designers and makers), could actively contribute to the socio-technological process of learning. Returning to my opening statement in this Foreword: the vision of apps as knowledge-amplifying tools can be realized if, and only if, we choose to use them with agency and participation in mind. Such choice would be a democratic choice, because as Feenberg (2012) pointed out, 'the fate of democracy is therefore bound up with our understanding of technology' (p. vii).

To conclude, the social and the technical are entangled in apps and we should not try to disentangle them in our observations as researchers, or in our pedagogies as teachers, or in our daily practices as users. Both aspects - the technological and the social - are moving and changing rapidly in the 21st century. This poses extreme challenges to our understanding of personal and collective values connected to technologies. The speed of socio-technical developments also pushes the limits of scientific capability to conduct studies that are relevant and at the same time, objective. In 2019, future vision for apps is uncertain. It can be either incremental, focused on amplifying knowledge through new access points, or it can be bigger and deeper, and persist as a step change that marks the beginning of a new technology era.

Natalia Kucirkova
University of Stavanger, Norway

REFERENCES

Common Sense Media. (2019). The Common Sense Census: Media Use by Tweens and Teens, 2019. *Common Sense Media*. Available online from: https://www.commonsensemedia.org/research/the-common-sense-census-media-use-by-tweens-and-teens-2019

Dumouchel, P. (1992). Gilbert Simondon's plea for a philosophy of technology. *Inquiry*, *35*(3-4), 407–421. doi:10.1080/00201749208602302

Eisen, S., & Lillard, A. S. (2016). Just Google it: Young children's preferences for touchscreens versus books in hypothetical learning tasks. *Frontiers in Psychology*, *7*, 1431. doi:10.3389/fpsyg.2016.01431 PMID:27713717

Feenberg, A. (2012). *Questioning technology*. London: Routledge. doi:10.4324/9780203022313

Giddens, A. (2003). *Runaway world: How globalization is reshaping our lives*. London: Profile Books.

Kucirkova, N. (2014). *iPads and tablets in the classroom: Personalising children's stories* (Vol. 41). UKLA.

Kucirkova, N., Messer, D., Sheehy, K., & Panadero, C. F. (2014). Children's engagement with educational iPad apps: Insights from a Spanish classroom. *Computers & Education, 71*, 175–184. doi:10.1016/j.compedu.2013.10.003

Latour, B. (1996). On actor-network theory: A few clarifications. *Soziale Welt*, 369–381.

Outhwaite, L. A., Faulder, M., Gulliford, A., & Pitchford, N. J. (2019). Raising early achievement in math with interactive apps: A randomized control trial. *Journal of Educational Psychology, 111*(2), 284–298. doi:10.1037/edu0000286 PMID:30774149

Picton, I., & Clark, C. (2015). *The Impact of Ebooks on the Reading Motivation and Reading Skills of Children and Young People: A Study of Schools Using RM Books. Final Report*. National Literacy Trust.

Richmond, J., Burn, A., Dougill, P., Raleigh, M., & Traves, P. (2017). *Curriculum and assessment in English 3 to 11: A better plan*. London: Routledge. doi:10.4324/9781315228464

Säljö, R. (2016). Apps and learning: A sociocultural perspective. In N. Kucirkova & G. Falloon (Eds.), *Apps, Technology and Younger Learners* (pp. 23–33). London: Routledge.

Simondon, G. (1958). *Du mode d'existence des objets techniques*. Paris: Aubier-Montaigne.

Simondon, G. (1964). *L'individu et sa genèse physico-biologuique: l'individuation à la lumière des notions de forme et d'information (No. 573.2)*. Presses universitaires de France.

Van Dijck, J. (2013). *The culture of connectivity: A critical history of social media*. Oxford University Press. doi:10.1093/acprof:oso/9780199970773.001.0001

Wegeriff, R. (2019). *Steps towards a theory of Educational Technology: Introducing Simondon*. Retrieved from http://www.rupertwegerif.name/blog/towards-a-theory-of-ed-tech-introducing-simondon

Preface

INTRODUCTION

In the last decade, there has been an increase in the use of smart mobile devices by young children around the globe (Kucirkova, Wells Rowe, Oliver, & Piestrzynski, 2017). The smart mobile device is a small-sized, portable computer that has a touch screen. Recently, these mobile devices have become increasingly powerful in terms of their processing, memory, and storage capabilities, allowing for sophisticated delivery of a range of digital content (McKenzie, Spence, & Nicholas, 2018).

Researchers worldwide believe that these interactive devices hold potential as powerful learning tools (Dorouka, Papadakis, & Kalogiannakis, 2019). A reason for that is their characteristics compared to the 'traditional' computers such as their intuitive interface, smart design, increased mobility, connectivity (GPS, Wi-Fi, 4G), multimodality and a huge range of applications among others (Falloon, 2014). Furthermore, the existence of a smart screen that responds to a range of tactile actions such as tap, swipe, and drag provides children a pleasant and playful experience (Kucirkova, 2017). Unlike mouse operated computers that require greater fine motor control, tablets remove this operational barrier and provide increased opportunities for learning even for children 2 years old and younger. It is not surprising that due to these user-friendly features, tablets are becoming an increasingly popular digital device amongst young age children in formal and informal settings (Kucirkova, 2019; Neumann, 2018).

For the reasons mentioned above, smart screen devices -and especially tablets- are making their way into western homes (Papadakis, Kalogiannakis & Zaranis, 2019), and they provide a new opportunity for young age children to learn and cultivate skills in a fun, engaging way (McQuiggan, McQuiggan, Sabourin, & Kosturko, 2015). In 2011, only 38% of children aged 0-8 in the United States had ever used a mobile device, but by 2017, that figure rose to 84% (Rideout, 2017). Taking into account the availability of these devices in the majority of homes in the western world as well as the abundance of apps that are often sold at a relatively low cost or for free, there are increasing opportunities for learners from diverse socioeconomic backgrounds to access them (Callaghan & Reich, 2018). Currently, over half of educational apps on the Apple and Android markets are advertised for young children (Vaala, Ly, & Levine 2015). In 2013, 58% of parents in the United States reported that they had downloaded apps for their children (Common Sense Media, 2013). Parents and educators (Kalogiannakis & Papadakis, 2019) have embraced educational apps and app development is targeting young children under the label of education (Papadakis et al., 2018).

Mobile apps were first developed in the early 1990s but did not gain widespread popularity until the early 2000s. The most famous digital app stores - iTunes and Google Play - are filled with an abundance of education-related apps for iOS and Android devices (Bouck, Satsangi, & Flanagan, 2016). An app is a software application that works, generally, on a smart mobile device. Apps can support a variety of functions, including games and education (Bouck et al., 2016). Increased student engagement, motivation, and independence are apparent benefits of using apps integrated into formal and informal settings (Lubniewski, Arthur, & Harriott, 2018). Furthermore, studies also suggest that these apps use can support the playful and creative practice as well as creative thinking through problem-solving in early years settings (Kalogiannakis, Nirgianaki, & Papadakis, 2018).

To realize the potential benefits of technology use in early childhood education (ECE), technology must be developmentally appropriate (Kalogiannakis, Ampartzaki, Papadakis, & Skaraki, 2018). That means that technology should be responsive to the ages and developmental levels of the children, to their individual needs and interests, and their social and cultural contexts (Flewitt, Messer, & Kucirkova, 2015; McManis & Gunnewig, 2012). Many apps are created but not necessarily evaluated by professionals in the field of education (Lubniewski et al., 2018). The majority of apps in both famous digital stores are presented in the form of games or quizzes, mostly designed for repeating number facts, and that only support input from a single-finger touch, as if it were the old fashioned mouse click, drag, or single taps on the keyboard (Baccaglini-Frank & Maracci, 2015). In a study of the 200 top-selling educational apps in the iTunes Store by the Joan Ganz Cooney Center at Sesame Workshop, it was found that the majority of apps for young children do not go beyond skill and drill and do not fully utilize the interactive features of an iPad (Shuler, Levine, & Ree, 2012).

The thousands of available apps require that parents can evaluate the appropriateness and quality of the app for their learners' characteristics, needs, and instructional goals (Bausch, Ault, & Tammaro, 2012; Bouck et al., 2016). At the same time, limited resources are available to help teachers and parents to evaluate apps pedagogical affordances or to make decisions about using them (Papadakis et al., 2017). Parents and teachers often make the mistake of using inappropriate apps; they select apps randomly and use them without examining their educational quality (Papadakis et al., 2018). For both teachers and parents, it is critical to look for applications that are created with evidence-based practices. But, the process of locating and evaluating apps can be difficult and time-consuming (Perry et al., 2016). Teachers and parents need reliable and easy to use tools that allow them to evaluate apps with relative ease but will also guide them in the effective use of the apps with their students (Lubniewski et al., 2018).

Smart mobile devices and available apps are becoming increasingly popular to expose preschool children to educational concepts - including STEM, Computational Thinking and coding - before they enter formal schooling (Sheehan, Pila, Lauricella, & Wartella, 2019). But, although technology promises to be transformative (Papadakis & Kalogiannakis, 2010) educational content must remain the main thrust of any educational mobile app development and not technology (Shing & Yuan, 2016). The choice of apps must be based on more than a suggestion from another person, and more than an appealing advertisement (Papadakis & Kalogiannakis, 2017a, 2017b). Furthermore, when used for instruction, they should not be chosen for entertainment value. Apps must meet specific predetermined academic criteria.

OVERVIEW

This book showcases recent development in exemplary uses of mobile learning in various disciplines, including the examination of the real value of the self-proclaimed educational apps, the usage of apps and smart mobile devices to promote learning in various disciplines such as natural sciences, in emergent literacy development, in STEM activities, in Computational Thinking concepts and Coding Skills, in

Linguistic Intelligence, and in the development of Oral Language. Moreover, this book presents the attitudes in mobile learning in the scope of parents, preservice teachers, in-service teachers and of course students. The book ends with an overview of the advances of mobile learning in terms such as IoT, in Early Childhood Social Development, and the instruction of English as a foreign language.

What follows is a brief overview of each chapter.

In the opening Chapter 1, Stamatios Papadakis and Michail Kalogiannakis present a comprehensive overview of the real value of self-proclaimed educational apps. They claim that although apps have the potential to positively transform the learning process in formal and informal learning environments, in fact, software developers are capitalizing on a growing market by creating a large variety of applications which are targeted at a young audience. For that reason, the authors tried to determine whether there are self-proclaimed 'educational' apps in digital app stores that offer great potential to provide a space for young children to play and learn. They looked for peer-reviewed articles between July 2010 and April 2019. The results of 22 studies revealed that most of the apps that were analyzed, although they are characterized by their developers as 'educational' have no real educational value. They conclude that educators and parents must carefully consider how to introduce and scaffold this type of technology as it is not a panacea.

Focusing on the development of educational mobile apps, Chapter 2 states the problem that although most research has explored the impacts of mobile devices, in particular, iPads, on preschool-aged children, less research has explored how to choose applications and even less on recommendations for application developers. This chapter will draw from two standards - the four-pillar model of Hirsh-Pasek et al. (2015) and the rubric for the evaluations of educational apps for preschool children (REVEAC) by Papadakis, Kalogiannakis, and Zaranis (2017) and analyze two applications designed for young children. Through the analysis, this chapter will propose the use of the REVEAC for educators and developers while also making specific suggestions when targeting children under 5 years old in the educational context

Following the six chapters discuss the impact of mobile learning on children's cognitive development in various disciplines in preschool education. Chapter 3 provides an idea of how tablets might prove useful to kindergarten students. It contributes to the relevant literature by presenting evidence that, in subjects related to animals' classification, kindergarten students, through tablets: (i) can learn more and retain more knowledge compared to printed material and (ii) are more motivated to learn and enjoy lessons more. In conclusion, the study's findings might prove useful to researchers and teachers in understanding the pros and cons of tablets and effectively using these devices in kindergarten education. Chapter 4 discusses the specific features of touch-screen tablets that facilitate their easy use even by very young children and can create exciting environments for learning and instruction in early childhood. These devices along with easily downloadable applications allow young children to handle multimodal features such as texts, pictures, and sounds. This chapter provides a review of the recent literature on children's use of touch-screen tablets and its impact in emergent literacy development at home and in the pre-school setting. Considerable evidence indicated that the use of touch-screen tablets has the potential to enhance the emergence of writing skills and to facilitate aspects of emergent literacy such as letter name/sound knowledge, print awareness, letter writing and name writing skills, and phonological awareness. However, the relation between tablets and emergent literacy seems to be reinforced by factors such as the quality of literacy apps, the type of touch-screen literacy activities and the type of scaffolding used by adults. Chapter 5 states that the use of developmentally mobile applications for preparing pre-service teachers to promote STEM activities in preschool classrooms has been recognized as an effective instructional approach to facilitating critical thinking and meaningful learning. Chapter 5 touches upon this topic by examining whether ScratchJr – a developmentally appropriate app- via a structured course influence pre-service preschool students' ability to teach STEM concepts as well as motivate them to use ScratchJr in their future teaching work. This chapter discusses that the study results showed that the use of ScratchJr environment for teaching STEM concepts can motivate the pre-service teachers and empower them to continue using smart mobile technology in their future daily teaching practice. The authors conclude that it is imperative the priority to prepare pre-service preschool teachers to integrate CT and STEM concepts into their daily teaching practice using touchscreen technologies and developmentally appropriate apps. Of course, this approach requires a multipronged approach that involves a diverse set of necessary skills to be developed so as the pre-service teachers to become computationally literate as well as the development of curricula that specifically consider the affordances of the smart mobile technology in the preschool classroom. Chapter 6, written by Stamatios Papadakis, examines the use of apps to promote computational thinking concepts and coding skills in children of preschool and pre-primary school age. The author highlights the fact that Computer Science (CS) and Computational Thinking (CT) must become a new imperative for the western societies, societies that call for a bit of innovation, entrepreneurship, teamwork, and creative thinking. Smart screen devices and apps due to their unique characteristics can be used to teach young children coding skills and CT concepts with game-like activities. The author tries to find out if there are apps to facilitate CT and coding skills among preschool and early elementary school-age children. The author categorized the coding apps in two different categories: apps that there are research studies that reinforce their claims and apps that there is no research to confirm their educational value. He confirms ScratchJr as a developmentally appropriate coding app to engage children in CT and coding activities, as well as to provide a space for them to encounter ideas from CS. He found 3 other apps (Kodable, Daisy the Dinosaur and the Lightbot Jr), that the empirical research highlights the fact that can help young children learn basic coding skills via structured app play, although the level of difficulty differs between the apps. Although he found other apps which claim to be educational the author couldn't find scientific research to certify that value and advice that we must be a skeptic for their ability to deliver an optimal educational experience. In Chapter 7, the authors discuss how parents in Indonesia teach children how to read, write, and understand Indonesian vocabularies using the Duolingo application. Duolingo is an application based on the Android platform which helps children mastering other languages that they can practice speaking, reading, listening, and writing through a play. The activity of playing naturally will stimulate the aspects of linguistic and symbolic that this development is closely related to learning to speak by involving the ability to vocalize, the use of this application will implement toward gamification method. Gamification uses game-based mechanics, aesthetics, and game thinking to engage people, motivate action, promote learning, and solve problems. This chapter illustrates that gamification can be used specifically in stimulating linguistic intelligence based on a case study in Indonesia. In the last chapter of this section, Chapter 8, the authors aim to develop an application for mobile devices called Gradual Interactive Language (GIL). For app creation, the authors used the five stages of Soft Systems Methodology (SSM) and the classic model of software development as well. The results of the pilot test indicated that it is alternative support to move from pre-language to language, not limited to children with ID, the only thing that would depend is the moment in which it will use.

Following the six chapters dealing with the usage of apps as tools to promote learning the next 4 chapters deal with the students, parents and educators' perceptions and attitude regarding the use of mobile learning and in general ICT for educational purposes. Chapter 9 tried to investigate the main factors of Greek kindergarten teachers against primary school teachers' attitudes towards ICT in the classroom. The participants of this research were drawn from teachers of public schools in the area of Athens and Crete. A questionnaire was designed to assess the attitudes of teachers toward computers and their computer knowledge in primary and kindergarten schools learning environment. A comparison was performed among the kindergarten and primary school teachers to see the interactions of factors on a) the ICT knowledge of teachers, b) the frequency of using ICT at home by the teachers and, c) the ICT use in the teaching process. The conclusion of these comparisons was summarized, and the changes of recent years were discussed in the hope that they may provide useful information on the education challenges of the kindergarten curriculum. In Chapter 10, George Zacharis investigated pre-service teachers' acceptance to use mobile devices as resources for learning in a university context. In the current study, a modified UTAUT model was chosen to investigate the factors affecting the acceptance of mobile devices for learning from pre-service students as potential educators in a Greek university. Based on the findings of the current study, using mobile devices for learning is an increasingly crucial learning method for students. When university students find the use of mobile devices engaging and easy to use for learning and consider them essential to improve their learning performance and effectiveness, their Behavioral Intention and Use Behavior toward using these are enhanced. Therefore, pre-service teachers as future educators will use mobile devices for learning in their educational practice. In Chapter 11, Kleopatra Nikolopoulou states the problem that although the popularity of tablets in the homes of young children is rapidly increasing, in Greece there is little empirical evidence on this issue. This chapter investigated young children's use of tablets at home and parents' views on tablet benefits and their concerns. A questionnaire was completed by the parents of 100 children aged 4-6. It was found that young children engage in a range of activities such as playing games, watching cartoons, listening to music, watching videos, and using educational apps while visiting websites and using email are seldom carried out. Gender and age had an occasional isolated impact on children's tablet activities. Most parents "agree and strongly agree" that tablets "teach basic technology skills", "teach foreign languages" and "can make learning fun", while their main concerns included dependence, reduction of communication and inappropriate content. Implications regarding links between home and kindergarten are discussed. The last chapter of this section, Chapter 12, focuses on primary school students' perceptions about utilizing mobile games in the classroom. The results from an analysis of 10.381 questionnaires indicated that students hold neutral to positive perceptions on the utilization of mobile games during lessons. They are more interested in games that enhance their knowledge and develop their thinking skills. Besides, students prefer to play games in association with lessons related to Physics, Mathematics, and Language Arts. The results also indicated that (i) boys hold more positive perceptions about mobile games than girls and (ii) students from higher grades hold more positive perceptions than students from lower grades. Furthermore, the findings from interviews with 18 students indicated that they believe that lessons will be more enjoyable and interesting using games and that learning will be easier and more effective.

The last section of this book discusses the role of new forms of technology and its implication on the methods of teaching. Chapter 13 discusses the role of IoT as a growing paradigm that combines education with new prospects: sensors and beacons may lead towards the pervasive provision of educational content to students and wearables may track the students' interaction with educational objects. Subsequently, learning is transforming and may take place anywhere, anytime and with any means; an evolution described under the term Ubiquitous learning. In the chapter 13 "Mobile Learning in the Era of IoT. Is Ubiquitous Learning the Future of Learning?" the authors describe the metamorphosis of traditional learning, to e-learning, mobile learning, and Ubiquitous learning and discuss the characteristics and applications of the latter. Furthermore, they shortly present the Experience API specification and its possible uses in adaptive learning applications to make Ubiquitous learning reality in Early Childhood learning and other stages of education.

In the following Chapter 14, Dr. Courtney Blackwell highlights that historically, new media technology brings with it both the promise to transform children's learning and development as well as fears over potential negative consequences of digital engagement, particularly in disrupting social skills and relationships. With young children's near-ubiquitous access to mobile devices, understanding the extent to which mobile media influences their social health and development are critical. As a starting place, this chapter reviews current theoretical and empirical work on how the integration of mobile media – particularly touchscreen tablets - in formal early childhood education environments enables or interferes with children's social skills and relationships. Implications for policy and practice are discussed in light of this review. In Chapter 15, the authors tried to investigate whether an Information Communication Technology (ICT)-oriented teaching approach that is based upon Van Hiele's levels of geometric thought influences students' comprehension of the English alphabet; Comprehension of the letters individually, in terms of form and pronunciation, as well as within the lexical context, in terms of their pronunciation. This teaching approach was designed upon the principles of Blended Learning. Other English Language Teaching (ELT) approaches were touched upon, such as the Translation Method, the Direct Method, the Audio-Lingual Method, the Cognitive Approach, Desuggestopedia, the Communicative Language Teaching and lastly Eclectism, and a link between mathematical understanding and English language comprehension was explored. The participating sample consisted of Greek second grade students attending a public primary school located in Crete. The sample was divided into two groups, which were distributed among an experimental and a control one. The control group was taught following the traditional instruction and the experimental group was taught through an ICT-oriented intervention which was based upon Van Hiele's levels and a Blended Learning design. The outcomes showed a significant difference between the two groups on students' performance. This book ends with Chapter 16 that provides an insight into a constructionist approach aimed at developing computational thinking and creativity by engaging students in programming and robotics activities in the early childhood classroom. The purpose of this chapter is to bring forward the main features of teaching interventions implemented in real conditions to kindergarten and elementary school students, as well as to provide the educational community with examples of practices for inspiration, implementation, and reflection. The paradigms concern the utilization of robotics systems as Bee-Bot for 5-6 years old students, as well as of Lego WeDo 1.0 and Lego WeDo 2.0 for 8-9 years old students. An exploratory case study has been conducted; the research framework, key aspects, and the findings are reported; solutions, recommendations and future research plans are discussed. The author argues that a combination of unplugged, robotics and programming developmentally appropriate activities better support student computational thinking and creativity development.

CONCLUSION

Mobile Learning Applications in Early Childhood Education examines the recent advancement of mobile technology in early childhood education, including the use of self-proclaimed educational apps to promote learning, explores the innovative ways that developmentally appropriate apps are used to enhance and transform teaching and learning, selects exemplary cases of mobile learning by adopting developmentally appropriate technology and pedagogy, and identifies challenges and future directions for practice, research, and theoretical development in using mobile learning in preschool education and pre-service and in-service teachers training. The 16 chapters together update the field with research-informed, evidence-based design and implementation recommendations in the use of mobile learning for higher levels of thinking and learning, which will help preschool teachers to correctly prepare their teachers for future challenges.

This book is useful for educators, education leaders, education researchers, and stakeholders in both early childhood, primary education as well as higher education institutions. Its theoretical and pedagogical frameworks, trends and best practices in using mobile learning, and challenges and future directions identified by the studies in this book will help the target audience gain a comprehensive and deep understanding of mobile learning in education and training.

REFERENCES

Baccaglini-Frank, A., & Maracci, M. (2015). Multi-touch technology and pre-schoolers' development of number sense. *Digital Experiences in Mathematics Education*, *1*(1), 7–27. doi:10.100740751-015-0002-4

Bausch, M. E., Ault, M. J., & Tammaro, M. (2012). Apps for all: Introduction. In M. J. Ault & M. E. Bausch (Eds.), Apps for all students: A teacher's desktop guide (pp. 1–5). Alexandria, VA: Technology and Media Division of the Council for Exceptional Children (TAM).

Bouck, E. C., Satsangi, R., & Flanagan, S. (2016). Focus on inclusive education: evaluating apps for students with disabilities: supporting academic access and success. Childhood Education, 92(4), 324-328.

Callaghan, M. N., & Reich, S. M. (2018). Are educational preschool apps designed to teach? An analysis of the app market. *Learning, Media, and Technology*, *43*(3), 280–293. doi:10.1080/17439884.2018.14 98355

Common Sense Media. (2013). *Zero to eight: Children's media use in America.* Retrieved from https:// www. commonsensemedia.org/research/zero-to-eight-childrens-media-use-in-america-2013

Dorouka, P., Papadakis, St., & Kalogiannakis, M. (2019). Tablets & apps for promoting Robotics, mathematics, STEM Education, and literacy in early childhood education. *International Journal of Mobile Learning and Organisation.* (forthcoming)

Falloon, G. (2014). What's going on behind the screens? Researching young students' learning pathways using iPads. *Journal of Computer Assisted Learning*, *30*(4), 318–336. doi:10.1111/jcal.12044

Flewitt, R., Messer, D., & Kucirkova, N. (2015). New directions for early literacy in a digital age: The iPad. *Journal of Early Childhood Literacy*, *15*(3), 289–310. doi:10.1177/1468798414533560

Kalogiannakis, M., Ampartzaki, M., Papadakis, St., & Skaraki, E. (2018). Teaching natural science concepts to young children with mobile devices and hands-on activities. a case study. *International Journal of Teaching and Case Studies*, *9*(2), 171–183. doi:10.1504/IJTCS.2018.090965

Kalogiannakis, M., Nirgianaki, G.-M., & Papadakis, St. (2018). Teaching magnetism to preschool children: The effectiveness of picture story reading. *Early Childhood Education Journal*, *46*(5), 535–546. doi:10.100710643-017-0884-4

Kalogiannakis, M., & Papadakis, St. (2019). Evaluating pre-service kindergarten teachers' intention to adopt and use tablets into teaching practice for natural sciences. *International Journal of Mobile Learning and Organisation, 13*(1), 113–127. doi:10.1504/IJMLO.2019.096479

Kucirkova, N. (2017). i RPD-A framework for guiding design-based research for iPad apps. *British Journal of Educational Technology, 48*(2), 598–610. doi:10.1111/bjet.12389

Kucirkova, N. (2019). Children's agency by design: Design parameters for personalization in story-making apps. *International Journal of Child-Computer Interaction, 21*, 112–120. doi:10.1016/j.ijcci.2019.06.003

Kucirkova, N., Wells Rowe, D., Oliver, L., Piestrzynski, L. E. (2017). Children's writing with and on screen(s): a narrative literature review. *COST ACTION ISI1410 DigiLitEY.*

Lubniewski, K. L., Arthur, C. L., & Harriott, W. (2018). Evaluating instructional apps using the app checklist for educators (ACE). *International Electronic Journal of Elementary Education, 10*(3), 323–329.

McKenzie, S., Spence, A., & Nicholas, M. (2018). Going on safari: the design and development of an early years literacy iPad application to support letter-sound learning. *The Electronic Journal of e-Learning, 16*(1), 1-16.

McManis, L. D., & Gunnewig, S. B. (2012). Finding the education in educational technology with early learners. *Young Children, 67*(3), 14–24.

McQuiggan, S., McQuiggan, J., Sabourin, J., & Kosturko, L. (2015). Engaging young users: apps for preschoolers. In S. McQuiggan, L. Kosturko, J. McQuiggan, & J. Sabourin (Eds.), *Mobile learning: A handbook for developers, educators, and learners*. Hoboken, NJ: John Wiley & Sons.

Neumann, M. M. (2018). Using tablets and apps to enhance emergent literacy skills in young children. *Early Childhood Research Quarterly, 42*, 239–246. doi:10.1016/j.ecresq.2017.10.006

Papadakis, S., & Kalogiannakis, M. (2010). eTwinning in the early childhood as starting line of innovative practices for the didactic of natural sciences. In M. Kalogiannakis, D. Stavrou, & P. G. Michaelidis (Eds.), *Proceedings of the 7th International Conference on Hands-on Science, Bridging the Science and Society Gap* (pp. 235-240). Academic Press.

Papadakis, S., & Kalogiannakis, M. (2017a). Mobile educational applications for children. What educators and parents need to know. *International Journal of Mobile Learning and Organisation, 11*(3), 256–277. doi:10.1504/IJMLO.2017.085338

Papadakis, S., & Kalogiannakis, M. (2017b). Evaluation of Greek Android mobile applications for preschoolers. *Preschool and Primary Education, 5*, 65–100. doi:10.12681/ppej.11208

Papadakis, S., Kalogiannakis, M., & Zaranis, N. (2017). Designing and creating an educational app rubric for preschool teachers. *Education and Information Technologies, 22*(6), 3147–3165. doi:10.100710639-017-9579-0

Papadakis, S., Kalogiannakis, M., & Zaranis, N. (2018). Educational apps from the Android Google Play for Greek preschoolers: A systematic review. *Computers & Education, 116*, 139–160. doi:10.1016/j.compedu.2017.09.007

Papadakis, S., Kalogiannakis, M., & Zaranis, N. (2019). Parental involvement and attitudes towards young Greek children's mobile usage. *International Journal of Child-Computer Interaction*. (forthcoming)

Perry, A. D., Thrasher, E. P., & Lee, H. S. (2016). High leverage iPad apps for the mathematics classroom. *Mathematics Teacher*, *107*(9), 706–711. doi:10.5951/mathteacher.107.9.0706

Rideout, V. (2017). *The common sense census: Media use by kids age zero to eight*. Retrieved from https://www.commonsensemedia.org/research/the-common-sense-census-media-useby-kids-age-zero-to-eight-2017

Sheehan, K. J., Pila, S. C., Lauricella, A. R., & Wartella, E. A. (2019). Parent-child interaction and children's learning from a coding application. *Computers & Education*, *140*, 103601. doi:10.1016/j.compedu.2019.103601

Shing, S., & Yuan, B. (2016). Apps developed by academics. *Journal of Education and Practice*, *7*(33), 1–9.

Shuler, C., Levine, Z., & Ree, J. (2012). iLearn II: An analysis of the education category of Apple's app store. New York: The Joan Ganz Cooney Center at Sesame Workshop.

Vaala, S., Ly, A., & Levine, M. H. (2015). *Getting a read on the app stores: a market scan and analysis of children's literacy apps. Full report*. New York: The Joan Ganz Cooney Center at Sesame Workshop.

Acknowledgment

The editors gratefully acknowledge the editorial advisory board, essential contributions from the chapter authors who are the faculty members, researchers, teachers, instructional designers, and thought leaders who have bravely entered the foray of the modern competency-based education movement in Preschool education. Their innovative spirits, scholarship, and leadership create a bright future for competency-based education and for the many students who benefit from alternative, learner-centered approaches to preschool education.

There are many reviewers to thank and we are grateful to those authors who so willingly reviewed more than one chapter. A particular thank you goes to those reviewers who were not authors but graciously shared their time and expertise to support this cause.

We also wish to thank Ms. Courtney Tychinski (IGI Special Projects Coordinator) and Ms. Maria Rohde (IGI Development Editor) for their extensive (and good-natured) editing support in preparation of the final manuscript and for their support in managing the many details associated with publishing a book.

We would like to thank our families for their on-going help, support and encouragement.

Stamatios Papadakis & Michail Kalogiannakis
October 14, 2019
Rethymno, Crete, Greece

Section 1
Apps Quality

Chapter 1
A Research Synthesis of the Real Value of Self-Proclaimed Mobile Educational Applications for Young Children

Stamatios Papadakis
Faculty of Education, Department of Preschool Education, University of Crete, Greece

Michail Kalogiannakis
https://orcid.org/0000-0002-9124-2245
Faculty of Education, Department of Preschool Education, University of Crete, Greece

ABSTRACT

Mobile devices are the preferred media choice for children, and their massive worldwide use has driven a new generation of software called mobile applications or apps for use in the educational context. But, while there has been an explosion of apps that are claimed to be educational for young children, there are large distinctions between the quality of the apps. Even though apps may provide an active, enjoyable, and engaging context, the question is whether they are attuned to children's educational needs. This study performed a meta-analysis and research synthesis of 22 studies published 2011 to 2019. In the study, journal articles and conference papers in the English language published during the period 2010-2019 were coded and analyzed. The findings provide evidence that very few of these so-called "educational" apps that have been evaluated and tested can promote children's intelligence and improve their learning performance.

INTRODUCTION

Interactive touch screen technology upended the smartphone market with the introduction of the iPhone and iTouch in 2007. But with the introduction of the Apple iPad on April 3, 2010 things changed. The general public had access to the first mass-produced touchscreen tablet with a 9.7-inch smooth glass

DOI: 10.4018/978-1-7998-1486-3.ch001

surface. In a very few years, touch screen devices such as tablets emerged as an alternative to 'traditional' Graphical User Interfaces (GUI) interfaces, providing children, even at a very young age, with a new way of interacting with the digital technology without the need for external input devices such as mouse or keyboard. By using natural ways to interact with touch screen devices, young age users can express themselves in a natural way, enhancing communication and collaboration between them (Roldán-Álvarez, Márquez-Fernández, Martín & Guzmán, 2016). Furthermore, compared to other connected devices such as desktop computers and/or laptops, tablets are cheaper, and the addition of applications, better known as apps, make tablets multipurpose and more versatile than other digital devices. Thus, although the first tablet type device (iPad) was introduced for entertainment purposes, emerged almost immediately as the dominant mobile learning self-education tool (Kalogiannakis & Papadakis, 2017a, 2017b).

Additionally, tablets become the device of choice for young children because of their large screen size, high mobility, offline and online media streaming ability, decreasing cost and increasing interactive capabilities (Kabali et al., 2015). As a result, there is an increasing trend towards using mobile devices from the youth population as their touchscreens are just the right size for a young child to handle and operate (Wohlwend, 2017). Even children, as young as 2 years of age, can effectively learn and perform with these devices (The Michael Cohen Group, 2011). In the United States, the nonprofit organization Common Sense Media reported that the percentage of children aged from birth to eight years who have access to a mobile device at home had sharply increased in the past few years from 52% in 2011 to 75% in 2013 (Common Sense Media, 2013).

Given the high access and usage of Android/iOS tablets, it is not surprising that applications related to the use of these devices that claim to provide educational experiences for young children have similarly high rates of growth (Godfrey & Reed, 2013; Papadakis & Kalogiannakis, 2017a, 2017b). Mobile applications or 'apps' as they are less formally called are software programs for tablets, smartphones and other mobile devices (Common Sense Media, 2013). With the appropriate content and context, apps have the potential to positively transform the learning, while the use of interactivity and personalization features offered, motivate young learners to invest more effort in the learning task (Calder, 2015; Papadakis, Kalogiannakis & Zaranis, 2017). The playful nature of these apps is often appealing to children (Callaghan, 2018). In fact, the app targeted parents of younger kids (ages 3-6) dominate the mobile app market. Comparing 2009 to 2011, apps for toddlers and preschoolers' apps in the Apple store saw the greatest growth, an increase of 23% (Shuler, Levine & Ree, 2012). The same researchers highlight the fact that with 58% of top 100 selling iOS apps targeted children in this age range. Another non-profit organization, Common Sense Media, reported that almost half the families in the United States had downloaded apps for their children to use (Rideout, 2017) in the form of games to encourage children to learn or to keep them busy at least (Dashti & Yateem, 2018). App developers claim that self-proclaimed educational apps can help children improve their learning skills and performance. In fact, developers are capitalizing on a growing market by creating a large variety of applications which are targeted at a young audience (Joe, 2018). Although these apps can engage students by making learning fun (Papadakis, 2018) in ways that are adaptive to different children needs, in fact, many of these 'educational' apps aren't the best way for young children to learn math, science and literacy skills (Callaghan, 2018).

As Donahoo (2012; 2013) highlights although the educational apps for young children in mobile apps stores are developed with the intention to obtain the best results, few really support children's cognitive development. Unfortunately, the vast majority of 'educational' apps available to download in the digital stores (Apple iTunes and Google Play) target only basic academic skills. This is because the software developers are not following the established curricula and/or often design the apps with little or no direction or guidance from developmental specialists (Martens, Rinnert & Andersen, 2018). Many self-proclaimed educational apps just replicate printed resources and outdated learning tools, like a series of flashcards (Donahoo, 2013).

In conclusion, apps targeting young children are continuously being released, with many of them claiming to be educational (proper for the teaching and learning process) (Kalogiannakis & Papadakis, 2017c). Many of these claims are unsubstantiated. Even though some apps may provide enjoyable, and/or engaging activities, the question is whether children can learn from these apps (Sari, Takacs & Bus, 2017). This article provides an overview of the research used to study the real value of 'educational' apps since the release of tablets in 2010. This study performed a meta-analysis and research synthesis trying to answer the following question: Do popular self-proclaimed 'educational' apps are designed to meet the developmental and learning needs of young children?

LITERATURE REVIEW

While the need for parents and educators to understand these newer smart screens, technologies are no doubt important, apps submitted to the Appstore and Google Play are often marketed as 'educational', with little to no assurances of educational value (Vaala et al., 2015). The Joan Ganz Cooney Center and New America have characterized the market for children's educational apps as a 'Digital Wild West' (Guernsey, Levine, Chiong & Severns, 2012). The digital app market is full of software but doesn't provide adequate information and support, to aid parents' and educators' decision-making around this topic (Vaala, Ly & Levine, 2015).

For parents and educators, selecting appropriate apps for young children is a daunting challenge (Papadakis, Kalogiannakis & Zaranis, 2016a, 2016b). In fact, determining what constitutes being an educational app is complicated, and up for debate. It is sometimes easier to determine what constitutes a lack of quality. For instance, Martens, Rinnert & Andersen, 2018 report characteristics such as in-app advertising, poor navigation design, non-functional gamified elements, privacy issues, etc. Kucirkova, Messer, Sheehy & Panadero (2014) state that researchers attempting to establish an educational framework for apps face many of the same challenges encountered in computer-based educational tools. Hirsh-Pasek and her colleagues describe the current app market as the 'first wave of app development' where the material that already exists in other non-digital form is transformed in a digital format (Hirsh-Pasek et al., 2015, p.5). Most designs are simple tasks that reflect familiar formats and only offer closed literacy experiences designed for passive recipients (Lynch & Redpath, 2014). Furthermore, Kucirkova et al (2014) note that apps with real educational value are focusing primarily on promoting education and not solely on providing entertainment.

For someone, a non-expert in the educational technology field, to figure out which app to select requires a great deal of effort. Furthermore, in two biggest digital stores (iTunes store and Play Google) the navigation system is problematic and doesn't provide enough information for someone to specify both the number and the quality of the 'educational' apps for effective decision-making. Martens, Rinnert & Andersen (2018) characteristically point out that a simple search in the App Store using the terms 'ABC' or 'Alphabet,' returned between 279 and 286 results. Most of the apps had no further information regarding the principles the designers and developers followed when developing them. Furthermore, there are few resources for evaluating them. There are already rubrics developed by content experts and researchers, but parents and educators whether ignoring their existence either it is difficult to use them and interpreter their results (Martens, Rinnert & Andersen, 2018). Although there is app information available in the digital stores it largely consists of an app marketing and thus cannot be used as an evaluation criterion of the app educational value (Larkin, 2013). Furthermore, this app information is provided by the developers and cannot be considered as accurate or reliable. Larkin (2015a, b) summarizes that the selection of apps is hindered not only by the 'sheer number of apps' and 'the poor structure of the iTunes app store user interface' (p. 7) but also from factors such as the poor app information and description, the search algorithms of the stores, etc.

Researchers such as Kucirkova (2016) and Hirsch-Pasek et al. (2015) highlight also the hard competition on the app market. Kucirkova (2016) states that developing an app is a costly endeavor, with an app production costs amounting to $10,000-70,000. At the same time an average app costs around $3. Thus fast-produced and fast-selling content for a broader audience is a tricky issue for developers for profits in a low-priced apps market in the digital stores. For that reason, many apps offer the same content but slightly different interface and media content (Hirsch-Pasek et al., 2015). Considering the fact that the app market is currently over-saturated with children's educational apps, the success or failure of a mobile app especially in terms of its economic side is not just a result of educational quality but rather how many users have installed the app, from the thousands of apps available in the digital stores and repositories. It is likely that the most best-selling or popular educational apps might not be the most effective from a pedagogical perspective (Kucirkova, 2016; Hirsch-Pasek et al., 2015). This situation can be misleading for parents and educators who purchase or download these products under the assumption they will be beneficial to their children and students (Flewitt, Messer & Kucirkova, 2014; Joe, 2018).

There are two broad approaches in the literature to evaluate the value of self-proclaimed educational apps. The first approach involves researchers targeting an app, or a small set of apps, for analysis for their educational characteristics (Kalogiannakis & Papadakis, 2019). A restriction of this type of app research is that some process of selection has already occurred prior to a more targeted investigation. The second approach involves experts looked at products in the iTunes App store, the Google Play, either 'paid' or free apps to provide a first-time snapshot, or a comprehensive survey, of what is most visible to parents and teachers. A restriction in this approach is the huge number of apps that need to be evaluated. In addition, both digital stores are in a constant state of flux, with educational apps added, obsoleted and modified continuously (Larkin, 2013; Larkin, Kortenkamp, Ladel & Etzold, 2018).

The Aim of the Present Study

Owing to the vast and continuously increasing number of 'educational' apps, there is a need for ongoing research to develop evidence for their real educational value (Calder, 2015; Hirsh-Pasek et al., 2015; Sari, Takacs & Bus, 2017). Various researchers over the past 10 years have tried to identify whether the 'educational' apps available for children in the digital app stores do not work as pure entertainment but have educational value. The aim of the present study is to determine whether there are self-proclaimed 'educational' apps in digital app stores that offer great potential to provide a space for young children to play and learn. Thus, the research question of this study is the following: Even though apps may provide an enjoyable approach, are really offer opportunities to foster young children's critical early-learning skills?

The Procedure

In respect to timeframe, this study include evidence gathered between July 2010 and April 2019. The process involved researchers using indexing databases (Scopus, Web of Knowledge, Google Scholar, ProQuest) and academic hosting websites such as ResearchGate to search for additional relevant studies. Sources were searched from April 1, 2019, through May 31, 2019, and were limited to peer-reviewed journal articles published in English. The results of these searches were downloaded to Zotero, and duplicates were removed. In some cases, both snowball sampling and independent research websites were also conducted.

The keywords used for the search were tablets, touchscreens, educational apps, preschool education and all variations of these terms (e.g., 'app' included mobile applications, applications, iPad software, Android software, mobile software, etc.) Furthermore, in order to ensure subject matter coverage of applications in education, manual searches were also conducted in the following journals: *Computers & Education, Journal of Computer Assisted Learning, Computers in Human Behaviour, British Journal of Educational Technology, Early Childhood Education Journal, Journal of Children and Media and the Journal of Educational Technology & Society.* In order to assess the academic body of knowledge, 'grey Literature' was excluded, e.g. technical writings, white papers, as well as feasibility studies that do not fully meet inclusion criteria. Non-academic sources such as news, and other non-peer reviewed articles were also excluded.

Are There Educational Apps With Real Educational Value?

The research synthesis revealed that since 2010 there are 22 studies with either big or small sample of self-proclaimed educational apps for Apple Store and/or Play Store targeted at preschool to elementary school-aged children.

Prior to 2010 research, it is worth mentioning the work by Shuler (2009) that found that of all the 100 top-selling apps in the iTunes App Store, (available on the iPhone and iPod Touch), only 47% were classified as being appropriate for preschool and elementary-aged kids.

Watlington (2011) reviewed the top 100 free iPod touch educational apps on the iTunes App Store for educational value and developmental appropriateness. She was one of the first researchers that notes that *'very little research is available on educational apps and consumers currently rely on app developers and iTunes ratings for their information'* (p. 3112). After applying specific criteria, she concluded that only 48% of the apps were 'developmentally appropriate' based on the study design. The researcher also found that language skills were the focus of these apps.

In 2011, in a study in Australia, a total of 202 'educational' apps across various disciplines such as English, Mathematics, etc. were tested for supporting students' learning and motivation. Depending on the study results, some apps were better than others at supporting students in their learning process, while almost half of all apps (53%) promoted ISTE.Nets Student Standards 3 and 4. The ISTE.Nets (2007) Standards provide considerations and implementation guidelines for the successful use of technologies in 21st-century education and were used in this study for external benchmarking for student learning with the use of iPads and educational apps. One of the main findings of the study was that the prevalence of the 'educational' apps was not suitable for Stage 3 learners, as their content was either too simplistic or significantly complicated for the students (p. 18) (Learning Exchange, 2011).

Goodwin and Highfield (2012) conducted a systematic content analysis of the 'top 10 education Apps' (240 apps) with particular attention paid to mathematics learning apps. In their work, they generally classified the apps into three types based on key pedagogical features; instructional, manipulable and constructive. Instructional apps are used as drill-and-practice environments for learning. There is simply no place for any deviation from the predefined activities requiring a simple answer. As a result, such apps don't put too much pressure on the learner and thus don't foster learner cognitive development. On the contrary, constructive apps have a more open-ended design that supports experimentation allowing learners to create their own content and may distribute it to other users. In their research, most of the apps were classified as being instructive (75%), 23% as manipulable, while only 2% as constructive. They conclude that the characterization of apps as being 'educational' is not correct and somewhat misleading, as most of the apps simply reinforce the basic knowledge and skills of the users through rote learning. These apps, in general, reproduce printed material e.g. flashcards in a digital format and consist mainly of drill & practice activities i.e. activities which require minimal effort for the learner.

Shuler (2012) from the Joan Ganz Cooney center conducted a content analysis of all available educational apps for children (iLearn II study) and included a list of 100 educational iPad and iPhone apps from the Apple iTunes store (a total of 200 apps). Schuler found that more than 80% of the top-selling paid apps in that category target children ranged in age from toddlers to high school. In this study, quantitative and qualitative variables such as age target and price, marketing description, name, images were analyzed. It was found that early learning was the most popular subject (47%), following by math apps (13%). The same study found that the majority of the most popular apps (86%) in the education category are not intended by the apps developers to be used in school as the target audience are mainly parents looking for apps to foster their young children's' early-learning skills mainly in informal learning environments such as family, community, etc.

Guernsey et al. (2012) looked for the top 20 both paid and free educational literacy apps in the two largest app stores (iTunes store and Google Play). In their analysis, they found that most of the apps targeted basic literacy skills, such as letter and word recognition. Very few apps attempt to target more advanced skills like comprehension and grammar. The same researchers highlight the fact that the app market in both digital stores 'felt a lot like a digital Wild West', with apps had little to no information on 'whether the developers had backgrounds in early literacy or whether the apps were vetted by reading experts or evaluated in any way'.

Byers and Hadley (2013) conducted a research on 150 specifically mathematics apps for children ages 3 to 8. Their results revealed that most of the apps organized in a question/answer format and/or traditional gaming layout. The researchers concluded that 'educational' apps features can actually hinder children's learning progressions while they are interacting with mathematics apps.

Falloon (2013) in his study investigated the impact of design and content features of 45 educational iPad apps for preschoolers. The apps were selected by an experienced teacher. Only 27 of these apps were considered as educational targeting various learning content such as solving math tasks, reinforcing spelling, etc. The other 18 apps were categorized as learning skills tools helping young children to develop various activities such as writing, reading skills, etc.

Handal, El-Khoury, Campbell and Cavanagh (2013) evaluated the 100 top rated mathematics apps for primary and secondary school students available from the Apple Store. The apps were categorized with respect to their instructional role and nine categories of content were identified. Unlike other studies, the researchers don't direct express concerns about the accuracy and richness of the mathematical app content although they warn teachers that it is important for them 'to keep a watchful but enthusiastic eye' in the area of mobile mathematics learning (p. 146).

Highfield and Goodwin (2013) reviewed 53 mathematics apps, in 3 countries (Australia, United States, and United Kingdom) in relation to age appropriateness, curriculum content, range of pedagogies, and pedagogic design for mathematics learning. The initial sample was the most popular educational paid apps on the App Store (n=360) in a six-month period. The examination of age and curriculum content revealed that '*apps catering to a range of age groups are presented in the ten top educational apps in the App Store*' (p. 381). The examination of curriculum content revealed 'a limited range of curriculum areas' (p. 381). Furthermore, the classification of educational design revealed a predominance of instructive apps (p. 382). Instructive apps focus on drill and practice activities, not allow for the creation of content. The researchers highlighted the limited representation of apps promoting open-ended learning activities that are essential to engage children in learning while they foster their critical thinking skills and creativity.

Larkin (2013) analyzed 142 apps in Apple store out of an initial pool of 4000 apps by content and knowledge focus, in terms of their relevance to the mathematics curriculum, their appropriateness for primary school classrooms, as well as their ability to develop conceptual and procedural epistemological understandings. Larkin observed that most of the apps were just digital flashcards encouraging rote learning. Only 40 of the 142 mathematics apps had features that supported and encouraged learning shifts. Like other researchers, Larkin highlighted the fact that apps had disparities between their name and the mathematics content, between targeted age levels and real age levels, and of course between the price of an app and the real educational value of a self-proclaimed educational app.

Cherner, Dix, & Lee (2014) reviewed 92 free apps from different sources such as the App Store, the Appitic and three online lists with content relative to the core curriculum for the public schools for PreK-12 students. They found that most of the apps didn't have a specific subject area as most of them could be used in multiple subject areas. Like other studies (Shuler, 2009) they found that apps focused on literacy, numeracy and general learning for toddlers, and preschool-aged children represented the majority of the educational category. They also conclude that educators must have developed a conceptual framework for an understanding of mobile learning prior to the use of tablets and apps in the classroom if they want the apps to meet students' instructional needs.

In 2015, the Joan Ganz Cooney Center and The New America Foundation released a new study entitled 'Getting a Read on the App Stores: A Market Scan and Analysis of Literacy Apps'. In that analysis of 183 popular and expert-awarded literacy-focused apps for children 0- to 8-years-old, the researchers found a considerable variability in the amount and distribution of information available to users in app descriptions within popular app markets (Apple App Store, Google Play, and Amazon Appstore) (Vaala, Ly & Levine, 2015). Particularly rare in app descriptions were what the researchers called benchmarks of educational quality. These benchmarks include containing a curriculum, providing feedback, imbedding scaffolding, using learning theory, and including educators or content experts in the app's development team. Furthermore, some of the study results were the following:

- The same apps weren't often recommended by multiple stores, repositories, etc.
- App descriptions are inconsistent in length.
- Apps for preschoolers outnumber other age groups.

In their research work, Cayton-Hodges, Feng & Pan (2015) examined a sample of 16 apps for mathematics education in the Apple App Store targeted at preschool to elementary school-aged children. The apps reference a wide range of mathematics topics: numbers, operations, algebra, geometrics, statistics, and probability. Their goal was to understand the instructional principles used in designing the apps for smart mobile devices. They concluded that either by design or by accident, the mathematics apps are equal in accuracy and richness of content of mathematics while most of them are characterized by inadequacies in accuracy and completeness.

Larkin (2015a), in another study, reviewed 53 geometry apps in the Apple store. It is found that most of the apps were limited in their ability to assist students to conceptualize the meaning of geometry concepts. Many of the 'educational' apps often include poor quality and/or inappropriate content. They tend to be limited to simple drill and practice format, focusing only on testing of basic while ignoring the development of higher levels of geometric thinking. This study also agrees with other studies results that the majority of the educational apps focus on one specific content area, while most of them are simply drill and practice, promoting rote learning and are not effective in promoting declarative and procedural knowledge acquisition.

Namukasa, Gadanidis, Sarina, Scucuglia & Aryee, (2016) reviewed 80 mathematics apps relevant to negative and positive integer numbers. The researchers found that the majority of the apps (93%), posed traditional, prescriptive practice tasks, which function in a manner similar to printed flashcards, with an almost identical right/wrong pattern of responses. Only 4 apps (5%) focused on building a conceptual understanding of mathematical concepts, either by introducing a new topic, or explaining how a procedure worked.

Yelland, Drake & Sadler (2017) according to the research commissioned by the 'Australian Government Department of Education and Training' investigated the effectiveness of STEM education apps for tablet technologies for preschool children (4-5 years of age). Following Shuler (2012) methodology, an analysis was conducted of the reported Top-100 paid and free apps. Using this approach, the researchers generated an initial list of 32 apps. Further analysis in the iTunes App store with various criteria returned a total of 45 apps for detailed analysis. The researcher's analysis revealed that although there are many apps that appear to be marketing themselves educational on the digital stores, only a limited number of apps can be considered as a useful educational tool for educators to support and enhance STEM teaching and learning in early childhood. Furthermore, the researchers state that the apps are not generally accompanied by any documentation to inform educators as to the most effective method of their use in order to improve the efficiency of learning using these technologies (p.22).

Sari, Takacs and Bus (2017) evaluated the most popular apps for children in Hungary, Turkey, Greece and the Netherlands (560 Dutch, 492 Hungarian, 532 Greek, and 494 Turkish apps) available on the Android and iOS app stores. In all four countries, the researchers found that most apps were advertised for the youngest age groups. Like the study of Guernsey & Levine (2015), they found that almost half (37%) of the best-selling apps that were categorized as educational offered training in basic academic skills. In their results, they highlight the fact that most apps in the four countries were not designed by local developers. Although this means that the apps had many similarities between the countries there was a serious consequence for local learners. Most of the apps didn't include local characters and typical visualizations, hindering children's development of language and literacy skills. They also found that there were many apps for young children containing any language. As the researchers note there were apps that didn't '*contain five spoken words with no restrictions on whether it is in a complete sentence or whether these words go beyond the most basic vocabulary*' (p.13). They conclude that from an educational perspective, these apps do not provide any language input and consequently additional benefits for younger children.

Papadakis, Kalogiannakis & Zaranis (2018), examined whether the self-proclaimed educational apps for Greek preschoolers are developmentally appropriate in terms of content and design and, thus, contribute to children cognitive development. The researchers found that only 40 apps fulfilled their criteria and then were evaluated using a specially developed rubric. In this study, almost all apps in the sample (95%) belonged in the 'drill & practice' category. As a result, the apps did nothing more than simply promote rote learning (p. 145). There were no open-ended type apps. Like other studies, it was also highlighted the fact that all apps were accompanied by a poor description according to their content, development teams, as well as the evaluation process. Finally, they noted the absence of correlation between the rubric and the app score according to the Google 'star' rating system. The similar mismatch was also found between the app rubric score and the user comments. In all these cases the apps seem to be overvalued in terms of their real educational value.

Bratitsis (2018) made a search for educational mobile apps in the Greek language for kindergarten students in Google Play. Although the first result returned 248 apps, he found that there were mistakes in the categorization of these apps according to the PEGI system. 39 of those apps – for various reasons- were not appropriate for the designated age group. Like Papadakis et al. (2018), the researcher found that the figures that accompany the app description in Google Play such as the number of downloads and evaluators, or the app stars are not necessarily the most appropriate source of information to guide both parents and teachers in their search for apps with real educational value. Considering the app discipline, Bratitsis (2018) found that almost half of the app content can be considered as related to language acquisition and mathematics learning. The rest of the apps were categorized in a game type format that had no proven real educational value. Bratitsis, (2018) concluded that very few apps had real, educational value and, thus, can be used in the classroom settings as they mostly belonged to the 'drill & practice' type, with the potential drawbacks.

Callaghan and Reich (2018) conduct a content analysis of 50 popular preschool literacy and math apps in the Apple App Store. The researchers found that the popular preschool category contain a mix of both low- and high-quality apps. Like other studies, they conclude that teachers cannot merely rely on app popularity in the digital store and it is up to them to identify engaging apps for use in the classroom.

In Turkey (Özeke, 2018), analyzed 44 apps for the Android mobile operating system for their educational value. The sample apps had Turkish content, targeted children five years old and under, and were either free, trial or freemium. The selected 44 apps had scored from 2.9 to 5 according to Google's 'star' rating system (min: 1; max: 5). In this work, the researcher by using an assessment tool found that the sample apps had a medium score as almost all apps promoted learning through repetitive practices. These scores revealed a medium to low educational value. The researcher also found that there was no statistically significant correlation between the scores of the rubric and a) Google Play scores; b) users' comments, and c) apps downloads.

Dubé, Kacmaz, Wen, Xu & Alam (2019) examined the basic information, educational content, and user ratings for the top 90 apps in the Apple store. The researchers found that there is a lack of transparency in how the Apple store selects the 'top' apps. Similar there is a lack of information from the developers considering the benchmarks of educational quality in their app descriptions. Only 32 of the sample apps had names to help educators find apps relevant to their lessons. Surprisingly, this percentage was smaller for the <5 age range, as compared to the 6-8 and 9-11 age ranges. Furthermore, 21 apps out of did not mention any math subjects.

Results

The key research question under investigation in this research synthesis is whether self-proclaimed educational apps are appropriate in terms of content and resources to support active learning and constructive learning activities. The conclusion of this study is that the app marketplace remains like the 'Digital Wild West' that Guernsey et al. (2012) documented 8 years earlier. Although there are apps that can provide children with experiences that are connected to learning both in and out of the classroom, almost the majority of self-proclaimed educational apps -either free or paid- since 2010, come with downsides (Hiniker, 2018). In fact, most of the apps analyzed in the present study, although they are characterized by their developers as 'educational' have no real educational value. Most apps marketed as educational are game-like in a form generally reflecting a behaviorist view of learning (Falloon, 2013).

Vaala et al. (2015) state that most apps follow testing-based approaches such as puzzle games, card games, and more. In line with the documented studies, the findings also suggest that few high-quality apps are available for basic literacy and numeracy skills – even with the most minimal requirements. Furthermore, many self-proclaimed educational apps promote activities which are not age appropriate as they target content covering a broad age range (0 to 5 years). Sari et al. (2017) also highlight the fact that many educational apps offer educational activities for young children that are not considered developmentally appropriate for their age.

While the potential of apps (either free or paid) for the children cognitive development is enormous, the market of high-quality children's apps is relatively narrow (Kalogiannakis & Papadakis, 2019). Early literacy and math are the most common app content (Kalogiannakis, Ampartzaki, Papadakis & Skaraki, 2018). Very few apps improve learning outcomes and teach '21st-century skills' such as critical thinking, curiosity, communication, teamwork, and creativity, to name a few (Donahoo, 2012). Most of the apps for young children rely solely on emotional and/or other approaches that remain relatively simplistic. As a result, the apps pay more attention to the entertainment aspects of the content than to cognitive values (Dubé, Alam, Xu, Wen & Kacmaz, 2019; Falloon, 2014). The self-proclaimed educational apps mostly include activities either based on simple digital flashcards, or on simple vocabulary, letter recognition and math exercises (Bratitsis, 2018).

Also, evidence has emerged that many of the app developers do not mention benchmarks of educational quality in their app store descriptions (Larkin, 2013; Vaala et al., 2015), which are needed for educators and even for parents to find good apps (Dubé, Kacmaz, Wen, Xu & Alam, 2019). Furthermore, the titles of most apps, as well as the written descriptions, do not provide any information about the content focus of the app. As Vaala, et al. (2015) highlights the Android and Apple digital app stores are offering plenty of apps to users but in fact both populated stores don't provide adequate information and support, to aid adults' decision-making.

Another trend that emerged, is that concerning children's exposure to their native language through 'educational' apps, the situation seems even more problematic and alarming in countries with a small language base like Greece or Hungary (Papadakis et al., 2018; Sari et al., 2017). Many language apps developed in a main language such as English or Spanish, etc. are not culturally adapted for use in other countries. In the best scenario, they simply translated into the local languages wherever required. In the worst-case scenario, the apps are developed without the support of oral language, in a developer effort to quickly and cheaply make the app available around the globe (Sari et al., 2017).

In conclusion, most 'educational' apps target only rote academic skills, are not based on established curricula and academic findings, and use little or no input from child developmental specialists, psychologists, and educators (American Academy of Pediatrics, 2016). All things considered, we must agree with Pelton and Pelton (2012) conclusion: the majority of existing 'educational' apps are primarily using flashcard drill instructional methods, do not provide effective support to decision-making, and fail to support active learning and creativity in a child-centered learning environment.

CONCLUSION

Discussion

In this study, like the other researches, we found that a relatively small number of apps are well designed and really support children to learn more easily and effectively. Apps developers don't have an understanding of how children learn and develop (Vaala et al., 2015). Thus, many apps simply replicating existing printed resources and learning tools, like digital flashcards (Donahoo, 2013). Given that some apps are of better quality and can be effective math and literacy learning tools, a logical next question is how educators and parents find these quality apps (Dubé, Kacmaz, Wen, Xu & Alam, 2019).

Unfortunately, teachers and parents must spend many hours and effort searching for appropriate apps to support their students' or children' learning experiences (Flewitt, Kucirkova & Messer, 2014; Gray, Dunn, Moffett & Mitchell, 2017; Papadakis & Kalogiannakis, 2019). Bratitsis (2018) considering his own experience, state that is a major frustration for the teacher or parent who has the willingness to incorporate tablet type devices and accompanied apps for use in the classroom or at home. A problem revealed during the research is the digital store's limited or problematic categorical organization which does not allow users to easily find apps that contain specific content or are intended for a specific age group (Joe, 2018). This situation adds value to various researchers claims that useful guidelines and frameworks for selecting appropriate apps are necessary as well as a significant redesign of digital stores with the use of more effective and efficient search algorithms for optimized results.

So, what teacher and parents can do when looking towards using self-proclaimed educational apps? A more nuanced approach should, therefore, take into considerations of adults than simply thinking about screen time limits and simply reducing children's screen time exposure. It is necessary for families and early educators to evaluate the appropriateness and relevance of an app to determine what is appropriate for a child age and development level (Lee, 2016).

Among the wide selection of apps available in the Android and iOS platform, educators and parents must carefully consider how to introduce and scaffold this type of technology (NAEYC & the Fred Rogers Center, 2012). The published joint position statement from NAEYC and the Fred Rogers Center (2012) highlights the critical adult role, 'in making certain that thoughtful planning, careful implementation, reflection, and evaluation all guide decision making about how to introduce and integrate any form of technology or media into the classroom experience' (p. 6) (Chau, 2014). It becomes the responsibility of everyone who is involved in the development of children apps to be thoughtful and purposeful. Developers but also educators and parents must realize how and why these digital and interactive tools can support children cognitive development (Donahoo, 2012).

In the best-case scenario, professionals that involve in app development must work closely with early childhood development experts to develop content to ensure that quality is incorporated into app content (Lee, 2016). Callaghan (2018) suggests consumers to be purposeful in educational app selection. Furthermore, the same researcher encourages parents to play apps with their children. Vaala, Ly & Levine

(2015) state that it is necessary for parents and educators to go beyond descriptions and comments in the app stores by visiting developers' websites. It is also imperative for adults to cross-check several reliable sites and databases for further guidance about a given app prior its use.

There is a plethora of different innovative technologies that can be used for young children cognitive development. Nobody can deny that tablet type devices and their accompanying apps are here to stay and already have transformed children digital experiences (Falloon, 2013). Thus, is necessary for experts to stop wondering whether tablets and apps are good education tools and instead should focus on identifying the most effective types of apps of contributing to children cognitive development and skills acquisition. Indeed, it is widely known that technology, whether used in a formal or informal setting, must be applied in approaches that promote children's learning and development (Lee, 2016).

COMPETING INTERESTS

The authors declare that they have no competing interests.

FUNDING

We are especially indebted to the Special Account of Research of the University of Crete, which financially supports our research effort, KA 1013, Mobile Educational Application Developmental Scale (MoEADS): A first attempt to evaluate mobile educational apps for young children in Greece.

REFERENCES

American Academy of Pediatrics. (2016). Media and young minds. *Pediatrics, 138*(5), e20162591. doi:10.1542/peds.2016-2591 PMID:27940793

Bratitsis, T. (2018). An Attempt for Critical Categorization of Android Applications Available for the Greek Kindergarten. In M. Auer & T. Tsiatsos (Eds.), *Interactive Mobile Communication Technologies and Learning. IMCL 2017. Advances in Intelligent Systems and Computing* (Vol. 725, pp. 56–68). Dordrecht: Springer. doi:10.1007/978-3-319-75175-7_7

Byers, P., & Hadley, J. (2013). Traditional and novel modes of activity in touch screen math apps. In J. P. Hourcade, N. Sawhney, & E. Reardon (Eds.), *Proceedings of the 12th International Conference on Interaction Design and Children*. New York: ACM.

Calder, N. (2015). Apps: Appropriate, Applicable, and Appealing? In T. Lowrie & R. Jorgensen (Zevenbergen) (Eds.), Mathematics Education in the Digital Era 4, Digital games and mathematics learning (pp. 233-250). Dordrecht: Springer.

Callaghan, M. (2018). *Educational apps designed for kids: Avoiding missed opportunities*. Retrieved from https://bold.expert/educational-apps-designed-for-kids-avoiding-missed-opportunities

Callaghan, M. N., & Reich, S. M. (2018). Are educational preschool apps designed to teach? An analysis of the app market. *Learning, Media and Technology*, *43*(3), 280–293. doi:10.1080/17439884.2018.1498355

Cayton-Hodges, G. A., Feng, G., & Pan, X. (2015). Tablet-based math assessment: what can we learn from math apps? *Journal of Educational Technology & Society, 18*(2), 3-20.

Chau, C. L. (2014). *Positive technological development for young children in the context of children's mobile apps* (Doctoral dissertation). Tufts University.

Cherner, T., Dix, J., & Lee, C. (2014). Cleaning up that mess: A framework for classifying educational apps. *Contemporary Issues in Technology & Teacher Education, 14*(2), 158–193.

Common Sense Media. (2013). *Zero to eight: Children's media use in America 2013.* Retrieved from https://www.commonsensemedia.org/research/zero-to-eight-childrens-media-use-in-america-2013

Dashti, F. A., & Yateem, A. K. (2018). Use of mobile devices: A case study with children from Kuwait and the United States. *International Journal of Early Childhood, 50*(1), 121–134. doi:10.100713158-018-0208-x

Donahoo, D. (2012). *The Future of Apps for Young Children: Beyond ABC & 123.* Retrieved from https://www.huffpost.com/entry/the-future-of-apps-for-yo_b_1771395

Donahoo, D. (2013). *Seeking better apps for young children.* Retrieved from https://www.fredrogerscenter.org/2013/03/seeking-better-apps-for-young-children/

Dubé, A. K., Alam, S. S., Xu, C., Wen, R., & Kacmaz, G. (2019). Tablets as Elementary Mathematics Education Tools: Are They Effective and Why. In K. Robinson, H. Osana, & D. Kotsopoulos (Eds.), *Mathematical Learning and Cognition in Early Childhood: Integrating Interdisciplinary Research into Practice* (pp. 223–248). Dordrecht: Springer. doi:10.1007/978-3-030-12895-1_13

Dubé, A. K., Kacmaz, G., Wen, R., Xu, C., & Alam, S. S. (2019). Mathematics apps in the App Store: Low transparency impedes educator's ability to find quality apps. *Proceeding of the American Educational Research Association (AERA).*

Falloon, G. (2013). Young students using ipads: App design and content influences on their learning pathways. *Computers & Education, 68*, 505–521. doi:10.1016/j.compedu.2013.06.006

Falloon, G. (2014). What's going on behind the screens? Researching young students' learning pathways using iPads. *Journal of Computer Assisted Learning, 30*(4), 318–336. doi:10.1111/jcal.12044

Flewitt, R., Kucirkova, N., & Messer, D. (2014). Touching the virtual, touching the real: iPads and enabling literacy for students experiencing disability. *Australian Journal of Language and Literacy, 37*(2), 107–116.

Godfrey, B. J., & Reed, M. (2013). *App store after five years.* Retrieved from http://actonline.org/2013/07/19/app-store-after-five-years/

Goodwin, K., & Highfield, K. (2012). *iTouch and iLearn: an examination of 'educational' Apps.* Paper presented at the Early Education and Technology for Children conference, Salt Lake City, UT.

Gray, C., Dunn, J., Moffett, P., & Mitchell, D. (2017). *Mobile Devices in Early Learning: Evaluating the Use of Portable Devices to Support Young Children's Learning*. Belfast, UK: Stranmillis University College.

Guernsey, L., & Levine, M. (2015). *Tap Click Read: Growing Readers in a World of Screens*. San Francisco, CA: Jossey Bass.

Guernsey, L., Levine, M., Chiong, C., & Severns, M. (2012). *Pioneering literacy in the digital wild west: Empowering parents and educators*. Washington, DC: Campaign for Grade-Level Reading.

Handal, B., El-Khoury, J., Campbell, C., & Cavanagh, M. (2013). A framework for categorising mobile applications in mathematics education. *Australian Conference on Science and Mathematics Education*. Retrieved from https://researchonline.nd.edu.au/edu_conference/70/

Highfield, K., & Goodwin, K. (2013). Apps for mathematics learning: A review of 'educational' apps from the iTunes App Store. In V. Steinele, L. Ball & C. Bardini (Eds.), *Mathematics education: Yesterday, today and tomorrow. Proceedings of the 36th annual conference of the Mathematics Education Research Group of Australasia*, (pp. 378-385). Melbourne, Australia: MERGA.

Hiniker, A. (2018). *Research Suggests Well-Designed Preschool Apps Can Encourage Family Engagement and Self-Regulation*. Retrieved from http://joanganzcooneycenter.org/2018/05/21/research-suggests-well-designed-preschool-apps-can-encourage-family-engagement-and-self-regulation/

Hirsh-Pasek, K., Zosh, J. M., Golinkoff, R. M., Gray, J. H., Robb, M. B., & Kaufman, J. (2015). Putting education in 'educational' apps: Lessons from the science of learning. *Psychological Science in the Public Interest*, *16*(1), 3–34. doi:10.1177/1529100615569721 PMID:25985468

ISTE. (2007). *Standards for Global Learning in the Digital Age*. Available: http://www.iste.org/standards.aspx

Joe, S. (2018). *Out of touch: An investigation of the disconnect between 'educational' app design and research* (Doctoral dissertation). Auckland University of Technology.

Kabali, H. K., Irigoyen, M. M., Nunez-Davis, R., Budacki, J. G., Mohanty, S. H., Leister, K. P., & Bonner, R. L. (2015). Exposure and use of mobile media devices by young children. *Pediatrics*, *136*(6), 1044–1050. doi:10.1542/peds.2015-2151 PMID:26527548

Kalogiannakis, M., Ampartzaki, M., Papadakis, St., & Skaraki, E. (2018). Teaching Natural Science Concepts to Young Children with Mobile Devices and Hands-on Activities. A Case Study. *International Journal of Teaching and Case Studies*, *9*(2), 171–183. doi:10.1504/IJTCS.2018.090965

Kalogiannakis, M., & Papadakis, St. (2017a). A proposal for teaching ScratchJr programming environment in preservice kindergarten teachers. In *Proceedings of the 12th Conference of the European Science Education Research Association (ESERA), Research, practice and collaboration in science education*. Dublin City University and the University of Limerick.

Kalogiannakis, M., & Papadakis, St. (2017b). Pre-service kindergarten teachers acceptance of 'ScratchJr' as a tool for learning and teaching computational thinking and Science education. In *Proceedings of the 12th Conference of the European Science Education Research Association (ESERA), Research, practice and collaboration in science education*. Dublin City University and the University of Limerick.

Kalogiannakis, M., & Papadakis, St. (2017c). Combining mobile technologies in environmental education: A Greek case study. *International Journal of Mobile Learning and Organisation, 11*(2), 108–130. doi:10.1504/IJMLO.2017.084272

Kalogiannakis, M., & Papadakis, St. (2019). Evaluating pre-service kindergarten teachers' intention to adopt and use tablets into teaching practice for natural sciences. *International Journal of Mobile Learning and Organisation, 13*(1), 113–127. doi:10.1504/IJMLO.2019.096479

Kucirkova, N. (2016). Trans-and intra-apps: innovating the app market and use. In *Apps, Technology and Younger Learners* (pp. 272–284). New York: Routledge. doi:10.4324/9781315682204

Kucirkova, N., Messer, D., Sheehy, K., & Panadero, C. F. (2014). Children's engagement with educational iPad apps: Insights from a Spanish classroom. *Computers & Education, 71*, 175–184. doi:10.1016/j.compedu.2013.10.003

Larkin, K. (2013). Mathematics Education: Is there an app for that? In V. Steinle, L. Ball, & C. Bardini (Eds.), *Proceedings of the 36th annual conference of the Mathematics Education Research Group of Australasia. Mathematics education: Yesterday, today and tomorrow*, (pp. 426–433). Melbourne, Australia: MERGA.

Larkin, K. (2015a). The search for fidelity in geometry apps: an exercise in futility? In M. Marshman, V. Geiger, & A. Bennison (Eds.), *Mathematics education in the margins. Proceedings of the 38th annual conference of the Mathematics Education Research Group of Australasia* (pp. 341–348). Sunshine Coast: MERGA.

Larkin, K. (2015b). An app! An app! My kingdom for an app: an 18 month quest to determine whether apps support mathematical knowledge building. In T. Lowrie & R. Jorgensen (Eds.), Digital games and mathematics learning: potential, promises and pitfalls. New York: Springer. doi:10.1007/978-94-017-9517-3_13

Larkin, K., Kortenkamp, U., Ladel, S., & Etzold, H. (2018). Using the ACAT Framework to Evaluate the Design of Two Geometry Apps: An Exploratory Study. *Digital Experiences in Mathematics Education, 5*(1), 59–92. doi:10.100740751-018-0045-4

Learning Exchange Parramatta. (2011). *iPads in Schools: Use Testing*. Parramatta, Australia: Catholic Education Diocese. Retrieved from http://www.learningexchange.nsw.edu.au/_resources/ipads-in-schools--usetesting-2.pdf

Lee, J. (2016). *Early Learning and Educational Technology Policy Brief. Office of Educational Technology*. US Department of Education.

Lynch, J., & Redpath, T. (2014). 'Smart' technologies in early years literacy education: A meta-narrative of paradigmatic tensions in iPad use in an Australian preparatory classroom. *Journal of Early Childhood Literacy, 14*(2), 147–174. doi:10.1177/1468798412453150

Martens, M., Rinnert, G. C., & Andersen, C. (2018). Child-Centered Design: Developing an Inclusive Letter Writing App. *Frontiers in Psychology*, *6*(9), 2277. doi:10.3389/fpsyg.2018.02277 PMID:30574104

Michael Cohen Group LLC. (2011). *Young Children, Apps & iPad.* Michael Cohen Group LLC Report. Retrieved from https://mcgroup.wpengine.com/wp-content/uploads/2015/03/ipad-study-cover-page-report-mcg-info_new-online.pdf

Namukasa, I. K., Gadanidis, G., Sarina, V., Scucuglia, S., & Aryee, K. (2016). Selection of apps for teaching difficult mathematics topics: An instrument to evaluate touch-screen tablet and smartphone mathematics apps. In *International perspectives on teaching and learning mathematics with virtual manipulatives* (pp. 275–300). Dordrecht: Springer. doi:10.1007/978-3-319-32718-1_12

National Association for the Education of Young Children & the Fred Rogers Center for Early Learning and Children's Media at Saint Vincent College. (2012). Technology and Interactive Media as Tools in Early Childhood Programs Serving Children from Birth through. *Age, 8*. Retrieved from https://www.naeyc.org/files/naeyc/file/positions/PS_technology_WEB2.pdf

Özeke, V. (2018). Evaluation of Educational Mobile Apps for Turkish Preschoolers from Google Play Store. *European Journal of Education Studies*, *4*(4), 238–259.

Papadakis, S. (2018). The use of computer games in classroom environment. *International Journal of Teaching and Case Studies*, *9*(1), 1–25. doi:10.1504/IJTCS.2018.090191

Papadakis, St., & Kalogiannakis, M. (2017a). Mobile educational applications for children. What educators and parents need to know. *International Journal of Mobile Learning and Organisation*, *11*(3), 256–277. doi:10.1504/IJMLO.2017.085338

Papadakis, St., & Kalogiannakis, M. (2017b). Evaluation of Greek Android mobile applications for preschoolers. *Preschool and Primary Education*, *5*, 65–100. doi:10.12681/ppej.11208

Papadakis, St., & Kalogiannakis, M. (2019). Evaluating a Course for Teaching Advanced Programming Concepts with Scratch to Preservice Kindergarten Teachers: A Case Study in Greece. In D. Farland-Smith (Ed.), *Early Childhood Education* (pp. 1-19). London: IntechOpen Limited. Retrieved from www.intechopen.com

Papadakis, St., Kalogiannakis, M., & Zaranis, N. (2016a). Comparing tablets and PCs in teaching mathematics: An attempt to improve mathematics competence in early childhood education. *Preschool and Primary Education*, *4*(2), 241–253. doi:10.12681/ppej.8779

Papadakis, St., Kalogiannakis, M., & Zaranis, N. (2016b). Developing fundamental programming concepts and computational thinking with ScratchJr in preschool education: A case study. *International Journal of Mobile Learning and Organisation*, *10*(3), 187–202. doi:10.1504/IJMLO.2016.077867

Papadakis, St., Kalogiannakis, M., & Zaranis, N. (2017). Designing and creating an educational app rubric for preschool teachers. *Education and Information Technologies*, *22*(6), 3147–3165. doi:10.100710639-017-9579-0

Papadakis, S., Kalogiannakis, M., & Zaranis, N. (2018). Educational apps from the Android Google Play for Greek preschoolers: A systematic review. *Computers & Education*, *116*, 139–160. doi:10.1016/j.compedu.2017.09.007

Papadakis, St., Kalogiannakis, M., & Zaranis, N. (2018). Educational apps from the Android Google Play for Greek preschoolers: A systematic review. *Computers & Education*, *116*, 139–160. doi:10.1016/j.compedu.2017.09.007

Pelton, F. L., & Pelton, T. (2012). Sharing strategies with Teachers: iPods in math class. In *Society for Information Technology & Teacher Education International Conference* (Vol. 2012, No. 1, pp. 4363-4366). Academic Press.

Rideout, V. (2017). *The Common Sense census: Media use by kids age zero to eight*. San Francisco, CA: Common Sense Media.

Roldán-Álvarez, D., Márquez-Fernández, A., Martín, E., & Guzmán, C. (2016). Learning experiences using tablets with children and people with autism spectrum disorder. In *European Conference on Technology Enhanced Learning* (pp. 640-643). Dordrecht: Springer. 10.1007/978-3-319-45153-4_81

Sari, B., Takacs, Z., & Bus, A. (2017). What are we downloading for our children? Best-selling children's apps in four European countries. *Journal of Early Childhood Literacy*; Advance online publication. doi:10.1177/1468798417744057

Shuler, C. (2009). *iLearn: A content analysis of the iTunes app store's educational section*. The Joan Ganz Cooney Center at Sesame Workshop.

Shuler, C., Levine, Z., & Ree, J. (2012). *iLearn II: An analysis of the education category of Apple's app store*. New York, NY: The Joan Ganz Cooney Center at Sesame Workshop. Retrieved from http://joanganzcooneycenter.org/upload_kits/ilearnii.pdf

Vaala, S., Ly, A., & Levine, M. H. (2015). Getting a Read on the App Stores: A Market Scan and Analysis of Children's Literacy Apps. Full Report. In *Joan Ganz Cooney Center at Sesame Workshop. Joan Ganz Cooney Center at Sesame Workshop*. Academic Press.

Watlington, D. (2011). Using iPod touch and ipad educational apps in the classroom. In M. Koehler & P. Mishra (Eds.), *Proceedings of Society for Information Technology & Teacher Education International Conference 2011* (pp. 3112-3114). Chesapeake, VA: AACE.

Wohlwend, K. E. (2017). Toddlers and touchscreens: Learning 'Concepts Beyond Print' with tablet technologies. In R. J. Meyer & K. F. Whitmore (Eds.), *Reclaiming early childhood literacies: Narratives of hope, power, and vision*. Mahwah, NJ: Lawrence Erlbaum.

Yelland, N., Drake, P., & Sadler, K. (2017). *Early Learning in STEM: Multimodal learning in the 21st century*. Melbourne: Victoria University.

ADDITIONAL READING

Dorouka, P., Papadakis, St., & Kalogiannakis, M. (2019). Tablets & apps for promoting Robotics, Mathematics, STEM Education and Literacy in Early Childhood Education. *International Journal of Mobile Learning and Organisation* (forthcoming article)

Kalogiannakis, M., Nirgianaki, G.-M., & Papadakis, St. (2018). Teaching magnetism to preschool children: The effectiveness of picture story reading. *Early Childhood Education Journal*, *46*(5), 535–546. doi:10.100710643-017-0884-4

Kazakoff, E. R., Sullivan, A., & Bers, M. U. (2013). The Effect of a Classroom-Based Intensive Robotics and Programming Workshop on Sequencing Ability in Early Childhood. *Early Childhood Education Journal*, *41*(4), 245–255. doi:10.100710643-012-0554-5

Orfanakis, V., Papadakis, St., Kalogiannakis, M., Ampartzaki, M., & Vassilakis, K. (2016). Digital Student Conference Platform Implementation: The case study of the "Research Project" course. *Open Education*, *12*(2), 5–23.

Papadakis, St., & Kalogiannakis, M. (2010). eTwinning in the early childhood as starting line of innovative practices for the didactic of natural sciences. In M. Kalogiannakis, D. Stavrou, & P.G. Michaelidis *(Eds.) Proceedings of the 7th International Conference on Hands-on Science, Bridging the Science and Society Gap*. Rethymno, 25-31 July 2010, 235-240.

Papadakis, S., & Kalogiannakis, M. (2018). Using Gamification for Supporting an Introductory Programming Course. The Case of ClassCraft in a Secondary Education Classroom. In A. Brooks, E. Brooks, & N. Vidakis (Eds.), *Interactivity, Game Creation, Design, Learning, and Innovation. ArtsIT 2017, DLI 2017. Lecture Notes of the Institute for Computer Sciences, Social Informatics and Telecommunications Engineering* (Vol. 229, pp. 366–375). Cham, Switzerland: Springer. doi:10.1007/978-3-319-76908-0_35

Papadakis, St., & Kalogiannakis, M. (2019). Evaluating the effectiveness of a game-based learning approach in modifying students behavioural outcomes and competence, in an introductory programming course? A case study in Greece, *International Journal of Technology Enhanced Learning*, Forthcoming article

Papadakis, St., Kalogiannakis, M., Orfanakis, V., & Zaranis, N. (2019). The appropriateness of scratch and app inventor as educational environments for teaching introductory programming in primary and secondary education. In Information Resources Management Association (IRMA) (Ed.), Early Childhood Development: Concepts, Methodologies, Tools, and Applications (pp. 797-819), USA: IGI-Global. doi:10.4018/978-1-5225-7507-8.ch039

Papadakis, S., Kalogiannakis, M., & Zaranis, N. (2019). Parental involvement and attitudes towards young Greek children's mobile usage. *International Journal of Child-Computer Interaction*. doi: . doi:10.1016/j.ijcci.2019.100144

Sullivan, A., & Bers, M. U. (2016). Robotics in the early childhood classroom: Learning outcomes from an 8-week robotics curriculum in pre-kindergarten through second grade. *International Journal of Technology and Design Education*, *26*(1), 3–20. doi:10.100710798-015-9304-5

Voogt, J., Fisser, P., Good, J., Mishra, P., & Yadav, A. (2015). Computational thinking in compulsory education: Towards an agenda for research and practice. *Education and Information Technologies*, *20*(4), 715–728. doi:10.100710639-015-9412-6

KEY TERMS AND DEFINITIONS

App: An app is a software application that works, generally, on a mobile technology such as a smartphone, tablet, or other similar device.

Early Childhood Education: Is a broad term used to describe any type of educational program that serves children in their preschool years, before they are old enough to enter kindergarten.

Mobile Device: The mobile device is a small-sized, portable computer that typically has touch-screen features.

Preschool Education: Is education that focuses on educating children from the ages of infancy until six years old.

Preservice Teachers: Students enrolled in an initial educator preparation program.

Smartphone: A smartphone is a term used to describe a category of mobile devices with computer-like functionality.

STEM: The term STEM is an acronym used by those relevant to the educational method concerning the fields of science, technology, engineering, and mathematics.

Chapter 2
Too Many Apps to Choose From:
Using Rubrics to Select Mobile Apps for Preschool

Rachel Ralph
https://orcid.org/0000-0001-8661-8343
Centre for Digital Media, Canada

Patrick Pennefather
https://orcid.org/0000-0002-1936-1872
University of British Columbia, Canada

Jillianne Code
https://orcid.org/0000-0002-3096-1289
University of British Columbia, Canada

Stephen Petrina
University of British Columbia, Canada

ABSTRACT

Substantive research investigates the effects and impacts of tablets, in particular iPads, on children's education, but few papers discuss support for teachers in deciding which iPad applications can be integrated into the classroom. Even fewer articles are directed towards application developers. This chapter explores two standards for choosing apps for children - the four-pillar model of Hirsh-Pasek et al. (2015) and the rubric for the evaluations of educational apps for preschool children (REVEAC) by Papadakis, Kalogiannakis, and Zaranis (2017). This chapter draws from two standards for choosing iPad applications for young children in the classroom and through analysis of two applications will propose the REVEAC for educators and developers while also suggesting specific features that developers could consider when targeting children under 5 years old in educational contexts.

DOI: 10.4018/978-1-7998-1486-3.ch002

INTRODUCTION

There exists a body of research that shows both the negative and the positive results of integrating technology into the classroom. Children model parent behavior (Feldman, Bamberger, & Kanat-Maymon, 2013; White et. al, 2014). Like it or not children are attracted to digital devices, particularly if new generations of parents provide access to them or children simply observe parents use of mobile devices as they age. In a recent survey children under five years old were found to spend two to three hours per day interacting with devices (Childwise, 2016; Common Sense Media, 2017; Karsenti, 2013). Another survey conducted in the United States found that over 98% of homes have at least one mobile device, of which 43% of children aged 2-4 own their own tablet device (Common Sense Media, 2017). The average time spent on devices continues to grow and the impact of these devices needs to be further explored if we are to better understand the implications of their use. As Common Sense Media (2017) states "the proportion of homes with a tablet has nearly doubled over the past four years (from 40 percent to 78 percent)" (p.23). Some consider this to be a challenging problem as "the average amount of time spent with mobile devices each day has tripled (again), going from 5 minutes a day in 2011 to 15 minutes a day in 2013 to 48 minutes a day in 2017" (Common Sense Media, 2017, p.3). Despite media sensationalism and some research provoking an insurgence of negative associations with mobile devices and other technologies (Cadoret et al, 2018; Kara, 2018), numerous positive outcomes have also emerged. For example, researchers have found that children exposed to media and technology at early ages has resulted in increased motivation for learning, higher level thinking skills such as problem solving as well as the ability to work more independently at an earlier age (Flewitt, Messer, & Kucirkova, 2014; Herodotou, 2018; Karsenti, 2013; Lynch & Redpath, 2014). Faced with pressures within and outside the classroom, educators are forced to make decisions regarding their use of emerging technologies in the classroom. While we could perform a review of the pros and cons of the use of the iPad in early childhood education, we believe that how the iPad is integrated into curriculum and in particular what content (iPad applications) children use, will be of more benefit to teachers.

While there is substantive research that investigates the effects and impacts of tablets, in particular iPads, on children's education, few papers discuss a process that can support teachers in deciding which iPad applications can be integrated into the classroom. Even fewer articles are directed towards application developers in order to better align them with standards that would help them target their software to teachers and students. Our goal is twofold: (1) to draw from two standards for choosing iPad applications for young children in the classroom; and (2) based on analysis of two applications, to propose specific features that developers could consider when targeting children under 5 years old in educational contexts. This chapter will explore two standards for choosing apps for children - the four-pillar model of Hirsh-Pasek et al. (2015) and the rubric by Papadakis, Kalogiannakis, and Zaranis (2017). While the former is simple, the latter has been validated. In this chapter, two apps, *Sock Puppets* and *ChatterPix Kids* were evaluated using the two frameworks. While the simple one makes for a quick uptake and provides a foundation of evaluation for educators, the validated framework is not only validated but provides a higher level of rigour across multiple categories. Before getting into the evaluation of each app, we will provide a background of the use of iPads in early education settings.

BACKGROUND

The Use of iPads

There are many mobile devices that children access at home. Every kind of tablet or smartphone seems to be at their fingertips. Mobile devices, in particular iPads, are also readily available at schools starting in Kindergarten, however not as often in early childhood education (ECE) settings. That said, imagine you are a preschool teacher. You have been told that your young students need to learn about digital literacy and all of a sudden, a bunch of iPads get dumped into your lap. It is now your job to figure out 1) what to do with them; 2) how to integrate them into your class; and 3) how to effectively integrate them through accessing research that discusses the pros and cons of doing so. This can be quite challenging to many ECE educators as media has suggested that iPads add to increased screen time, which may promote negative social behaviours (AAP, 2014). On the other hand, iPads may intuitively provide teachers with a tool that can replace passive use of computers with a more touch-based interactive system.

But why iPads? iPads entered the market only in the last eight or nine years, and more recently in schools, teachers have access to an increasing amount of research documents that reveal their pros and cons. iPad devices have become more portable, affordable, and efficient (Flewitt et al., 2014; Karsenti, 2013; Lynch & Redpath, 2014). These devices also allow for material, physical, and virtual productions, as well as multiple opportunities to blend these together (Rowsell & Harwood, 2015). Additionally, iPads are one of the most "cutting-edge, culturally powerful yet enigmatic technological tools" for young learners that are available (Flewitt et al., 2014, p. 3). Apple's education program encourages the purchase of iPads for schools as they offer high volume discount prices of devices and apps (Apple, 2015). There are over 400 million iPads sold offering over 2.2 million apps and in particular over 300,000 for children (Alper, 2013; Apple, 2015, 2017; Hendela, 2014; Protalinkski, 2018).

Based on market trends, the influence of iPads in the classroom continues to grow and there are many benefits in choosing to integrate iPads into current curriculum. These include: increased social learning, communication and collaboration, creativity, motivation, concentration, and independence. Children's social learning is fostered by communication and collaboration allowing for cognitive growth through social interaction (Beschorner & Hutchison, 2013; Plowman, Stevenson, Stephen, & McPake, 2012). The use of iPads fostered communication and collaboration with peers, but also between teachers and students (Karsenti, 2013; Roskos et al., 2012). Moreover, iPad integration supports multiple forms of communication through various modes of media that allow oral, written, and graphic communication that enable the majority of children to participate in multiple activities that may be restrictive in other devices or low-tech options (Flewitt et al., 2014; McPake, Plowman, & Stephen, 2013). The multimodal approach encourages creative endeavors.

The literature attracts teachers to using iPads in reports that show children use a variety of apps to create multimedia products including voice recordings, video recordings, graphics, and text (Arnott, Grogan, & Duncan, 2016; Flewitt et al., 2014; Karsenti, 2013; McPake et al., 2013). Some research shows that creativity encourages independent thinking and the ability to produce and think on an independent level is fostered through iPad use (Flewitt et al., 2014; Lynch & Redpath, 2014; Masek, Murcia, & Morrison, 2012). Educators can use iPads to allow learners to work at different paces and styles (Masek et al., 2012). Children become independent users much more quickly than they did with previous devices. iPad use further inspires increased motivation and concentration for students (Axelsson, Andersson, & Gulz, 2016; Flewitt et al., 2014; Lynch & Redpath, 2014; Masek et al., 2012). Children are able to make their own choices (Beschorner & Hutchison, 2013) and they enjoy working beyond skill and drill apps. Additionally, children are highly motivated with heightened concentration that allows for more advanced literacy skills and information technology skills (Flewitt et al., 2014; Karsenti, 2013).

Touch interfaces of iPads have influenced cognitive abilities, as touch is a primary form of interaction for young children and it extends their understanding (Crescenzi, Jewitt, & Price, 2014; Mangen, 2010; Neuman & Neuman, 2014). Even though early touch interactions may be referring to physical contact with other people, the touch interaction with iPads is still an influential part of extending understanding. Moreover, haptic perception through tactile senses takes children beyond scanning and browsing into more interactive behaviours (Mangen, 2010; Roskos, Burstein, & You, 2012). Preschool-aged children's fine motor skills are less developed, but they are able to use their fingers in various apps (Moyer-Packenham et al., 2015). These applications continue to help refine children's fine motor skills. It is important to note that learning isn't fostered by the iPads alone. Curriculum, structured lessons that are guided by the teacher, and continuous management of student interactions can support increased benefits with these devices.

Several potential benefits of the use of the iPad in the classroom have been indicated in the section above. However, an iPad is only as effective as their applications. The challenge that teachers face is to sift through an overwhelming number of applications. A teacher can also get quite exhausted filtering through over 300,000 applications targeting young children in the apple store).To support educators, two models have been developed by researchers to support this daunting task: the four pillar model and the rubric for the evaluations of educational apps for preschool children. We will describe both and then analyze two iPad applications using the REVEAC model.

The Four Pillar Model

The four pillar model, designed by Hirsch-Pasek et al. (2015), is based on Chi (2009) taxonomy of learning and is a child centric approach and how some apps are designed to engage children (or not). The four pillars include: active, engaged, meaningful, and socially interactive learning.

Based on these four categories, we developed a 4 point score sheet (reflective of a rubric) (see Table 1). Each category has a broad description to which we provided a 4 point scale of whether the app matches that category.

Rubric for the EValuations of Educational Apps for Preschool Children (REVEAC)

The REVEAC was designed with several pedagogical and technological considerations that can validate app selection. It is divided into four main categories (educational content, functionality, technical characteristics, and design) and 18 subcategories (knowledge package appropriateness, learning provision, levelling, motivation/engagement; child-friendliness, autonomy, instructions existence, configuration ability; performance and reliability, advertisements/electronic transactions, social interactions, error correction/feedback provision, progress monitoring/sharing, bias free; graphics, sound, layout/scenery, app/menu design). Within each category there are 4 levels of evaluation, including: unsatisfactory/poor, needs improvement, good quality, and exemplary. For full rubric see Papadakis, Kalogiannakis, and Zaranis (2017).

Table 1. Hirsch-Pasek et al four pillars as a 4 level rubric

Category	Description	unsatisfactory/ poor *1*	needs improvement *2*	good quality *3*	Exemplary *4*
Active learning	"minds on" activity rather than physical activity suggests that children are not just copying what is going on in their lives rather they are active knowledge builders. Hirsh-Pasek et al. (2015) use this idea to identify app features where children are actively participating in the activity and not just passive viewers. They suggest that some apps encourage this behaviour through touching the screen, moving the device, talking into the microphone, or interacting with the camera. Hirsh-Pasek et al. (2015) suggests that "children can be actively involved with an app when they use it as a platform to discover new information about a content area" (p. 10).				
Engaged learning	Preschool-aged children are predisposed to distractor that can disrupt the primary objective and understanding for children. It is also interesting to note how distractions made by advertisements can make the learning even more challenging, especially when the children might click on the ad and go into the app store and potentially purchase the app if it is linked to a credit card. Further, educational apps should engage children cognitively, which is a challenge when designing for young children whose developmental levels can be quite distinct.				
Meaningful learning	"learning with a purpose, learning new material that is personally relevant, and linking learning to pre-existing knowledge" (Hirsh-Pasek et al., 2015, p. 13). This means moving beyond disconnected facts or rote learning. Finding meaning can be challenging for designers, as meaning is very contextual. However, Hirsh-Pasek et al. (2015) ask "does the app ask the child to go beyond rote learning?; Does the app experience tap into the child's personal history, activate prior knowledge of a subject, or build a rich narrative?" (p. 15). The idea is to expand the learning beyond just learning one simple piece and connect it back to the children's lives to make it meaningful.				
Socially interactive learning	In television, we call this co-viewing where a teacher has dialogue with children to promote active engagement. Also, apps that encourage joint media engagement can support social interaction. Some apps do this better than others by asking for children to provide an immediate tap or swipe response, but not all apps are socially interactive. App design can incorporate social interaction with face-to-face interactions (i.e., collaborating on a project or game), mediated and support social relations with on-screen character (i.e., characters responding to children's speech.				

The REVEAC, compared to the 4 pillars, is very rigorous. This rubric is based on several other frameworks including the four pillars of Hirsch-Pasek et al. (2015) where Papadakis, Kalogiannakis, and Zaranis (2017) as a guide for "researchers, educators and designers in evidence-based app development and secondly to create an evidence-based guide" (p. 3150). The REVEAC takes this further to consider educational research, app rating scales but targeting preschool-aged children (that many other frameworks do not consider), and evaluation from researchers. Further, the REVEAC applies worldwide rubric standards that consider criteria size (not too small) and detailed performance level descriptions.

Methods

Since there is value in using both models, we will analyze two iPad applications through the lens of each. Both will support teachers in evaluating applications they want to integrate into their classroom curriculum. To model how teachers could interrogate an iPad application they wish to use in the classroom, we will evaluate *Sock Puppets by Smith Micro Software* and *ChatterPix Kids by Duck Duck Moose*.

Sock Puppets is an app for ages 4+ that allows users to create "lip-synced videos…[with] puppets, props, scenery, and backgrounds" (Sock Puppets App, 2016). In the app, there are multiple puppets that each child can choose from (see Figure 1). The user can also choose from multiple backgrounds or settings and add other props or effects to the scene. Once these decisions are made, users now animate and record their sock puppets to create a story (see Figure 2).

ChatterPix Kids is also for ages 4+ that is sponsored by Duck Duck Moose and Khan Academy that lets users "take any photo, draw a line to make a mouth, and record your voice" (ChatterPix Kids app, 2019) to animate a still image (see Figure 3).

To evaluate each app, two researchers downloaded the apps onto their iPads. They "played" each app once to go through the experience to reflect the experience of the child or teacher using the app for the first time. They completed the process from onboarding to final product. After the "play" round, each researcher evaluated the apps, independently, applying a score out of 4 for each model (unsatisfactory/poor, needs improvement, good quality, and exemplary). It is important to note that the 4 pillar model does not have these evaluation categories, however, we gave each of the four pillars a score indicator to reflect the REVEAC in order to compare the evaluations of each framework.

Figure 1 . Select characters in Sock Puppets app

Figure 2 . Recording an animation in Sock Puppets app

RESULTS

As noted, each reviewer independently evaluated each app using the two frameworks. Scores for each framework will be presented for each app. The scores are out of 4: 1= unsatisfactory/poor, 2= needs improvement, 3= good quality, and 4=exemplary. Occasionally, reviewers scored the item differently, the average of each score will be presented which may include a score in between each of the 4 (such as 2.5 that needs improvement but is approaching good quality).

Four Pillar Model Evaluation of Two Apps

Both *Sock Puppets* and *ChatterPix Kids* were overall satisfactory for learning with no score lower than 2 (see Figure 4). While both apps need improvement on their active engagement and meaningful learning levels (between 2 and 2.5), they both would be considered socially active (a score of 3 and 3.5 respectively). This can suggest that these two apps provide social learning opportunities for educators.

At a quick glance in Figure 5, one can see the level of rigour provided by the REVEAC that the four pillar model does not. Even though some educators may prefer the four pillar model as it is much simpler and quicker for analysis, the detail provided by the REVEAC is much better. More categories may be more complex but it seems to provide an overall better evaluation for the app.

The first seven categories indicating educational content scores rate *ChatterPix Kids* either needing some improvement or good quality with bias approaching exemplary (between 2 and 3.5), whereas *Sock Puppets* scores ranged between unsatisfactory/poor or needing improvement and good quality (between 1 and 3). The next four categories focused on design, both apps scored similarly of needing improvement and approaching good quality (scores 2.5-3). For functionality, the next four categories, *ChatterPix Kids* was approaching good quality (scores 2.5-3) while *Sock Puppets* mostly remained needing improvement (scores 2-3, with more below 2.5). The final three categories focusing on technical characteristics were mostly similar (scores 2.5-4), with both apps receiving exemplary on no advertisements.

Figure 3 . Adding a mouth to animate a picture of a teddy bear in ChatterPix Kids app

ANALYSIS

Cohen's Kappa was calculated to determine the inter-rater reliability. Cohen's Kappa represents moderate agreement value at .5, good agreement value at .7, and very good agreement above .8 (Pallant, 2010).

For the four-pillar model, there was very good agreement for engaged learning, meaningful learning, and socially interactive learning (Cohen's Kappa =.85, $p < .0005$). For active learning, there was good agreement (Cohen's Kappa =.75, $p < .0005$).

For the REVEAC, inter-rater reliability was measured for each major evaluation sector: educational content, functionality, technical characteristics, and design. For educational content there is good agreement (Cohen's Kappa =.71, $p < .0005$). Design and technical characteristics were moderate agreement (Cohen's Kappa =.50, $p < .0005$). Functionality was very good agreement (Cohen's Kappa =.90, $p < .0005$).

Figure 4. Four Pillar model average scores for Sock Puppets and ChatterPix Kids

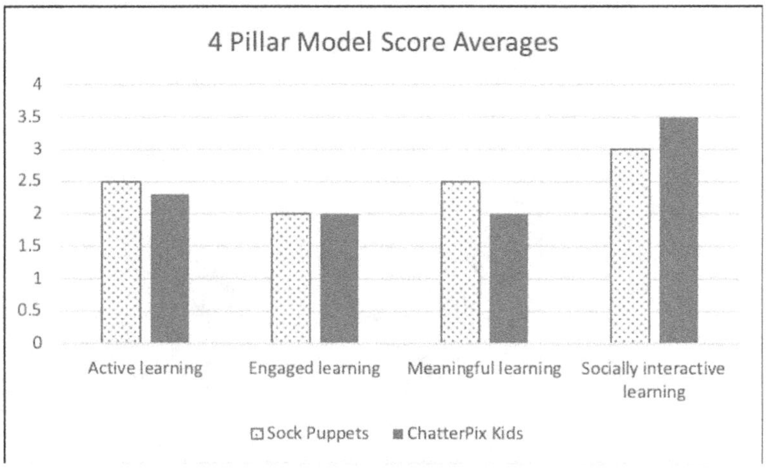

Figure 5. REVEAC average scores for Sock Puppets and ChatterPix Kids

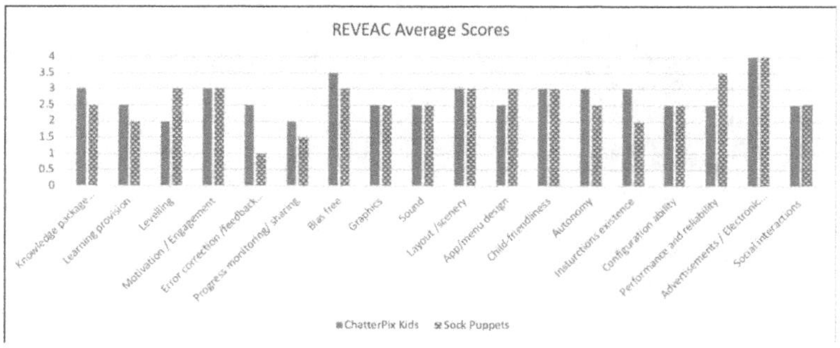

DISCUSSION

The increased use of mobile devices in classrooms has prompted a shift in the pedagogical approach to media and technology. Mobile devices and in particular iPads, have become pervasive in 21st century classrooms. Devices have become more portable, affordable, efficient, and prominently used (Common Sense Media, 2017, Flewitt et al., 2014; Karsenti, 2013; Lynch & Redpath, 2014). Benefits and detriments of iPad use have been reported in the research literature, however app use and selection need continuous evaluation. Teachers need access to appropriate frameworks for app selection validated by researchers. Moreover, we need more context specific research as the ubiquitous nature of these devices is undeniable.

Matching Models, Pillars, Rubrics, and Frameworks with ECE Goals

How to choose appropriate apps can be difficult and time consuming, but this selection can be supported by using validated models. Researchers and teachers know that:

humans learn best when they are actively involved ('minds-on'), engaged with the learning materials and undistracted by peripheral elements, have meaningful experiences that relate to their lives, and socially interact with others in high-quality ways around new material, within a context that provides a clear learning goal. (Hirsh-Pasek et al. 2015, p. 7).

Based on this understanding, we know that we want digital tools and activities to also support these ideas. Even though the four pillars aren't as rigorous as the REVEAC, it identifies important aspects of learning that many teachers would support. It is essential to select media and technology that "allow children opportunities to discover make choices…to explore, imagine and problem-solve" (Beschorner & Hutchison, 2013, p. 17). In ECE, these choices are often based upon pre-existing program goals. While researchers and teachers would benefit from integrating REVEAC, for example, the model offers different components that would still be useful in applying in order to make decisions as to which apps to use. A pedagogical purpose to media and technology integration should align with ECE goals (Aronin & Floyd, 2013; Morgado, 2008). By keeping ECE goals top of mind, researchers can further filter applications they wish to use to conduct inquiry.

In some parts of Canada, the Early Learning Advisory Group established an early learning framework that "refers to the emerging and expanding of young children's physical, intellectual, emotional, social, and creative capacities" (Early Learning Advisory Group, 2008, p. 2). The early learning framework written for parents, teachers, or caregivers, is meant to support the child, as they are a person with "complex identities, grounded in their individual strengths and capacities, and their unique social, linguistic, and cultural heritage" (Early Learning Advisory Group, 2008, p. 4). Recognizing unique social and cultural landscape, this framework also reflects the Convention of the Rights of a Child (The United Nations, 1989) as well as regulations on early childcare practices. Most importantly the framework identifies the five key areas of development, including:

1. aesthetic and artistic development
2. emotional and social development
3. intellectual development
4. physical development and well-being, and
5. the development of social responsibility.

In the Language and Literacy area, the learning goal related to media and technology is: "communicate thoughts and experiences creatively using many different forms of expression" (Early Learning Advisory Group, 2008, p. 29). More specifically for preschool-aged children: "what opportunities do children have to use technology to explore their thoughts and ideas (e.g., computer software, websites, cameras, camcorders)?" (Early Learning Advisory Group, 2008, p. 30). Since the Early Learning Framework was published in 2008, media and technology has evolved dramatically (iPads did not even exist on the market until 2010) and therefore this early learning framework needs updating to current and present and potential future technological practices.

Sociability and Controlled Access in ECE

Even though some research suggests the detrimental effects of media and technology, especially on young children, a nuanced approach to tech integration is slowly becoming the consensus. In particular, the AAP has revised their initial ban on devices statement. They do recommend parents to prioritize unplugged time for infants and toddlers, but recognize that "some media can have educational value for children…[and] that this be high quality programming" (American Academy of Pediatrics (AAP), 2016). They continue to make positive recommendations towards *Sesame Workshop* and Public Broadcasting Service (PBS) programming. The AAP, *Sesame Workshop*, and PBS want to help children understand what they are seeing and have critically reflective discussions.

Similarly, some theories and traditions of early childhood learning, such as Montessori and Waldorf, recommend cautious introductions of digital media and technology (Dunn, 2000; Tosco, 2015). In particular, the research describing the detrimental effects of media and technology focuses on screen-time, where children are passive observers rather than participants in interactive media and technology (Drew et al., 2019; Mares & Woodward, 2005). The inclusion of media and technology has been integrated into ECE by referring to the National Association for the Education of Young Children (NAEYC) position statement on media and technology. The NAEYC alongside the Fred Rogers Center support the AAP in discouraging obsessive screen use under the age of 2 as early brain development occurs; however, they also understand there may be appropriate times for screen use at this young developmental age (American Academy of Pediatrics (AAP), 2014, 2016; "Fred Rogers Center," 2014; NAEYC, 2012). For example, appropriate screen time would be a Skype or FaceTime call with grandparents or trying an interactive app or watching *Sesame Street* and participating in a *Sesame Workshop*. The NAEYC and the Fred Rogers Centre are aware of concerns and have developed the following position statement:

Technology and interactive media are tools that can promote effective learning and development when they are used intentionally by early childhood educators, within the framework of developmentally appropriate practice, to support learning goals established for individual children. (NAEYC, 2012, p. 5)

Intentionally designed use implies that the use of interactive media like iPads be integrated into curricular design and that access to interactive technologies be mediated through parents. Future studies would use the REVEAC to choose apps not only because of its validated metrics, but also the detailed criteria, the design specifically for preschool-aged children, but also as it addresses four key areas (educational content, design, functionality, and technical characteristics).

SOLUTIONS AND RECOMMENDATIONS

The two models analyzed in this chapter focus on selecting apps. While we have identified some of the potential use of these two models, we have not addressed the latter part of our goal: to propose specific features that developers could consider when targeting children under 5 years old in educational contexts. Developers and app designers could use these models alongside other design pieces to make more appropriate decisions for the 0-4 age range.

In particular, we can identify the 4 subsections of "design" in the REVEAC (see Table 2): graphics, sound, layout/scenery, and app/menu design.

For the first category, designers should consider high quality graphics. This does not mean that it needs to be three dimensional with hi-poly counts, 2D or low polycounts can be just as aesthetically pleasing. The attractiveness of the graphics is quite important as many children ages 0-4 are pre-literate and messages need to be conveyed via graphics alone. Sound is also as important in this respect. Hirsh-Pasek

Table 2. Design section of the REVEAC

Evaluation sector	unsatisfactory/ poor *1*	needs improvement *2*	good quality *3*	Exemplary *4*
Graphics	Poor quality images, not attractive graphics	Medium quality images, not so attractive graphics	Good quality images, attractive graphics.	High-quality images, very attractive graphics.
Sound	Poor sound quality. Unnecessary sounds, such as background music which it distracts child attention.	Moderate sound quality. Slightly distracting.	Good sound quality.	Excellent sound quality. Enhancing the learning experience.
Layout/ scenery	Layout elements are confusing. Is often unreasonable and unpredictable. Incorrect use of color and contrast for text and/ or graphics.	Layout / scenery. The layout is not intuitive or often is unreasonable. Text or graphics are hard to read or distinguish as they are often incompatible with the background	The layout, background, text, and graphics are mostly intuitive. The child may need further assistance on locating elements.	Clearly visible elements, easy to use, efficient and responsive layout.
App/menu design	Incorrect menu design, which complicates app handling. There is a lack of design consistency the different levels / screens of the app.	App/menu design is partially correct. There are inconsistencies or errors in the design.	App/menu design is correct. There is still some design errors or inconsistencies.	Proper menu design, which facilitates app handling. There is a design consistency between the different levels / screens of the app.

et al. (2015) also discuss the importance of limited distractors through engaged learning as children are prone to short attention spans. Meaningful choices for sound are critical when designing apps for this age. Is it just music? Is it music and narration? Is it music, narration, and sound effects? App developers should be user testing all of these features. Further, the layout/scenery play an important role for young children. By designing apps that children can find the features and use on their own, we are encouraging young children to become independent learners. Too many features and the app becomes too saturated and confusing for young children, not enough and it becomes boring or too text focused. It is the goldilocks problem to get it just right. Finally, in the app/menu design having consistency is very important. Similar to the layout, menus need to be consistent and simple so that children can play independently.

For some designers, these models are helpful, but they might need more than these 4 categories. Another model to consider is Malone's (1980) essential characteristics for designing computer games. These characteristics are divided into 3 major categories and several sub-categories (see Table 3). Each of these categories (and subcategories) are put into many apps made today and something that designers should consider. Even though Malone wrote about the design of computer games, it can be applied across multiple systems and devices. Malone intended to explore what makes things fun to learn. At an early age, we want to encourage young children to learn and by having a bit of fun, we can hopefully continue to encourage rather than discourage learning in future classroom settings.

Overall, using frameworks can assist in influencing designers of apps, but also teachers selecting apps to use with their students.

Table 3. Malone's (1980) essential characteristics for designing computer games

Guideline	Specific Characteristic	Detailed description of design
Challenge	Goal (objective of the game)	· Simple games = obvious goal · Complex games with built in goals or structured with level of difficulty · Some sort of feedback
	Uncertain outcome (all you know is that you will win or lose)	· Variable difficulty level: automatic, chosen by player, determined by skill · Multiple level goals (there is a meta goal like score keeping, or speed – do a task as fast as possible to win) · Hidden info with selective revealing · randomness
	Self-esteem for player	Game should encourage your self-esteem and not damage it (it can be difficult but not damaging)
Fantasy	Intrinsic	Skill is dependent on fantasy, fantasy is dependent on skill More interesting More explorative More 'learning'
	Extrinsic	Fantasy is only dependent on if skill is used correctly (right or wrong answer) Domain independent
	Emotional aspects	Can be completely different or reflect real life ex. War, destruction, competition
Curiosity	Sensory	Audio/visual effects: as decoration, to enhance fantasy, as a reward, as a representation system
	Cognitive	Completeness, consistency, parsimony: connect to existing knowledge to build understanding/skill
	Informative feedback	Surprising/random Educational/constructive

FUTURE RESEARCH DIRECTIONS

Researchers need to continue to test and validate both frameworks presented in this chapter. Researchers and educators using apps should use the frameworks before using apps. Even though we had good levels of inter-rater reliability, more testers for more apps should be studied. Also, app developers could be included in research. Researchers could work with developers and use the design section of the REVEAC framework and compare it to Malone's framework. Another assistance for teachers would be for app developers to put their REVEAC/four pillar/Malone scores in the app score. Providing a transparent score not just based on if people liked the app but with a more rigourous score affiliated to what teachers need.

Teachers as researchers need to develop and test more rapid approaches to support the choices they make when integrating new applications into their course plans. The REVEAC framework is one such approach. The framework can also be used to draw comparisons between two applications if a teacher is unsure as to which their students will benefit more from. Further validation can be achieved with a more experimental approach to an application's implementation in the classroom. For example, why not let the students decide which application is the right one for them? Simple survey questions or post-app reflection with the students can confirm or challenge a teacher's preconceived notions of what application would be most beneficial to the learning process.

CONCLUSION

Children now enter schools within curricula and policies framed for 21st century learning (Beschorner & Hutchison, 2013). Some of these policies advocate for social learning, intentional use of computers to facilitate learning, and personal learning approaches. Amidst changing policies, teachers are striving to find a balance on the integration of emerging technologies to support learning at home and within the classroom. Teachers would benefit from existing research models that might lead to a better understanding of how, when and why to engage children with technologies like the iPad. In addition, keeping ECE goals top of mind will help teachers and researchers identify mobile applications that will benefit their young students. Using validated methods like the REVEAC would support these methods. Children now enter schools within curricula and policies framed for 21st century learning (Beschorner & Hutchison, 2013) and teachers are striving to find a balance of appropriate apps for "plugged-in" activities. Furthermore, app developers could benefit with access to similar frameworks. By working with teachers and industry professionals we can connect industry, education, theory and research to provide validated structures for young children entering schools in the third decade of the 21st century. Continuous discourse and dissemination of validated methods can encourage teachers as they continue to move forward.

ACKNOWLEDGMENT

This research received no specific grant from any funding agency in the public, commercial, or not-for profit sectors.

REFERENCES

Alper, M. (2013). Developmentally appropriate new media literacies: Supporting cultural competencies and social skills in early childhood education. *Journal of Early Childhood Literacy*, *13*(2), 175–196. doi:10.1177/1468798411430101

American Academy of Pediatrics (AAP). (2014). The benefits of limiting TV. Retrieved from http://www.healthychildren.org/English/family-life/Media/Pages/The-Benefits-of-Limiting-TV.aspx

American Academy of Pediatrics (AAP). (2016). American Academy of Pediatrics announces new recommendations for children's media use. Retrieved from https://www.aap.org/en-us/about-the-aap/aap-press-room/Pages/American-Academy-of-Pediatrics-Announces-New-Recommendations-for-Childrens-Media-Use.aspx

Apple. (2015). Retrieved from http://www.apple.ca

Apple. (2017). App Store shatters records on New Year's Day. Retrieved from http://www.apple.com/newsroom/2017/01/app-store-shatters-records-on-new-years-day.html

Arnott, L., Grogan, D., & Duncan, P. (2016). Lessons from using iPads to understand young children's creativity. *Contemporary Issues in Early Childhood*, *17*(2), 157–173. doi:10.1177/1463949116633347

Aronin, S., & Floyd, K. K. (2013). Using an iPad in inclusive preschool classrooms to introduce STEM concepts. *Teaching Exceptional Children, 45*(4), 34–39. doi:10.1177/004005991304500404

Axelsson, A., Andersson, R., & Gulz, A. (2016). Scaffolding executive function capabilities via play-&-learn software for preschoolers. *Journal of Educational Psychology, 108*(7), 969–981. doi:10.1037/edu0000099

Beschorner, B., & Hutchison, A. (2013). iPads as a literacy teaching tool in early childhood. *Education Publications, 26*, 16–24.

Cadoret, G., Bigras, N., Lemay, L., Lehrer, J., & Lemire, J. (2018). Relationship between screen-time and motor proficiency in children: A longitudinal study. *Early Child Development and Care, 188*(2), 231–239. doi:10.1080/03004430.2016.1211123

ChatterPix Kids App. (2019). Retrieved from http://www.duckduckmoose.com/educational-iphone-itouch-apps-for-kids/chatterpixkids/

Chi, M. T. H. (2009). Active-constructive-interactive: A conceptual framework for differentiating learning activities. *Topics in Cognitive Science, 1*(1), 73–105. doi:10.1111/j.1756-8765.2008.01005.x PMID:25164801

Childwise. (2016). *The monitor pre-school report: Key behaviour patterns among 0-4 year olds*. Retrieved from http://www.childwise.co.uk/

Cingel, D. P., Sumter, S. R., Stoeten, E., & Mann, S. (2019). Can television help to decrease stigmatization among young children? The role of Theory of Mind and general and explicit inserts. *Media Psychology*, 15213269.2019.1601570. doi:10.1080/15213269.2019.1601570

Common Sense Media. (2017). Zero to eight: Children's media use in America: A common sense research study. Retrieved from https://www.commonsensemedia.org/sites/default/files/uploads/research/csm_zerotoeight_fullreport_release_2.pdf

Crescenzi, L., Jewitt, C., & Price, S. (2014). The role of touch in preschool children's learning using iPad versus paper interaction. *Australian Journal of Language and Literacy, 37*(2), 86–95.

Dunn, S. E. (2000). Technology: Where is it taking us? A "writing" perspective. *Montessori Life, 12*(1), 34–35.

Early Learning Advisory Group. (2008). *British Columbia early learning framework*. British Columbia, Canada: Ministry of Health and Ministry of Children and Family Development.

Feldman, R., Bamberger, E., & Kanat-Maymon, Y. (2013). Parent-specific reciprocity from infancy to adolescence shapes children's social competence and dialogical skills. *Attachment & Human Development, 15*(4), 407–423. doi:10.1080/14616734.2013.782650 PMID:23544455

Flewitt, R., Messer, D., & Kucirkova, N. (2014). New directions for early literacy in a digital age: The iPad. *Journal of Early Childhood Literacy, 15*(4), 1–22.

Fred Rogers Center. (2014). Retrieved from http://www.fredrogerscenter.org/

Hendela, T. (2014). App Store sales top $10 billion in 2013. *Apple Press Info*. Retrieved from http://www.apple.com/ca/pr/library/2014/01/07App-Store-Sales-Top-10-Billion-in-2013.html

Herodotou, C. (2018). Mobile games and science learning: a comparative study of 4 and 5 years old playing the game Angry Birds. *British Journal of Educational Technology*, *49*(1), 6–16. doi:10.1111/bjet.12546

Hirsh-Pasek, K., Zosh, J. M., Golinkoff, R. M., Gray, J. H., Robb, M. B., & Kaufman, J. (2015). Putting education in "educational" apps lessons from the science of learning. *Psychological Science in the Public Interest*, *16*(1), 3–34. doi:10.1177/1529100615569721 PMID:25985468

Kara, H. G. E. (2018). A case study on reducing children's screen time: the project of screen free week. *World Journal of Education*, *8*(1), 100–110. doi:10.5430/wje.v8n1p100

Karsenti, T., & Fievez, A. (2013). *The iPad in education: uses, benefits, and challenges - A survey of 6,057 students and 302 teachers in Quebec, Canada*. Montreal, Canada: CRIFPE.

Lynch, J., & Redpath, T. (2014). 'Smart' technologies in early years literacy education: A metanarrative of paradigmatic tensions in iPad use in an Australian preparatory classroom. *Journal of Early Childhood Literacy*, *14*(2), 147–174. doi:10.1177/1468798412453150

Malone, T. W. (1980). *What makes things fun to learn? Heuristics for designing instructional computer games*. In *Proceedings of the 3rd ACM SIGSMALL symposium and the first SIGPC symposium on Small systems*, Palo Alto, CA. 10.1145/800088.802839

Mangen, A. (2010). Point and click: Theoretical and phenomenological reflections on the digitization of early childhood education. *Contemporary Issues in Early Childhood*, *11*(4), 415–431. doi:10.2304/ciec.2010.11.4.415

Mares, M.-L., & Woodard, E. (2005). Positive effects of television on children's social interactions: a meta-analysis. *Media Psychology*, *7*(3), 301–322. doi:10.1207/S1532785XMEP0703_4

Masek, M., Murcia, K., & Morrison, J. (2012). Getting serious with iPads: The intersection of game design and teaching principals. *Australian Educational Computing*, *27*(2), 34–38.

McPake, J., Plowman, L., & Stephen, C. (2013). Pre-school children creating and communicating with digital technologies in the home. *British Journal of Educational Technology*, *44*(3), 421–431. doi:10.1111/j.1467-8535.2012.01323.x

Morgado, L. (2008). TEL practices in preschool and kindergarten education: Integrating computer use and computer programming in off-computer activities. In M. D. Lytras, D. Gasevic, P. Ordonez de Pablos, & W. Huang (Eds.), *Technology enhanced learning: Best practices*. Hershey, PA: IGI Global. doi:10.4018/978-1-59904-600-6.ch002

Moyer-Packenham, P. S., Shumway, J. F., Bullock, E., Tucker, S. I., Anderson-Pence, K. L., Westenskow, A., ... Jordan, K. (2015). Young children's learning performance and efficiency when using virtual manipulative mathematics iPad apps. *Journal of Computers in Mathematics and Science Teaching*, *34*(1), 41–69.

NAEYC. (2012). *Technology and young children – ages 3 through 8. A position statement of the National Association for the Education of Young Children.* Washington, DC: National Association for the Education of Young Children.

Neuman, M., & Neuman, D. (2014). Touch screen tablets and emergent literacy. *Early Childhood Education Journal, 42*(4), 231–239. doi:10.100710643-013-0608-3

Papadakis, S., Kalogiannakis, M., & Zaranis, N. (2017). Designing and creating an educational app rubric for preschool teachers. *Education and Information Technologies, 22*(6), 3147–3165. doi:10.100710639-017-9579-0

Plowman, L., Stevenson, O., Stephen, C., & McPake, J. (2012). Preschool children's learning with technology at home. *Computers & Education, 59*(1), 30–37. doi:10.1016/j.compedu.2011.11.014

Protalinski, E. (2018). Apple passes 400 million iPads sold. *Venture Beat.* Retrieved from https://venturebeat.com/2018/10/30/apple-passes-400-million-ipads-sold/

Roskos, K., Burstein, K., & You, B.-K. (2012). A typology for observing children's engagement with eBooks at preschool. *Journal of Interactive Online Learning, 11*(2), 47–66.

Rowsell, J., & Harwood, D. (2015). "Let it go": Exploring the image of the child as a producer, consumer, and inventor. *Theory into Practice, 54*(2), 136–146. doi:10.1080/00405841.2015.1010847

Sock Puppets App. (2016). Retrieved from https://apps.apple.com/ca/app/sock-puppets/id394504903

The United Nations. (1989). Convention on the rights of the child. *Treaty Series, 1577*, 3.

Tosco, J. E. (2015). *The effects of technology on engagement and retention among upper elementary Montessori students.* (Unpublished Master's Thesis), St. Catherine University.

White, N., Ensor, R., Marks, A., Jacobs, L., & Hughes, C. (2014). "It's mine!" Does sharing with siblings at age 3 predict sharing with siblings, friends, and unfamiliar peers at age 6? *Early Education and Development, 25*(2), 185–201. doi:10.1080/10409289.2013.825189

ADDITIONAL READING

Busschaert, C., Cardon, G., Van Cauwenberg, J., Maes, L., Van Damme, J., Hublet, A., ... De Cocker, K. (2015). Tracking and predictors of screen time from early adolescence to early adulthood: A 10-year follow-up study. *The Journal of Adolescent Health, 56*(4), 440–448. doi:10.1016/j.jadohealth.2014.11.016 PMID:25636817

Gentile, D. A., Berch, O. N., Choo, H., Khoo, A., & Walsh, D. A. (2017). Bedroom Media: One Risk Factor for Development. *Developmental Psychology, 53*(12), 2340–2355. Doi.org/10.1037/dev0000399. doi:10.1037/dev0000399 PMID:28945440

Hesketh, K. D., Crawford, D. A., Abbott, G., Campbell, K. J., & Salmon, J. (2015). Prevalence and stability of active play, restricted movement and television viewing in infants. *Early Child Development and Care*, *185*(6), 883–894. doi:10.1080/03004430.2014.963066

Hinkley, T., Salmon, J., Okely, A. D., Crawford, D., & Hesketh, K. (2012). Preschoolers' physical activity, screen time, and compliance with recommendations. *Medicine and Science in Sports and Exercise*, *44*(3), 458–465. doi:10.1249/MSS.0b013e318233763b PMID:21900847

Lin, L., Cherng, R., Chen, Y., Chen, Y., & Yang, H. (2015). Effects of television exposure on developmental skills among young children. *Infant Behavior and Development*, *38*, 20–26. doi:10.1016/j.infbeh.2014.12.005 PMID:25544743

Meziane-Fourati, A. (2017). Tunisian Mothers' Perception of Their Young Children's Interactive Screen Time at Home. *International Journal of Early Childhood Education and Care*, *6*, 1–10.

Neumann, M. M., & Neumann, D. L. (2017). The Use of Touch-Screen Tablets at Home and Pre-School to Foster Emergent Literacy. *Journal of Early Childhood Literacy, 17*(2), 203-220. DOI org/doi:10.1177/1468798415619773

Plowman, L., Stephen, C., & McPake, J. (2010). *Growing up with technology: Young children learning in a digital world.* New York: Routledge. doi:10.4324/9780203863619

Rideout, V. J., Foehr, U. G., & Roberts, D. F. (2010). Generation M2: Media in the lives of 8- to 18-year-olds *Henry J. Kaiser Family Foundation*. Retrieved from January 16, 2017 https://kaiserfamilyfoundation.files.wordpress.com/2013/04/8010.pdf

Tandon, P. S., Zhou, C., Lozano, P., & Christakis, D. (2010). Preschoolers' Total Daily Screen Time at Home and by Type of Child Care. *The Journal of Pediatrics*, *158*(2), 297–300. doi:10.1016/j.jpeds.2010.08.005 PMID:20980020

KEY TERMS AND DEFINITIONS

Four Pillar Model: Created by Hirsh-Pasek, K. et al to provide a framework for choosing appropriate educational apps on tablets. The four pillars are: active learning, engaged learning, meaningful learning, and socially interactive learning.

iPad(s): Apple's tablet mobile device first on the market in 2010.

Mobile Devices: A variety of devices including but not exclusive to: smartphones, tablets (iPads and other brands), and other touch devices (like iPods).

Montessori: A non-traditional school model where children have more "freedom", self-regulated/directed, hands-on, and multi-aged.

National Association For The Education of Young Children (NAEYC): a professional membership organization for children ages 0-8 that focuses on promoting early learning, including practice, policy and research.

Rubric For The Evaluations of Educational Apps For Preschool Children (REVEAC): Developed by Papadakis, Kalogiannakis, and Zaranis as a rigorous method for app selection based on four key areas: educational content, functionality, technical characteristics, and design.

Screen time: The amount of time spent in front of a screen (TV, computer, mobile device, etc.) more often referring to passive viewing versus interactive experiences.

Waldorf: From Rudolf Steiner, this education model promotes students artistic and practical skills holistically focusing on development of creativity and imagination.

Section 2
Apps as Tools to Promote Learning

Chapter 3
Teaching Natural Sciences to Kindergarten Students Using Tablets:
Results From a Pilot Project

Emmanuel Fokides
https://orcid.org/0000-0003-3962-0314
University of the Aegean, Greece

Dimitra Zachristou
University of the Aegean, Greece

ABSTRACT

The chapter presents the results of a project in which tablets were used for teaching natural sciences to kindergarten students. The classification of animals depending on certain characteristics was the subject matter. Forty-five students participated, divided into three groups. The first used printed material, the second used computers and webpages, and the third used tablets and AR applications. Bybee's 5Es provided the teaching framework for all groups. Data were collected using evaluation sheets and structured interviews. The students in the tablets' group performed better in all the evaluation sheets compared to the ones who were taught using printed material, but there were no statistically significant differences compared to the computers' group. A positive impact on motivation and enjoyment was noted in the tablets' group. Thus, it can be concluded that tablets are an interesting alternative teaching tool for very young students. Implications for research and practice are also discussed.

INTRODUCTION

In recent years, the importance of introducing concepts related to natural sciences in early childhood education has been acknowledged, as it contributes to the cognitive development of very young students (Eshach, 2006; Harlen, 2018; Trundle, 2010). Indeed, the exploration of the natural world is the source

DOI: 10.4018/978-1-7998-1486-3.ch003

of children's primary experiences; their innate curiosity and effort to comprehend the world that surrounds them compels them to build their original understanding/perceptions for it. As a result, children establish a body of ideas, in the form of interpretative models, before they even enter formal education (Akerson, Flick, & Lederman, 2000; Driver, Squires, Rushworth, & Wood-Robinson, 2014). Then again, studies have confirmed that many concepts related to this field are difficult to understand; both preschool and primary school children face serious problems. The plethora of their misconceptions, as recorded in the relevant literature, serves as a proof of the above statement (e.g., Driver et al., 2014). An area in which students face significant difficulties is the classification of living organisms on the basis of scientific criteria (Allen, 2015; Braund, 1998; Chen & Ku, 1998; Cinici, 2013; Gelman & Meyer, 2011; Kattmann, 2001; Kubiatko & Prokop, 2007; Papadopoulou & Athanasiou, 2015). Usually, their classifications are arbitrary, random, uncorrelated, and lack hierarchical reasoning (Driver, 1985). It has to be noted that most of the above studies involved primary or high-school students; kindergarten students were not that well studied.

Contemporary views for learning and teaching, such as mobile learning, highlight the role of digital technologies and introduce new instruments into instruction. Tools such as smartphones and tablets combine game and learning while offering students a better visualization of phenomena related to natural sciences. This, in turn, allows them to have a better understanding of these concepts. Children come into contact and are attracted to these devices from a very young age (Shuler, 2009). Since very young children view the school environment as a natural extension of their family environment, the use of mobile devices at school is self-evident, at least in their own eyes. Additionally, their interest, as well as their enthusiasm and enjoyment for lessons are amplified (Blackwell, 2013; Fokides, 2018; Fokides & Mastrokoukou, 2018). The learning outcomes of preschool children who participated in tablet-assisted projects were encouraging, to say the least (Al-Zu'bi, Omar-Fauzee, & Kaur, 2017; Bebell & Pedulla, 2015; McManis & McManis, 2016; Papadakis, Kalogiannakis & Zaranis, 2016; Zaranis, Kalogiannakis, & Papadakis, 2013; Zomer & Kay, 2016). At the same time, other advantages that emerged from tablets' use included opportunities for the development of fine motor skills and practical training of functional capacities (Bebell & Pedulla, 2015; Blackwell, 2013), fostering of students' creativity and imagination, and development of problem-solving skills (Blackwell, 2013). Kindergarten teachers can benefit from the use of mobile devices as well. They can record the teaching process for evaluating it at a later time, develop students' digital portfolios indicative of their progress, and save important incidents/observations that could help them to plan and organize their teaching (Parnell & Bartlett, 2012).

Taking into account that: (a) although kindergarten students face quite a lot of problems in animals' classification, (b) the issue is understudied and (c) tablets are considered effective teaching tools, a pilot project was designed and implemented in order to study whether tablets can indeed be effective tools for teaching kindergarten students subjects related to animals' classification. The rationale, methodology, and the results of the project are presented and analyzed in the coming sections.

BACKGROUND

Animals' Classification and Kindergarten Students

Children have an innate interest in animals. Indeed, even before entering formal education, preschoolers have already developed various perceptions for animals, either from their direct experiences (e.g. through contact with pets) or from other sources such as TV series, books, and the Internet (Kubiatko & Prokop, 2007; Patrick et al., 2013). These perceptions, although they reflect children's true beliefs (Carey, 2000), are far from being scientifically correct (Chyleńska & Rybska, 2018) and are persistent to change even after organized instruction (Kubiatko, 2012). Quite a large number of studies examined children's ideas about living organisms. Children did not find it difficult to identify animals as living organisms, although, when asked, they gave examples of large mammals, such as elephants and cows (Chen & Ku, 1998). Interestingly, children did not classify humans as animals (Papadopoulou & Athanasiou, 2015). On the other hand, they classified "living things" based on morphological and functional characteristics that are readily recognizable to humans (Martínez-Losada, García-Barros, & Garrido, 2014). For example, movement was the most commonly reported trait that children used for identifying humans and animals as being alive, but also for justifying why plants are not (Cardak, 2009).

Moreover, most children face serious problems when it comes to animals' classification (Braund, 1998). Usually, they form groups of animals that are arbitrary or illogical (Driver, 1985), using criteria derived from the so-called "Folklore Biology" (Patrick et al, 2013). The persistence of biological thinking based on an instinctive understanding of biological issues does not change even when children enter puberty (Coley, Arenson, Xu & Tanner, 2017). As Kattmann (2001) noted, young students classified animals according to the place they live and the way they move; morphological or anatomical characteristics were, mostly, ignored. As they grow older and having learned a few things related to Biology, students' misconceptions exist side-by-side with scientific knowledge, meaning that they classify animals using both scientific and unscientific criteria. Similar were the findings of Kubiatko and Prokop (2007) regarding the classification of mammals by six-years-old students and Cinici's (2013) findings regarding the classification of vertebrates and invertebrates by high-school students. In an older study, Chen and Ku (1998) noted that young students classified animals on the basis of their differences rather than on their similarities.

As far as kindergarten students are concerned, little research has been carried out, as it appears to be an age group that researchers did not include in their studies. Nevertheless, the results of relevant studies indicated that preschool children based their animals' classification in accordance to the level of their cognitive development and on the basis of their past experiences (Allen, 2015; Gelman & Meyer, 2011). The classifications they carry out involve criteria of similarity and correlation, but they tend to use causal relationships when they cannot find similarities. Also, as they cannot fully understand the classification process or the scientific criteria that are used, children use inductive reasoning for "labeling" animals (e.g. a fish is not a mammal) (Gelman & Meyer, 2011).

The archetypal representations of animals, which were established through their prior experiences, are one of the criteria they used. Classification problems arise when toddlers have not developed archetypes for some animals (e.g., for insects). Moreover, archetypal representations may lead either to right conclusions (e.g., a crab is not a fish although it lives underwater) or wrong ones as some animals do not fit the general appearance of the species they belong to (e.g. the hippocampus is not a fish since it does not look like a fish) (Allen, 2015). The appearance, motion, behavior, nutrition, physiology and anthropomorphic elements were also the criteria most commonly encountered in the animals' classifications by preschool children (Allen, 2015; Papadopoulou & Athanasiou, 2015). As observed in older students, preschool children find it difficult to make hierarchical categorizations (i.e., in phyla, classes, orders, families, genera, and species), as they seem unable to focus on multiple elements, but rather use one or two individual and unrelated criteria that often contradict each other (Allen, 2015). Cultural elements and language also offer children of young age directions for their categorization without using visual observations (e.g. a goldfish is immediately perceived as a fish since part of its name is "fish") (Gelman & Meyer, 2011).

Tablets and Kindergarten Students

Children, from very early on, come into contact with mobile devices, as parents often share their smartphones and tablets with their offsprings (Shifflet, Toledo & Mattoon, 2012). Such devices are omnipresent and ready to use in almost every place where small children dwell. So, children, when entering pre-school education, are already skilled users of mobile devices (Shuler, 2009). Especially tablets' large touchscreens, ease of use, responsiveness to voice commands, easy to navigate menus, portability, and access to the Internet, are some of the features that render them an ideal learning tool in the hands of young children (Cavus & Ibrahim, 2009). In the eyes of toddlers, the use of tablets in the school environment is self-evident, since the school is viewed as an extension of their family environment. Thus, the educational exploitation of these devices is a legitimate (and feasible) objective (McManis & McManis, 2016).

Tablets offer preschool children unique opportunities to develop functional and fine motor skills (Bebell & Pedulla, 2015; Bedford, Urabain, Irati, Cheung, Karmiloff-Smith, & Smith, 2016; Blackwell, 2013). Tapping, double-tapping, dragging, and other gestures are performed with ease and almost instinctively (Shifflet et al., 2012); more complex ones (e.g., shaking and resizing) can be learned by mimicking others or by following instructions (Hiniker, Sobel, Hong, Suh, Irish, Kim & Kientz, 2015). The above, enable them to successfully perform a variety of tasks using tablets. For example, children aged three to six can practice skills necessary for the development of writing (Vatavu, Cramariuc, & Schipor, 2015). What is more, tablets are considered easier to use compared to a keyboard or mouse, as the use of the latter requires already developed motor skills (Blackwell, 2013; Hirsh-Pasek, Zosh, Golinkoff, Gray, Robb, & Kaufman, 2015).

The early familiarization of toddlers with tablets promotes their autonomy and independence from adult supervision. Their self-esteem and self-confidence can also benefit. The confidence that they feel when they become skilled users of these devices, motivates them to experiment and further explore their competences (Couse & Chen, 2010). Preschool children who participated in programs in which tablets were used, showed more interest (Strouse & Ganea, 2017), enthusiasm, and enjoyment for the learning activities (Blackwell, 2013). At the same time, they became more persistent in achieving their goals (Couse & Chen, 2010; Flewitt, Messer & Kucirkova, 2014), especially when the activities were presented in the form of games (Papadakis et al., 2016). Preschoolers' social and collaboration skills, as well as their responsibility toward others, were also fostered (Blackwell, 2013). Studies indicated that toddlers rarely use tablets by themselves. They prefer to share ideas with their classmates, help each other, and ask or answer questions (Shifflet et al., 2012), at least more than when working with paper and pencil (Bebell & Pedulla, 2015).

At the same time, kindergarten students, through the use of tablets and by conducting activities of escalating difficulty, displayed signs of cognitive enhancement and readiness in fields such as linguistic development (Zomer & Kay, 2016), attention (Strouse & Ganea, 2017), recognition of letters and words, and preliminary writing skills (Bebell & Pedulla, 2015; McManis & McManis, 2016; Neumann & Neumann, 2013). Promising results were achieved regarding the development of reading skills (McManis & McManis, 2016), creativity, critical thinking (Flewitt et al., 2014), imagination (Laidlaw & Wong, 2016), and metacognitive skills (Al-Zu'bi et al., 2017). The development of all the above can be accelerated if similar activities are conducted at home, under the supervision or help of parents or older siblings. This has also been proven to be an effective strategy for the smooth and successful transition to formal education (Wong, 2015).

Correspondingly, positive were the results regarding the understanding of mathematical concepts, numbering, measurements, spatial orientation, as well as in geometry (Bebell & Pedulla, 2015; McManis & McManis, 2016; Zaranis et al., 2013, Zaranis, 2019). That is because tablets support personalized learning and incorporate a multitude of learning strategies (e.g., blended learning) (Highfield & Goodwin, 2013); thus, supporting children with different needs (Papadakis et al., 2016). Even when tablets were used for recreation, in transitional periods between learning activities in kindergarten, they created opportunities for practice, so that even this period of time can be productive (Blackwell, 2013). Finally, tablets' educational impact seems to be correlated with students' age; students aged five to six benefited more than students aged three to four (Al-Zu'bi et al., 2017; Huber, Tarasuik, Antoniou, Garrett, Bowe, Kaufman, & Team, 2016; Vatavu et al., 2015; Zomer & Kay, 2016).

Although the relevant literature is rather limited, studies concluded that the use of tablets for teaching subjects related to natural sciences, helped kindergarten students to initiate the process of understanding more abstract concepts, the teaching of which was avoided in the past, as they were considered advanced (and unsuitable) for students of this age. For example, Cascales, Laguna, Pérez-López, Perona, and Contero (2013) in their study regarding the teaching of animals, reported increased learning gains with the use of tablets compared to conventional material. Students were also able to recognize more new animals (Strouse & Ganea, 2017) or name them (Silawati & Rachmania, 2016). To the best of the authors' knowledge, there are no studies in which animals' classification was examined using tablets and having kindergarten students as a target group.

Given the problems kindergarten students face in animals' classification, the lack of relevant studies, and the educational potential of tablets, the following research hypotheses were formulated:

H1. The use of tablets for teaching kindergarten students subjects related to animals' classification, produces better learning outcomes compared to the use of other teaching tools such as computers and printed material.

H2. When tablets are used, students find their use easy, are more motivated to learn, enjoy lessons more, and collaborate better.

Development and Evaluation of the Project

For examining the study's hypotheses, a pilot project was designed and implemented. A quasi-experimental design with one experimental and two control groups was applied, as data were to be collected from intact classrooms. Students in the experimental group used tablets, in the first control group they used printed material and in the second they used computers. The project's theme was the classification of animals divided into three teaching units/sessions. These sessions presented the classification of animals according to (i) their habitat (jungle, forest, desert, etc.), (ii) diet (carnivores, herbivores, and omnivores), and (iii) body characteristics (teeth, paws-nails, ears, eyes, etc.). Details for the procedure that was followed, for the materials, and instruments used, are presented in the coming sections.

Participants and Duration of the Project

The target group was kindergarten students (aged four to six) who: (i) never before used tablets as part of their teaching and (ii) were not taught subjects related to animals' classification (or similar subjects). As a result, a total of forty-nine students were recruited from three kindergarten schools in the city of Agrinio, Greece. In each school, a teaching tool (printed material, computers, and tablets) was randomly assigned. As the research involved minors, permission from the University's ethical committee was granted. The parents of the participating students were also contacted and their written consent was obtained. The teachers of the participating schools were briefed for the study's objectives and methods. Also, they were asked to strictly follow the teaching method that was assigned to each, as described in the "Procedure" section. The project lasted for about a month (three sessions in each school, from mid-October to late November 2018). Each session lasted for three teaching hours, so as students to have enough time at their disposal for conducting the sessions' activities.

Materials

The teaching material for each session's subject matter (animals' habitat, diet, and body characteristics) was developed in three different forms, as there were three teaching tools/students' groups. Maps, charts, images, photos, animations, diagrams, and videos of animals, animal habitats, diet-feeding behaviors, and characteristics, were gathered and edited. Because students were very young and could not read (at least fluently), audio files were recorded, including narrations and descriptions. It has to be noted that for facilitating students' understanding of animals' classification, the animals that were selected had easily identifiable characteristics (e.g., sharp teeth and claws). Insects, birds, and amphibians were excluded.

For the development of tablets' applications Blippbuilder (https://www.blippar.com/) was used, which enables the rapid development of applications for mobile devices that fall into the category of Augmented Reality (AR) applications. In short, AR is a technology that merges the real with the digital world by presenting to the user, in real-time, a combination of real and virtual objects, multimedia elements, and information, while allowing his/her interaction with the above (Billinghurst, Clark, & Lee, 2015). A marker (an image) is used for triggering/starting an AR application. The markers were printed and handed to students who were going to use the tablets. A total of nine applications were developed, three for each session. For example, in the session discussing the body characteristics of animals, a crocodile was chosen for presenting the characteristics of carnivores. After triggering the relevant application, students were able to watch videos describing crocodiles, tap on thumbnail images of their teeth and claws (and simultaneously hear the relevant narrations), and see a comparison of these characteristics with the corresponding characteristics of other animals (Figure 1). It has to be noted that the applications were not developed by a group of experts but by one of the researchers who had no previous experience in the development of such applications. The development of the applications required around fifty hours.

Following the same logic and using the same material, a website was developed. The website was later packed into a self-contained application, consisting of a single, installable bundle including all the web-pages and necessary files. Therefore, the website could be locally executed to any computer as a stand-alone application without the need for an Internet connection. For the group that used printed material, the website's material was printed on paper in the form of three booklets. Extensive screenshots from the videos replaced the video files. The audio files were transcribed on documents so as to be read/narrated by the class's teacher.

Figure 1. Screenshot from the development of tablets' applications Source: The authors

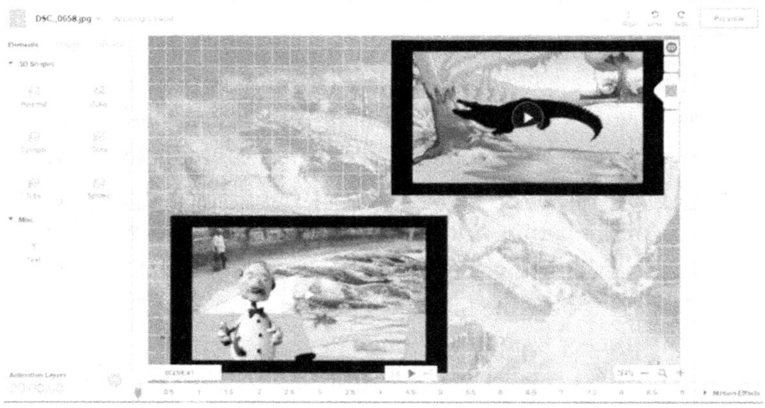

Procedure

For science subjects, it is recommended students to work in small groups (Harlen, 2018). Accordingly, it was decided students (of all groups) to work in pairs (Figure 2). Bybee's 5Es (Bybee, Taylor, Gardner, Van Scotter, Powell, Westbrook, & Landes, 2006) is an instructional model commonly used when teaching science-related subjects. Consequently, it was followed, though its five phases were slightly modified so as to suit the study's needs:

- The purpose of the "Engage" stage is to excite students' interest, to encourage them to get personally involved in the lesson, and sets the groundwork for the activities to follow. The teachers provided the necessary stimuli, by making a short introduction and by engaging in conversations with students related to the session's theme.
- During the "Explore" stage students explore ideas through activities and the "Explain" stage allows students to communicate what they have learned. These stages were merged and were conducted using either the printed material or the webpages or the tablets' applications. For example, in the session in which animals were classified according to their habitat, the first screen of the corresponding tablets' application presented an assortment of animals. A similar webpage was presented to students who used computers and the website. As for the group that used the printed material, animals' photos were given to each pair of students. Students discussed what they might have in common. Each pair of students "recorded" their views by sticking animals' pictures in a piece of cardboard. Students were then asked to proceed to the next pages of the application (or the next webpages), in which each habitat was presented. Students in the printed material group were given the corresponding booklet and the teacher was given the task to narrate/read the accompanying text. Following that, students revisited their "recorded" views, re-discussed, and, if necessary, they made corrections.
- The "Extend" stage allows students to further explore the implications of what they have learned. The purpose of the "Evaluation" stage is to determine how much learning has taken place. Both stages were also merged. In the above example, the next screen of the tablets' application (or the following webpage or the next pages of the booklet) presented yet more animals, but this time, a

Figure 2. Screenshots from students' groups Source: The authors

world map was also presented (in which jungles, deserts, forests, and the polar regions were specified using images). Once again, students "recorded" their views by sticking animals' pictures in a printed word map. During the "Evaluation" stage, each pair presented their recorded views to the whole class and engaged in discussions until a common consensus was reached.

- It has to be noted that during all stages and most of the time, students were free to work by themselves. The teachers acted as facilitators of the process by starting or joining in students' discussions, by drawing their attention to important aspects of their work, and by providing guidelines (but without enforcing their views).

To summarize, in all groups of students: (i) the same teaching procedure was followed, (ii) the same subjects were taught, and (iii) the number of sessions was the same. The only difference the three groups had, was the teaching tool that each used.

Instruments

For data collection purposes, five evaluation sheets (pre- and delayed post-tests, and one for each of the three teaching units), as well as structured interviews were used. The pre-tests were administered prior to the beginning of the project for recording students' previous knowledge on animals and their classification. The delayed post-tests were administered two weeks after the end of the project and checked knowledge retention, while the other three evaluation sheets were administered at the end of each session and examined what students were able to understand. All evaluation sheets consisted of yes-no and multiple-choice questions and were structured so as: (i) to fully cover the content of each unit, and (ii) questions to be of escalating difficulty. Then again, having as target group kindergarten students, imposed the same problem as in the project's material, text could not be used. Instead, it was decided to use images, to voice, and verbally explain each question. To avoid questions' oversimplicity, in most, there was more than one correct answer and, in others, there was no correct answer at all (Figure 3). Moreover, in all the evaluation sheets, students were asked to draw animals highlighting certain

Figure 3. Example of a question in an evaluation sheet Source: The authors

characteristics, depending on the session's subject matter. For example, in the session in which animals were classified according to their body characteristics, students were asked to draw a carnivore, using black color for indicating why it is a carnivore (Figure 4). The drawings were later examined for their accuracy and were graded accordingly.

As a substitute for questionnaires, structured interviews were used for recording students' views regarding the three teaching tools. The interviews' questions (presented in the Appendix) were drawn from the Learning Experience Questionnaire, a validated modular scale for examining users' views for educational software (Fokides, Kaimara, Deliyiannis, & Atsikpasi, 2019). Although this scale examines a total of twelve factors, four were selected: motivation, enjoyment, collaboration, and ease of use (not applicable for printed material). The method for obtaining quantitative data from students' responses is presented in the coming section.

Data Screening and Data Processing

Out of the initial forty-nine students, four had to be excluded from the study because they were absent in one or more sessions, reducing the final sample size to forty-five students. As already mentioned in the "Instruments" section, for examining students' views and attitudes for the tools that were used, structured interviews were utilized, with questions coming from a validated scale. In order to statistically analyze students' responses, the qualitative raw data for each question were transformed into quantitative ones,

Figure 4. Example of a student's drawing Source: The authors

namely, to Likert-type scales (anchored at 1-low and 5-high). For that matter, a multi-step procedure was followed. First, a thematic coding analysis was applied (Saldaña, 2015). In short, this method allows the indexing of the text into categories/codes and the establishment of a framework of thematic ideas/themes (Gibbs, 2007). The recordings were transcribed verbatim on three documents (one for each group). Two individuals with expertise in educational software acted as coders and Atlas Ti was used for extracting/labeling the codes. The coders were trained and their reliability was assessed: (a) in a pilot test using a randomly selected quarter of the responses and (b) formally during the coding of the full sample. For determining consistency among raters an interrater reliability analysis was conducted using Cohen's kappa coefficient and it was found to be very good [$\kappa = .890, p < .001$, 95% CI (.878, .902)] (Landis & Koch, 1977). Responses were viewed once, for identifying the main ideas. Then, they were re-viewed in more detail and the ideas were labeled with codes. This process was repeated one more time to reduce overlap and redundancy of codes and a small set of themes emerged. As all responses were labeled with themes, the two coders viewed them once again, assigned ranking scores, and an average of the two scores was calculated. For example, the response "It was so much fun playing with tablets" was labeled as "high enjoyment" and received a score of 5. Finally, as there was more than one question examining each factor (motivation, enjoyment, ease of use, and collaboration), a composite score for each factor was calculated, by averaging the scores of its corresponding questions.

In order to determine if the three groups of students had any statistically significant differences in their scores (in both the evaluation sheets and the interviews), one-way ANOVA tests were to be conducted. Prior to conducting these tests, it was checked whether the assumptions for ANOVA testing were violated. It was found that: (i) all groups had the same number of participants ($N = 15$), (ii) there were no outliers, (iii) the data were normally distributed, as assessed by Q-Q plots and the Shapiro-Wilk test, and (iv) the homogeneity of variance, as assessed by Levene's Test of Homogeneity of Variance, was violated in one case (first evaluation sheet). As in this case, heteroscedasticity was a concern, it was decided to proceed using the Brown-Forsythe test (1974).

Results

As already mentioned, the final sample size was forty-five students (twenty boys and twenty-five girls), divided into three groups of fifteen students each (students' mean age was almost identical in the three groups and the same applied for the ratio of boys and girls). Each group was taught using a different tool (printed material-Group1, computers-Group2, and tablets-Group3). Mean scores and standard deviations per group of participants and per evaluation sheet are presented in Table 1.

For determining whether the scores of the three groups had any statistically significant differences, one-way ANOVA tests were conducted. It was found that in the pre-test there were no statistically significant differences [$F(2, 42) = .076, p = .927$], while to the rest of the tests there were statistically significant differences [ES1: Brown-Forsythe $F(2, 36.219) = 9.577, p < .001$; ES2: $F(2, 42) = 10.444$, $p < .001$; ES3: $F(2, 42) = 16.071, p < .001$; delayed post-test: $F(2, 42) = 5.573, p = .007$]. Post-hoc comparisons were conducted on all possible pairwise contrasts in cases where statistically significant differences were noted. The results of these test are presented in Table 2.

Taken together, the above results suggested that Group3 (tablets group) outperformed Group1 (printed material group) in all cases, including the delayed post-test. Then again, Group3 did not outperform Group2 (computers group) in any case. Also, Group2 outperformed Group1 in two out of four cases. Given the above, H1 is partially confirmed; the use of tablets for teaching kindergarten students subjects

Table 1. Means and standard deviations for the evaluation sheets

Test	Groups					
	Group1 (*N* = 15)		Group 2 (*N* = 15)		Group 3 (*N* = 15)	
	M	*SD*	*M*	*SD*	*M*	*SD*
Pre-test (max = 30)	14.80	2.48	14.67	2.19	15.00	2.39
ES1 (max = 30)	13.20	1.97	14.93	2.19	16.13	1.25
ES2 (max = 30)	14.80	2.57	17.80	2.48	18.40	1.81
ES3 (max = 30)	17.80	1.66	20.13	1.92	21.20	1.42
Delayed post-test (max = 35)	22.33	2.19	23.73	1.87	24.60	1.50

Notes: ES = evaluation sheet; maximum scores for each ES are presented in parenthesis

related to animals' classification, produces better learning outcomes compared to the use of printed material, but compared to the use computers, the results are the same.

Mean scores and standard deviations per group of participants and per questionnaire's factor are presented in Table 3. As previously, for determining whether the scores in the questionnaire were statistically significantly different, one-way ANOVA tests were conducted. It was found that in collaboration and ease of use there were no statistically significant differences [$F(2, 42) = 1.41$, $p = .255$; $F(1, 28) = 0.71$, $p = .673$ respectively]. On the other hand, there were statistically significant differences in the other two factors [motivation: $F(2, 42) = 13.59$, $p < .001$; enjoyment: $F(2, 42) = 9.35$, $p = .001$]. Again, post-hoc comparisons were conducted on all possible pairwise contrasts, as presented in Table 4.

Taken together, the questionnaire's results suggested that students in Group3 were more motivated to learn and enjoyed lessons more than students in the other two groups. As for collaboration between students and tablets' ease of use, there were no differences with the other two groups. Thus, H2 can be

Table 2. Pairwise comparisons for evaluation sheets

Evaluation sheet	Groups	Mean Difference	Sig.	Interpretation
ES1	1-2	-1.73	.075	NS
	1-3	-2.93	< .001	Group3 outperformed Group1
	2-3	-1.20	.178	NS
ES2	1-2	-3.00	.003	Group2 outperformed Group1
	1-3	-3.60	< .001	Group3 outperformed Group1
	2-3	-.60	.758	NS
ES3	1-2	-2.33	.001	Group2 outperformed Group1
	1-3	-3.40	< .001	Group3 outperformed Group1
	2-3	-1.07	.203	NS
Post test	1-2	-1.40	.114	NS
	1-3	-2.27	.005	Group3 outperformed Group1
	2-3	-.87	.423	NS

Table 3. Means and standard deviations for the interviews

Factor	Groups					
	Group1 (N = 15)		Group 2 (N = 15)		Group 3 (N = 15)	
	M	*SD*	*M*	*SD*	*M*	*SD*
Motivation	3.02	1.01	3.74	0.72	4.49	0.50
Enjoyment	3.45	0.87	3.51	1.12	4.62	0.28
Ease of use	NA	NA	3.67	0.90	3.89	0.74
Collaboration	3.76	0.76	3.31	1.18	3.14	1.14

Note: NA = not applicable

partially accepted; when tablets are used, students are more motivated to learn and enjoy lessons more. On the other hand, collaboration and easiness of use are not that different compared to other tools.

SOLUTIONS AND RECOMMENDATIONS

The study sought to examine the learning outcomes from the use of tablets by kindergarten students when teaching them the classification of animals. For that matter, three groups of students were formed; one used printed material, the second used computers and webpages, while in the third tablets were used. As far as the learning outcomes are concerned, the results indicated that the use of tablets produced better results compared to the use of printed material. On the other hand, compared to the use of computers and webpages, the two groups did not have any statistically significant differences, meaning that the learning outcomes did not differ that much. The same applied for the delayed post-tests, which examined knowledge retention. Consequently, the study's results are in support of the finding of previous studies which concluded that the use of tablets results in better learning outcomes compared to conventional tools (e.g., Al-Zu'bi et al., 2017; Bebell & Pedulla, 2015; McManis & McManis, 2016; Zomer & Kay, 2016). At the same time, however, the results are in line with those studies that reported neutral results when tablets were compared to other -digital- media (e.g., Dündar & Akcayir, 2012).

A number of factors may have contributed to the above results. A fact that should be taken into account is the study's learning subject. Students, of all ages, have trouble classifying animals; their mis-

Table 4. Pairwise comparisons for the interview's factors

Factors	Groups	Mean Difference	Sig.	Interpretation
Motivation	1-2	0.72	.038	Group2 outperformed Group1
	1-3	2.16	< .001	Group3 outperformed Group1
	2-3	0.75	.029	Group3 outperformed Group2
Enjoyment	1-2	0.06	.979	NS
	1-3	1.17	.001	Group3 outperformed Group1
	2-3	1.11	.002	Group3 outperformed Group2

conceptions are notable (Coley et al., 2017; Martínez-Losada et al., 2014; Patrick et al, 2013) and are hardly lifted even after systematic teaching (Kubiatko, 2012). This was confirmed by the results in the evaluation sheets of all groups. In both the pre-test and the first evaluation sheet (animals' classification according to their habitat) about half of the students' answers were wrong (see Table 1). Slightly better were the results in the second evaluation sheet (animals' classification according to their diet), but still rather disappointing. On the other hand, it would be unrealistic to expect dramatic -positive- changes, in such a difficult learning subject, after a small number of interventions.

A second thing to consider is the teaching method. It has to be reminded that it was based on Bybee's 5Es, that students worked in pairs, that they had quite an increased autonomy, and that the teacher's role was active and supportive. A large number of researchers support the view that the satisfactory learning outcomes resulting from the use of tablets are due to students' increased autonomy and control over their learning pace (e.g., van Deurse, ben Allouch, & Ruijter, 2016). Other studies concluded that the learning outcomes can also be attributed to the reinforcement of collaborative learning through the use of tablets (e.g., Blackwell, 2013; Shifflet et al., 2012). The above cannot be supported by the findings of the present study. Indeed, by examining Table 3, it becomes clear that collaboration in the tablets group was the worst among the three groups, although not statistically significantly different. The short duration of the project together with the fact that students were not used in working with tablets are plausible explanations for this result. Another finding that contradicts the findings of previous research is related to tablets' ease of use. While researchers suggested that tablets are easier to use compared to a keyboard or mouse (Blackwell, 2013; Hirsh-Pasek, Zosh, Golinkoff, Gray, Robb, & Kaufman, 2015), in this study students regarded both computers and tablets as equally easy to use. Again, the project's short duration might have not allowed students to familiarize themselves with the use of tablets (and the applications that were used).

The relevant literature suggested that tablets allowed the development of an attractive and pleasant learning environment which made learning an enjoyable process (e.g., Blackwell, 2013). Indeed, this was confirmed by the interviews' results; the enjoyment students felt during lessons was far greater in the tablets' group than in the other two groups (see tables 3 and 4). The interviews' analysis also revealed that students were more motivated to learn, thus, confirming research pointing to this direction (e.g., Strouse & Ganea, 2017).

To summarize and on the basis of the study's results, fun/enjoyment and motivation proved to be the advantages of tablets over other teaching tools. Then again, these advantages were not enough for producing better learning outcomes compared to the use of other digital tools, namely, computers and webpages. Furthermore, the use of tablets resulted in clearly better learning outcomes compared to printed material (as the tablets' group outperformed the printed material group in all cases), while the use of computers and webpages also produced good results (as the computers' group outperformed the printed material group in two out of four cases). Thus, a legitimate conclusion is that digital learning tools, in general, are more preferable than conventional material, as they are expected to produce better learning outcomes.

Implications for Research and Practice

The study's implications for research are, up to a certain degree, related to the teaching method that was followed, which was a modified version of Bybee's 5Es. It has to be reminded that this framework was developed before tablets became mainstream. As others suggested, we are still in need of a robust pedagogy in order to take full advantage of tablets' educational potential (e.g., Clarke & Svanaes, 2014). Consequently, it is necessary either to update the existing or to develop new teaching methods. Toward this end, the study's teaching method might provide some ideas.

The study's results also have implications for software developers and educators. As already mentioned in the "Materials" section, the applications were developed by one of the researchers; thus, they were far from reaching professional standards. Furthermore, it is quite probable that their flaws had a negative impact on the learning outcomes. On the other hand, commercial applications, suitable for the study, were simply not available. This problem is not uncommon in countries in which the software industry is not well-developed. Two are the possible solutions to the above issue. The first is teachers to become producers of educational applications. The feasibility of this solution is questionable, given the time and effort needed for the development of such applications. The second solution is software developers and education professionals to work hand in hand, as suggested by Shuler, Levine, and Ree (2012). The close collaboration between these two groups is probably the ideal arrangement for the development of technically, as well as educationally sound applications.

In this project, each session lasted for three teaching hours. Two were the underlying reasons. Given that students were very young, settling down, booting the devices, and loading the applications, was time-consuming; valuable teaching time was lost. Second, it was of utmost importance to provide students with enough time so as to study at their own pace and conduct all the activities. Indeed, on the basis of the results, three-hour sessions proved to be sufficient. Therefore, kindergarten teachers, for successfully integrating tablets (or any other ICT tool for that matter) into their teaching, are advised to carefully plan the activities and allocate enough time for conducting them.

FUTURE RESEARCH DIRECTIONS

Although the study's results were interesting, there are limitations that should be acknowledged but also provide several paths for future research. The sample size, although acceptable for the type of statistical analysis that was conducted, could have been larger. Moreover, it was limited to one city in Greece. The above might raise concerns for the generalizability of the results. The long-term retention of knowledge is also unknown. In future research, the sample, besides being larger, can be more diverse, encompassing even younger or slightly older students. A larger variety of subjects (not only related to natural sciences) will help to determine the disciplines in which tablets are more effective. Additional research tools can also be utilized; observations and unstructured interviews with students and teachers will allow an in-depth understanding of tablets' educational value. Professionally developed applications can also be used and examine whether there is a significant variation in the results. Other devices (e.g., smartphones) or other technologies (e.g., virtual reality) can be used and compare the results. Finally, it would be interesting to conduct research maximizing or minimizing the teacher's role. By doing so, it would be easier to determine the exact impact tablets have on knowledge acquisition.

CONCLUSION

In sum, despite the aforementioned limitations, the study provided an idea about how tablets might prove useful to kindergarten students. What is more, it contributes to the relevant literature by presenting evidence that, in subjects related to animals' classification, kindergarten students, through tablets: (i) can learn more and retain more knowledge compared to printed material and (ii) are more motivated to learn and enjoy lessons more. In conclusion, the study's findings might prove useful to researchers and teachers in understanding the pros and cons of tablets and effectively using these devices in kindergarten education.

REFERENCES

Akerson, V. L., Flick, L. B., & Lederman, N. G. (2000). The influence of primary children's ideas in science on teaching practice. *Journal of Research in Science Teaching: The Official Journal of the National Association for Research in Science Teaching, 37*(4), 363-385. doi:10.1002/(sici)1098-2736(200004)37:4<363::aid-tea5>3.0.co;2-#

Al-Zu'bi, M. A. A., Omar-Fauzee, M. S., & Kaur, A. (2017). The effect of iPad apps on creative thinking among preschoolers in Jordan. *International Journal of Multidisciplinary Education and Research, 2*(4), 29-35.

Allen, M. (2015). Preschool children's taxonomic knowledge of animal species. *Journal of Research in Science Teaching, 52*(1), 107–134. doi:10.1002/tea.21191

Bebell, D., & Pedulla, J. (2015). A quantitative investigation into the impacts of 1: 1 iPads on early learner's ELA and math achievement. *Journal of Information Technology Education: Innovations in Practice, 14*(1), 191–215. doi:10.28945/2175

Bedford, R., de Urabain, S., Irati, R., Cheung, C. H., Karmiloff-Smith, A., & Smith, T. J. (2016). Toddlers' fine motor milestone achievement is associated with early touchscreen scrolling. *Frontiers in Psychology, 7*, 1108. doi:10.3389/fpsyg.2016.01108 PMID:27531985

Billinghurst, M., Clark, A., & Lee, G. (2015). A survey of augmented reality. *Foundations and trends in human-computer interaction, 8*(2-3), 73–272. doi:10.1561/1100000049

Blackwell, C. (2013). Teacher practices with mobile technology integrating tablet computers into the early childhood classroom. *The Journal of Educational Research, 7*(4), 231–255.

Braund, M. (1998). *The development of children's ideas on animal classification, form and function; Is school experience becoming increasingly impoverished?* (Report No. 143). East Lansing, MI: National Center for Research on Teacher Learning. (ERIC Document Reproduction Service No. ED436423).

Brown, M. B., & Forsythe, A. B. (1974). Robust tests for the equality of variances. *Journal of the American Statistical Association, 69*(346), 364–367. doi:10.1080/01621459.1974.10482955

Bybee, R. W., Taylor, J. A., Gardner, A., Van Scatter, P., Carlson-Powell, J., Westbrook, A., & Landes, N. (2006). *BSCS SE instructional model: Origins and effectiveness. A report prepared for the Office of Science Education, National Institutes of Health.* Colorado Springs, CO: BSCS.

Cardak, O. (2009). Science students' misconceptions about birds. *Scientific Research and Essays*, *4*(12), 1518–1522.

Carey, S. (2000). Science education as conceptual change. *Journal of Applied Developmental Psychology*, *21*(1), 13–19. doi:10.1016/S0193-3973(99)00046-5

Cascales, A., Laguna, I., Pérez-López, D., Perona, P., & Contero, M. (2013). An experience on natural sciences augmented reality contents for preschoolers. *Proceedings of the International Conference on Virtual, Augmented and Mixed Reality*, 103-112. Berlin, Germany: Springer. 10.1007/978-3-642-39420-1_12

Cavus, N., & Ibrahim, D. (2009). m-Learning: An experiment in using SMS to support learning new English language words. *British Journal of Educational Technology*, *40*(1), 78–91. doi:10.1111/j.1467-8535.2007.00801.x

Chen, S. H., & Ku, C. H. (1998). Aboriginal children's alternative conceptions of animals and animal classification. *Proceedings of the National Science Council Republic of China Part D Mathematics Science and Technology Education*, *8*(2), 55–67.

Chyleńska, Z. A., & Rybska, E. (2018). Understanding students ideas about animal classification. *EURASIA Journal of Mathematics. Science and Technology Education*, *14*(6), 2145–2155. doi:10.29333/ejmste/86612

Cinici, A. (2013). Turkish high school students' ideas about invertebrates: General characteristics and classification. *International Journal of Environmental and Science Education*, *8*(4), 645–661.

Clarke, B., & Svanaes, S. (2014). *An updated literature review on the use of tablets in education. Tablets for Schools*. UK: Family Kids & Youth.

Coley, J. D., Arenson, M., Xu, Y., & Tanner, K. D. (2017). Intuitive biological thought: Developmental changes and effects of biology education in late adolescence. *Cognitive Psychology*, *92*, 1–21. doi:10.1016/j.cogpsych.2016.11.001 PMID:27865155

Couse, L. J., & Chen, D. W. (2010). A tablet computer for young children? Exploring its viability for early childhood education. *Journal of Research on Technology in Education*, *43*(1), 75–96. doi:10.1080/15391523.2010.10782562

Driver, R. (1985). *Children's ideas in science*. UK: McGraw-Hill Education.

Driver, R., Squires, A., Rushworth, P., & Wood-Robinson, V. (2014). *Making sense of secondary science: Research into children's ideas*. Routledge.

Dündar, H., & Akçayır, M. (2014). Implementing tablet PCs in schools: Students' attitudes and opinions. *Computers in Human Behavior*, *32*, 40–46. doi:10.1016/j.chb.2013.11.020

Eshach, H. (2006). *Science literacy in primary schools and pre-schools* (Vol. 1). Springer Science & Business Media; doi:10.1007/1-4020-4674-X

Flewitt, R., Messer, D., & Kucirkova, N. (2015). New directions for early literacy in a digital age: The iPad. *Journal of Early Childhood Literacy*, *15*(3), 289–310. doi:10.1177/1468798414533560

Fokides, E. (2018). Tablets in primary schools: Results of a study for teaching the human organ systems. *International Journal of Smart Education and Urban Society*, *9*(3), 1–16. doi:10.4018/IJSEUS.2018070101

Fokides, E., Kaimara, P., Deliyiannis, I., & Atsikpasi, P. (2019). Development of a scale for measuring the learning experience in serious games. Preliminary results. *Proceedings of the Digital Culture & Audiovisual Challenges International Conference*. Corfu, Greece: Ionian University.

Fokides, E., & Mastrokoukou, A. (2018). Results from a study for teaching human body systems to primary school students using tablets. *Contemporary Educational Technology*, *9*(2), 154–170. doi:10.30935/cet.414808

Gelman, S. A., & Meyer, M. (2011). Child categorization. *Wiley Interdisciplinary Reviews: Cognitive Science*, *2*(1), 95–105. doi:10.1002/wcs.96 PMID:23440312

Gibbs, G. R. (2007). Thematic coding and categorizing. In Analyzing qualitative data, 38-56. London, UK: Sage. doi:10.4135/9781849208574.n4

Harlen, W. (2018). *The teaching of science in primary schools*. David Fulton Publishers. doi:10.4324/9781315398907

Highfield, K., & Goodwin, K. (2013). Apps for mathematics learning: A review of "educational" apps from the iTunes app store. *Proceedings of the 36th Annual Conference of the Mathematics Education Research Group of Australia*, 378- 385. Melbourne, Australia.

Hiniker, A., Sobel, K., Hong, S., Suh, H., Irish, I., Kim, D., & Kientz, J. A. (2015). Touchscreen prompts for preschoolers: Designing developmentally appropriate techniques for teaching young children to perform gestures. *Proceedings of the 14th International Conference on Interaction Design and Children*, 109-118. ACM. 10.1145/2771839.2771851

Hirsh-Pasek, K., Zosh, J. M., Golinkoff, R. M., Gray, J. H., Robb, M. B., & Kaufman, J. (2015). Putting education in "educational" apps: Lessons from the science of learning. *Psychological Science in the Public Interest*, *16*(1), 3–34. doi:10.1177/1529100615569721 PMID:25985468

Huber, B., Tarasuik, J., Antoniou, M. N., Garrett, C., Bowe, S. J., Kaufman, J., & Team, S. B. (2016). Young children's transfer of learning from a touchscreen device. *Computers in Human Behavior*, *56*, 56–64. doi:10.1016/j.chb.2015.11.010

Kattmann, U. (2001). Aquatics, flyers, creepers and terrestrials-students' conceptions of animal classification. *Journal of Biological Education*, *35*(3), 141–147. doi:10.1080/00219266.2001.9655763

Kubiatko, M. (2012). Kindergarten Children's perception of animals focusing on the look and fear of animals. *Educational Sciences: Theory and Practice*, *12*(4), 3181–3186.

Kubiatko, M., & Prokop, P. (2007). Pupils' misconceptions about mammals. *Journal of Baltic Science Education*, *6*(1), 5–14.

Laidlaw, L., & Wong, S. S. H. (2016). Literacy and complexity: on using technology within emergent learning structures with young learners. *Complicity: An International Journal of Complexity and Education*, *13*(1), 30–42. doi:10.29173/cmplct24008

Landis, J. R., & Koch, G. G. (1977). The measurement of observer agreement for categorical data. *Biometrics, 33*(1), 159–174. doi:10.2307/2529310 PMID:843571

Martínez-Losada, C., García-Barros, S., & Garrido, M. (2014). How children characterise living beings and the activities in which they engage. *Journal of Biological Education, 48*(4), 201–210. doi:10.1080/00219266.2013.849281

McManis, M. H., & McManis, L. D. (2016). Using a touch-based, computer-assisted learning system to promote literacy and math skills for low-income preschoolers. *Journal of Information Technology Education, 15*, 409–429. doi:10.28945/3550

Neumann, M. M., & Neumann, D. L. (2014). Touch screen tablets and emergent literacy. *Early Childhood Education Journal, 42*(4), 231–239. doi:10.100710643-013-0608-3

Papadakis, S., Kalogiannakis, M., & Zaranis, N. (2016). Comparing tablets and PCs in teaching mathematics: An attempt to improve mathematics competence in early childhood education. *Preschool and Primary Education, 4*(2), 241–253. doi:10.12681/ppej.8779

Papadopoulou, P., & Athanasiou, K. (2015). Children ideas for "animal" compared to teachers' conceptions. *International Journal of Innovation and Research in Educational Sciences, 2*(6).

Parnell, W. A., & Bartlett, J. (2012). iDocument: How smartphones and tablets are changing documentation in preschool and primary classrooms. *Technology and Young Children, 67*(3), 50–59.

Patrick, P., Byrne, J., Tunnicliffe, S. D., Asunta, T., Carvalho, G. S., Havu-Nuutinen, S., ... Tracana, R. B. (2013). Students (ages 6, 10, and 15 years) in six countries' knowledge of animals. *Nordic Studies in Science Education, 9*(1), 18–32. doi:10.5617/nordina.624

Saldaña, J. (2015). The coding manual for qualitative researchers. Atlanta, GA: *Sage*.

Shifflet, R., Toledo, C., & Mattoon, C. (2012). Touch tablet surprises: A preschool teacher's story. *Young Children, 67*(3), 36.

Shuler, C. (2009). *Pockets of potential: Using mobile technologies to promote children's learning*. New York, NY: The Joan Ganz Cooney Center.

Shuler, C., Levine, Z., & Ree, J. (2012). iLearn II: An analysis of the education category of Apple's app store. New York: The Joan Ganz Cooney Center at Sesame Workshop.

Silawati, E., & Rachmania, S. (2016). Character building of early young learners through gadget as learning media in kindergarten. *Proceedings of the 3rd International Conference on Early Childhood Education (ICECE 2016)*. Atlantis Press. doi:10.2991/icece-16.2017.26

Strouse, G. A., & Ganea, P. A. (2017). A print book preference: Caregivers report higher child enjoyment and more adult–child interactions when reading print than electronic books. *International Journal of Child-Computer Interaction, 12*, 8–15. doi:10.1016/j.ijcci.2017.02.001

Trundle, K. C. (2010). Teaching science during the early childhood years. Best Practices in Science Education. *Pridobljeno, 18*(10), 2016.

van Deurse, A. (2016). Tablet use in primary education: Adoption hurdles and attitude determinants. *Education and Information Technologies*, *21*(5), 971–990. doi:10.100710639-014-9363-3

Vatavu, R. D., Cramariuc, G., & Schipor, D. M. (2015). Touch interaction for children aged 3 to 6 years: Experimental findings and relationship to motor skills. *International Journal of Human-Computer Studies*, *74*, 54–76. doi:10.1016/j.ijhcs.2014.10.007

Wong, S. S. H. (2015). Mobile digital devices and preschoolers' home multiliteracy practices. *Language and Literature*, *17*(2), 75–90. doi:10.20360/g2cp49

Zaranis, N. (2018). Comparing the effectiveness of using tablet computers for teaching addition and subtraction. In L. Daniela, & M. Lytras (Eds.), *Learning strategies and constructionism in modern education settings*. Hershey, PA: IGI Global. doi:10.4018/978-1-5225-5430-1.ch009

Zaranis, N., Kalogiannakis, M., & Papadakis, S. (2013). Using mobile devices for teaching realistic mathematics in kindergarten education. *Creative Education*, *4*(7A1), 1-10. doi:10.4236/ce.2013.47a1001

Zaranis, N., & Valla, V. (2019). Tablets in learning mathematics for kindergarten students. In *Didactics of smart pedagogy* (pp. 267–284). Cham, Switzerland: Springer; doi:10.1007/978-3-030-01551-0_14

Zomer, N. R., & Kay, R. H. (2016). Technology use in early childhood education: A review of literature. *Journal of Educational Informatics*, *1*, 1–25.

ADDITIONAL READING

Akçayır, M., & Akçayır, G. (2017). Advantages and challenges associated with augmented reality for education: A systematic review of the literature. *Educational Research Review*, *20*, 1–11. doi:10.1016/j.edurev.2016.11.002

Bidin, S., & Ziden, A. A. (2013). Adoption and application of mobile learning in the education industry. *Procedia: Social and Behavioral Sciences*, *90*, 720–729. doi:10.1016/j.sbspro.2013.07.145

Chu, H. C. (2014). Potential negative effects of mobile learning on students' learning achievement and cognitive load-A Format assessment perspective. *Journal of Educational Technology & Society*, *17*(1), 332–344.

Coley, J. D., Arenson, M., Xu, Y., & Tanner, K. D. (2017). Intuitive biological thought: Developmental changes and effects of biology education in late adolescence. *Cognitive Psychology*, *92*, 1–21. doi:10.1016/j.cogpsych.2016.11.001 PMID:27865155

Fokides, E. (2018). Teaching basic programming concepts to young primary school students using tablets: Results of a pilot project. *International Journal of Mobile and Blended Learning*, *10*(1), 34–47. doi:10.4018/IJMBL.2018010103

Fokides, E., & Atsikpasi, P. (2017). Tablets in education. Results from the initiative ETiE, for teaching plants to primary school students. *Education and Information Technologies*, *22*(5), 2545–2563. doi:10.100710639-016-9560-3

Rikala, J., Vesisenaho, M., & Mylläri, J. (2013). Actual and potential pedagogical use of tablets in schools. *Human Technology*, *9*(2), 113–131. doi:10.17011/ht/urn.201312042736

Wilkinson, K., & Barter, P. (2016). Do mobile learning devices enhance learning in higher education anatomy classrooms? *Journal of Pedagogic Development*, *6*(1), 14–23.

KEY TERMS AND DEFINITIONS

Animals' Classification: A methodological grouping of animals that allows scientists to study the relationships in animal groups and to see the whole animal family tree as it has developed through time.

Augmented Reality: A technology that merges the real with the digital world by presenting to the user, in real-time, a combination of real and virtual objects.

Constructivism: A learning theory supporting the view that people construct their own understanding and knowledge of the world, through experiencing things and reflecting on those experiences.

Kindergarten Education: A level of education preceding primary education. Depending on each country's educational system, it is either a mandatory or optional for a child to attend this level.

Natural Sciences: A branch of science which deals with the physical world (e.g. physics, chemistry, geology, and biology).

Quasi-experimental Design: An empirical interventional study used to estimate the causal impact of an intervention on the target population without random assignment.

Tablet: A mobile device, with a mobile operating system, a touchscreen display, a processing circuitry, and a rechargeable battery, in a thin, flat package.

APPENDIX

Table 5. The interviews' questions (as rephrased in order to suit the study's needs)

Factor	Item
Enjoyment	I think the lessons were fun I felt bored during lessons* I enjoyed lessons I really enjoyed studying It felt good to successfully complete the lessons' tasks I felt frustrated*
Collaboration	I was displeased because it was impossible to collaborate with others; everyone had a mind of his own* Doing things together with my fellow students was interesting. With my fellow students, we were able to jointly decide what to do
Perceived ease of use	I think it was easy to learn using the applications/tablets I found the applications/use of tablets unnecessarily complex* I imagine that most students will learn to use the applications/tablets very quickly I felt that I needed help from someone else in order to use the applications/tablets* It was easy for me to become skillful at using the applications/tablets
Motivation	The lessons did not hold my attention* During lessons, I did not have the impulse to learn more about the learning subject* The lessons did not motivate me to learn*

Note. * = Negatively worded question; responses and the corresponding scores were reversed

Chapter 4
The Impact of Mobile Learning via Touch-Screen Tablets in Emergent Literacy Development

Ioannis Grigorakis
Faculty of Education, University of Crete, Greece

ABSTRACT

Children's use of touch-screen tablets is increasing as these mobile devices become increasingly available. The interactive, tactile touch-screen interface and easily downloadable applications make tablets especially popular with pre-schoolers. This chapter's literature review provides an overview of recent research into tablets and emergent literacy development at home and in the pre-school setting. A multi-faceted synthesis of research on children's use of tablets and its impact in emergent literacy skills was conducted. The evidence indicated that mobile learning via touch-screen tablets has the potential to broadly enhance emergent writing and may facilitate the development of letter name/sound knowledge, print awareness, letter writing and name writing skills, and phonological awareness. The relation between the use of tablets and emergent literacy development seems to be complex as it is mediated by factors such as the type of multisensory experiences through literacy applications and the type of scaffolding used by adults. Recommendations and directions for future research are discussed.

INTRODUCTION

Emergent literacy is referred to the early development of a number of oral language and code-related skills (see Storch & Whitehurst, 2002) that are foundational skills upon which children build future skilled reading and writing (Kirby, Parrila, & Pfeiffer, 2003; Landerl & Wimmer, 2008). The provision of high-quality literacy experiences at home and in the pre-school setting as well as the quality of the interactions between adults and young children contribute to the development of children's early literacy skills (Niklas & Schneider, 2013; Dickinson & McCabe, 2001). With new technological advances continually emerging, the literacy practices of young children and their families are characterized, more than ever before, by the everyday use of digital technologies (Flewitt, Messer, & Kucirkova, 2015). In

DOI: 10.4018/978-1-7998-1486-3.ch004

fact, the entry of digital technologies primarily into the home environment and to a lesser extent into the school environment (Flewitt et al., 2015) has exposed children, parents and educators to new forms of literacy combining active learning and pleasure (Shamir & Korat, 2015).

Particularly, the rapid entry and use of portable technologies and efficient mobile devices such as touch- screen tablets brought new habits to the level of everyday life and new tools at the educational level (Kokkalia, Drigas, & Economou, 2016). These devices offer the opportunity for independent learning and easy access to information as well as they provide motivation and stimulus that promote the development of digital literacy (Zaranis, Kalogiannakis, & Papadakis, 2013). Touch-screen tablets are becoming rapidly popular among pre-schoolers because of their technological features (Papadakis, Kalogiannakis, & Zaranis, 2016) and this trend is rising rapidly (Rideout, 2017). They are able to provide all the advantages of mobile learning in all educational levels, not only for one-to-one learning but also as an entertaining, attractive, and potentially powerful learning tool for work in different learning environments (e.g. inside and outside the classroom) (Zaranis et al., 2013). In addition, they offer the potential for accompanying educational applications (apps) (Terras & Ramsay, 2012) that can create exciting environments for learning and instruction even in early childhood (Papadakis, Kalogiannakis, & Zaranis, 2017). The engaging features of tablets could afford a range of opportunities for promoting emergent literacy development (Neumann & Neumann, 2017). Their integration in family literacy practices is also changing patterns of parent/adult-child interactions and the way children learn from such interactions (Kucirkova, Messer, Sheehy, & Flewitt, 2013). Due to the increasing popularity of touch-screen tablets as digital devices for pre-schoolers it is essential to explore the potential effects tablets have on emergent literacy. In this direction, the present chapter reviews recent research that has examined the use of tablets at home and in the pre-school setting to foster emergent literacy skills.

In selecting the studies for this review, searches were conducted on PsycInfo, ERIC, and Google Scholar databases. The key words searched were ‘touch-screen’ or ‘tablets’ or ‘iPads’ or ‘applications’ or ‘mobile learning’ or ‘mobile device’ along with ‘emergent literacy’ or ‘early literacy’ as well as along with ‘pre-school’ or ‘home’. All abstracts (N=163) were then screened to include studies that conducted the period 2013-2018 and examined emergent literacy development or aspects of emergent literacy in young children (<6 years old) who used touch-screen tablets at home and/or the pre-school setting. Studies with children experiencing disabilities were not included. This generated 9 home-based studies (8 published studies and one master’s thesis), 11 pre-school-based studies (8 published studies and 3 dissertations) and 2 review studies. Fifteen studies on more general early childhood use of tablets and apps at home and in the pre-school setting were also reviewed. Recommendations will be discussed next to literature review.

BACKGROUND

Emergent literacy skills include a range of knowledge and understandings about language, letter sounds and names, phoneme grapheme correspondence, print concepts, linguistic awareness, early reading and early writing behaviors (Bowman & Treiman, 2004; Clay, 1998; Mann & Foy, 2003; Storch & Whitehurst, 2002; Whitehurst & Lonigan, 1998). Through the family literacy environment, young children are involved in a variety of formal and informal reading and writing experiences, become familiar with environmental print, develop literacy habits, interact with writing tools in order to write spontaneously, explore the literacy activities of people around them and participate in literacy activities with family members (Sénéchal & LeFevre, 2002; Manolitsis, 2016). Studies on home-based formal or informal literacy experiences have shown that the literacy habits of family members, the parent mediation during joint activities, as well as, the frequency and the quality of shared literacy interactions are decisive factors upon the development of emergent literacy skills (Curenton, & Justice, 2008; Neumann, Hood, & Neumann, 2009; Neumann, Hood, & Ford, 2012). Just as home, in the pre-school environment, young children participate in a variety of literacy activities (see Manolitsis, 2016) and are exposed to a variety of materials that contain printed letters, words, symbols, numerals and icons ((Neumann, 2014). This exposion is further facilitated by the several features afforded by digital technology (Patchan, & Puranic, 2016).

The introduction of Information and Communication Technologies (ICT) in the educational reality is without doubt dynamic. Computer aided learning is more realistic than ever before (Zaranis et al., 2013). However, in our days the content of the term ICT has been expanded to deal not only with computers but well as digital media, mobile technologies and online resources (see Zaranis et al., 2013). Computers and digital media have been shown to build basic emergent literacy skills through the use of interactive platforms (Hisrich & Blanchard, 2009). There have been prior studies on technology benefits revealing the usability of computer assisted instruction, as a learning tool, in improving early literacy skills (e.g. early reading and writing abilities, phonological awareness) in pre-school classrooms in different educational systems (see for a review Vernadakis, Avgerinos, Tsitskari, & Zachopoulou, 2005). Also, prior empirical research in early childhood education has indicated a significant impact of different digital media on literacy development (see for a review Zomer & Kay, 2016).

Mobile learning is defined as any kind of learning that takes place in learning environments taking into consideration the mobility of technology, the mobility of learners, and the mobility of learning processes (see Sharples, Taylor, & Vavoula, 2007). Most of the descriptions for mobile learning are based on learning theories "that focus mobility, efficiency, technology, that consider social structures and infrastructure, and that describe mobility as mobility of expectations" (Seipold, 2014, p. 31). Mobile learning is considered to denote learning involving the use of the mobile touch-screen devices (e.g. tablets and smartphones) and provides insight to the educational affordances of learning with these devices "as learning is untethered, happening across contexts, time, subjects, people, and technologies" (Crompton & Burke, 2018, pp. 53-54). Research on mobile learning reflects positive correlations between student engagement and effective use of technology (Chen, Lambert, & Guidry, 2010; Falloon & Khoo, 2014) perceived learning while using tablets and comfort levels and prolonged use of the tablet (Diemer, Fernandez, & Streepey, 2012). It is argued that the tablet is a game changer even in the preschool education and there is a need for more ways to harness it as a powerful teaching and learning tool (Kucirkova et al., 2013).

Learning through touch-screen technologies presents new dimensions for interactive digital experiences in different ways compared with other devices (Lam & Duan, 2012 ; Papadakis & Kalogiannakis, 2017). Tablets have three features that facilitate their easy use even by very young children: (a) they are portable and light-weight, (b) they do not require the child's fine motor skills to handle conventional computer peripherals such as mice and keyboards, and (c) they are designed in such a way to accommodate apps that have a child friendly intuitive design (Kucirkova, 2014; Papadakis, Kalogiannakis, & Zaranis, 2018). Specifically, touch-screen tablet provide children the opportunity to handle multimodal features such as texts, images and sounds through various gestures such as swipe, tap, touch, slide (Kucirkova, 2014; Lynch & Redpath, 2012). The nature of these devices allows for a child-centered approach to learning, giving pre-school children access to applications that respond to their individual needs, interests and abilities (Kucirkova, 2014). The use of tablets can make a significant contribution to the learning process at home and in the pre-school setting as it motivates the child's interest through interactive and multimedia content management while facilitating the development of social and individual skills (Judge, Floyd & Jeffs, 2015). Learning and cognitive development achieved by these devices is much more natural and more combatible for young children as it depends less on cognitive skills that they have not yet developed (Flewitt et al., 2014).

From very early age children are exposed visually to their parent's mobile touch-screen interfaces through exploring and playing with mobile touch-screen phones. As their fine motor skills develop they begin to explore the touch-screen displays and therefore to interact with visually attractive print (e.g. icons, symbols, letters, words, and numerals) (Neumann & Neumann, 2014). The touch-screen environment provides motivations for young children to draw, make symbols, write, create signs, messages and lists and generally to communicate in a digital environment in the same way as adults do (Beschorner & Hutchison, 2013). Additionally, touch-screen tablets are in the form of a writing/note pad and can rotate their surface. This feature may help children's orient in use of the print environment (Orrin & Olcese, 2011). Besides the advantages of the features of touch-screen interfaces, tablets are in favour of young children mostly due to the existence of mobile apps designed for these devices (Zaranis et al., 2013). The flexibility of these apps provides children the opportunity and motivation to develop digital skills and to create their own material with many media (e.g. images, texts, recordings) (Flewitt et al, 2015). These may include taking photos, watching videos, playing games, talking to family members on video chats or text messaging, mark making, and reading story books (Neumann, Merchant, & Burnett, 2018). As mobile technology becomes more important and its use may be influencing young children's emerging ideas about literacy, Beschorner and Hutchison (2013) note that "the conceptions of literacy and instruction students will receive may be changing to include the integration of technology" (p. 16).

THE USE OF TOUCH-SCREEN TABLETS AT HOME IN RELATION WITH EMERGENT LITERACY SKILLS

Young children use tablets at home in order to locate data on the internet, to play educational as well as entertainment apps and to interact through reading e-books (Livingstone et al., 2014; Salmon, 2013). The children's involvement with tablets may promote their understanding about the functions of print interface (Neumann & Neumann, 2014). For example, it has been observed that pre-schoolers identify basic functional words (e.g. Play) on tablet screens (Livingstone et al., 2014). Their early interaction with digital print via the use of touch-screen tablets and apps may aid in fostering aspects of emergent literacy skills (Neumann & Neumann, 2014). At home this interaction may occur through the children's own exploration of tablets or the support of their parents. A number of studies have broadly examined the use of tablets by young children for: (a) emergent literacy outcomes in the home context (Alliagas & Margallo, 2017) as well as its relationship to emergent literacy skills (Neumann, 2014, 2016), (b) literacy activities through parental scaffolding (Kucirkova et al., 2013; Neumann, 2018b; Petkovski, 2014), and (c) the impact of tablet-based e-book features in respect of emergent literacy aspects when joint reading (Krcmar and Cingel, 2014; O'toole & Kannass, 2018; Piotrowski & Krcmar, 2017).

Neumann (2014) conducted a study based on a parent survey (access and use of touch-screen tablets at home by their children). She found that pre-school children's (N=109; age 3-5 years) access to tablets was positively correlated with letter sound knowledge and name writing skill. In this respect tablets may be a potential resource for emergent literacy skills. On the contrary, duration could not be linked to any literacy measurement. The development of emergent literacy skills may be enhanced through the quality of children's interaction with tablet literacy apps rather than through the time spent. In this direction, Neumann (2016) carried out a second study based again on a parent survey and examined children's (N=57; age 2-4 years) participation in a variety of digital and non-digital literacy activities at home. Children's access to a larger number of apps was positively correlated with print knowledge. Also, the frequency of tablet use for writing was positively correlated with print awareness, print knowledge, and sound knowledge. According to the author, it seems that tablets can be effective digital writing tools for young children at home. Through writing and drawing apps they are engaging in activities such as scribbling, writing and copying letters and words. Yet, no associations occurred with regards to the frequency of e-book reading.

Tablet-based e-books, like print books, have the potential to foster pre-literacy skills (O'toole & Kannass, 2018). The tablet e-books that have additional features (e.g. read-to-me options and hotspots) and that can support highly interactive, multimedia experiences, offer notably different co-reading experiences than print books and simple digitized versions of print books do (Chiong, Ree, Takeuchi, & Erickson, 2012). To date, there is evidence demonstrating that tablet-based e-books and print books are both effective in improving a range of emergent literacy outcomes (for a review refer to Reich, Yau, & Warschauer, 2016).

However, tablet-based e-books should be examined whether they actually promote or hinder parent-child conversation about the book content and its structure (Reich et al., 2016). Krcmar and Cingel (2014) investigated the difference between e-books and print books comparing parent-child (N=70; children's age 2-5 years) shared reading of print-based book with an enhanced tablet-based e-book. Whether children recalled the stories was assessed by running a comprehension test. The results indicated that children scored significantly lower in comprehending the tablet-based e-book. Yet, in the tablet format, parent-child conversations focused less on the narrative content and more on e-book format and environment. Specifically, parents were observed to engage in more distractive talk with their child about the e-book format and environment. This non–content-related talk may explain the lower comprehension scores of the e-book condition. Conversely, with regards to the traditional book, parent-child verbal interactions actually focused more on the narrative of the book. The authors argue that the additional cognitive load required by the features of the tablet-based e-book may have influenced parent-child conversation and interaction, decreasing the child's ability to focus upon the narrative content.

The e-book reading apps for tablet may need to be better designed to reduce children's potential distraction for learning and to promote parent-child shared reading activities associated with the narrative of a story at home. As Reich et al. (2016) indicated, the well-designed tablet-based e-books that support learning seem to be equivalent, and occasionally better, than print books for pre-schoolers. Piotrowski and Krcmar (2017) employed a two group (hotspot on vs. hotspot off) randomly assigned design to examine how Dutch preschoolers (N=78; age 2-5 years) behave and respond to a story read to them by a tablet when that story either contains or does not contain hotspots. Results indicated that children in the hotspot condition made significantly more relative comments to the narrative rather than their peers in the non-hotspot condition. Ultimately, both groups performed very well on comprehension/recall test, recalling the majority of the central narrative elements, with no significant difference detected between the groups. The authors suggest that tablet-based e-book developers should try to ensure the double function of hotspots (activated or deactivated) so that parents or even children can determine what best meets their unique needs.

Also, the narrative style during reading may influence young children's conception of the story content. O'toole and Kannass (2018) compared preschoolers' (N= 100; 4 years old) word learning and story comprehension from print and tablet e-book controlling for narration source. Children were randomly assigned to one of four conditions: an adult reading out loud a print book, an e-book read aloud by an adult, a print book narrated by an audio device, and an e-book narrated by an audio device. The results indicated that tablet e-books and audio devices promoted word learning. On the other hand, there were no effects of book type or narration style on story comprehension. However, children in all cases did very well on this assessment. Interestingly, when controlling attention, the story comprehension scores tended to be higher for children with the live reader experience rather than the audio device. The authors suggest that the presence of an adult is essential to scaffold children's early literacy skills in all book reading contexts (e.g. print, electronic, read aloud, and narrated).

Qualitative studies have examined children's more general engagement with e-books, which present a unique context for documenting how children practice their digital literacies in the home setting (see for a review Kucirkova, 2018). For example, Aliagas and Margallo (2017) studied children's literacy experiences with storybook tablet app during family shared book reading and focused on the effects of screen-based interactivity. Some e-books' interactivity relies on identifying text key points and teaching new words, but it is also associated with children's engagement in the story (Kucirkova, 2018). The researchers studied for two years tablet-based share reading events of four Spanish families (children's age 1,5 – 5 years; N=4) and analysed children's type of interactivity with regards to storybook apps. Their study showed that there was interactivity between the storybook app and the children's responses which facilitated emergent readers to be part of the story as collaborators, storytellers, authors or internal characters, that is, to adopt active reading roles. Alliagas and Margallo (2017) argue that these interactive, intellectual and emotionally immersive experiences provide a map that carries the potential to change children's reading experiences with tablet storybook apps. According to Kucirkova (2018), although there isn't a staunch delimitation of the term interactivity in children's actual use of e-books, "interactivity is more than just children tapping on hotspots that trigger a certain sound effect" (p. 27).

Parent-child general interactions with tablets certify that these devices are positive and engaging learning tools for young children (Kucirkova & Sakr, 2015; Wood et al., 2016). Kucirkova et al. (2013) used a case-study approach to inspect the shared tablet stories' interactions between a mother and her 33-month-old daughter in the home setting as well as how they used the Our Story app to jointly create a personalised story based on their past holiday. The mother was sitting closely together with her daughter holding the tablet with her right arm and hugging the young girl whilst using her left-hand index to point out pictures on the screen accompanied with written and spoken words. At the same time, the mother asked her daughter questions about their past holidays experience. Observing the session showed the mother's support facilitated the child's early understanding of the printed word, i.e. that the written words on the tablet screen had a relationship with real-life events and objects. The researchers argue that the shared use of tablet apps such as Our Story provide parents with the opportunity to build on their children's experiences with texts, sounds and pictures.

Petkovski (2014) explored the types of parental supportive strategies when tablets are used by their young children. In particular, she examined parental scaffolding by asking mothers (N=72) and fathers (N=32) to play for 10 minutes with seven reading apps and five math apps with their 2-6 year old children. The 104 parent-child couples were free to select any of the apps as well as any typical app/function, which appears on the tablets. The self-report (survey and interview) and observational data showed that parents provided their children with four different types of scaffolding: verbal (e.g. reading out a list of items), physical (e.g. pointing to the touch screen), emotional-verbal (e.g. positive reinforcement and providing confidence), and emotional-physical (e.g. physical expressions of praise). However, conclusions about the effects of literacy apps on emergent literacy cannot be drawn from this interactive session.

The type of parental support during a child's interactions with tablets and literacy apps may influence emergent literacy development (Neumann & Neumann, 2017). A more nuanced approach is therefore to examine the association between the type of parental scaffolding and aspects of emergent literacy. Neumann (2018b) how maternal behaviors differ when using a touch-screen tablet as a writing tool compared to traditional writing tools. Mother-child dyads (N=47; children's age 1.98-4.59 years) were asked to complete two joint-writing tasks. Mothers were provided with the instruction to help their child in any way that they liked, to write two words by using the touch-screen tablet and the Draw Buddy app and subsequently to note the same words on a paper using a pencil. Video-recorded observations showed that across both types of writing tools mothers mediated their children's letter and word writing using similar scaffolding strategies (e.g. writing the word for the child to copy and dictating the letter names) and used similar directional language to guide and extend children's writing efforts. Also, all mothers used positive affective feedback (e.g. "That's beautiful") to encourage their child's efforts. Correlational analyses indicated that maternal print mediation during tablet use was positively related to letter name knowledge, letter-sound knowledge, letter writing, and name writing. Neumann (2018b) argues that parental scaffolding of preschoolers' writing with tablets may provide a roadmap for helping children develop letter shaping skills.

Overall, the research reviewed, although precludes any causation, suggests that writing activities with interactive literacy tablet apps in the home setting can support aspects of emergent literacy learning such as letter sound knowledge, name writing (Neumann, 2014), print awareness, print knowledge, and sound knowledge (Neumann, 2016) particularly when these activities are supported and encouraged by parental mediation and scaffolding strategies (Neumann, 2018b). Also, suggests that young children's active engagement in shared tablet-based literacy events can support early literacy outcomes (e.g. early understanding of the printed word) (Alliagas & Margallo, 2017; Kucirkova et al., 2013). Tablet-based e-book reading could be effective to story comprehension (Piotrowski and Krcmar (2017), especially when this activity is a shared experience between an adult and child (O' toole & Kannass, 2018). Further empirically designed studies will help to clarify the impact of the use of touch-screen tablets in emergent literacy skills in the home setting.

USING TOUCH-SCREEN TABLETS IN SUPPORTING EMERGENT LITERACY SKILLS IN PRE-SCHOOL CLASSROOMS

In the pre-school setting touch-screen tablets, like many other classroom tools, have been used independently, in small groups and for whole class instruction to support young children's development of emergent literacy skills (Beschorner & Hutchison, 2013). The way on how touch-screen tablets support emergent literacy skills in early childhood classrooms has been examined both by qualitative studies (Beschorner & Hutchison, 2013; Bigelow, 2013; Crescenzi, Jewitt & Price, 2014 ; Fink, 2018; Flewitt et al., 2015; Isgett & Mann, 2018) and by experimental or semi-experimental studies (Cubelic, 2013; Neumann, 2018a; Patchan & Puranic, 2016; Reeves, Gunter, & Lacey, 2017; Vatalaro et al., 2018).

The qualitative studies have mainly described how the use of tablets can support the development of early literacy skills, such as emergent writing (Beschorner & Hutchison, 2013; Bigelow, 2013; Crescenzi et al., 2014), emergent reading and writing (Flewitt et al., 2015), letter/letter sound knowledge (Isgett & Mann, 2018) and early literacy learning (Fink, 2018). Beschorner and Hutchison (2013) conducted a study across two pre-school classrooms (N= 35; age 4-5 years) providing teachers with six tablets for individual, group, and class activities. Children were free to use a variety of tablet apps that were presented to them twice a week over a total period of seven weeks. Data analysis showed that children's emergent writing was facilitated by specific tablet writing apps (e.g. iWrite), which allowed pre-schoolers to actually write using letters, symbols and/or drawings formed on the screen using their finger and to write typed text using the keyboard. Even children with a limited letter shaping ability could use the on-screen keyboard to identify and write letters and touch them on the screen. Pre-schoolers were excited about the communicational purpose of their written efforts (constructing emails and messages for family members), thus understanding even more the function and the importance of writing. The creation of messages is a common practice in early childhood classrooms. In this direction, Bigelow (2013) examined preschoolers' (N=15; age 43-57 months) digital message making practices and products over a period of six months. Specifically, she explored how young children interact with touch-screen tablets during the digital composing process and what kind of digital products they create when composing and sending emails to their parents. The data analysis showed that through their involvement with the iWrite activities children willingly engaged in and began to understand more about emails and their purpose as well as to compose digital messages. Children engaged in on-screen interactions resulted in a variety of on-screen actions (e.g. tapping, stamping, typing) for their written efforts. A range of digital products were observed many of which, according to author, exhibited characteristics reported in print-based emergent literacy research (e.g. the use of multiple colours when drawing and one colour when writing, the use of familiar names in both multimodal and typed mail products).

Digital technology such as tablets seems to offer different ways to interact with literacy than pen and paper (Bigelow, 2013). For example, the pop-up keyboard feature enabled children in the above two studies to create their own digital stories or messages, which may be difficult tasks if using pen and paper_(Neumann & Neumann, 2017). Crescenzi et al. (2014) compared over a five-day period the creation of messages via finger painting processes in digital (tablet) versus physical (paper) learning environments by very young children (N=7; age 27-37 months). Both digital (e.g. Doodle Buddy) and non-digital environments (i.e. sheets of white papers with or without a pre-drawn image, paint palette) offered a restricted range of colours and touch-based interaction. Considering as semiotic resources tapping touches, straight and circular strokes, these were mostly used in the tablet environments while press touch occurred most often in the paint-paper environment. Touch repetition as well as sequences of continuous touch prevailed with the tablets rather than on paper. According to the researchers tablets may be helpful when continuous touch movements are the primary action. It furthermore enhances an early stage of drawing and supports the formation of digital skills in tandem with touch experiences. Tablets appear less disruptive than paints as children's attention remains centered on the creation of the message, whilst the movement between paper and paint palettes is distractive. On the other hand, some tablet features may restrict children's non-digital activities especially in terms of sensory experiences: a digital tap is always the same, coloring on paper may vary (Crescenzi et al., 2014).

Also, it seems that touch-screen based literacy activities lead teachers to re-evaluate children's literacy competence. They encourage children to construct positive images of themselves in the literacy classroom. Flewitt et al. (2015) lent tablets to a children's centre nursery (N= 16; age 3-4 years) and to a primary school reception class (N=30; age 4-5 years). They discussed over a two-month period their potential uses with staff in pre- and post-semi-structured interviews and observed how tablets were integrated into the classroom activities. Their post findings showed that the integration of tablets offered many opportunities for the development of early literacy skills (emergent writing/reading) in both autonomous and collaborative learning environments. Teachers emphasized the creative use of touch-screen tablets and recognized their benefits in developing children's self-esteem and enthusiastic involvement in multiple reading and writing activities. Specifically, they underlined that for some children, the tablet offered gateways for revealing their true reading potential. Furthermore, they noted that young children's involvement in digital literacy activities with tablets changed the ways in which children were perceived by peers and teachers (e.g. "good spellers" or "more capable readers"), forming a new relational framework in the classroom.

In relation with the nature of tablet apps used, the 'closed content' apps position children as recipients of narrowly defined literacy knowledge, while the 'open content' apps engage children to create original personal stories in multiple media (Flewit et al., 2015). Isgett and Mann (2018) examined over a six-week period the effect of using a 'closed content' touch screen app (e.g. Phonics Island) to kindergarden aged children's (N=26) knowledge of letters and letter sounds in two classrooms. Children were given a tablet for approximately fifteen minutes per day. They showed that this tablet app was a great resource to use along with teacher led lessons to motivate young children's knowledge of uppercase/lowercase letters and letter sounds. However, they did not determine if the increasing knowledge of letters/letter sounds was due only to the use of the tablets as there were many teacher led and student led independent activities that happened over the course of the six weeks and did not include technology. In another study with kindergarden aged children (N=13) Fink (2018) examined over a course of fifteen week the integration of tablets and 'open content' apps into early literacy instruction. She indicated that creation based learning tasks, which utilize the tablet and related 'open content' apps (e.g. Show Me) can be effectively integrated into pedagogy to facilitate early literacy learning in the kindergarden classroom. The 'open content' tablet apps provided digital experiences where children had the opportunity to create meaningful connections between the words they stated, the images they used, and the product that each artifact represented. Particularly, her findings indicated that young children utilized the tablet and related apps to create a variety of digital products (e.g. interactive presentation, talking picture, and digital comic book story) that demonstrated their early literacy learning.

In order to evaluate causal effects for the use of tablets in supporting emergent literacy skills in preschool classrooms a number of studies have used a pre-test and post-test control group methodology. These studies have mainly focused on the effects of tablet use on specific emergent literacy skills such as phonemic awareness, alphabetic principle (Cubelic, 2013), vocabulary development (Vatalaro et al., 2018) letter name/sound knowledge, print concepts, name writing (Neumann, 2018a), capital letter writing (Patchan & Puranic, 2016), phonological awareness (Reeves et al., 2017).

In a large intervention study that involved 16 kindergarten classrooms Cubelic (2013) investigated the relation between achievement scores in the areas of phonemic awareness and alphabetic principle of children who received supplemental literacy instruction using tablet apps and those children who did not. Children in the treated classrooms (N=144) used tablet literacy apps (e.g. ABC Phonics) that were selected and monitored by their teachers to foster phonemic awareness and the alphabetic principle for approximately sixty minutes per week over the course of a school year. Children in the control classrooms (N=147) engaged in the existing literacy framework with their teachers. Data for letter naming, first sound fluency, phoneme segmentation fluency, and nonsense word fluency showed a large pre-post-test effect for both groups. Yet, children in the treatment group outperformed their peers in the control group in phoneme segmentation fluency (phonemic awareness) and nonsense word fluency (alphabetic principle) only across the different socio-economic status categorizations. By simply incorporating the use of tablets in pre-school classrooms, achievement gains upon emergent literacy skills are difficult to interpret. Technical scaffolding and feedback by the tablet app itself, is likely to support aspects of emergent literacy skills. Scaffolding and feedback are seen crucial components to the learning process (Shute, 2008; Pentimonti & Justice, 2010).

Vatalaro et al. (2018) carried out an 8-week intervention study to examine the efficacy of using tablets and different intentionally chosen apps to increase the receptive and expressive vocabulary development of pre-school children aged 3 to 5 years old. It included four classrooms, which served as two intervention groups and two comparison groups. In the first intervention group (N=16) the teachers exposed children to scaffolding-like interactive vocabulary apps (e.g. Endless Alphabet) that labeled, defined, and animated vocabulary for an average of 1.8 hours per day. In the second intervention group (N=15) children were exposed to open-ended vocabulary apps (e.g. Beck and Bo) in which they more often had to produce vocabulary for an average of 1.3 hours per day. In the two comparison groups (N=18; N=14) children were free to use various chosen apps (e.g. Letter School) without instruction for approximately 1 hour per day. Children in all groups were pre- and post-tested in terms of receptive and expressive vocabulary. The children who used the scaffolding-like apps had higher gains in receptive vocabulary than their peers who used open-ended apps. None of the intervention groups differed significantly from the comparison groups on this measure. On the other hand, although there were no significant differences between the groups with relevance to the expressive vocabulary measures, children using the scaffolding-like apps had greater means to increase in their expressive vocabulary than those using open-ended apps.

It seems that the type of apps, regardless of the teacher's guidance, may be primary vehicle for scaffolding aspects of emergent literacy skills. Neumann (2018a) using a pre-test post-test randomised control group design explored the effects of using literacy tablet apps with multimodal, interactive and in-built support features on a range of emergent literacy skills in children aged 2 to 5 years old. Children in the tablet group (N=24) were given a nine week intervention program (30 min/week) in small groups using literacy apps (e.g. Letter school) focusing on three new alphabet letters each week. The control group (N=24) participated in joint non-digital activities with their teachers. Children in both groups were pre- and post-tested individually on specific emergent literacy skills (print concepts, letter name and sound knowledge, name writing and letter writing). The results showed that while there were no significant differences at pre-test between the two groups, the tablet group showed significantly higher scores at post-test in letter name and sound knowledge, print concepts, and name writing when compared to the control group. Neumann (2018a) argues that young children can independently and competently use tablets and apps for literacy learning as well as that the supporting features of literacy apps may positively stimulate children's interactions and experiences with letters and words.

Despite the importance of scaffolding-like apps, the feedback by touch-screen itself may be helpful to the development of early literacy skills. Patchan and Puranic (2016) tested the effectiveness of using tablets to teach pre-school children (N=46; age 41-65 months) how to write capital letters. They further examined the influence of two factors afforded by the tablet device itself (i.e. extrinsic and intrinsic feedback) on the development of children's writing ability. The extrinsic feedback was provided while the child was still thinking about the form of the letter. On the other hand, the intrinsic feedback was about the enriched tactile experience resulting from direct contact of the child's finger during writing. Children were divided into small groups and were taught to write capital letters for 8 weeks (3 times per week). Each group was randomly assigned to one of the three instructional conditions: tablet finger, tablet stylus, and paper and pencil. Children in all groups were pre- and post-tested individually on letter writing using paper and pencils. The findings indicated that although children from the tablet stylus condition wrote a similar number of letters correctly at post-test, as those from paper and pencil condition, the children from the tablet finger condition wrote correctly more letters than those from the tablet stylus condition. Therefore, the extrinsic feedback did not provide an additional benefit. In contrast, the intrinsic feedback made salient by the enriched tactile experience did enhance children's learning. That is, the enhancement of touch experiences during learning was more beneficial for letter writing.

Reeves et al. (2017) using a two-group quasi-experimental design investigated the effect of mobile learning with the use of informal feedback on enhancing emergent literacy skills of pre-kindergarten students aged 4-5 year. Children in the experimental group (N=20) practiced early literacy and math skills using a variety of tablet apps two days a week for approximately fifteen minutes of guided instruction, coupled with informal feedback from students for each app used, over the course of a school year. The high degree of engagement and success of children in one challenging but creative app it was signaling the usability of this app and therefore its importance for enhanced instruction on a subject area. Children in the control group (N=8) were taught with the traditional curriculum without tablets. Both groups were pre- and post-tested on print knowledge, phonological awareness, mathematics, and oral language. The results revealed significantly higher phonological awareness for the experimental group. The researchers note that many of the tablet apps used by children specifically focused on phonics. The experimental group outperformed the control group at post-test for print knowledge and oral language but the results were not significant. However, both groups improved these two skills significantly. Reeves et al. (2017) support the notion that the integration of mobile technology in content-specific areas with the use of student-centered informal feedback and innovative teaching practices could enhance early literacy learning.

On the whole, the research reviewed suggests that the use of touch-screen tablets in the pre-school classrooms can support aspects of emergent literacy learning. Experimental studies showed that tablets and literacy apps do significantly improve skills such as phonological awareness (Reeves et al., 2017), letter writing (Patchan & Puranic, 2016), letter name/sound knowledge, print concepts, and name writing (Neumann, 2018a). However, the picture with regard to vocabulary, phonemic awareness and alphabetic principle was unclear (Cubelic, 2013; Vatalaro et al., 2018). Further experimental designed studies should take into account factors such as: sample size, randomised allocation of participants to experimental groups, time of instruction, digital and non-digital cognitive and language instructional strategies, educational tablet-based practices, app-based features, and home use of tablets. On the other hand, quantitative studies indicated that tablets and apps are a positive educational tool that can support emergent writing (Beschorner & Hutchison, 2013; Flewitt et al., 2015). Although, it seems that the use of tablets can have a positive impact upon aspects of emergent literacy skills more research is needed.

RECOMMENDATIONS

The features of touch-screen tablet provide a platform to support children's emerging understanding of literacy. A number of studies on tablets and emergent literacy in the home and pre-school setting have shown that these devices can be used in multiple ways and may facilitate emergent literacy skills such as letter name/sound knowledge (Neumann, 2014, 2016, 2018a), print awareness (Neumann, 2016, 2018a), letter writing (Patchan & Puranic, 2016), name writing (Neumann, 2014, 2018a), phonological awareness (Reeves et al., 2017) and emergent writing (Beschorner & Hutchison, 2013; Flewitt et al., 2015).

In fact, the relation between tablets and emergent literacy development may be complex (Neumann & Neumann, 2017). From the research reviewed, some key themes have been raised.

Children's access of touch-screen tablets at home may promote the emergence of some aspects of literacy but raises questions concerning the type and quality of digital activities in which children engage when use tablets at home. In Neumann's (2016) study young children aged 2-4 years who used tablets more often for writing had greater print awareness and letter/sound knowledge. Children used writing and drawings apps and they had the opportunity to express their early writing efforts. Thus, they were engaging in activities such as scribbling, writing and copying letters and words throughout a constructive process. It seems that the touch-based features of a tablet provide further valuable opportunities for young children who have not yet fully their fine motor skills to easily make marks with their finger (Neumann, 2016). Research has shown that young children's multisensory experiences contribute to learning to write (Labat, Ecalle, Baldy, & Magnan, 2014).

Creative literacy experiences through touch-screen tablets that engage and scaffold children's learning may be an important factor in developing emergent literacy skills. These experiences show off the importance of the quality of literacy apps. "Tablet experiences and emergent literacy skills may be depended upon the quality of the literacy apps" (Neumann & Neumann, 2017, p. 213). It is predominantly argued that literacy learning with tablets is likely to be influenced by the types of multisensory features as well as by the presence of in-built support features of apps (Neumann, 2018a). The constructive apps which are designed for creative expression and are characterized by an 'open-ended' design that allows young children to create their own content (Papadakis & Kalogiannakis, 2017) may promote early literacy learning (Fink, 2018). Thus, literacy apps should have specific features in their design and should provide children with rapid feedback and multiple avenues for engaging, learning and communicating with their world (Neumann & Neumann, 2014). With such a multifunctional technology young children's learning has the potential to be enriched (Wood et al., 2016). In any case, literacy apps need to be linked to psychological (e.g. behaviors, thoughts, needs, interests and abilities of young children), pedagogical (e.g. how to deliver the content to children through tablets), and technological dimensions (e.g. how the features of tablet itself increase children's engagement) (Mohamad, Lakulu, & Samsudin, 2016).

On the subject of tablet-based e-books, they often contain distractive features (e.g. hotspots) that may divert children's attention to the story line, thus reducing their story recall ability (Northrop & Killeen, 2013). Such distractive features may increase the cognitive load and impact upon the narrative content (Krcmar & Cingel, 2014). This may be the reason why the research in which early literacy outcomes during tablet-based e-book and paper book reading have been compared has provided mixed evidence (Richter & Courage, 2017). There are four main factors that should be emphasized. First, the challenge of comparative studies is to ensure that researchers compare like with like, which is not always possible with e-books as there are many types and kinds of them (Kucirkova, 2018). Second, the importance of parent-child joint reading and conversation about the story content. According to Kucirkova (2018) "so long as electronic books provide space for parents to have a conversation with their child about the story content, they may be a viable alternative to print books" (p. 19). Third, the narrative style during reading as findings suggest that the shared book reading, regardless of book type, should be between a child and an adult reader (O'toole & Kannass, 2018) and the narration function should be turned off (Hoffman & Paciga, 2013). Fourth, if an e-book contains unrelated and not well designed hotspots and games, thus taxing the working memory (Zipke, 2017). Tablet-based e-book developers should try to ensure the double function of hotspots (activated or deactivated) so that parents or even children can determine what best meets their unique needs (Piotrowski & Krcmar, 2017).

The parent-child interaction with a tablet is without a doubt of vital importance as it provides young children with scaffolded instruction and helps to strengthen skill sets in various domains (Petkowski, 2014). Parents have been observed to provide their children with a range of tablet-scaffolded strategies such as physical (e.g. pointing to words), verbal (e.g. asking questions) and emotional (e.g. positive reinforcement) to build their children's learning during interactions with literacy apps (Petkowski, 2014) or during the creation of a personalised story (Kucirkova et al., 2013). On the other hand, children are self-motivated to interact with tablets independently and play apps for entertainment purposes that are not age or developmentally appropriate and lack learning features (Neumann, 2014; Neumann & Neumann, 2017). Also, a small proportion of parents believe that the use of tablets would not help literacy learning and may decrease their children's appreciation of traditional literacy activities (Livingstone et al.,2014). So the question is how to best help parents manage children's use of tablets at home and to adopt a pedagogical approach to support emergent literacy. Early childhood educators could help parents to create positive tablet-based early literacy learning experiences and to supervise tablet use.

The research reviewed showed that the integration of touch-screen tablets in the repertoire of pre-school classroom-based activities offers promising educational opportunities for early writing (Beschorner & Hutchison, 2013; Flewitt et al., 2015). However, the integration of these devices into classroom literacy practice requires a great deal of thought and commitment from teaching staff as it presupposes the development of a pedagogy that supports their creative use (Flewitt et al., 2015). Teacher beliefs about tablets as an educational tool may influence children's engagement, interaction, and learning about these devices (Neumann, Merchant & Burnett, 2018). Thus, pre-school teachers need support in integrating tablets into their classrooms and mainly in understanding about the use of tablets at pre-school settings for early literacy learning. Studies have shown that they can be used independently (Lu et al., 2017) and in small groups (Northrop & Killeen, 2013) as a tool for encouraging social interaction, communication, play, and creating (Geist, 2014). The mobility of these devices allows children to freely change positions into the preschool classroom by using tablet at various body postures (Neumann, 2018a). This content offers many opportunities in both autonomous and collaborative learning environments. Qualitative studies showed that young children can independently and in small groups use tablets and apps in pre-school classrooms for literacy learning (Beschorner & Hutchison, 2013; Flewitt et al., 2015; Fink, 2018). In Neumann's (2018a) experimental study children in the tablet group attended a weekly session in small groups of 3–4 children and showed significantly higher letter name, sound knowledge, print concepts and name writing skills than children in the control group. The research reviewed showed that the impact of the use of tablets in emergent literacy skills was more clear in studies in which children could independently and in small groups use tablets and apps for literacy learning (Neumann's, 2018a; Patchan & Puranic, 2016) rather than in studies with whole classroom instruction (Cubelic, 2013; Reeves et al., 2017; Vatalaro et al., 2018).

Beschorner and Hutchison (2013) have noted the important role pre-school teachers play in supporting children during literacy activities on touch-screen tablet by providing prompts, encouragement and technical help. As Neumann and Neumman (2014) mention, "the optimal use of tablets for early literacy learning may be dependent upon the type of scaffolding used by parent or teacher" (p. 231). It is possible that the use of directional language in supporting children during literacy activities on touch-screen tablets reinforces the development of aspects of emergent literacy skills. The use of higher levels of mediation and the focus on cognitive-based questions may support children in a literacy activity that they would unable to do on their own (Neumann, 2018a). Lynch and Redpath (2014) argue that the true impact of tablets and apps depends greatly on how they are utilized.

FUTURE RESEARCH DIRECTIONS

The image emerging from the reviewed literature suggests that the use of tablets and apps at home may have potentially positive effects on children's emergent literacy development but does not imply causation. Thus, further empirically designed research is needed. As the home literacy environment in our days consists of both traditional and digital literacy tools the development of emergent literacy skills in such an environment should be firmly on the agenda of early literacy researchers in order to provide information for understanding how children's early literacy experiences are changing. However, although many parents do allow their children to use mobile devices (Petkowski, 2014) others do not or not so systematically. It might be important in future research to take into consideration the young child's level of ability to use tablet and it's relation with emergent literacy skills (e.g. novice users vs. more advanced users). Another important factor is the interaction that takes place between parents and children when introducing their child to tablet-based literacy apps. Future research should take into consideration the child's behavior to parental supporting attempts. It is interesting to explore whether parents' support when their children use tablets for literacy activities is efficient and applicable.

Moreover, it is critical to evaluate how tablets can be used as an educational tool in the pre-school classrooms. The research reviewed showed that the use of touch-screen tablets can support aspects of emergent literacy. However, there have been a wide range of methods, procedures, and materials used to study the questions of interest. In some studies the age range of the participant samples was wide and the language skill variability might was present. Thus, future research examining tablets and early literacy skills should control for early language variability. In addition, another factor that was not adequately addressed is the influence of children's home use of tablets and literacy apps on emergent literacy. This specific child-level characteristic may be decisive in early childhood for children who are behind in tablet-based literacy experiences and requires more investigation. Also, there is a need to consider how tablets and literacy apps can be integrated into interventional studies using stronger control methods (e.g. regular classroom activities vs. small groups). Classroom level factors such as peer interactions, teacher's knowledge and ability in using tablets for instruction may impact upon learning. In most of studies reviewed there were used different but specific literacy apps. It is likely the design level and the cognitive prerequisites of these apps to differ. Future research could also examine young children's interactions with 'closed-content' and 'open-content' literacy apps in order to gain insight into specific cognitive strategies that they are apply to build early literacy skills.

CONCLUSION

It seems that touch screen-tablets can be used to foster young children's emergent literacy skills at home and in the pre-school setting. These mobile devices can provide multisensory structured experiences that may facilitate aspects of emergent literacy although further research is needed. Tablet experiences and emergent literacy skills may be dependent upon the quality of the literacy apps. These apps should have specific features in their design in order to provide a high level of interactivity, to connect children with the printed screen symbols and to offer opportunities for constructivist learning and regular feedback. The association between touch-screen literacy activities and the development of emergent literacy seems to be reinforced by the type of scaffolding used by adults. It is important to provide parents and teachers with evidence-based strategies in order to create positive tablet-based literacy experiences in early childhood.

REFERENCES

Aliagas, C., & Margallo, A. M. (2017). Children's responses to the interactivity of storybook apps in family shared reading events involving the iPad. *Literacy*, *51*(1), 44–52. doi:10.1111/lit.12089

Beschorner, B., & Hutchison, A. (2013). iPads as a literacy teaching tool in early childhood. *International Journal of Education in Mathematics, Science, and Technology*, *1*(1), 16–24.

Bigelow, E. C. (2013). *iWrite: Digital Message Making Practices of Young Children*. (PhD Thesis, Vanderbilt University). Available at https://etd.library.vanderbilt.edu

Bowman, M., & Treiman, R. (2004). Stepping stones to reading. *Theory into Practice*, *43*(4), 295–303. doi:10.120715430421tip4304_8

Chen, P. S. D., Lambert, A. D., & Guidry, K. R. (2010). Engaging online learners: The impact of web-based technology on college student engagement. *Computers & Education*, *54*(4), 1222–1232. doi:10.1016/j.compedu.2009.11.008

Chiong, C., Ree, J., Takeuchi, L., & Erickson, I. (2012). *Print Books vs e-Books*. New York, NY: The Joan Ganz Cooney Center.

Clay, M. M. (1998). *By different paths to common outcomes*. Portland, ME: Stenhouse.

Crescenzi, L., Jewitt, C., & Price, S. (2014). The role of touch in preschool children's learning using iPad versus paper interaction. *Australian Journal of Language and Literacy*, *37*, 87–95.

Crompton, H., & Burke, D. (2018). The use of mobile learning in higher education: A systematic review. *Computers & Education*, *123*, 53–64. doi:10.1016/j.compedu.2018.04.007

Cubelic, C. J. (2013). *iPad 2 applications and emergent literacy: do they have an impact on the acquisition of early literacy skills?* (PhD Thesis, Youngstown State University). Available at http://rave.ohiolink.edu/etdc/

Curenton, S. M., & Justice, L. M. (2008). Children's preliteracy skills: Influence of mothers' education and beliefs about shared-reading interactions. *Early Education and Development*, *19*(2), 261–283. doi:10.1080/10409280801963939

Dickinson, D. K., & McCabe, A. (2001). Bringing it all together: The multiple origins, skills, and environmental supports of early literacy. *Learning Disabilities Research & Practice*, *16*(4), 186–202. doi:10.1111/0938-8982.00019

Diemer, T. T., Fernandez, E., & Streepey, J. W. (2012). Student perceptions of classroom engagement and learning using iPads. *Journal of Teaching and Learning with Technology*, *1*(2), 13–25.

Falloon, G., & Khoo, E. (2014). Exploring young students' talk in iPad-supported collaborative learning environments. *Computers & Education*, *77*, 13–28. doi:10.1016/j.compedu.2014.04.008

Fink, M. A. (2018). *Effective iPad integration in the kindergarten literacy curriculum through creation-based literacy tasks: an action research study*. (PhD Thesis, University of Pittsburgh). Available at http://d-scholarship.pitt.edu/35369/1/

Flewitt, R., Messer, D., & Kucirkova, N. (2015). New directions for early literacy in a digital age: The iPad. *Journal of Early Childhood Literacy*, *15*(3), 289–310. doi:10.1177/1468798414533560

Geist, E. (2014). Using tablet computers with toddlers and young pre-schoolers. *Young Children*, *69*, 58–63.

Hisrich, K., & Blanchard, J. (2009). Digital media and emergent literacy. *Computers in the Schools*, *26*(4), 240–255. doi:10.1080/07380560903360160

Hoffman, J. L., & Paciga, K. (2013). Click, Swipe, and Read: Sharing e-Books with Toddlers and Preschoolers. *Early Childhood Education Journal*, *42*(6), 379–388. doi:10.100710643-013-0622-5

Isgett, J., & Mann, L. (2018). *Using iPads to increase students' letters and letter sounds knowledge.* Retrieved from https://sophia.stkate.edu/ maed/253

Judge, S., Floyd, K., & Jeffs, T. (2015). Using mobile media devices and apps to promote young children's learning. In K. L. Heider & M. Renck Jalongo (Eds.), *Young children and families in the information age* (pp. 117–131). Springer Netherlands. doi:10.1007/978-94-017-9184-7_7

Kirby, J. R., Parrila, R. K., & Pfeiffer, S. L. (2003). Naming speed and phonological awareness as predictors of reading development. *Journal of Educational Psychology*, *95*(3), 453–464. doi:10.1037/0022-0663.95.3.453

Kokkalia, G., Drigas, A. S., & Economou, A. (2016). Mobile learning for preschool education. *International Journal of Interactive Mobile Technologies*, *10*(4), 57–64. doi:10.3991/ijim.v10i4.6021

Krcmar, M., & Cingel, D. P. (2014). Parent-child joint reading in traditional and electronic formats. *Media Psychology*, *17*(3), 262–281. doi:10.1080/15213269.2013.840243

Kucirkova, N. (2014). iPads in early education: Separating assumptions and evidence. *Frontiers in Psychology*, *5*, 715. doi:10.3389/fpsyg.2014.00715 PMID:25071678

Kucirkova, N. (2018). *How and why to read and create children's digital books.* London, UK: UCL Press. Available at www.ucl.ac.uk/ucl-press

Kucirkova, N., Messer, D., Sheehy, K., & Flewitt, R. (2013). Sharing personalised stories on iPads: A close look at one parent–child interaction. *Literacy*, *47*(3), 115–122. doi:10.1111/lit.12003

Kucirkova, N., & Sakr, M. (2015). Child–father creative text-making at home with crayons, iPad collage & PC. *Thinking Skills and Creativity*, *17*, 59–73. doi:10.1016/j.tsc.2015.05.003

Labat, H., Ecalle, J., Baldy, R., & Magnan, A. (2014). How can low-skilled 5-year-old children benefit from multisensory training on the acquisition of the alphabetic principle? *Learning and Individual Differences*, *29*, 106–113. doi:10.1016/j.lindif.2013.09.016

Lam, J., & Duan, G. (2012). A review of mobile learning environment in higher education sector of Hong Kong: Technological and social perspectives. ICHL (pp. 165-173). Springer.

Landerl, K., & Wimmer, H. (2008). Development of word reading fluency and spelling in a consistent orthography: An 8-year follow-up. *Journal of Educational Psychology*, *100*(1), 150–161. doi:10.1037/0022-0663.100.1.150

Livingstone, S., Marsh, J., Plowman, L., ... (2014). *Young children (0-8) and digital technology: a qualitative exploratory study national report.* UK: Joint Research Centre, European Commission, Luxembourg. Retrieved from http://eprints.lse.ac.uk/60799/1/__lse.ac.uk_storage_LIBRARY

Lu, Y., Ottenbreit-Leftwich, A. T., Ding, A., & Glazewski, K. (2017). Experienced iPad-using early childhood teachers: Practices in the one-to-one iPad classroom. *Computers in the Schools: Interdisciplinary Journal of Practice, Theory, and Applied Research, 34.* doi:. 2017.1287543 doi:10.1080/07380569

Lynch, J., & Redpath, T. (2012). Smart technologies in early years literacy education: A Meta-narrative of paradigmatic tensions in iPad use in an Australian preparatory classroom. *Journal of Early Childhood Literacy, 14*(2), 147–174. doi:10.1177/1468798412453150

Mann, V. A., & Foy, J. G. (2003). Phonological awareness, speech development, and letter knowledge in preschool children. *Annals of Dyslexia, 53*(1), 149–173. doi:10.100711881-003-0008-2

Manolitsis, G. (2016). Emergent literacy in early childhood education: New issues and educational implications. *Preschool and Primary Education, 4*(1), 3–34. doi:10.12681/ppej.9970

Mohamad, A. J., Lakulu, M., & Samsudin, K. (2016). The development of mobile applications for kindergarten early reading: challenges and opportunities. *Journal of Engineering and Applied Sciences (Asian Research Publishing Network), 11*(3), 380–383.

Neumann, M. M. (2014). An examination of touch screen tablets and emergent literacy in Australian preschool children. *Australian Journal of Education, 58*(2), 109–122. doi:10.1177/0004944114523368

Neumann, M. M. (2016). Young children's use of touch screen tablets for writing and reading at home: Relationships with emergent literacy. *Computers & Education, 97,* 61–68. doi:10.1016/j.compedu.2016.02.013

Neumann, M. M. (2018a). Using tablets and apps to enhance emergent literacy skills in young children. *Early Childhood Research Quarterly, 42,* 239–246. doi:10.1016/j.ecresq.2017.10.006

Neumann, M. M. (2018b). Maternal scaffolding of preschoolers' writing using tablet and paper-pencil tasks: relations with emergent literacy skills. *Journal of Research in Childhood Education, 32*(1), 67–80. doi:10.1080/02568543.2017.1386740

Neumann, M. M., Hood, M., & Ford, R. (2012). Mother-child joint writing in an environmental print setting: Relations with emergent literacy. *Early Child Development and Care, 182*(10), 1349–1369. do i:10.1080/03004430.2011.615928

Neumann, M. M., Hood, M., & Neumann, D. L. (2009). The scaffolding of emergent literacy skills in the home environment: a case study. *Early Childhood Education Journal, 36*(4), 313–319. doi:10.100710643-008-0291-y

Neumann, M. M., Merchant, G., & Burnett, C. (2018). Young children and tablets: The views of parents and teachers. *Early Child Development and Care, 1550083.* doi:10.1080/03004430

Neumann, M. M., & Neumann, D. L. (2014). Touch screen tablets and emergent literacy. *Early Childhood Education Journal, 42*(4), 231–239. doi:10.100710643-013-0608-3

Neumann, M. M., & Neumann, D. L. (2017). The use of touch screen tablets at home and pre-school to foster emergent literacy. *Journal of Early Childhood Literacy*, *17*(2), 203–220. doi:10.1177/1468798415619773

Niklas, F., & Schneider, W. (2013). Home literacy environment and the beginning of reading and spelling. *Contemporary Educational Psychology*, *38*(1), 40–50. doi:10.1016/j.cedpsych.2012.10.001

Northrop, L., & Killeen, E. (2013). A framework for using iPads to build early literacy skills. *The Reading Teacher*, *66*(7), 531–537. doi:10.1002/TRTR.1155

O'Toole, K. J. (2015). *Emergent literacy skills in print and electronic contexts: The influence of book type, narration style, and attention* (Doctoral dissertation). Available at http:// ecommons.luc.edu/luc_diss/1652

O'Toole, K. J., & Kannass, K. N. (2018). Emergent literacy skills in print and electronic contexts: The influence of book type, narration source, and attention. *Journal of Experimental Child Psychology*, *173*, 100–115. doi:10.1016/j.jecp.2018.03.013 PMID:29702378

Orrin, M. T., & Olcese, N. R. (2011). Teaching and learning with iPads, ready or not? *TechTrends*, *55*(6), 42–48. doi:10.100711528-011-0540-6

Papadakis, S., & Kalogiannakis, M. (2017). Mobile educational applications for children: What educators and parents need to know. *International Journal of Mobile Learning and Organisation*, *11*(3), 256–277. doi:10.1504/IJMLO.2017.085338

Papadakis, S., Kalogiannakis, M., & Zaranis, N. (2016). Comparing tablets and PCs in teaching mathematics: an attempt to improve mathematics competence in early childhood education. *Preschool and Primary Education*, *4*(2), 241–253. doi:10.12681/ppej.8779

Papadakis, S., Kalogiannakis, M., & Zaranis, N. (2017). Improving mathematics teaching in kindergarten with realistic mathematical education. *Early Childhood Education Journal*, *45*(3), 369–378. doi:10.100710643-015-0768-4

Papadakis, S., Kalogiannakis, M., & Zaranis, N. (2018). Educational apps from the Android Google Play for Greek preschoolers: A systematic review. *Computers & Education*, *116*, 139–160. doi:10.1016/j.compedu.2017.09.007

Patchan, M. M., & Puranic, C. S. (2016). Using tablet computers to teach preschool children to write letters: Exploring the impact of extrinsic and intrinsic feedback. *Computers & Education*, *102*, 128–137. doi:10.1016/j.compedu.2016.07.007

Pentimonti, J. M., & Justice, L. M. (2010). Teachers' use of scaffolding strategies during read alouds in the preschool classroom. *Early Childhood Education Journal*, *37*(4), 241–248. doi:10.100710643-009-0348-6

Petkovski, M. (2014). *Parent-child interaction in shared-computer activities: an exploratory study examining parent–child interaction with a mobile device. (Master's Thesis*, Wilfrid Laurier University, Canada). Available at http://scholars.wlu.ca/etd

Piotrowski, J. T., & Krcmar, M. (2017). Reading with hotspots: Young children's responses to touchscreen. *Computers in Human Behavior*, *70*, 328–334. doi:10.1016/j.chb.2017.01.010

Reeves, J. L., Gunter, G. A., & Lacey, C. (2017). Mobile learning in pre-kindergarten: using student feedback to inform practice. *Journal of Educational Technology & Society*, *20*(1), 37–44.

Reich, S. M., Yau, J. C., & Warschauer, M. (2016). Tablet-based eBooks for young children: what does the research say? *Journal of Developmental and Behavioral Pediatrics*, *37*(7), 585–591. doi:10.1097/DBP.0000000000000335 PMID:27575440

Richter, A., & Courage, M. L. (2017). Comparing electronic and paper story books for preschoolers: Attention, engagement, and recall (2017). *Journal of Applied Developmental Psychology*, *48*, 92–102. doi:10.1016/j.appdev.2017.01.002

Rideout, V. (2017). *The common sense census: Media use by kids age zero to eight*. San Francisco, CA: Common Sense Media; Available at http://www.commonsensemedia.org

Salmon, L. G. (2013). Factors that affect emergent literacy development when engaging with electronic books. *Early Childhood Education Journal*, *42*(2), 85–92. doi:10.100710643-013-0589-2

Seipold, J. (2014). Mobile learning: Structures, concepts and practices of the British and German mobile learning discusssion from a media education perspective. *MedienPädagogik*, *24*, 30–52.

Sénéchal, M., Le Fevre, J., Smith-Chant, B. L., & Colton, K. V. (2001). On refining theoretical models of emergent literacy. The role of empirical evidence. *Journal of School Psychology*, *39*(5), 439–460. doi:10.1016/S0022-4405(01)00081-4

Sharples, M., Taylor, J., & Vavoula, G. (2007). A theory of learning for the mobile age. In R. Andrews & C. Haythornthwaite (Eds.), *The Sage handbook of e-learning research* (pp. 221–247). London, UK: Sage. doi:10.4135/9781848607859.n10

Shute, V. J. (2008). Focus on formative feedback. *Review of Educational Research*, *78*(1), 153–189. doi:10.3102/0034654307313795

Storch, S. A., & Whitehurst, G. J. (2002). Oral language and code-related precursors to reading: Evidence from a longitudinal structural model. *Developmental Psychology*, *38*(6), 934–947. doi:10.1037/0012-1649.38.6.934 PMID:12428705

Terras, M., & Ramsay, J. (2012). The five central psychological challenges facing effective mobile learning. *British Journal of Educational Technology*, *43*(5), 820–832. doi:10.1111/j.1467-8535.2012.01362.x

Vatalaro, A., McDonald Culp, A., Hahs-Vaughn, D. L., & Barnes, A. C. (2018). A quasi-experiment examining expressive and receptive vocabulary knowledge of preschool head start children using mobile media apps. *Early Childhood Education Journal*, *46*(4), 451–466. doi:10.100710643-017-0877-3

Vernadakis, N., Avgerinos, A., Tsitskari, E., & Zachopoulou, E. (2005). The use of computer assisted instruction in preschool education: making teaching meaningful. *Early Childhood Education Journal*, *33*(2), 99–104. doi:10.100710643-005-0026-2

Whitehurst, G. L., & Lonigan, C. J. (1998). Child development and emergent literacy. *Child Development*, *69*(3), 848–872. doi:10.1111/j.1467-8624.1998.tb06247.x PMID:9680688

Wood, E., Petkovski, M., De Pasquale, D., Gottardo, A., Evans, M. A., & Savage, R. S. (2016). Parent scaffolding of young children when engaged with mobile technology. *Frontiers in Psychology*, *7*, 690. doi:10.3389/fpsyg.2016.00690 PMID:27242603

Zaranis, N., Kalogiannakis, M., & Papadakis, S. (2013). Using mobile devices for teaching realistic mathematics in kindergarten education. *Creative Education*, *4*(07), 1–10. doi:10.4236/ce.2013.47A1001

Zipke, M. (2017). Preschoolers explore interactive storybook apps: The effect on word recognition and story comprehension. *Education and Information Technologies*, *22*(4), 1695–1712. doi:10.100710639-016-9513-x

Zomer, N. R., & Kay, R. H. (2016). Technology use in early childhood education: a review of literature. *Journal of Educational Informatics*, *1*, 1–25.

ADDITIONAL READING

Bonneton-Botte, N., Beucher-Marsal, C., Bara, F., & ... . (2019). Teaching cursive handwriting: A contribution to the acceptability study of using digital tablets in French classrooms. *Journal of Early Childhood Literacy*, *0*(0), 1–24. doi:10.1177/1468798419838587

Fallon, G. (2013). Young students using iPads: App design and content influences on their learning pathways. *Computers & Education*, *68*, 505–521. doi:10.1016/j.compedu.2013.06.006

Flewitt, R., Kucirkova, N., & Messer, D. (2014). Touching the virtual, touching the real: iPads and enabling literacy for students experiencing disability. *Australian Journal of Language and Literacy*, *37*(2), 107–116.

Homer, B. D., Kinzer, C. K., Plass, J. L., Letourneau, S. M., Hoffman, D., Bromley, M., ... Kornak, Y. (2014). Moved to learn: The effects of interactivity in a Kinect-based literacy game for beginning readers. *Computers & Education*, *74*, 37–49. doi:10.1016/j.compedu.2014.01.007

Marsh, J. (2016). The digital literacy skills and competences of children of pre-school age. *Medical Education*, *7*(2), 197–214.

McKenzie, S., Spence, A., & Nicholas, M. (2018). Going on Safari: The Design and Development of an Early Years Literacy iPad Application to Support Letter-Sound Learning. *The Electronic Journal of e-Learning Volume 16*(1), 1-16. (Available online at www.ejel.org)

Oliemat, E., Ihmeideh, F., & Alkhawaldeh, M. (2018). The use of touch-screen tablets in early childhood: Children's knowledge, skills, and attitudes towards tablet technology. *Children and Youth Services Review*, *88*, 591–597. doi:10.1016/j.childyouth.2018.03.028

Rowe Wells, D., & Miller, M. E. (2016). Designing for diverse classrooms: Using iPads and digital cameras to compose eBooks with emergent bilingual/ biliterate four-year-olds. *Journal of Early Childhood Literacy*, *16*(4), 425–472. doi:10.1177/1468798415593622

KEY TERMS AND DEFINITIONS

Constructive Applications: Applications that maintain an emphasis on inquiry and social processes as well as on distributed cognition designs.

Digital Literacy: The ability to use ICT to find, evaluate, create, and communicate information, requiring both cognitive and technical skills.

Emergent Writing: The progression of stages a young child moves through beginning with random marks and ending with the conventional spelling of a language.

Feedback: A process in which learners make sense of information about their performance and use it to enhance the quality of their work or learning strategies.

Literacy: The ability to read and write.

Phonological Awareness: The ability to identify and manipulate units of spoken sound.

Scaffolding: A learning process in which teachers and/or parents model or demonstrate how to solve a problem, and then step back, offering support as needed.

Chapter 5
The Use of Developmentally Mobile Applications for Preparing Pre-Service Teachers to Promote STEM Activities in Preschool Classrooms

Michail Kalogiannakis
https://orcid.org/0000-0002-9124-2245
Faculty of Education, Department of Preschool Education, University of Crete, Greece

Stamatios Papadakis
https://orcid.org/0000-0003-3184-1147
Faculty of Education, Department of Preschool Education, University of Crete, Greece

ABSTRACT

Studies suggest that the exposure to STEM learning opportunities early in life is important because the development of STEM skills can further students' interest and educational attainment in STEM, expanding their career choices later in life. Smart mobile devices have become ubiquitous in schools and have been transforming educational practices at all ages and levels and almost all over the world. At the same time, there is evidence that teacher education departments lack the knowledge and skill to teach pre-service teachers about using these devices in their daily teaching practice. The findings of this chapter underline the need to develop teaching and learning processes that go beyond a mere transmission of the technical knowledge required to use mobile technologies with educational purposes, focusing instead on raising students' awareness about the educational benefits that the integration of mobile technologies can bring to formal education.

DOI: 10.4018/978-1-7998-1486-3.ch005

INTRODUCTION

It is widely recognized that technological innovation, over the past decade, has influenced all our lives and that as such, formal schooling should have changed to reflect that fact (Murphy,& Headley, 2018; Papadakis & Kalogiannakis, 2017; Yelland & Gilbert, 2018). Children can understand more science during the preschool and early school education than was believed in the previous years (Moomaw & Davis, 2010; Kalogiannakis, Ampartzaki, Papadakis, & Skaraki, 2018). Over the last decade, STEM (Science, Technology, Engineering, and Math) topics have become a focus of interest for researchers and educators. A reason is that STEM education is considered as a key enabler to further equip students with the skills they'll need for the future workplace (Psycharis, 2018; Saxton et al., 2014). STEM refers to evidence-based knowledge and practices in the fields of science, mathematics, engineering, and technology (Gonzalez & Kuenzi, 2012) by engaging students in inquiry-driven learning practices to solve real-world problems in their daily lives through authentic learning experiences (Daugherty, 2013). STEM education includes student use of math and science concepts to facilitate interdisciplinary collaborations among engineering and technology. Although four disciplines are included in the acronym S.T.E.M., science and mathematics are the most familiar to teachers of young children although math and science disciplines, instead of being taught in isolation, they are used to solve several problems arising in different spaces (Chamberlin & Pereira, 2017; Psycharis, 2018; Dorouka, Papadakis & Kalogiannakis, 2019; Margot & Kettler, 2019).

Tablets and smartphones permeate children's lives (Zaranis, Kalogiannakis & Papadakis, 2013). Young age children use tablet type devices on a daily basis as the larger screen of a tablet allows for easier use (Eisen & Lillard, 2017). Furthermore, research results have shown that the touchscreen devices could serve as a tool for in-service and pre-service teachers to better meet the individual learner needs in today's and future's (preschool) classrooms. Thus, tablet-type devices have become ubiquitous in modern schools and have been transforming educational practices of all ages, of all educational levels, of all social classes around the globe (OECD, 2015).

Children can learn lessons and gain experience from interactive screen media, provided that media (a) considers content as means to improving teaching and learning, (b) are absorbed in a context that stimulates and scaffold children's engagement, and (c) are tailored on the developmental needs of the individual child (Hurwitz, 2018). In recent years, the potential of interactive mobile technology to provide innovative solutions for the transformation of education has raised high expectations for researchers and educators. It is precise because of this flexibility and creativity of tablet type devices that can promote new and effective teaching practices (Lynch & Redpath, 2014; McTavish, 2014; Dunn, Gray, Moffett & Mitchell, 2018). For instance, smart mobile devices can act as an active source of data receivers (e.g. data loggers) from a set of quite different experimental situations (Zaranis et al., 2013). In addition, a smart mobile device can act as learning hubs promoting the integration of formal and informal learning contexts, enabling the creation and maintenance of a broad range of artefacts associated with multiple learning contents such as concept maps, photos, animations, etc. (Looi et al., 2011).

The importance of exposure and experience in STEM learning opportunities for young children (as an integrated approach of the curriculum) is well recognized as it can promote a continuous and independent learning process for the students in STEM disciplines both in the academic and professional sector (Papadakis, 2018a; Sheehan, Hightower, Lauricella & Wartella, 2018). Also, given the ubiquity of interactive media and especially smart mobile devices, amongst young children, there is a need for schools in providing opportunities for their students to develop their confidence and abilities in STEM skills by using new interactive smart screen technologies (Dunn, Gray, Moffett & Mitchell, 2018). One potential way to promote earlier students experience in STEM content is through the use of interactive media, which can be a powerful tool for supporting young children interest in STEM prior to formal school entry (Sheehan et al., 2018). It is also known, that developmentally appropriate mobile applications (or apps) could potentially be transformative for children's early development (Papadakis, 2018b). Rather than the myriad iPad and Android apps that have flooded the digital market, all children should have the opportunity either individually or in groups to engage with appropriate apps that will help them develop problem-solving skills that are foundational for students' interest in STEM fields and for later academic success (Kwan, 2017; Papadakis et al., 2017a; 2017b). With developmentally appropriate, appealing tools and approaches, young age children can learn basic scientific concepts in early childhood (Wong & Jiang, 2018).

In early education, children can be creative and use their enthusiasm and ideas while making progress on something they are interested in. Furthermore, it is known that young children develop their creativity and problem-solving skills while taking part in Computational Thinking (CT) activities (Bers, 2018b). In conjunction with the importance of CT methods in early childhood, coding is not considered as an issue of technical skills, rather than an effective way to build the '21st-century skills', like the 3 R's of reading, writing, and arithmetic were considered the fundamental skills in the previous centuries (Bers, 2018b). The last decade, there is an explosion of coding activities for very young age children as new technological visual and interactive resources have offered new opportunities for effective and engaging learning experiences (Kalogiannakis & Papadakis, 2017; Papavlasopoulou, Giannakos & Jaccheri, 2019). But even if we accept that early childhood education classrooms provide a powerful and friendly environment for an effective STEM implementation (Estapa & Tank, 2017) to amplify the students exposure in STEM-related areas, it is essential that universities should educate pre-service teachers in STEM thinking practices regardless of academic discipline (Kalogiannakis & Papadakis, 2019; Papadakis & Kalogiannakis, 2019; Dorouka et al., 2019).

STEM education requires a pedagogical design to flexible constructivist teaching approaches that foster student's engagement (Margot & Kettler, 2019). Pre-service teachers need to be equipped with the necessary knowledge and support to be good curriculum facilitators. The scope is that they should be able to co-design subject-specific curriculum materials facilizing the classroom enactment of lessons supported by mobile technology. But, a curriculum reorganization in terms of mobile learning integration requires an integrative view of how this redesign is structured and how the educational scenarios are adapted so as student learning is situated in contexts and activities via smart mobile technology across various concepts (Looi et al., 2011; Papadakis, 2018b).

Teachers that have been sufficiently trained in the STEM subjects with the use of mobile devices will harness these advances to transform their lessons into a 'mobilized' curriculum that makes science learning fun and motivating via the use of mobile technologies by the students (Looi et al., 2011). Students in universities are now considered as 'digital natives' who daily use smart mobile technology for entertainment and solely for education purposes (Barnes, Marateo & Ferris, 2007; So et al., 2012). But, by the time they will perceive that the use of this technology as a learning tool adds a sense of playfulness and interactivity into the classroom they might be more inclined to use it in the future classroom (Sánchez-Prieto, Hernández-García, García-Peñalvo, Chaparro-Peláez & Olmos-Migueláñez, 2019). Numerous studies have examined the motives of teachers-in-training experiences in the way of adopting technology in the classroom. Teachers, prior experience with technology affects the amount of cognitive effort needed for a technology-driven implementation in the classroom (Pamuk, 2011). There is also evidence that educational university departments lack the curriculum designs to guide pre-service teachers according to the facilitation of STEM experiences into the classroom via touch screen technologies (Yadav et al., 2011, 2014). Educational departments should train pre-service teachers in STEM teaching approaches to develop strategies for effective implementation of mobile-based learning activities in their classrooms. Learning how to use new forms of technology may be demanding as it acquires knowledge and skills that encourage innovation and knowledge transfer in new and efficient ways (Papadakis, 2017). In early childhood education, the encouragement of exploration and science requires active teacher involvement in the design of learning environments through play-based approaches appropriate to young children's needs and learning (Yelland & Gilbert, 2018).With this increased interest for STEM activities even in preschool education, however, there are key questions about how pre-service teacher's education programs prepare them to effectively integrate STEM concepts with the use of mobile technology in their future preschool classrooms.

Our study builds on what is known from previous research and involves the use of a developmentally appropriate mobile application for preschoolers, ScratchJr, to educate pre-service preschool teachers in developmentally appropriate multidisciplinary learning activities to use the specific app to teach STEM concepts in their future daily practice. This study proposes the use of apps such as ScratchJr as an effective instructional method to familiarize pre-service preschool teachers with STEM activities. The current study aimed to develop a better understanding of how pre-service teachers experiences of using a developmentally appropriate mobile app affect the extent to which those teachers are able and willing to use mobile technologies to integrate STEM concepts into their future preschool classroom.

LITERATURE REVIEW

STEM Education and Pre-Service Teachers

In order to address the ongoing need for more STEM-literate workers, classrooms at all educational levels are integrating STEM curriculum (Margot & Kettler, 2019). The reason for that is that many international initiatives refer to the importance of the STEM education as one of the best academic investments as it fosters broad-based scientific literacy, in efforts to improve undergraduate science, math and engineering education in elementary, middle, and secondary schools (English, 2016).

However, successful integration and implementation of STEM curriculum is difficult, mainly, because educators need to develop a comprehensive understanding of the integration concepts and tools (Psycharis, 2018; Kalogiannakis et al., 2018). Technology with new educational kits and methods brings rapid advances in education for teachers to design learning contents adapted to the demanding needs of the students. This can present a challenge as it requires a great effort for preschool and elementary teachers who often have limited content knowledge within STEM (Estapa & Tank, 2017).

Research shows that both from an economic and an educational prospect, early educational interventions have lower costs but durable effects (Cunha & Heckman, 2007; Bers, 2018a). International research suggests that children who are exposed to STEM concepts and coding activities at an early age are faced with less gender-based stereotypes regarding their future selection in STEM careers (Bers, 2018b).

One of the most effective ways to provide support for the inclusion of STEM in preschool classrooms is through systematic and targeted professional development that can produce learning opportunities for teachers to understand how to successfully implement the new STEM practices or content (Estapa & Tank, 2017).

The key to preparing STEM educators is to first begin by developing their conceptual framework of an integrated STEM approach in the classroom (Kelley & Knowles, 2016; Psycharis, 2018; Moore et al., 2014). Basically, adjustment for STEM education can be described as educational approaches between two or more subject STEM areas, and/or between a STEM subject and any other school discipline (Sanders, 2009). In fact, preschool and elementary school teachers often have limited STEM content knowledge, limited access to materials and resources, required for the effective implementation of STEM instruction (Brophy et al. 2008; Moore et al. 2014; Estapa & Tank, 2017). O'Brien et al. (2014) characteristically state 'a lack of STEM subject matter expertise and content development experiences, coupled with high anxiety and low self-efficacy can lead to excess stress, and lack of effectiveness as well as a lack of interest in, and an absence of motivation in STEM subjects' (p. 280). Numerous studies have examined the motives and experiences of teachers-in-training in conjunction with technology adoption into their teaching practice. These studies have revealed that prior exposure to relevant technologies affects teachers' perceived difficulty or the amount of effort needed to effectively integrate technology into their classrooms (Pamuk, 2011). Other studies revealed that teachers who perceive technology as having a positive effect on learning support, are more likely to adopt technology, regardless of its complexity (Chen, 2010).

Thus, educational institutions need to provide pre-service teachers during their studies with opportunities to adopt in conjunction with the adequate technical support, supervision and training to adopt the use of digital technologies, especially the mobile-centric ones (Reid, 2014; Lucas, 2018; Sánchez-Prieto et al., 2019). Nonetheless, even when the adequate tools and kits are available (external barriers) their presence does not guarantee that teachers will immediately start using certain technology in widespread ways. There are internal barriers, such as motivation, that play a major motivation role.

These terms in general, reflect on how the educators transform their own teaching practices (McLoughlin, Wang & Beasley, 2008), and therefore these barriers are often associated with motivational, action-relevant characteristics.

Hence, the training a pre-service teacher accept during his/her studies plays a major role (Sang et al., 2010) as this particular training influences a pre-service teacher professional identity, in terms of the teaching subject and the school-life practice (So et al., 2012; Sánchez-Prieto et al., 2019).

A key factor that influences teacher's decision to adopt the use of digital equipment to the teaching-learning process is the perceived perception that such 'innovation' is going to improve his/her learning practice significantly (Mac Callum, Jeffrey & Kinshuk, 2014). Undoubtedly, pre-service teachers perceived the usefulness of technology is one of the main determinants of the use of this technology in the future classroom (Sánchez-Prieto et al., 2019).

Children and Learning Technologies

Learning technologies are increasingly being used in children's homes around the globe. At the same time, children are spending an increasing amount of time in front of screens, while many of them, are learning from smart screen experiences, such as educational apps (Donohue, 2016). According to the National Association for the Education of Young Children (NAEYC) technology is acceptable when used intentionally to promote learning and development. Thus, technology must be suitable to the needs, interests, ages and cognitive development of the children, while considering their social and cultural contexts with which it is used (Copple and Bredekamp, 2009; NAEYC 2012).

Since a decade, researchers, highlight the fact that the rise of new interactive technology, including smart mobile devices and apps, provide teachers and students with a variety of digital choices in formal and informal contexts (Stevenson, 2010). Young children can use smart mobile devices in active, playful and developmentally appropriate explorations which enable them to better communicate their ideas via multimodal formats. This fact leads teachers to rethink and therefore transform their pedagogy to ensure that children with the use of new interactive technologies can extend their ideas in conjunction with real-world experiences to meet the 21st-century students' learning styles and expectations (Yelland & Gilbert, 2018).

Researchers note that as children are provided with personalized interactive learning experiences through the usage of smart mobile devices, they are able to improve their ability in various cognitive tasks while expanding the skills needed to succeed in demanding work environments. Digital technology can promote unique affordances for learning when encouraging general open-ended exploration, provide scaffolding experiences, and promote co-creation models (Hirsh-Pasek et al., 2015; Vaala et al., 2015).

In addition, apps quality is also an important factor to consider is whether an app is based on rote memorization or creative and problem-solving tasks can improve children's learning outcomes (Neumann, 2016). Since 1980, Papert highlighted the fact that open-ended learning software that follows a scaffolding approach allowing students to actively construct their own knowledge through their interests is preferred than instructor-directed alternatives (Papert, 1980).

These experiences are immensely important for children learning (Hershman et al., 2018). These factors will affect the ways in which teachers are using apps, as they must move away from using the 'traditional' skills-based apps, to more creative apps that allow for an open-ended exploration of knowledge (Yelland & Gilbert, 2018).

With the advent of smart screen technology, in early childhood education has begun the integration of learning experiences that expose children to hands-on projects including STEM. Furthermore, programming and robotics kits specifically for young age children have become widespread over the last decade (Cheng, 2019).

These new developmentally appropriate tools offer the opportunity for young age children to take part in activities that help them develop and integrate knowledge and understanding about Computational Thinking (CT) (Bers, 2018a). Young children are encouraged to learn by experimenting, to express themselves creatively through projects creation, as well as to collaborate with family and peers while also having fun (Leidl, Bers & Mihm, 2017).

The use of touchscreen technologies is almost imperative as their integration is considered as an integral part of preschool and elementary classrooms to help students to become multiliterate learners in the 21st-century as they can facilitate multimodal learning experiences in linguistic, spatial, visual, gestural, aural, communicative modes (Yelland & Gilbert, 2018). Smart mobile devices are considered as ideal candidates for the development of new learning STEM experiences as they integrate several technological components in a platform, such as digital audio and video recorders, high-quality photo cameras, a virtual keyboard and access to the internet.

Furthermore, children's engagement with these devices can be highly mobile, increasing children's engagement with their learning, especially if it involves children action with apps that require external input, collaboration, creativity and physical movement (Kucirkova & Falloon, 2018).

ScratchJr

In the last decades, the significant role that Computational Thinking (CT) plays in educational and professional contexts has been widely discussed in the literature (Papadakis, Kalogiannakis, Zaranis & Orfanakis, 2016). Wing (2006, p.33) states that 'CT represents a universally applicable attitude and skill set everyone, not just computer scientists, would be eager to learn and use'. CT can be described as a general approach that incorporates problem-solving, designing systems, and understanding human behaviors by employing the central concepts of Computer Science (CS) (Wing, 2006). CT, lately it is also seen as an important element of STEM disciplines. Characteristically, National Research Council highlights the fact that CT along with mathematical thinking is one of the keys to success in science and engineering (Wong & Jiang, 2018).

Furthermore, nowadays, the importance of learning to code has become equally important as math and reading (Papavlasopoulou et al., 2019; Psycharis & Kalia, 2017). However, if a coding introduction is going to be introduced in the early years, there is a need to explore technological tools and pedagogical approaches that take into consideration the cognitive and physical limitations, as well as experiences and emotional readiness of young children (Bers, 2018b). The programming approach based on building blocks on Logo-like programming languages, such as Scratch, is used for decades to introduce to young students various CT skills and CS concepts, such as algorithmic thinking, problem-solving, etc. In these programming environments for novices, a user creates projects by simply putting visual command blocks together on the screen, which represent simple instructions: a control structure, an operator, a variable, etc. (Papadakis, Kalogiannakis, Orfanakis & Zaranis, 2017). These elements and components that support object-oriented and event programming can be combined on a drag-and-drop interface in a certain way that enables the novice programmer to make less cognitive efforts. These programming environments belong to a certain category that characterized as 'low floor' (beginners can learn how to use it quickly and easily), 'high ceiling' (beginners can create both simple but also advanced projects) as well as 'wide walls' (beginners can extend the functionality of their projects in novel ways) while they engage in rich, playful and interactive programming exercises.

ScratchJr was born out of the need for developmentally appropriate environments that facilitate code learning outcomes for young learners (Bers, 2018b). ScratchJr was developed as a result of an academic collaboration between the DevTech Research Group at Tufts University, the MIT Lifelong Kindergarten Group, and the Playful Invention Company (Bers, 2018a). It is a mobile programming environment for introducing children between 5 to 7 years in computational thinking activities while coding, providing a space in which powerful ideas can be played out (Leidl et al., 2017). ScratchJr is based on Scratch and thus enables children to create projects in the form of interactive stories and games by snapping together graphical pieces on the screen to give life on digital characters (sprites) by making them move, jump, dance, etc. (Bers, 2018a).

Block-based programming environments (like Scratch and App Inventor) have the advantage of using visual command interfaces that fit properly only when they make a logical sequence of events and/or commands. That characteristic reduces novice programmer cognitive load as they don't have to cope with the strenuous syntactic rules and problems associated with textual programming languages (Wilson & Moffat, 2010, p. 70; Papavlasopoulou et al., 2019). This open-ended design approach makes also ScratchJr ideally for younger ages, as it is very easy for them to code simply by moving pieces of command blocks (Leidl et al., 2017).

Researchers claim that the use of open-ended programming environments enables children to easily learn the basics of CS, such as debugging, testing, and modularity (Leidl et al., 2017). According to ScratchJr research team, during the app development, the scope was the design of an introductory programming environment to act as an intellectual and educational tool for novice programmers than to prepare them for a career in Computer Science (Bers, 2018b). ScratchJr design features reduce unnecessary cognitive effort by keeping the challenges at a level adaptable to young children cognitive development while at the same time engage children in high-level thinking processes (Bers, 2018b). Similarly, to Scratch, in ScratchJr, there is an embedded paint editor, a sound-recording system, and a predefined collection of sprites and backgrounds (Bers, 2018a). Another advantage of ScratchJr is that it is freely available for the two famous mobile Operating Systems, iOS, and Android e.g. for Apple and Android smart devices. Due to its popularity, ScratchJr, in 2018 had been downloaded over 9.5 million times and had been used in almost every country in the world (except North Korea and Western Sahara) (Bers, 2018a).

METHODS

Sample

Seventeen pre-service preschool students took part in a semester course at a Greek university participated in the study after research ethics approvals were received. All participants had basic computer literacy skills and had completed, as part of their university studies, a semester course on teaching with multimedia technology in preschool education. The participants had no previous experience with neither CT concepts and coding skills nor with the use of ScratchJr app. As STEM concepts we define interdisciplinary concepts that are transversal to the different cognitive areas included in STEM disciplines.

Research Questions

The present study sought to answer the following two research questions:

Hypothesis One: How did engaging with ScratchJr influence pre-service teachers' ability to teach STEM concepts?

Hypothesis Two: How did learning with ScratchJr influence pre-service teachers' motivation to use ScratchJr in their future teaching work?

Procedure

The study duration was from October to December 2018. The total number of teaching hours was 39, spread over 13 weeks. The educational activity was called 'Software development'. The students were required to design 3 different activities that could use to teach STEM concepts in preschoolers. In this course with the use of ScratchJr, we focused on projects development, so that pre-service teachers could learn the process of turning an initial idea into something that can be used into the preschool classroom. As they created projects, pre-service teachers learned not only how to work with code, but how to find new ideas, how to develop new strategies, and how to creatively refine their ideas over time and of course to work collaboratively.

Throughout the learning sessions, the students were initially presented the ScratchJr environment to get familiarized with that. Subsequently, they had to create simple projects using the ScratchJr app. The choice of using ScratchJr instead another programming environment for novices (e.g. Scratch, App Inventor) was based on our own previous teaching experience but also in international research results that highlight the benefits of ScratchJr as a very simple and helpful programming environment. In short, with ScratchJr the CT concepts and coding activities become challenging to an appropriate degree (Kelleher & Pausch, 2005), so novices can focus on creative problem-solving (Lin & Liu, 2012; Papavlasopoulou et al., 2019). Furthermore, these constructionism-based coding activities, are fostering in an environment in which novice programmers -in our case the pre-service teachers - are stimulated and inspired by the environment, that positively affects their overall learning experience (Papavlasopoulou et al., 2019). During the first weeks, the students were intended to make games, animations, and interactive stories. Then the students were asked to design more sophisticated projects to teach STEM concepts in preschoolers.

The course-specific objectives were to help participants:

- Familiarize themselves with the ScratchJr user interface.
- Understand the purpose of ScratchJr as a programming environment.
- Design, develop and debug ScratchJr projects.

Considering the results from previous studies (Bers et al., 2014), we adopted a learning-by-doing training approach during the course. The aim was to motivate participants to become engaged in the process of developing CT concepts and coding skills through ideas that could be used in the preschool classroom. The students were free either to work alone or in dyads to create their projects orientated in STEM concepts. During each lesson, participants were encouraged to follow the step-by-step illustrated instructions as were described in an e-book created by the researchers to able to create their own learning pathway. Of course, they could also seek advice and assistance from others, such as the instructor, or peers. The instructor during the course encouraged the ongoing interaction, collaboration, and exchange of ideas between the pre-service teachers.

Years of research have fortunately provided much insight into the problem's novice programmers have to face when dealing with coding activities. In general, coding activities can be very demanding as if they are now well organized, they can produce a high cognitive effort for novices, affecting their performance, and overall experience with the related tasks. Thus, an appropriate organization, scheduling, and integration of the learning sources in conjunction with an appropriate representation of the related tasks are necessary to avoid learners' cognitive overload and improve task execution effectiveness (Papavlasopoulou et al., 2019). As learning sources, we used teaching material which can be found on the ScratchJr website (scratchjr.org). This is designed by the ScratchJr team to facilitate the use of ScratchJr in both formal and informal learning environments. The learning material is freely available and has been designed with a focus on learning ScratchJr by emphasizing interdisciplinary connections with other disciplines such as literacy, numeracy, etc. (Bers, 2018b). Additionally, we also used the 'Collaborative ScratchJr Projects Guide' (https://sites.tufts.edu/devtech/files/2018/07/Collaborative-ScratchJr-Project-Guide.pdf), developed by the DevTech Research Group. This guide has been created to help users in the planning of collaborative projects (Bers, 2018b). In addition to the ScratchJr app and freely available teaching material, the 'Official ScratchJr Book', written by Marina Umaschi Bers and Mitchel Resnick (Bers & Resnick, 2015) was also used as well as videos relevant to course content from the YouTube channel.

Classroom Observations

Data from classroom observations revealed that the pre-service teachers enjoyed the freedom of self-directed learning procedure while researching new CT topics and coding skills with the ScratchJr app. Furthermore, most of them during the lessons enjoyed searching and watching relevant videos online. The fact that students asked their own questions and looked for the material is considered as very significant cultural change considering that they boosted their confidence and changed their mindset while increased their comfort with the ScratchJr programming environment. Another factor, also noted in the classroom, was the fact that the students implemented the tasks on the mobile device without immediate guidance from the teacher. Furthermore, we noticed an increased social interaction as part of the students YouTube video sharing experience, the sharing of interesting things they found on the Internet (such as ScratchJr tips, notes, etc.), the willingness to help one another to solve technical problems and conflicts, etc. In summary, we can conclude that the use of ScratchJr encouraged a face-to-face interaction and dialogue while the participants' engaged on the tasks, improving their overall learning experience.

Attitudes Towards the Use of Mobile Devices for Learning

We collected both qualitative and quantitative data regarding students' beliefs about ScratchJr to achieve a more holistic picture of their intrinsic motivation and engagement. In terms of attitudes towards the use of mobile technology for learning, pre- and post-questionnaire surveys were administered to the students to compare baseline opinions at the beginning and after the completion of the course. Most of the pre-service teachers felt the ScratchJr environment was easy to use and quick to learn. As a result, the pre-service teachers held positive and favorable views towards the use of ScratchJr as a tool for teaching STEM concepts in their classrooms. Note that almost 80% of the participants thought that by using the ScratchJr, will help them understood the science concepts learned better.

Paired samples t-tests were conducted to detect significant differences in the mean scores of the pre-service teacher's attitudes towards ScratchJr for teaching STEM concepts between the pre and post-survey measures. We considered a significance level of 0.05 for all tests. First, as shown in Table 1, there was a significant improvement in student attitudes towards the use of smart mobile technology for teaching from the pre- to post- survey. The items were measured on a 5-point Likert scale where 1 = Strongly Agree, 2 = Agree, 3 = Neutral, 4 = Disagree and 5 = Strongly Disagree. Thus, the lower the score, the more positive were pre-service teacher's attitudes (or higher acceptance) towards the usage of mobile technologies for learning. After the teaching intervention, the students were found to have more positive attitudes towards mobile learning use in the classroom [$t(16) = -1.876, p < .01$]. Also the participants liked the learning activities using mobile learning devices and the ScratchJr app more than before [$t(16) = -1.902, p < .01$] and they perceived as more effective and collaborative when working in groups rather than working alone [$t(17) = -2.122, p < .05$].

Semi-Structured Interview

We utilized semi-structured interviews with the pre-service teachers who attended the course. Each interview lasted approximately 15 to 30 minutes and followed a structured questionnaire. The protocol was designed to further explore pre-service teachers' opinions on questions that were not covered - or covered poorly - in the quantitative questionnaire survey. These data may provide a better understand-

Table 1. Paired sample t-test comparing the pre-service teachers' attitudes towards mobile devices for learning

		M	*N*	*SD*	*t*
The mobile devices will help me teach STEM subjects-concepts in my class.	Post-survey	1.29	17	.399	-2.954**
	Pre-survey	1.86		.687	
The mobile devices will help me in my daily teaching practice.	Post-survey	1.32	17	.342	-2.352**
	Pre-survey	1.98		.698	
I like the learning activities using mobile devices and associated apps.	Post-survey	1.09	17	.313	-2.156*
	Pre-survey	2.02		.598	
I intend to use mobile devices in the classroom.	Post-survey	1.28	17	.204	-3.112*
	Pre-survey	2.05		.406	
Teaching STEM concepts to my students is easy for me.	Post-survey	1.44	17	.318	-2.583**
	Pre-survey	2.06		.607	

*$p < .05$; **$p < .01$

ing of participants' views on the benefits and problems of ScratchJr app. The findings could also shed light on the factors associated with their behavioral intentions to use ScratchJr to teach STEM concepts.

The interview protocol consisted of the following three questions:

- What are your perceptions of ScratchJr and STEM concepts?
- What are the benefits and problems of ScratchJr?
- What will enable you to further use ScratchJr?

The findings of the interviews are consistent with the results of the quantitative research confirming the positive effects of perceived usefulness and ease of use regarding ScratchJr app usage as well as students' intention to use ScratchJr to teach STEM concepts. When questioned about their beliefs of the potential benefits and problems of ScratchJr, the participants recognized that ScratchJr was a useful tool for developing useful projects to teach young children STEM concepts. They also said that ScratchJr was easy to use, due to its draggable blocks visual coding environment. Participants were asked what would motivate them to continue to use the ScratchJr in the future. Most of the people indicated that the easiness of the project's creation can be a major motivating factor for pre-service teachers. This qualitative analysis is consistent with quantitative findings that confirm pre-service teachers' behavioral intention to use ScratchJr. Furthermore, pre-service teachers believed that the type and quality of learning experiences promoted using ScratchJr were not only appropriate but would provide a strong foundation for the teaching of STEM subject areas in the preschool classroom. They also perceived as positive the plethora of instructional resources related to ScratchJr.

Teachers also noted that they believe that the kinesthetic activities provided by ScratchJr will motivate their students to participate in STEM activities. They also mentioned that ScratchJr can easily adapt to the individual needs of each child so that they can dynamically change their learning approaches in a manner that meets their students' needs. Of course, pre-service teachers mentioned that they lack detailed knowledge concerning STEM content and they ask for more knowledge to feel fully prepared to integrate STEM subjects in the classroom. According to the course itself from their responses the pre-service students found that was interesting:

- The course structure
- The course activities
- Collaboration with partners
- Learning new things.

In fact, if one considers that the pre-service teachers had no prior experiences and knowledge with the use of ScratchJr, as well as with coding skills and computational thinking concepts the obtained results are very encouraging. We can consider that the use of ScratchJr as a teaching strategy had a positive effect on pre-service teachers' attitudes toward mobile technology and technological inquiry abilities.

DISCUSSION

Perspectives

In this study, we presented motivation results in teaching the ScratchJr environment at pre-service kindergarten teachers. It is widely known, that in the educational settings the introduction of a new technology usually involves additional cognitive cost and workload for the teaching staff, and this potential workload determine whether the teaching staff consider the impact of these new technologies as confusing or difficult (Sánchez-Prieto et al., 2019).

Our results show that the use of ScratchJr environment for teaching STEM concepts can motivate the pre-service teachers and empower them to continue using smart mobile technology in their future daily teaching practice (Papadakis, Kalogiannakis & Zaranis, 2018). This is considered very important as if pre-service teachers perceived during the course that the use of smart technology and accompanying apps to develop new educational approaches related to STEM concepts was not easy or effective would most likely discourage them from using these tools in their future practice.

However, the findings of this study suggest that the ScratchJr, as a 'low floor', 'high ceiling' and 'wide walls' app, enabled pre-service teachers to learn the foundational of the coding skills and computational concepts quickly and easily without imposing a too-high cognitive load on them. ScratchJr allowed pre-service teachers to implement stories and games. They became familiar with the fundamentals of programming and computational thinking without worrying about syntax. Also, the analysis of the ScratchJr projects revealed statistically significant changes in pre-service teachers' self-efficacy to design new methods of teaching STEM concepts indicating that the participants developed skills and gained confidence while enacting with interactive touchscreen environments during the course (Psycharis & Kotzampasaki, 2019). But, to capitalize on the potential benefits of the integration of smart screen technology and relative apps, educational institutions must streamline STEM education. Furthermore, pre-service teachers need to be introduced to computational thinking concepts in ways that match their specific instructional goals and of course, best fit their needs. Pre-service teachers who are aware of mobile technology and accompanying apps and possess positive attitudes toward the use of this new technology on teaching and learning are more likely to incorporate such technology choices into their future routine (Chen, 2010). This is considered very important as various studies surrounding the under-representation of women in workforce related to STEM careers have suggested that gender differences in career choices could be due to intrinsic differences in domain-specific skill in STEM (Kersey, Braham, Csumitta, Libertus, & Cantlon, 2018). The findings of this study underline the need of the education departments for developing a framework for teaching and learning explanation that goes beyond a mere technical proficiency to integrate smart mobile technology in the classroom. Indeed, it requires a focus on raising students' awareness of the benefits that smart mobile technology in conjunction with the appropriate apps can bring to early education (Sánchez-Prieto et al., 2019). These new approaches involve the adaptation of existing curriculum and teaching methods in terms of fostering and or applying scenarios that enable the creative use of mobile technologies in educational settings (Sánchez-Prieto et al., 2019).

However, the results of our study should be interpreted with caution as there are some limitations associated with its design. First, the study was based on a very small number of female pre-service teachers and therefore may not have accurately represented the disparities between genders, regarding perception and response to work with ScratchJr for STEM concepts integration as well as the development of their knowledge according to fundamental coding skills and CT concepts. Given this methodological limitation, however, further studies should be carried out across different educational university departments that consider the diversity in demographics of the participants.

The last decade, there is a wave of enthusiasm for STEM education for the improvement of the early childhood education programs as research has indicated that it would be helpful for children's development if children could be exposed to STEM education at a young age (Kazakoff, Sullivan, & Bers, 2013). Thus, it is imperative the priority to prepare pre-service preschool teachers to integrate CT and STEM concepts into their daily teaching practice using touchscreen technologies and developmentally appropriate apps. This approach, of course, requires a multipronged approach that involves a diverse set of necessary skills to be developed so as the pre-service teachers to become computationally literate as well as the development of curricula that specifically consider the affordances of the smart mobile technology in the preschool classroom.

ACKNOWLEDGMENT

The authors would like to thank all the students for their participation and assistance during this study.

COMPETING INTERESTS

The authors declare that they have no competing interests.

FUNDING

We are especially indebted to the Special Account of Research of the University of Crete, which financially supports our research effort, KA 1013, Mobile Educational Application Developmental Scale (MoEADS): A first attempt to evaluate mobile educational apps for young children in Greece.

REFERENCES

Bers, M. U. (2018a). Coding, playgrounds and literacy in early childhood education: The development of KIBO robotics and ScratchJr. In *Global Engineering Education Conference (EDUCON)*, 2018 *IEEE* (pp. 2094-2102). IEEE. 10.1109/EDUCON.2018.8363498

Bers, M. U. (2018b). Coding and computational thinking in early childhood: the impact of ScratchJr in Europe. *European Journal of STEM Education, 3*(3), 08.

Chen, R. (2010). Investigating models for preservice teachers' use of technology to support student-centered learning. *Computers & Education, 55*(1), 32–42. doi:10.1016/j.compedu.2009.11.015

Cheng, G. (2019). Exploring factors influencing the acceptance of visual programming environment among boys and girls in primary schools. *Computers in Human Behavior, 92*, 361–372. doi:10.1016/j.chb.2018.11.043

Dorouka, P., Papadakis, St., & Kalogiannakis, M. (2019). ((Forthcoming). Tablets & apps for promoting robotics, mathematics, STEM education and literacy in early childhood education. *International Journal of Mobile Learning and Organisation*. Retrieved from https://www.inderscience.com/info/ingeneral/forthcoming.php?jcode=ijmlo

Eisen, S., & Lillard, A. S. (2017). Young children's thinking about touchscreens versus other media in the US. *Journal of Children and Media, 11*(2), 167–179. doi:10.1080/17482798.2016.1254095

English, L. D. (2016). STEM education K-12: Perspectives on integration. *International Journal of STEM Education, 3*(3), 1–8.

Estapa, A. T., & Tank, K. M. (2017). Supporting integrated STEM in the elementary classroom: a professional development approach centered on an engineering design challenge. *International Journal of STEM education, 4*(1), 6.

Hershman, A., Nazare, J., Sysoev, I., Fratamico, L., Buitrago, J., Soltangheis, M., ... Roy, D. (2018). Family learning coach: engaging families in children's early literacy learning with computer-supported tools. In *Proceedings of the 26th International Conference on Computers in Education. Philippines: Asia-Pacific Society for Computers in Education*, 637-646.

Kalogiannakis, M., Ampartzaki, M., Papadakis, St., & Skaraki, E. (2018). Teaching natural science concepts to young children with mobile devices and hands-on activities. a case study. *International Journal of Teaching and Case Studies, 9*(2), 171–183. doi:10.1504/IJTCS.2018.090965

Kalogiannakis, M., & Papadakis, St. (2017). Combining mobile technologies in environmental education: A Greek case study. *International Journal of Mobile Learning and Organisation, 11*(2), 108–130. doi:10.1504/IJMLO.2017.084272

Kalogiannakis, M., & Papadakis, St. (2019). Evaluating pre-service kindergarten teachers' intention to adopt and use tablets into teaching practice for natural sciences. *International Journal of Mobile Learning and Organisation, 13*(1), 113–127. doi:10.1504/IJMLO.2019.096479

Kelley, T. R., & Knowles, J. G. (2016). A conceptual framework for integrated STEM education. *International Journal of STEM Education*, *3*(11). doi:10.118640594-016-0046-z

Kersey, A. J., Braham, E. J., Csumitta, K. D., Libertus, M. E., & Cantlon, J. F. (2018). No intrinsic gender differences in children's earliest numerical abilities. *NPJ Science of Learning, 3*(1), 12.

Kucirkova, N., & Falloon, G. (2018). iPad apps and visual methodologies: Empirical and ethical issues in achieving authentic data. *Research in Learning Technology*, *26*(2029), 1–18.

Leidl, K. D., Bers, M. U., & Mihm, C. (2017). Programming with ScratchJr: a review of the first year of user analytics. In *Conference Proceedings of International Conference on Computational Thinking Education 2017*. Hong Kong: The Education University of Hong Kong.

Looi, C. K., Zhang, B., Chen, W., Seow, P., Chia, G., Norris, C., & Soloway, E. (2011). 1: 1 mobile inquiry learning experience for primary science students: A study of learning effectiveness. *Journal of Computer Assisted Learning*, *27*(3), 269–287. doi:10.1111/j.1365-2729.2010.00390.x

Margot, K. C., & Kettler, T. (2019). Teachers' perception of STEM integration and education: A systematic literature review. *International Journal of STEM Education*, *6*(2). doi:10.118640594-018-0151-2

Moomaw, S., & Davis, J. (2010). STEM Comes to Preschool. *Young Children*, *65*(5), 12–18.

Moore, T. J., Stohlmann, M. S., Wang, H. H., Tank, K. M., Glancy, A. W., & Roehrig, G. H. (2014). Implementation and integration of engineering in K-12 STEM education. In J. Strobel, S. Purzer, & M. Cardella (Eds.), *Engineering in precollege settings: synthesizing research, policy, and practices* (pp. 35–60). Lafayette: Purdue University Press. doi:10.2307/j.ctt6wq7bh.7

Murphy, S., & Headley, M. (2018). The role of digital technology in teen mothers' and their children's literacy. *Journal of Early Childhood Literacy*, Online First. doi:10.1177/1468798418783326

Neumann, M. M. (2016). Young children's use of touch screen tablets for writing and reading at home: Relationships with emergent literacy. *Computers & Education*, *97*, 61–68. doi:10.1016/j.compedu.2016.02.013

OECD. (2015). Computers and learning: making the connection. (PISA, OECD Publishing, 2015).

Pamuk, S. (2011). Understanding preservice teachers' technology use through TPACK framework. *Journal of Computer Assisted Learning*, *28*(5), 425–439. doi:10.1111/j.1365-2729.2011.00447.x

Papadakis, S. (2017). Creativity and innovation in European education. 10 years eTwinning. Past, present and the future. *International Journal of Technology Enhanced Learning*, *8*(3-4), 279–296.

Papadakis, S. (2018a). Gender stereotypes in Greek computer science school textbooks. *International Journal of Teaching and Case Studies*, *9*(1), 48–71. doi:10.1504/IJTCS.2018.090196

Papadakis, S. (2018b). The use of computer games in classroom environment. *International Journal of Teaching and Case Studies*, *9*(1), 1–25. doi:10.1504/IJTCS.2018.090191

Papadakis, S., & Kalogiannakis, M. (2017). Mobile educational applications for children. What educators and parents need to know. *International Journal of Mobile Learning and Organisation*, *11*(3), 256–277. doi:10.1504/IJMLO.2017.085338

Papadakis, S., & Kalogiannakis, M. (2019). Evaluating a course for teaching advanced programming concepts with scratch to preservice kindergarten teachers: a case study in Greece. In D. Farland-Smith (Ed.), *Early Childhood Education* (pp. 1–19). London, UK: IntechOpen Limited. doi:10.5772/intechopen.81714

Papadakis, S., Kalogiannakis, M., Orfanakis, V., & Zaranis, N. (2017). The appropriateness of Scratch and App Inventor as educational environments for teaching introductory programming in primary and secondary education. [IJWLTT]. *International Journal of Web-Based Learning and Teaching Technologies, 12*(4), 58–77. doi:10.4018/IJWLTT.2017100106

Papadakis, S., Kalogiannakis, M., & Zaranis, N. (2016a). Comparing tablets and PCs in teaching mathematics: An attempt to improve mathematics competence in early childhood education. *Preschool & Primary Education Journal, 4*(2), 241–253. doi:10.12681/ppej.8779

Papadakis, S., Kalogiannakis, M., & Zaranis, N. (2016b). Developing fundamental programming concepts and computational thinking with ScratchJr in preschool education. A case study. *International Journal of Mobile Learning and Organisation, 10*(3), 187–202. doi:10.1504/IJMLO.2016.077867

Papadakis, S., Kalogiannakis, M., & Zaranis, N. (2017a). Designing and creating an educational app rubric for preschool teachers. *Education and Information Technologies, 22*(6), 3147–3165. doi:10.100710639-017-9579-0

Papadakis, S., Kalogiannakis, M., & Zaranis, N. (2017b). Improving mathematics teaching in kindergarten with realistic mathematical education. *Early Childhood Education Journal, 45*(3), 369–378. doi:10.100710643-015-0768-4

Papadakis, S., Kalogiannakis, M., & Zaranis, N. (2018). Educational apps from the Android Google Play for Greek preschoolers: A systematic review. *Computers & Education, 116*, 139–160. doi:10.1016/j.compedu.2017.09.007

Papadakis, S., Kalogiannakis, M., Zaranis, N., & Orfanakis, V. (2016). Using Scratch and App Inventor for teaching introductory programming in secondary education. A case study. *International Journal of Technology Enhanced Learning, 8*(3/4), 217–233. doi:10.1504/IJTEL.2016.082317

Papavlasopoulou, S., Giannakos, M. N., & Jaccheri, L. (2019). Exploring children's learning experience in constructionism-based coding activities through design-based research. *Computers in Human Behavior, 99*, 415–427. doi:10.1016/j.chb.2019.01.008

Psycharis, S. (2018). STEAM in education: A literature review on the role of computational thinking, engineering epistemology, and computational science. Computational STEAM Pedagogy (CSP). *Scientific Culture, 4*(2), 51–72.

Psycharis, S., & Kalia, M. (2017). The Effects of Computer Programming on high school students' problem solving, reasoning skills and self-efficacy in Mathematics. *Instructional Science, 45*(5), 583–602. doi:10.100711251-017-9421-5

Psycharis, S., & Kotzampasaki, E. (2019). The impact of a STEM Inquiry Game Learning scenario on Computational Thinking and Computer self-confidence. *Eurasia Journal of Mathematics, Science & Technology Education, 15*(4), em1689, 1-18, doi:10.29333/ejmste/103071

Sánchez-Prieto, J.-C., Hernández-García, H., García-Peñalvo, F.-J., Chaparro-Peláez, J., & Olmos-Migueláñez, S. (2019). Break the walls! second order barriers and the acceptance of mLearning by FIRST-YEAR PRE-SERVICE TEAChers. *Computers in Human Behavior*, *95*, 158–167. doi:10.1016/j.chb.2019.01.019

Sanders, M. (2009). STEM, STEM education, STEMmania. *Technology Teacher*, *68*(4), 20–26.

Sheehan, K. J., Hightower, B., Lauricella, A. R., & Wartella, E. (2018). STEM Media in the Family Context: The effect of STEM career and media use on preschoolers' science and math skills. *European Journal of STEM Education*, *3*(3), 17. doi:10.20897/ejsteme/3877

Wing, J. M. (2006). Computational thinking. *Communications of the ACM*, *49*(3), 33–35. doi:10.1145/1118178.1118215

Wong, G. K., & Jiang, S. (2018, December). Computational thinking education for children: algorithmic thinking and debugging. In *2018 IEEE International Conference on Teaching, Assessment, and Learning for Engineering (TALE)* (pp. 328-334). IEEE.

Yelland, N., & Gilbert, C. (2018). Transformative technologies and play in the early years: Using tablets for new learning. *Global Studies of Childhood*, *8*(2), 152–161. doi:10.1177/2043610617734985

Zaranis, N., Kalogiannakis, M., & Papadakis, S. (2013). Using mobile devices for teaching realistic mathematics in kindergarten education. *Creative Education (Special Issue in Preschool Education)*, *4*(7A1), 1-10.

ADDITIONAL READING

Bers, M. U. (2008). *Blocks to robots: Learning with technology in the early childhood classroom*. New York, NY: Teachers College Press.

Jimoyiannis, A., & Komis, V. (2007). Examining Teacher's beliefs about ICT in Education. *Teacher Development*, *11*(2), 149–173. doi:10.1080/13664530701414779

Kalogiannakis, M., Nirgianaki, G.-M., & Papadakis, St. (2018). Teaching magnetism to preschool children: The effectiveness of picture story reading. *Early Childhood Education Journal*, *46*(5), 535–546. doi:10.100710643-017-0884-4

Kalogiannakis, M., & Papadakis, St. (2017). Combining mobile technologies in environmental education: A Greek case study. *International Journal of Mobile Learning and Organisation*, *11*(2), 108–130. doi:10.1504/IJMLO.2017.084272

Kazakoff, E. R., Sullivan, A., & Bers, M. U. (2013). The Effect of a Classroom-Based Intensive Robotics and Programming Workshop on Sequencing Ability in Early Childhood. *Early Childhood Education Journal*, *41*(4), 245–255. doi:10.100710643-012-0554-5

Orfanakis, V., Papadakis, St., Kalogiannakis, M., Ampartzaki, M., & Vassilakis, K. (2016). Digital Student Conference Platform Implementation: The case study of the "Research Project" course. *Open Education, 12*(2), 5–23.

Papadakis, St., & Kalogiannakis, M. (2010). eTwinning in the early childhood as starting line of innovative practices for the didactic of natural sciences. In M. Kalogiannakis, D. Stavrou, & P.G. Michaelidis (Eds.) *Proceedings of the 7th International Conference on Hands-on Science, Bridging the Science and Society Gap Rethymno*, 25-31 July 2010, 235-240.

Papadakis, S., & Kalogiannakis, M. (2018). Using Gamification for Supporting an Introductory Programming Course. The Case of ClassCraft in a Secondary Education Classroom. In A. Brooks, E. Brooks, & N. Vidakis (Eds.), *Interactivity, Game Creation, Design, Learning, and Innovation. ArtsIT 2017, DLI 2017. Lecture Notes of the Institute for Computer Sciences, Social Informatics and Telecommunications Engineering* (Vol. 229, pp. 366–375). Cham, Switzerland: Springer. doi:10.1007/978-3-319-76908-0_35

Papadakis, St., & Kalogiannakis, M. (2019). Evaluating a Course for Teaching Advanced Programming Concepts with Scratch to Preservice Kindergarten Teachers: A Case Study in Greece. In D. Farland-Smith (Ed.), *Early Childhood Education,* 1-19, UK-London: IntechOpen Limited https://www.intechopen.com/online-first/evaluating-a-course-for-teaching-advanced-programming-concepts-with-scratch-to-preservice-kindergart/

Papadakis, St., Kalogiannakis, M., Orfanakis, V., & Zaranis, N. (2019). The appropriateness of scratch and app inventor as educational environments for teaching introductory programming in primary and secondary education. In Information Resources Management Association (IRMA) (Ed.), Early Childhood Development: Concepts, Methodologies, Tools, and Applications (pp. 797-819), USA: IGI-Global*. doi:10.4018/978-1-5225-7507-8.ch039

Papadakis, S., Kalogiannakis, M., & Zaranis, N. (2019). Parental involvement and attitudes towards young Greek children's mobile usage. *International Journal of Child-Computer Interaction.* doi: . doi:10.1016/j.ijcci.2019.100144

Psycharis, S. (2015). The Impact of Computational Experiment and Formative Assessment in Inquiry Based Teaching and Learning Approach in STEM Education. *Journal of Science Education and Technology, 25*(2), 316–326. doi:10.100710956-015-9595-z

Sullivan, A., & Bers, M. U. (2016). Robotics in the early childhood classroom: Learning outcomes from an 8-week robotics curriculum in pre-kindergarten through second grade. *International Journal of Technology and Design Education, 26*(1), 3–20. doi:10.100710798-015-9304-5

Voogt, J., Fisser, P., Good, J., Mishra, P., & Yadav, A. (2015). Computational thinking in compulsory education: Towards an agenda for research and practice. *Education and Information Technologies, 20*(4), 715–728. doi:10.100710639-015-9412-6

KEY TERMS AND DEFINITIONS

App: An app is a software application that works, generally, on a mobile technology such as a smartphone, tablet, or other similar device.

Early Childhood Education: Is a broad term used to describe any type of educational program that serves children in their preschool years, before they are old enough to enter kindergarten.

Mobile Device: The mobile device is a small-sized, portable computer that typically has touch-screen features.

Preschool Education: Is education that focuses on educating children from the ages of infancy until six years old.

Preservice Teachers: Students enrolled in an initial educator preparation program.

Smartphone: A smartphone is a term used to describe a category of mobile devices with computer-like functionality.

STEM: The term STEM (Science, Technology, Engineering and Mathematics) is an acronym used by those relevant to the educational method concerning the fields of Science, Technology, Engineering and Mathematics.

Chapter 6

Apps to Promote Computational Thinking Concepts and Coding Skills in Children of Preschool and Pre-Primary School Age

Stamatios Papadakis
https://orcid.org/0000-0003-3184-1147
Faculty of Education, Department of Preschool Education, University of Crete, Greece

ABSTRACT

A decade ago, Computational Thinking (CT) and coding were typically considered part of the secondary education programs, as the focus was on programming and algorithm development. The early childhood classroom was not exactly the area expected to find students-developed coding skills. But as has been the case lately, CT and coding have been characterized as fundamental skills of the 21st century, not only for computer scientists but for all citizens. Yet, through the application of developmentally appropriate technologies, the development of coding skills is increasingly possible, and the result may be the advancement of CT fluency or at least familiarity in young age children. Given the enormous success of smart mobile devices and accompanying mobile apps the rationale for this chapter is to investigate if there are apps that provide the children of preschool and pre-primary school age with opportunities to cultivate their foundational coding and CT skills.

INTRODUCTION

Until recently, Information and Communications Technology (ICT) acceptance and adoption focused primarily on the effectiveness of ICT as an educational tool, to be understood through the development of information literacies. Areas such as Computer Science (CS) or Computational Thinking (CT) were typically isolated into post-secondary education curriculum or below, with a focus on program development (Falkner, 2015). However, CS is quickly redefining what it means to be a literate citizen of the 21st century (Harel, 2016). The need to include CS in every school curriculum has arisen, with the belief

DOI: 10.4018/978-1-7998-1486-3.ch006

that young people should have the opportunity to develop their CT and coding skills, to be creators, not just consumers of technology (Falkner, 2015). The CT has been implemented (in school curricula) in several western countries, as well as related concepts like logical thinking, problem-solving, abstraction, etc. (Minchillo, Vellozo, Borin & Borin, 2018).

And there is a reason for that. Researchers worldwide call for the need that CS and CT must become a new imperative for the western societies, societies that call for a bit of innovation, entrepreneurship, teamwork, and creative thinking. In the 19th century, the '3 Rs' - reading, writing, and arithmetic – were declared as the foundational skills of modern Western education. For Generation Z, also called Post Millennials (Moschella & Basso, 2018) the '3 Rs' aren't going to be nearly enough to succeed in the 21st century economy. Since the last two decades, in the current literature of the '5th C' of 21st-century skills, critical thinking, creativity, collaboration, communication and Computational Thinking have witnessed a growing interest as an essential component of the school curriculum as core skills that must be taught to all students (Grover, 2018). Wing (2006) said that CT is regarded as an essential skill that was ignited by the rise and widespread of computers, just as that the traditional '3 Rs' of reading, writing, and arithmetic serve as the basic skills students need to succeed by the advent of printing (Chen et al., 2017). CT can help get young people to engage with powerful ideas, to unleash their own creativity and express themselves in new and interesting ways, as well as to understand the rapidly changing world around them (Yu & Roque, 2019). It has been pointed out by many scholars that there exists the relationship between understanding and practice of coding and the development of higher order thinking skills including problem-solving, critical-thinking and collaboration competencies (Kaldor, 2017).

To explain what CT is, and what is its relationship to programming (or coding), Grover (2018) defines CT as the thinking skills that are employed in understanding a problem and formulating a solution before coding. CT is not the same as programming, but rather a mix of skills, programmer uses in order to solve different types of problems. Coding is more than simply a technical matter of codes and protocols; Coding is the literacy of the 21st century (Bers, 2018). Coding is an excellent, fun, and useful context for developing CT skills (Pila, 2019). This is especially true in early childhood when opportunities are especially promising to cultivate an interest in computing as well as for the development of social, emotional, physical and cognitive skills (Yu & Roque, 2019). In fact, in many countries across the world, digital game-based learning applications, designed to support the more effective development of coding and CT skills in children aged 4 to 7 years old, have begun to appear.

Since 2012, the National Association for the Education of Young Children (NAEYC) and the Fred Rogers Center for Early Learning and Children's Media at Saint Vincent College—had already recognized the use of technology for high-quality instruction as a gap in the education system affecting teachers across grade levels and published a joint position statement on technology and interactive media as early childhood educational tools. In this statement they increased early childhood educators' awareness of this gap in their practice as they clearly state that educators who are innovating in the use of digital media in learning that can significantly enrich the learning experience in almost any knowledge area, particularly when their use of technology and media is child-centered, play-oriented, hands-on, relationship-building, and aligned with curriculum goals (NAEYC & the Fred Rogers Center, 2012). Furthermore, Bers (2018) confirms previous research that has shown that children as young as four years old can adequately understand, program their own robot to perform specific objectives while also building their CT and higher-order skills.

Consequently, using programming to teach CT is common in postgraduate and continuing education (Minchillo et al., 2018). New curricula around the globe, points to the need to improve for digital literacy,

CT and programming (or coding), to support a generation of children who will build the future of the civilized countries (Falkner, 2015). This shift has prompted an uptake in pedagogies and frameworks in the education systems worldwide. Coding and CT can often meet difficulties to fit into a formal early childhood program while some educators are being discouraged to teach young learners how to code for various reasons such as the lack of prior training or studies and/or extensive experience with coding (Herdzina, 2018). For that reason, learning initiatives to fostering computational literacy, are now actively being developed such as the 'hour of code' in the United States and the 'code week' in European Union (Strawhacker, Lee & Bers, 2018). Furthermore, several computational tools and kits are currently available to help younger children, with different coding backgrounds or levels of expertise, to grasp the most basic elements of coding through the use of interactive games and activities (Kyriakides, Meletiou-Mavrotheris & Prodromou, 2016). Digital applications or apps for smart mobile devices that are used to teach young children coding skills and CT concepts with game-like activities have gained great interest recently (e.g. ScratchJr, Kodable, etc.) (Pila, 2019). Research has shown that the most developmentally appropriate of them offer opportunities for interactions and collaborative learning (Kalogiannakis & Papadakis, 2019; McLennan, 2017) while at the same time through playful ways encourage children to explore coding. Digital kids are using smart screen devices and apps at increasing rates, while many apps are already being advertised as 'educationally valuable' (Schroeder & Kirkorian, 2016). Thus, it is critical to identify and evaluate whether and how these apps are designed in order to maximize educational benefits.

In this paper, we present a review of coding apps either for free or with a fee, that are designed to enable children to learn basic programming concepts via smart mobile devices. This paper is intended to be a starting point for further research useful both to academics, developers but also to educators and parents of children of preschool and pre-primary school age e.g. children from 3 up to 6 years. In addition to peer-reviewed research articles, we also explore other sources such as, for example, blog posts, as the most recent information regarding coding apps for young children is not merely published in academic journals.

LITERATURE REVIEW

The Need for Young Age Children to Get in Touch with Ct and Coding

In response to the digital age in which we live, there is a growing need for employees to keep pace with the changing trends and technologies (Bers, González-González & Armas–Torres, 2019) as there are more jobs available than there are skilled people to fill them (Pila, Aladé, Sheehan, Lauricella, & Wartella, 2018). For instance, the US Department of Labor projects that 1 million coding positions will go unfilled by 2020. These are jobs that would allow employers to earn a decent living and anticipate future developments (Harel, 2016). Coding and CT are the foundation of fast economic development and an increasingly technological society—a society that calls for innovation and entrepreneurship that can help spark needed economic growth (Harel, 2016). There has been growing recognition that being able to solve problems computationally is rapidly becoming a prerequisite competency for a career in various fields (Grover, 2018). Lee (2019) highlights the fact that technology-based systems are everywhere in society, and most children either experienced or recognized their presence positively. Consequently, to be creators and not just consumers of this powerful technology is increasingly seen as an essential credential for the digitalized workplace (Sáez-López, Román-González & Vázquez-Cano, 2016).

In Europe, the European Commission has taken an active role in promoting 'a digital agenda for Europe' which says that coding is the literacy of the 21st century (Bers et al., 2019). The notion is that coding accelerates learning and new ways of thinking and even young age children exposed in a range of tools that are inherent to the field of computer science (Bers, 2018; Kalogiannakis, Ampartzaki, Papadakis, & Skaraki, 2018; Sullivan & Bers, 2016). Furthermore, in July 2017, Informatics Europe published a report detailing how important is for pupils to get in touch with digital literacy and CS in the early school years, although is not generally regarded as a scientific subject (Moschella & Basso, 2018). In general, around the globe, various educational programs and initiatives are focusing on STEM literacy aiming at making CT and coding a priority for youth education (Bers et al., 2019). Various countries have begun introducing computer science (CS) to all students as either a part of new CS curricula or integrated into existing subjects (Grover, 2018). Those countries include United Kingdom, Canada, Singapore, Finland, Estonia, Ireland, Italy, South Korea, China, Australia, New Zealand, Brazil, United States to name a few (Bers et al., 2019; Walsh & Campbell, 2018). Numerous scholars have argued that technology – the 'T' in STEM – is vital in helping schools to ensure that young children are learning to keep up with the reality of the 21st-century (Pila, 2019).

Currently, there is a focus in schools to prepare students to productively participate in a 21st-century society that is increasingly ruled by digital technology (Aranda & Ferguson, 2018). In countries such as the United States and in Australia, there is a growing, government-led trend highlighting the significant interest involving children and youth in the areas of science, technology, engineering, and mathematics - or the STEM subjects (Highfield, Paciga & Donohue, 2018). For instance, in the United States, the US Department of Education - Office of Educational Technology (Lee, 2016) states that technology can be an effective tool for supporting students' construction of knowledge in STEM. At the same time, it highlights the fact that although there are many resources online and offline that offer tools for STEM to make them more applicable for young children, the 'T' in STEM is often confused with technological tools such as laptops, PCs and recently tablet type devices or with the broad field of educational technology. But as Lee (2016) notes the 'T' in STEM, however, is intended to introduce children and explain key concepts in the field of information technology. This includes CT, which is the basic logic underlying the interaction of the CS knowledge and code and is beginning to be applied in early childhood educational settings.

Additionally, Sullivan & Bers (2016) highlight the fact that the 'T' of technology and the 'E' of engineering, which have been systemically neglected in the past in early childhood, has gotten a lot more discussion in recent years with the establishment of new approaches for integrating media with educational content. New curricula introduced in several countries such as England, Australia, New Zealand, United States, have identified the need to educate both adults and youth for digital literacy and CS, as well as the need to promote deeper content learning and learning experiences on these areas since the years of early childhood (Falkner, 2015). As these tools, practices, and approaches are universally applicable, they can be taught, not only through CS courses but as part of an integrated way with other curricular inputs at school and from an early age (Bers, 2018).

Early coding, or precoding, can be fun and exciting. Over three decades of research has led to the discovery that the aforementioned tools are beneficial in giving children the practice they need (Highfield et al., 2018) for improving learning, memory, executive functioning, general cognitive problems, etc. (Strawhacker & Bers, 2019). Although CT concepts and coding skills for young age groups have grown in popularity among academic researchers and industrial practitioners (Strawhacker & Bers, 2019) this research area is not new: there's already a great deal of interest of these issues and a strong evidence base including tools such as the Logo programming language with a turtle as its starring character. Research has proven that children as young as four or five years of age can learn code-related skills such as sequencing and conditions along with literacy and numeracy (Pila et al., 2018). Furthermore, the sooner the children learn the basics of coding from a young age, the easier they will be able to learn, and expose themselves to computer language processing in the future (McLennan, 2017).

In early childhood education settings, it is widely known that 'developmentally appropriate practices must guide decisions about whether and when to integrate technology and interactive media' (NAEYC & the Fred Rogers Center, 2012). Programming is a creative activity in itself. Thus, at the preschool level, children are not taught CT and coding as a separate subject but focus on abilities while engaging in coding activities such as algorithms, as well as the opportunity to use programming to make things happen (Heikkilä & Mannila, 2018). In early childhood education, evidence suggests the positive effects of coding activities on attitudes and skills such as problem-solving and critical thinking (Bers et al., 2014; Sullivan and Bers, 2016). Platforms like ScratchJr engage children in meaningful learning experiences that fit in their cognitive needs, while nonprofit organizations like Code.org (http://code.org/) integrate CT and coding approaches into interactive lessons for students of all ages, starting from kindergarten (Strawhacker & Bers, 2019). Pila (2019) highlights the fact that even 'Elsa and Anna from Disney's Frozen have gotten in on the coding fun as 'The Hour of Code' project on code.org allows users to join Anna and Elsa as they help make a winter wonderland by creating snowflakes and more through code' (Pila, 2019). Furthermore, robotics kits such as KIBO provide students a screen-free opportunity to code a robot with sequences of symbols as well as 'If, Then' statements (Kaldor, 2017). There is also a growing number of interactive apps designed to teach young children about coding. Some of them, such as ScratchJr, 'Daisy the Dinosaur' and the Kodable can significantly enhance learning outcomes. Some others claim that teach young children coding skills, but they don't have still proven their real educational value. Furthermore, various scholars highlight the fact that research suggests that coding experience can help to overcome gender stereotypes associated with STEM subjects. Especially, early exposure to computational thinking activities can help mitigate barriers to participation in computing (Lee, 2019; Sullivan & Bers, 2016; Yu & Roque, 2019; Ching, Hsu & Baldwin, 2018).

COMPUTATIONAL THINKING AND CODING FOR CHILDREN

'Coding is the literacy of the 21st century' (Bers et al., 2019). The acquisition of Computational Thinking (CT) knowledge is also becoming recognized as a core competency that must be taught and acquired by the new generations of children (Román-González, Pérez-González & Jiménez-Fernández, 2017). What is coding and computational thinking and why are many countries increasingly focused on helping to introduce coding to younger users? Coding is writing the code and is an efficient tool to teach CT (Bers et al., 2019). Coding should be considered as a necessary starting point for developing concepts and skills related to CT and CS, rather than simply being an 'efficient' communication tool between human and machine, related to the ICT sector (Buitrago Flórez et al., 2017).

Definitions of CT vary across the literature, but all have their roots in Papert's 'constructionism' that he began to explore in the early 1980s with his book, 'Mindstorms – Children, computers and powerful ideas' (Papert, 1980 as cited in Aranda & Ferguson, 2018). Dr. Seymour Papert was envisioned young children start learning the basic concepts of coding as he developed the Logo computer programming language. Since then, researchers have found that young age children exposed to Logo performed significantly better on various tasks compared to a group of children that were not exposed to the programming environment. In addition, recent research on the effects of teaching coding to young children has shown that children as young as four or five years of age can successfully code a robot to perform a specific task or activity while most can actively in age and developmentally appropriate programming activities (Pila, 2019). Bers et al. (2019) study results agree with the literature which demonstrates that children can begin learning code at an early age.

Computational Thinking is a concept that has been defined in many ways (Bers et al., 2014) while encompasses a broad range of learning and problem-solving skills, patterns and approaches used in computer science. The current term of CT was largely popularised by Janette Wing who in 2006, was one of the first to define the term computational thinking as 'solving problems, designing systems and understanding of human behavior, using the fundamental concepts of computer science' (Wing, 2006, p. 33). In 2011 Wing revisited her first definition clarifying that CT 'is the thought processes involved in formulating problems and their solutions so that the solutions are represented in a form that can be effectively carried out by an information-processing agent' (Wing, 2011). In 2011, International Society for Technology in Education (ISTE) and the Computer Science Teachers Association (CSTA) (ISTE & CSTA, 2011) with contributions from leading figures in higher education, industry, and K–12 education developed an operational definition of CT. ISTE & CSTA (2011) defines CT as a problem-solving process that includes at least the following characteristics:

- Formulating problems
- Logically organizing and analyzing data
- Representing data
- Automating solutions through algorithmic thinking
- Identifying, analyzing, and implementing possible solutions
- Generalizing and transferring this problem-solving process to a wide variety of problems

Programming is not only a fundamental skill of CS and a key tool for integrating the various cognitive tasks involved in CT but the application of acquired competencies as well (Grover & Pea, 2013). It involves the ability of someone to create step-by-step instructions of computer code (McLennan, 2017). As the digitalization of society is one of the most critical issues of our time, coding, is seen as the new modern literacy that soon could be as important as grammar and spelling (Pila, 2019). Coding offers children 21st-century skills experiences including communication, thinking, and problem-solving, skills that are essential for future success in the marketplace (McLennan, 2017). As Gomes, Falcão & Tedesco (2018) note already in the eighties, researchers had already pointed out the relationship between coding and thinking skills: according to them, coding helps students to analyse and reflect on their thought process concisely.

One popular assumption about CT is that it is the same as coding. That's not quite accurate. While coding is a popular vehicle to teach CT, CT is much more than programming. Coding is considered as an excellent, fun, and useful context for developing CT skills (Grover, 2018). CT is the thinking skills that are employed in understanding a problem and formulating a solution before coding (Grover, 2018). CT is the process of problem analysis and problem decomposition that precedes (in addition to happening during) coding and programming. CT should also not be confused with 'digital literacy' which is often the 'technology' subject in schools that focus on how to use the software, digital tools, and the internet. CT is not about how to use technology; rather, it lifts the hood on how digital solutions are designed using computational tools.

Some basic CT concepts that children and/or novice programmers tend to use when they are coding are: Sequence, Iteration, Conditional statements, Threads, Event handling, User interface design, Input (Brennan & Resnick, 2012). According to Minchillo et al (2018), CT concepts can be grouped by school level as follows:

- **Preschool Ages 3 to 5:** Understand a problem, break down a bigger problem into smaller problems and take specific steps to solve it.
- **Elementary school Ages 6 to 10:** Problem understanding, problem decomposition, and solution specification. Represent algorithms by using a visual programming environment.

Table 1 shows Computational Thinking concepts associated with a target age range and Table 2 shows a short description of these concepts.

APPS

Over the intervening years, technology has become increasingly prevalent in the education sector, based on the growing interest and recognition that technology can be used to improve how young children learn (Hernandez, et al., 2015). With the advent of the iPad, the youth's engagement with digital devices such as tablet-type devices has grown exponentially since 2010 (Highfield et al., 2018). Instead, the exponential rate of growth of smart mobile devices has dramatically increased the number of children of

Table 1. Computational Thinking modules per age (Adapted from Minchillo et al., 2018)

Level	Ages	Prerequisites	Concepts
1	4 to 6	Reading (basic)	Sequence, loop, event, problem solving
2	6 or more	Reading, math	Conditional, algorithm, debugging

Table 2. Computational concepts (Adapted from Yu & Roque, 2018)

Concept	Description
Sequence	Identifying a series of steps for a task
Loops	Running the same sequence multiple times
Events	One thing causing another thing to happen
Parallelism	Making things happen at the same time
Conditionals	Making decisions based on conditions
Operators	Identifying a series of steps for a task
Data	Storing, retrieving, and updating values

all ages in the civilized world having access to these devices (Kyriakides et al., 2016). These devices are particularly suited for the young population as they are light and very mobile and do not rely on specific motor skills as they do not require the respondents to use a keyboard or a mouse (Kucirkova, 2014).

Young children are increasingly heavy users of applications (or 'apps') for these devices (Papadakis, Kalogiannakis & Zaranis, 2016a, 2016b). Early adopters of this type of technology like to actively engage in fun and play activities such as drawing and games (Lee, 2016). According to Notari, Hielscher & King (2016), the term 'mobile application' or 'app' can be defined as a 'self-contained program' or 'piece of software' designed for a specific purpose or to fulfill a particular objective. The apps can be downloaded by a user on a variety of smart mobile devices from an online store such as Google Play - the app store for Android devices - or Apple store - the app store for Apple devices -.

Outhwaite, Faulder, Gulliford, & Pitchford (2019) note that mobile apps, created for touch-screen devices are particularly suited for the young population, as children found them attractive, easy and intuitive to use. These apps build on gamification, that is, implementing game features in non-game contexts, aimed at, for instance, increasing user's intentions to use the app, while they try teaching young students the fundamental concepts of CT and coding in a playful mood (Heikkilä & Mannila, 2018; Papadakis, 2018). Indeed, educationally sound apps, supporting the development of coding skills, problem-solving and creativity for young people, (e.g. ScratchJr, Hopscotch, Lightbot, Daisy the Dinosaur, Kodable, etc.) have begun to appear around the globe (Kyriakides et al., 2016; Pila, 2019).

Many researchers claim that the early introduction of coding skills requires technologies and practices that are developmentally appropriate for young children and that take into consideration the cognitive limitations, maturity, and stage of development (Bers, 2018). Yet while the developers claim that their apps have real educational value, researchers currently know little whether these apps are essential ingredients to promote learning (Papadakis, Kalogiannakis & Zaranis, 2017, 2018). This is particularly true for apps that target the areas of STEM (Schroeder & Kirkorian, 2016). Though a huge number of apps in the market claim to be educational, there is little research evidence to support that, make it difficult for even the professionals to know what type of app would be appropriate and effective (Lee, 2016; Papadakis, Kalogiannakis & Zaranis, 2018).

Thus, a question arises due to the widespread of mobile technology and accompanied apps: How someone determines what is developmentally appropriate for young children when it comes commercially aggressive apps? (Lee, 2016). The first step is to define what is an educational mobile app. According to Notari et al. (2016), an educational app has a great potential for learning and training purposes either to help users connect and collaborate with others or with ordinary tasks in his or her daily school life. Hirsch-Pasek and her colleagues suggest that children under the age of 8 years learn best when they are able to engage in active learning, when learning experiences are socially and cognitively prepared, and they guided by a specific, achievable goal with a focus on app design and development (Hirsch-Pasek et al., 2015, p. 7).

THE STUDY

Until recently, an early childhood classroom was not the kind of place where someone expected to find out students engaged with CT and/or coding tools. But while leading politicians like former president of the United States Barack Obama, and top business figures like Bill Gates (Microsoft) and Mark Zuckerberg (Facebook) promote learning to code as a necessary skill in the 21st century, coding is increasingly viewed as a meaningful and important skill needed to expand the traditional literacy of reading and writing (Walsh & Campbell, 2018). The advances in technologies, in the last ten years, have given birth to new forms of educational tools designed to engage young age children in coding and CT activities in a fun and interactive way (Ching, Hsu & Baldwin, 2018).

The early studies have found a positive impact on young children cognitive development and the use of the Logo programming language. However, very young children find it incredibly difficult to use the Logo as it is a computer-based language and children may not have yet developed the motor skills necessary to control an external device, such as a mouse. Even the famous programming environment Scratch (version 2 or older) is not suitable for very young age learners. For these reasons, researchers except using 'touch gestural interfaces' with robots have also developed coding apps for kindergarten and first-grade students (Pila et al., 2018).

However, the application of developmentally appropriate techniques and tools has led to a substantial incline in the technological fluency of young children of kindergarten ages (Bers et al., 2014). Coding apps for mobile devices are a potential way for very young children to learn relative skills as these apps are screen-based like coding on a computer screen while they support visual block programming interfaces (Papadakis & Kalogiannakis, 2017a). This approach makes programming more accessible to younger children with less fine motor skills as it is not necessary to use external devices such a keyboard and/or a mouse for the input (Ching et al., 2018). Furthermore, access to these devices and accompanying apps is almost ubiquitous for families throughout the civilized world rather than PCs (Pila et al., 2018). Education researchers and teachers tend to believe that apps may be a great way to introduce young children to foundational coding skills while providing a fun, easy and enjoyable path to practice these skills (Pila et al., 2018).

Over the last years, several apps have been developed to encourage young children engagement with coding and CT related concepts and ideas (Aladé et al., 2016). At the same time, the research on the educational value of these apps is relatively new. Little prior work has focused on the educational impact of these apps (Aladé et al., 2016). There are already apps that help young children improved their knowledge of coding concepts and learn foundational coding skills (Pila, 2019; Pila et al., 2018). Most of the current research focuses on ScratchJr, while 'Daisy the Dinosaur' and 'Kodable' are mentioned fewer times. These results show that coding apps can teach young children basic general coding skills, provided them with a fun and flexible platform on which to work (Ching et al., 2018). In most apps, the use of visual blocks of code where children drag and drop is the most frequently used form (Yu & Roque, 2018). Furthermore, ScratchJr offers children experiences that are open-ended, giving them the opportunity to create interactive stories and animations (Yu & Roque, 2018).

Whilst research with young children and coding has focused on computer programming and robotics kits (Pila et al., 2018) research on the effect of apps on younger children coding and CT skills development is rare. The present study looks at the advent of new technologies in the form of apps as a way to facilitate CT and coding skills among preschool and early elementary school-age children. For this study, we categorize the coding apps in two different categories: apps that there are research studies that reinforce their claims (such as ScratchJr, Lightbot, Daisy the Dinosaur and Kodable) and apps that there is no research to confirm their educational value.

TABLET BASED APPS WITH EMPIRICAL RESEARCH

ScratchJr

ScratchJr (https://www.scratchjr.org/) is a free open-ended environment that allows children to create their own interactive games, animations, games, and stories (Strawhacker et al., 2018). The app has a simple, block-by-block interface, a library of characters and background scenes as well as a paint editor with drawing and editing controls. Children drag-and-drop prefabricated code blocks from the palette into the scripting area, such as puzzle pieces fitting together.

The ScratchJr app launched in 2014 and has been downloaded over 10 million times. Also, app statistics show that ScratchJr for both mobile operating systems (iOS and Android) is used in every country in the world (except North Korea and Western Sahara) (Bers, 2018). The DevTech Research Group at the Tufts University has also provided teaching guides and material (Strawhacker et al., 2018) which are freely available on the ScratchJr website (Bers & Sullivan, 2018). The advantage of the usage of ScratchJr and the accompanying material is that it helps educators to make connections with other content areas such as literacy and numeracy (Bers & Sullivan, 2018). In addition to digital materials, there also exists printed material such as the 'The Official ScratchJr Book: Help Your Kids Learn to Code' (Bers & Resnick, 2015) as well as a set of coding cards (Bers & Sullivan, 2018). These 75 activity cards involve three different kind of activities: off-the-screen games; on-screen activities; and ScratchJr challenges.

Kodable

Kodable (https://www.kodable.com) is an app for children in kindergarten to fifth grade and aims to teach young learners to program while they attempt to make their way through 45 levels of a maze. The app also comes with lesson plans that are developed according to the standards and recommendations of the U.S. Computer Science to guide educators and/or parents in the effective integration of the app to help young children through the first steps of programming approach (Ching et al., 2018). Using directional arrows (up, down, right, and left) as command blocks in a drag-and-drop interface, the app helps the child to roll the 'fuzz' through the different maze levels. As the child completes the challenges on each level, it learns to recognize the meaning of different concepts such as loops, conditions, and sequences.

There are two published studies that evaluate Kodable app. In the first study, in Turkey, the researchers found that the 25 five-year-old children enjoyed playing with the app while they became able to master the challenges of the sequence by the end of the week (Karadeniz, Samur & & Özden, 2014). The participants improved their knowledge of sequencing and conditional (if-then-else) statements and they were able to develop basic algorithms through the game. Indeed, the researchers noted that more guidance was needed to support the loop statement. However, it is worth mentioning that Flannery & Bers (2013) note that it remains unclear, how exactly the researchers measured such mastery.

In a second study, in the United States, Pila et al. (2018) conducted a naturalistic study of young children using two common apps: 'Daisy the Dinosaur' and 'Kodable'. After a week, the researchers found that children improved their coding skills such as sequencing, conditions, and loops. They also improved their knowledge of 'Daisy the Dinosaur' commands and in 'Kodable' gameplay. The researchers concluded that the children were able to learn and acquire basic coding skills, but 'they did not improve their ability to verbally explain what coding is'. They also found that that app was appealing to children helping them to learn foundational coding skills.

Daisy the Dinosaur

'Daisy the Dinosaur', which is only for Apple devices, is an app that teaches children the basics of coding from a young age (https://apps.apple.com/us/app/daisy-the-dinosaur/id490514278). On the app side, it uses a drag-and-drop interface with written commands, to help children learn coding skills such as sequence, loops, and conditional events.

There is only study evaluating the effectiveness of Daisy the Dinosaur as a coding app. As already mentioned, Pila et al. (2018) in their study tried to teach basic coding skills to young children using two different apps: 'Daisy the Dinosaur' and 'Kodable'. Children improved their knowledge of Daisy commands although they didn't know what coding was. The researchers also found that the app was also appealing to children.

Lightbot Jr

Lightbot and Lightbot Jr (https://lightbot.com/) belong to the same apps' family but with a different audience. Lightbot Jr. is suitable for ages four to eight while Lightbot for ages nine and up. Both are coding apps that try to teach young learners how to properly plan, test and debug procedures and loops (Ching et al., 2018). Lightbot Jr is a puzzle game that asks players to control the movement of a robot to light up all the blue squares in a scene (Yu & Roque, 2018). As the game progresses the levels will keep getting harder as more sophisticated combinations of commands are needed to achieve the task (Gouws et al., 2013). Lightbot Jr uses a similar layout with drag-and-drop icon-based blocks (Lin, 2015; Rose et al., 2017). The model of Light-Bot Jr, a visual representation of programming, makes the whole procedure significantly more attractive (Gouws et al., 2013).

Rose and colleagues in a study with Light-Bot Jr app in northern England included 20 boys and 20 girls with an age range of 6 to 7 years, found that the participants had used less manipulation. Also, programs in Lightbot Jr were faster to create compared to ScratchJr. Also, the Lightbot Jr interface naturally helped players to decompose complex problems into smaller and manageable sections (Rose et al., 2017). The same researchers also found that the participants interacted differently with the Lightbot Jr interface compared to the ScratchJr.

TABLET BASED APPS WITHOUT EMPIRICAL RESEARCH

Move the Turtle

Move the Turtle (http://movetheturtle.com/) is a paid app (for iPhone and iPad) that tries to teach children (ages 5+) the basics about coding through a game type interface. Children can program the motion of a turtle sprite to complete different tasks. Like other apps, the children can get familiar with loops, procedures, variables and conditional instructions. Although the first tasks are simple and easy, the next tasks are demanding and not suitable for the very young age children.

Run Marco!

Run Marco! is an app (https://runmarco.allcancode.com/) in the form of an adventure game designed for 6 to 12 years old children. It uses a visual set of instructions to guide Marco or Sophia across different levels of difficulty. The first challenges try to teach basic coding skills while the next levels try to teach various types of iteration and conditional logic. The game is divided into three sections, which comprise 36 levels. For instance, at level 1 the skills students will practice are the sequence of commands such as forward, turn, jump.

The Foos

The Foos app is a commercial app to teach programming for ages 5+ and can be considered as an extended version of The Foos Coding, which includes various types of games that let kids design, program and share their own games, or 'remixes' of games that other children created using the platform. The Foos, try to teach basic computer programming skills with no reading necessary so it is a learn-to-code platform for young age children. Children can learn to code by arranging animated icons in an order that results in a character moving through a series of 44 different puzzles.

Kidlo Coding

Kidlo Coding is a commercial app with monthly or yearly subscription which tries to teach children the very basics of programming with more than 50 fun games. Kidlo Coding helps children between 6 to 8 years old to improve their problem-solving skills and learn basic programming concepts such as sequences, loops, functions, etc.

Bee-Bot App

This app is based on the original Bee-Bot floor robot. The app provides the child with the opportunity to program a virtual robot to help it move across different pathways. The challenges get harder as children work through them. As with the physical Beebot toys, there are lots of opportunities for children to communicate how they think they need to get the robot to where he needs to be.

Code Adventures: Coding Puzzles For Kids

It is a commercial coding app suitable for children aged 6 and over. The app claims that children will improve their coding skills while they learn basic principles such as: order of operations, loops, conditionals, etc. The children must help an utterly lovable fuzzball to get back home by solving tricky spatial puzzles using only programming commands.

Freggers Play

Freggers Play (https://www.freggersplay.com/) is a free app that allows children up to 3 years old to animate 3D characters and objects in a playful way while they learn to code in an intuitive and entertaining way.

DISCUSSION

When evaluating any new approach or technology for use in early educational practice, it is important to consider the developmentally appropriate use of technology in terms of content, context, and children active involvement (Lee, 2016). The recent years, in both digital stores – Apple iTunes & Android Play Google-, educationally sound game-based apps (Papadakis & Kalogiannakis, 2017b, 2019), supporting the development of critical coding skills for young people, have begun to appear. While programming in developmentally appropriate coding apps, children think creatively, logically, and sequentially and learn the building blocks of CS, such as debugging and coding. Coding apps for young children should be developed to encourage them to engage in computational thinking while coding. Following Bers (2018) ideas, in coding apps, children must be able to choose which is the most appropriate for a particular activity while they use their skills and imagination during the project creation procedure. Through this procedure, young age children can develop fundamental coding skills such as abstract and sequential thinking as well as problem-solving strategies while supporting achievement.

It is widely known that educators can confidently introduce and support various forms of learning when they have access to substantial computational resources and information. Although, NAEYC and the Fred Rogers Center (2012) state that 'developmentally appropriate practices must guide decisions about whether and when to integrate technology and interactive media into early childhood programs' the problem remains: are there developmentally appropriate apps to support CT and coding activities in children of preschool and pre-primary school age?

The findings of this study provide evidence that young age children can facilitate early STEM learning and foundational coding skills from playful learning experiences on smart mobile devices. This study contributes to a small but growing body of literature that investigates how children process and learn fundamental coding skills from touchscreen devices. This study agrees with other studies results confirming ScratchJr as a developmentally appropriate coding app to engage children in CT and coding activities, as well as to provide a space for them to encounter ideas from CS (Leidl, Bers & Mihm, 2017). Various researchers agree that ScratchJr is an app that can teach young children to code while design their own stories, animations and games (Portelance, Strawhacker, & Bers, 2016; Walsh & Campbell, 2018). The design characteristics behind ScratchJr interface has been to create an environment, which is easily 'communicated' where young children can create their projects (Heikkilä & Mannila, 2018). Therefore, we can conclude that teachers and parents must look for developmentally appropriate coding apps such as ScratchJr to engage children in CT and foundational coding skills (Leidl et al., 2017).

The review also found 3 other apps (Kodable, Daisy the Dinosaur and the Lightbot Jr), which are available either in the Google and/or the Apple's app stores, that the empirical research highlights the fact that can help young children learn basic coding skills via structured app play, although the level of difficulty differs between the apps. All these apps were also appealing to children. This study also revealed apps which claim to be educational. In fact, some of them in their websites promote awards and distinctions that verify their educational value such as 'Rated Best of the Best' or 'Certified by …'. In this study, we don't claim that these apps don't have educational value, but as we couldn't found scientific research to certify that value, we must be skeptic for their ability to deliver an optimal educational experience. Also, there is another category of coding apps such as Bitsbox or Osmo Coding Awbie, etc. These apps use hands-on physical blocks to control character or to create projects, so we didn't include these types of apps in this review.

CONCLUSION

Given the fact that the quality early learning experiences predict future educational attainment, it has become increasingly important to use innovative pedagogical approaches to engage learners in STEM learning and foundational coding skills in the early years (Aladé et al., 2016). Due to the great appeal attractiveness of touchscreens and apps for children, there has been a shifting focus of organizations and governments to use these interactive technologies to support learning activities at very young ages (Aladé et al., 2016).

Yet when well designed, the technology can improve the teaching and learning practice offering meaningful opportunities for young children to learn in inclusive settings, and engage in a new, meaningful, positive and powerful way (Aladé et al., 2016). Educational apps that feature STEM content are becoming more widely available but, as our research suggests, these are clearly not being utilised. It is widely known that not all modern technology is designed in a way that is appropriate and useful for early learners (Lee, 2016). So, teachers and parents need to be thoughtful how they employ these new interactive technologies. (Lee, 2016).

Various studies have shown that young children can engage in different levels of computational literacies, such as abstraction, algorithms, and debugging, through engineering tasks with support from adults (Yu & Roque, 2019). In incorporating mobile learning into daily teaching and learning, educators can create a technology-rich environment that promotes the development of coding and CT skills (Walsh & Campbell, 2018). Hence, there is a need for frameworks that can properly integrate all these aspects in a coherent formulation. From our perspective, early year educators need pedagogical perspectives to integrate and incorporate CT, STEM, and coding concepts into their classrooms (Bers et al., 2019; Kalogiannakis & Papadakis, 2017a, 2017b, 2017c).

There is a great need for high-quality professional development programs that can empower teachers with new tools that can be integrated the use of educational technology into their practice (NAEYC & the Fred Rogers Center, 2012). These issues need to be addressed in order to make CT integration useful and sufficient in early childhood education. In concluding, there is a need for new and innovative technologies as well as rigorous designs and interfaces, specifically designed for young people to promote the development of CT concepts and coding skills in young children's life (Sullivan & Bers, 2016).

COMPETING INTERESTS

The authors declare that they have no competing interests.

FUNDING

We are especially indebted to the Special Account of Research of the University of Crete, which financially supports our research effort, KA 1013, Mobile Educational Application Developmental Scale (MoEADS): A first attempt to evaluate mobile educational apps for young children in Greece.

REFERENCES

Aladé, F., Lauricella, A. R., Beaudoin-Ryan, L., & Wartella, E. (2016). Measuring with Murray: Touchscreen technology and preschoolers' STEM learning. *Computers in Human Behavior*, *62*, 433–441. doi:10.1016/j.chb.2016.03.080

Aranda, G., & Ferguson, J. P. (2018). Unplugged programming: The future of teaching computational thinking? *Pedagogika*, *68*(3), 279–292. doi:10.14712/23362189.2018.859

Bers, M. U. (2018). Coding and computational thinking in early childhood: the impact of ScratchJr in Europe. *European Journal of STEM Education*, *3*(3), 1–13. doi:10.20897/ejsteme/3868

Bers, M. U., Flannery, L., Kazakoff, E. R., & Sullivan, A. (2014). Computational thinking and tinkering: Exploration of an early childhood robotics curriculum. *Computers & Education*, *72*, 145–157. doi:10.1016/j.compedu.2013.10.020

Bers, M. U., González-González, C., & Armas–Torres, M. B. (2019). Coding as a playground: Promoting positive learning experiences in childhood classrooms. *Computers & Education*, *138*, 130–145. doi:10.1016/j.compedu.2019.04.013

Bers, M. U., & Resnick, M. (2015). *The official ScratchJr book*. San Francisco, CA: No Starch Press.

Bers, M. U., & Sullivan, A. (2018). *ScratchJr coding cards: creative coding activities*. San Francisco, CA: No Starch Press.

Brennan, K., & Resnick, M. (2012). New frameworks for studying and assessing the development of computational thinking. In *Proceedings of the 2012 annual meeting of the American Educational Research Association*, (Vol. 1, p. 25). Vancouver, Canada.

Buitrago Flórez, F., Casallas, R., Hernández, M., Reyes, A., Restrepo, S., & Danies, G. (2017). Changing a generation's way of thinking: Teaching computational thinking through programming. *Review of Educational Research*, *87*(4), 834–860. doi:10.3102/0034654317710096

Chen, G., Shen, J., Barth-Cohen, L., Jiang, S., Huang, X., & Eltoukhy, M. (2017). Assessing elementary students' computational thinking in everyday reasoning and robotics programming. *Computers & Education*, *109*, 162–175. doi:10.1016/j.compedu.2017.03.001

Ching, Y. H., Hsu, Y. C., & Baldwin, S. (2018). Developing computational thinking with educational technologies for young learners. *TechTrends*, *62*(6), 563–573. doi:10.100711528-018-0292-7

Falkner, K. (2015). *Computational thinking as the 'new literacy': professional development opportunities*. Retrieved from https://www.scisdata.com/connections/issue-95/computational-thinking-as-the-new-literacy-professional-development-opportunities/

Flannery, L. P., & Bers, M. U. (2013). Let's dance the 'robot hokey-pokey!' Children's programming approaches and achievement throughout early cognitive development. *Journal of Research on Technology in Education*, *46*(1), 81–101. doi:10.1080/15391523.2013.10782614

Gomes, T. C. S., Falcão, T. P., & Tedesco, P. C. D. A. R. (2018). Exploring an approach based on digital games for teaching programming concepts to young children. *International Journal of Child-Computer Interaction, 16*, 77-84.

Gouws, L. A., Bradshaw, K., & Wentworth, P. (2013). Computational thinking in educational activities. *Proceedings of the 18th ACM Conference on Innovation and Technology in Computer Science Education. ACM, New York, NY.*

Grover, S. (2018). *The 5th 'C' of 21st century skills? Try computational thinking (not coding).* Retrieved from https://www.edsurge.com/news/2018-02-25-the-5th-c-of-21st-century-skills-try-computational-thinking-not-coding

Grover, S., & Pea, R. (2013). Computational thinking in K–12: A review of the state of the field. *Educational Researcher, 42*(1), 38–43. doi:10.3102/0013189X12463051

Harel, I. (2016). *American schools are teaching our kids how to code all wrong*. Retrieved from https://qz.com/691614/american-schools-are-teaching-our-kids-how-to-code-all-wrong/

Heikkilä, M., & Mannila, L. (2018). Debugging in programming as a multimodal practice in early childhood education settings. *Multimodal Technologies and Interaction*, *2*(3), 42. doi:10.3390/mti2030042

Herdzina, J. (2018). *TEC play and lesson plans: T+M In STEM: Coding and computational thinking*. Retrieved from http://teccenter.erikson.edu/early_childhood_educators/tecmentors_coding/

Hernandez, M. W., Estrera, E., Markovitz, C. E., Muyskens, P., Bartley, G., Bollman, K., . . . Silberglitt, B. (2015). Uses of technology to support early childhood practice. OPRE Report 2015-38. Washington, DC: Office of Planning, Research and Evaluation, Administration for Children and Families, U.S. Department of Health and Human Services.

Highfield, K., Paciga, K. A., & Donohue, C. (2018). Supporting whole child development in the digital age. In *Digital childhoods, international perspectives on early childhood education and development 22* (pp. 165–182). Singapore: Springer. doi:10.1007/978-981-10-6484-5_11

Hirsh-Pasek, K., Zosh, J. M., Golinkoff, R. M., Gray, J. H., Robb, M. B., & Kaufman, J. (2015). Putting education in 'educational' apps: Lessons from the science of learning. *Psychological Science in the Public Interest, 16*(1), 3–34. doi:10.1177/1529100615569721 PMID:25985468

International Society for Technology in Education (ISTE) and the Computer Science Teachers Association. (CSTA). (2011). *Computational Thinking in K–12 Education teacher resources second edition.* Retrieved from https://csta.acm.org/Curriculum/sub/CurrFiles/472.11CTTeacherResources_2ed-SP-vF.pdf

Kaldor, T. (2017). *Toying with tech: early coding and computational thinking in a museum setting.* Retrieved from http://teccenter.erikson.edu/in-practice/toying-with-tech/

Kalogiannakis, M., Ampartzaki, M., Papadakis, S., & Skaraki, E. (2018). Teaching natural science concepts to young children with mobile devices and hands-on activities. a case study. *International Journal of Teaching and Case Studies*, *9*(2), 171–183. doi:10.1504/IJTCS.2018.090965

Kalogiannakis, M., & Papadakis, St. (2017a). A proposal for teaching ScratchJr programming environment in preservice kindergarten teachers. In *Electronic Proceedings of the ESERA 2017 Conference. Research, Practice and Collaboration in Science Education, Part 15/Strand 15* (pp. 2095-2105). Dublin, Ireland: Dublin City University.

Kalogiannakis, M., & Papadakis, S. (2017b). Pre-service kindergarten teachers' acceptance of "ScratchJr" as a tool for learning and teaching computational thinking and Science education. In *Proceedings of the 12th Conference of the European Science Education Research Association (ESERA), 'Research, practice, and collaboration in science education'* (pp. 21-25). Dublin City University and the University of Limerick, Dublin, Ireland.

Kalogiannakis, M., & Papadakis, St. (2017c). Combining mobile technologies in environmental education: A Greek case study. *International Journal of Mobile Learning and Organisation, 11*(2), 108–130. doi:10.1504/IJMLO.2017.084272

Kalogiannakis, M., & Papadakis, S. (2019). Evaluating pre-service kindergarten teachers' intention to adopt and use tablets into teaching practice for natural sciences. *International Journal of Mobile Learning and Organisation, 13*(1), 113–127. doi:10.1504/IJMLO.2019.096479

Karadeniz, S., Samur, Y., & Özden, M. Y. (2014). Playing with algorithms to learn programming: A case study on 5 years old children. In *9th International Conference on Information Technology and Applications (ICITA2014)* (pp. 1-4). Sydney, Australia.

Kucirkova, N. (2014). iPads in early education: Separating assumptions and evidence. *Frontiers in Psychology, 5*, 715. doi:10.3389/fpsyg.2014.00715 PMID:25071678

Kyriakides, A. O., Meletiou-Mavrotheris, M., & Prodromou, T. (2016). Mobile technologies in the service of students' learning of mathematics: The example of game application ALEX in the context of a primary school in Cyprus. *Mathematics Education Research Journal, 28*(1), 53–78. doi:10.100713394-015-0163-x

Lee, J. (2016). *Early learning and educational technology policy brief.* Office of Educational Technology, US Department of Education. Retrieved from https://tech.ed.gov/files/2016/10/Early-Learning-Tech-Policy-Brief.pdf

Lee, J. (2019). Coding in early childhood. *Contemporary Issues in Early Childhood.* doi:10.1177/1463949119846541

Leidl, K. D., Bers, M. U., & Mihm, C. (2017). Programming with ScratchJr: a review of the first year of user analytics. In *Proceedings of International Conference on Computational Thinking Education 2017* (pp. 116-121). Hong Kong: The Education University of Hong Kong.

Lin, V. J. (2015). *Computational thinking and technology toys.* (Honors Thesis Collection, 307, Wellesley College, MA).

McLennan, D. M. P. (2017). Creating coding stories and games. *Teaching Young Children, 10*, 3. Retrieved from https://www.naeyc.org/resources/pubs/tyc/feb2017/creating-coding-stories-and-games

Minchillo, L. V., Vellozo, A., Borin, E., & Borin, J. F. (2018). Towards better tools and methodologies to teach computational thinking to children. In *26º Workshop sobre Educação em Computação (WEI 2018) (Vol. 26, No. 1/2018). SBC.* Universidade do Estado do Rio Grande do Norte, Brazil.

Moschella, M., & Basso, D. (2018). *Computational thinking at primary school: a research proposal for post millennial teachers.* Retrieved from https://eera-ecer.de/ecer-programmes/conference/23/contribution/44245/

National Association for the Education of Young Children & the Fred Rogers Center for Early Learning and Children's Media at Saint Vincent College. (2012). Technology and interactive media as tools in early childhood programs serving children from birth through. *Age, 8.* Retrieved from https://www.naeyc.org/files/naeyc/file/positions/PS_technology_WEB2.pdf

Notari, M. P., Hielscher, M., & King, M. (2016). Educational apps ontology. In D. Churchill, J. ... (Eds.), *Mobile learning design, lecture notes* (pp. 83–96). Singapore: Springer. doi:10.1007/978-981-10-0027-0_5

Outhwaite, L. A., Faulder, M., Gulliford, A., & Pitchford, N. J. (2019). Raising early achievement in math with interactive apps: A randomized control trial. *Journal of Educational Psychology, 111*(2), 284–298. doi:10.1037/edu0000286 PMID:30774149

Papadakis, S. (2018). The use of computer games in classroom environment. *International Journal of Teaching and Case Studies, 9*(1), 1–25. doi:10.1504/IJTCS.2018.090191

Papadakis, S., & Kalogiannakis, M. (2017a). Mobile educational applications for children. What educators and parents need to know. [Special Issue on Mobile Learning Applications and Strategies]. *International Journal of Mobile Learning and Organisation, 11*(3), 256–277. doi:10.1504/IJMLO.2017.085338

Papadakis, S., & Kalogiannakis, M. (2017b). Evaluation of Greek Android mobile applications for preschoolers. *Preschool and Primary Education, 5*, 65–100. doi:10.12681/ppej.11208

Papadakis, S., & Kalogiannakis, M. (2019). Evaluating a course for teaching advanced programming concepts with scratch to preservice kindergarten teachers: a case study in Greece. In D. Farland-Smith (Ed.), Early Childhood Education, 1-19. London, UK: IntechOpen Limited. doi:10.5772/intechopen.81714

Papadakis, S., Kalogiannakis, M., & Zaranis, N. (2016a). Comparing tablets and PCs in teaching mathematics: An attempt to improve mathematics competence in early childhood education. *Preschool and Primary Education, 4*(2), 241–253. doi:10.12681/ppej.8779

Papadakis, S., Kalogiannakis, M., & Zaranis, N. (2016b). Developing fundamental programming concepts and computational thinking with ScratchJr in preschool education: A case study. *International Journal of Mobile Learning and Organisation, 10*(3), 187–202. doi:10.1504/IJMLO.2016.077867

Papadakis, S., Kalogiannakis, M., & Zaranis, N. (2017). Designing and creating an educational app rubric for preschool teachers. *Education and Information Technologies, 22*(6), 3147–3165. doi:10.100710639-017-9579-0

Papadakis, S., Kalogiannakis, M., & Zaranis, N. (2018). Educational apps from the Android Google Play for Greek preschoolers: A systematic review. *Computers & Education, 116*, 139–160. doi:10.1016/j.compedu.2017.09.007

Papert, S. (1980). *Mindstorms: Children, computers, and powerful ideas.* Basic Books.

Pila, S. (2019). *Young children: What's code got to do with it? TEC Center at Erikson Institute.* Retrieved from http://teccenter.erikson.edu/what-were-reading/pila_coding/

Pila, S., Aladé, F., Sheehan, K. J., Lauricella, A. R., & Wartella, E. A. (2018). Learning to code via tablet applications: An evaluation of Daisy the Dinosaur and Kodable as learning tools for young children. *Computers & Education, 11*, 52–62.

Portelance, D. J., Strawhacker, A. L., & Bers, M. U. (2016). Constructing the ScratchJr programming language in the early childhood classroom. *International Journal of Technology and Design Education, 26*(4), 489–504. doi:10.100710798-015-9325-0

Román-González, M., Pérez-González, J. C., & Jiménez-Fernández, C. (2017). Which cognitive abilities underlie computational thinking? Criterion validity of the Computational Thinking Test. *Computers in Human Behavior, 72*, 678–691. doi:10.1016/j.chb.2016.08.047

Rose, S., Habgood, J., & Jay, T. (2017). An exploration of the role of visual programming tools in the development of young children's computational thinking. *Electronic Journal of e-Learning, 15*(4), 297-309.

Sáez-López, J. M., Román-González, M., & Vázquez-Cano, E. (2016). Visual programming languages integrated across the curriculum in elementary school: A two year case study using 'Scratch' in five schools. *Computers & Education, 97*, 129–141. doi:10.1016/j.compedu.2016.03.003

Schroeder, E. L., & Kirkorian, H. L. (2016). When seeing is better than doing: Preschoolers' transfer of STEM skills using touchscreen games. *Frontiers in Psychology, 7*, 1377. doi:10.3389/fpsyg.2016.01377 PMID:27679590

Strawhacker, A., & Bers, M. U. (2019). What they learn when they learn coding: Investigating cognitive development and computer programming in young children. *Educational Technology Research and Development, 67*(3), 541–575. doi:10.100711423-018-9622-x

Strawhacker, A., Lee, M., & Bers, M. U. (2018). Teaching tools, teachers' rules: Exploring the impact of teaching styles on young children's programming knowledge in ScratchJr. *International Journal of Technology and Design Education, 28*(2), 347–376. doi:10.100710798-017-9400-9

Sullivan, A., & Bers, M. U. (2016). Robotics in the early childhood classroom: Learning outcomes from an 8-week robotics curriculum in pre-kindergarten through second grade. *International Journal of Technology and Design Education, 26*(1), 3–20. doi:10.100710798-015-9304-5

Walsh, C., & Campbell, C. (2018). Introducing coding as a literacy on mobile devices in the early years. In Mobile technologies in children's language and literacy: innovative pedagogy in preschool and primary education (pp. 51-66). Emerald Publishing. doi:10.1108/978-1-78714-879-620181004

Wing, J. M. (2006). Computational thinking. *Communications of the ACM, 49-3*, 33-35.

Wing, J. M. (2011). Computational thinking: What and why. Retrieved from http://www.cs.cmu.edu/~CompThink/resources/TheLinkWing.pdf

Yu, J., & Roque, R. (2018). A survey of computational kits for young children. In *Proceedings of the 17th ACM Conference on Interaction Design and Children* (pp. 289-299). ACM, New York, NY. 10.1145/3202185.3202738

Yu, J., & Roque, R. (2019). A review of computational toys and kits for young children. *International Journal of Child-Computer Interaction, 21*, 17–36. doi:10.1016/j.ijcci.2019.04.001

ADDITIONAL READING

Dorouka, P., Papadakis, St., & Kalogiannakis, M. (2019). ((forthcoming). Tablets & apps for promoting Robotics, Mathematics, STEM Education and Literacy in Early Childhood Education. *International Journal of Mobile Learning and Organisation*. article https://www.inderscience.com/info/ingeneral/forthcoming.php?jcode=ijmlo and https://www.inderscience.com/info/ingeneral/forthcoming.php?jcode=ijtcs

Kalogiannakis, M., Nirgianaki, G.-M., & Papadakis, St. (2018). Teaching magnetism to preschool children: The effectiveness of picture story reading. *Early Childhood Education Journal, 46*(5), 535–546. doi:10.100710643-017-0884-4

Kazakoff, E. R., Sullivan, A., & Bers, M. U. (2013). The Effect of a Classroom-Based Intensive Robotics and Programming Workshop on Sequencing Ability in Early Childhood. *Early Childhood Education Journal, 41*(4), 245–255. doi:10.100710643-012-0554-5

Orfanakis, V., Papadakis, St., Kalogiannakis, M., Ampartzaki, M., & Vassilakis, K. (2016). Digital Student Conference Platform Implementation: The case study of the "Research Project" course. *Open Education, 12*(2), 5–23.

Papadakis, St., & Kalogiannakis, M. (2010). eTwinning in the early childhood as starting line of innovative practices for the didactic of natural sciences. In M. Kalogiannakis, D. Stavrou, & P.G. Michaelidis *(Eds.) Proceedings of the 7th International Conference on Hands-on Science, Bridging the Science and Society Gap* Rethymno, 25-31 July 2010, 235-240.

Papadakis, S., & Kalogiannakis, M. (2017). Using Gamification for Supporting an Introductory Programming Course. The Case of ClassCraft in a Secondary Education Classroom. In *Interactivity, Game Creation, Design, Learning, and Innovation* (pp. 366–375). Cham: Springer.

Papadakis, St., & Kalogiannakis, M. (2019). Evaluating a Course for Teaching Advanced Programming Concepts with Scratch to Preservice Kindergarten Teachers: A Case Study in Greece. In D. Farland-Smith (Ed.), *Early Childhood Education* (pp. 1–19). London, UK: IntechOpen Limited. doi:10.5772/intechopen.81714

Papadakis, St., & Kalogiannakis, M. (2019). Evaluating the effectiveness of a game-based learning approach in modifying students behavioural outcomes and competence, in an introductory programming course? A case study in Greece, *International Journal of Technology Enhanced Learning*, Forthcoming article

Papadakis, St., Kalogiannakis, M., Orfanakis, V., & Zaranis, N. (2019). The appropriateness of scratch and app inventor as educational environments for teaching introductory programming in primary and secondary education. In Information Resources Management Association (IRMA) (Ed.), Early Childhood Development: Concepts, Methodologies, Tools, and Applications (pp. 797-819), USA: IGI-Global. doi:10.4018/978-1-5225-7507-8.ch039

Papadakis, S., Kalogiannakis, M., & Zaranis, N. (2019). Parental involvement and attitudes towards young Greek children's mobile usage. *International Journal of Child-Computer Interaction*. doi: . doi:10.1016/j.ijcci.2019.100144

Sullivan, A., & Bers, M. U. (2016). Robotics in the early childhood classroom: Learning outcomes from an 8-week robotics curriculum in pre-kindergarten through second grade. *International Journal of Technology and Design Education*, *26*(1), 3–20. doi:10.100710798-015-9304-5

Voogt, J., Fisser, P., Good, J., Mishra, P., & Yadav, A. (2015). Computational thinking in compulsory education: Towards an agenda for research and practice. *Education and Information Technologies*, *20*(4), 715–728. doi:10.100710639-015-9412-6

KEY TERMS AND DEFINITIONS

App: An app is a software application that works, generally, on a mobile technology such as a smartphone, tablet, or other similar device.

Early Childhood Education: Is a broad term used to describe any type of educational program that serves children in their preschool years, before they are old enough to enter kindergarten.

Mobile Device: The mobile device is a small-sized, portable computer that typically has touch-screen features.

Preschool Education: Is education that focuses on educating children from the ages of infancy until six years old.

Preservice Teachers: Students enrolled in an initial educator preparation program.

Smartphone: A smartphone is a term used to describe a category of mobile devices with computer-like functionality.

STEM: The term STEM (Science, Technology, Engineering and Mathematics) is an acronym used by those relevant to the educational method concerning the fields of Science, Technology, Engineering and Mathematics.

Chapter 7
Gamifying Children's Linguistic Intelligence With the Duolingo App:
A Case Study From Indonesia

Muhibuddin Fadhli
https://orcid.org/0000-0002-4935-0701
Universitas Muhammadiyah Ponorogo, Indonesia

Saida Ulfa
Educational Technology Department, State University of Malang, Indonesia

Sukirman Sukirman
Institut Agama Islam Negeri Palopo, Indonesia

Happy Susanto
Universitas Muhammadiyah Ponorogo, Indonesia

Aldo Redho Syam
https://orcid.org/0000-0002-5979-6555
State University of Malang, Indonesia

ABSTRACT

This chapter discusses the use of Duolingo App to enhance children's linguistic intelligence. Linguistic intelligence is expertise in applying vocabularies effectively and efficiently. Duolingo is an application based on the Android platform which helps children mastering other languages that they can practice speaking, reading, listening, and writing through a play. Children in early childhood are in the period of 'playing' spontaneously. They will do activities of playing without any instruction from others. The activity of playing naturally will stimulate the aspects of linguistic and symbolic that this development is closely related to learning to speak by involving the ability to vocalize. The use of this application will implement toward gamification method. Gamification uses game-based mechanics, aesthetics, and game thinking to engage people, motivate action, promote learning, and solve problems. This chapter provides an illustration that gamification can be used specifically in stimulating linguistic intelligence based on a case study in Indonesia.

DOI: 10.4018/978-1-7998-1486-3.ch007

INTRODUCTION

The term gamification started to exist around 2010 as part of a long process. In the beginning, the term was applied when office workers were getting bored with their routines and needed activities to refresh their bodies and souls. Academic researchers have held that "gamification is a term originated in the digital media industry". The first documented uses date back to 2008, but gamification entered widespread adoption only in the second half of 2010, when several industry players and conferences popularised it (Deterding, Khaled, Nacke, & Dixon, 2011). The use of gamification spread through the education field, until Kapp employed the term in a learning context in his book explaining how gamification can motivate learners. He defines gamification as "using game-based mechanics, aesthetics, and game-thinking to engage people, motivate action, promote learning, and solve problems" (Kapp, 2014). In gamification, while elements of games such as points, badges, freedom to fail, and challenge are used, the intent is not to create a self-contained unit—not to create a game. The intent is to use elements from games to encourage the learners to engage with the content and to progress toward a goal. The argument is supported by Nicholson's observation that gamification is like a reward. Scoot states: "Rewards have been used for centuries to change behaviour; children and pets are trained through rewards and punishments, soldiers are rewarded for achievements through ranks and badges, and schools use grades to entice students to do schoolwork" (Neeli, Reiners, & Wood, 2015). The rewards referred to in this article are used to change behaviour, so that children previously uninterested in something will become interested, and will want to join in certain activities. Reward is one of elements in game which can be used to attract attention. This article provides an illustration in which gamification is used specifically to stimulate intelligence, in the sense stated by Gardner, who identifies eight intelligences in his book *Frames of Mind: the Theory of Multiple Intelligences* (Gardner, 2011). In this writing, the authors emphasise linguistic intelligence. Linguistic intelligence is the ability to understand and use spoken and written language. This can include expressing yourself effectively through speech or the written word as well as showing a facility for learning foreign tongues. Writers, poets, lawyers, and speakers are among those that Gardner sees as having high linguistic intelligence (Gardner, 2011). The example of famous people with high linguistic intelligence are William Shakspeare and J.K Rowling.

The use of smartphones in Indonesia has recently increased sharply, due to ease of access and affordability (BPS, 2017). In 2019, the number of smartphone users in Indonesia increased to 92 million (Katadata, 2019). The use of these devices affects the pattern of educating children. In the past, parents taught children how to read by giving them notebooks or letting them scribble on the wall. This shift in the pattern of education is a necessity that must be faced. If parents forbid children to use smartphones, they will become a generation closed to the outside world. Yet, allowing their children to interact with smartphones without clear objectives and boundaries will create unruly characters, highly dependent on their gadgets. Wise use of smartphones creates a generation of children who are independent and have broad knowledge. 'Wise' means obeying time restraints, obeying the rules, and being balanced. One of the wise things that can be done is to provide the right application for the child's age. Planting characters through smartphones is also a wise solution to their use. The phenomenon of education using smartphones also occurs in Indonesia, with parents installing applications to teach children to read, write, and understand vocabularies. One such application that is often used and installed is Duolingo. Irmayanti in her research argued the roles of the parents in assisting the use of gadget in pre-school children is extremily important, the factors which are supporting to guide children use the smarthphone are the knowledge of the technology and the impacts of using smarthphone to teach children (Irmayanti, 2018). One of efforts of parents to maintain the postive effects of the Gadget is by mentoring the children, parents can supervise the children and directly guide the children to access the only positive contents in their gadget .

In this chapter, we will discuss how parents in Indonesia teach children how to read, write, and understand Indonesian vocabularies using the Duolingo application. This Android-based game is generally employed to teach children to recognise letters, words and sentences through interesting games; the case study approach was therefore employed to explore both the pattern of parents teaching their offspring how to read and write using the application, and the development of reading skills of these children. This case study lasted for approximately a year, with the authors as the research instrument observing a child from five to six years old. An approach of interview and observation is employed to explore data and the facts of how parents use this application in learning.

Background

Gamification has been much discussed by some experts, initially in the economics sector. Activities such as training of trainers, employees' motivation, and briefings before starting work are activities related to gamification. Torsten's book *Gamification in Education and Business* explains how gamification "draws from multiple areas and (at present) from multiple theories borrowed from multiple disciplines, providing a medley of theoretical foundations for research in this area" (Neeli et al., 2015). These disciplines include, for example, the sciences of business, economics, and education. Gamification is part of game-based learning, Prensky argued that game-based learning can be use to figure out and invent ways to include reflection and critical thinking (either built into the game or through a process of instructor-led debriefing) with the learning and still make it a fun game (Prensky, 2000). However, the main difference between the gamification and game-based learning is the integration of game mechanics with training content. Game-based learning fully integrates the two, so the game is the training. On the other hand, gamification uses game elements as a reward for completing existing training modules (Findlay, 2016).

In the education field, gamification can be used thoroughly at both basic and upper levels – in practice, up to the level of senior high school or university. Recent research, however, indicates that gamification can be applied at the level of younger children. The strongest reason why this gamification is applied to children is that it comprises elements of game-based instruction. Research on gamification for children is very limited, with only a few literatures discussing the implementation of gamification methods for early childhood learning. In a book entitled *Gamification for Human Factors Integration: Social, Education, and Psychological Issues*, Tootell's chapter about the implementation of gamification provides an illustration of gamification as one of learning methods which can maximise learning outcomes in accordance with the curriculum. Curriculums from various countries, such as The Australian Early Years Learning Framework (EYLF), US National Association for the Education of Young Children (NAEYC), New Zealand Te Whariki, and UK Early Years Foundation Stage (EFYS), all agree on six learning outcomes which must be maximized in early childhood education. These are: child centred; active learning; plenty of time for children to pursue interests; studies or projects; reading with dialogue, questions and discussions; and creative, open-ended experiences (Bishop, 2014).

Naturally, when young children in early childhood are in a period of 'playing' spontaneously, they will carry out play activities without any instruction from others. Mary D. Sheridan advises: "Give the opportunity for children playing 'spontaneously'". In other words, they provide their own motivation to play and act without prompting or intervention by an adult. The type and duration of the play in which they engage are entirely determined by them, and activities can be taken up and stopped at will.

"The child, playing is an end in itself, and to an observer there may not seem to be any obvious goal or conclusion" (Sheridan, Howard, & Alderson, 2010). Synodi adds: "Playing is considered to be a part of young children's life, which is recognized as a contributor to a child's social, personal, linguistic, physical, cognitive, moral, creative and artistic development" (Synodi, 2010). According to him, playing is closely related to the parts of children's life which can contribute to social, emotional, linguistic, physical, cognitive, moral, creative, and aesthetic development.

When playing, children experience growth in accordance with their age (Allen, 1989). Various aspects of development are ready to be stimulated, such as physical aspects, so that when playing, children can control their movement. The age of 3 to 6 years offen reffered to as the "preschool period". Children's language, social emotional and cognitive skills are rapidly expanding. During this period the stimulation and learning that come from play (UNICEF, 2018). The activity of playing naturally will stimulate the motor skills of children, both fine and gross. The second aspects are cognitive and symbolic, and the development of both is closely related to acquiring knowledge and skill through processing and using information meaningfully. This skill includes imagination and creativity, abstract thought, logical reasoning, problem solving, and action. The third aspect refers to linguistic and symbolic, an area of development closely related to learning to speak by involving the ability to vocalise (or to pronounce words), to understand, and to obtain vocabularies which continue to develop. Children learn by using language in more complex ways, for example by giving opinions and exploring ideas. The fourth aspect which can be developed through play is related to social and emotional growth. Babies greatly depend on their carers, but when they play they become independent children, developing skills of self-help and understanding of social and cultural perspectives. Besides, control over emotion develops along with the way they express opinions and feelings. The last aspect is moral and spiritual. When playing, children will be examined on how tough they are in facing problems, particularly about honesty. Moral development generally copes with the understanding of values such as honesty, justice and respect, the concept of 'right' and 'wrong', responsibility, and the consequences of a person's actions.

Playing is closely related to intelligence. Research on play in early childhood has been increasing, supported by some of the academic journals, and in Norway playing activity is included in to the curriculum (Guss, 2005). In this journal article, Guss states: "A new understanding is arrived at not by asking what is play, but by asking how is playing?" This is important, because by asking only "what" play leaves children stuck only in a definition; asking "how" they play suggests involvement in the game. The statement is strengthened in a journal article entitled "Governing Early Childhood Education through Play" (Ailwood, 2003) which specifically discusses how to manage a game for early childhood education. This idea about playing is also implemented in language/literacy teaching. The research explores teaching in early childhood, with reading ability in different languages developed through playing activity (Moon & Reifel, 2008).

Intelligence for some people is defined by the ability to accept and to manage something. But, is intelligence limited only to this definition? The article discusses intelligences according to some experts. The traditional perspective suggests that intelligence is closely related to IQ. Yet Swiss psychologist Jean Piaget casts doubt on this, pointing out that when children take IQ tests, mistakes occur, making data about the intelligence of the children taking the test inaccurate. Piaget states that "Intelligence/cognitive development is when the scheme is obtained about how someone perceives his environment in the stages of development, when someone obtains new ways in presenting information mentally"(Piaget, n.d.). Piaget never criticises theories of intelligence using IQ, but states his opinion that: "The IQ movement is blindly empirical. It is based simply on tests with some predictive power about success in school and, only marginally, on a theory of how the mind works" (Gardner, 2011). An IQ is empirical, and therefore cannot be proved by a predictive test. Moreover, it addresses only success at school. Piaget's statement is quoted by Gardner, who defines intelligence as "a skill, with process of completeness, which is able to handle specific problem content in the world" (Gardner, 2008).

Gardner closely related to the theory of multiple intelligences (MI). To Gardner, all children are born with each talent of intelligence, so that we cannot say a child is intelligent from only one perspective of intelligence. Gardner introduces ideas about multiple intelligences in his book *Frames of Minds: The Theory of Multiple Intelligences*, proposing that there are seven intelligences (Gardner, 1988). Gardner refines his idea to eight intelligences in the third edition (Gardner, 2011). Those eight intelligences are:

Linguistic Intelligence (the ability to manage language); Musical Intelligence (the ability which is closely related to music, according to Gardner, this intelligence is the earliest intelligence seen in the process of children's development); Logical-Mathematical Intelligence (the intelligence which is related to logics and numbers); Spatial-Intelligence (the intelligence dealing with the sensitivity of someone to perceive visual spatial world); Bodily-Kinesthetic (the ability to control body movement and the proficiency in managing object); Interpersonal Intelligence (this intelligence is marked with the ability to absorb and to respond moods appropriately); Intrapersonal Intelligence (the ability to understand self-moods and the ability to differentiate emotion); Natural Intelligence (the ability to recognize members of species; to recognize the existence of other species, and to map the relationship among some species, both formally and informally) (Fadhli, 2016)

The above arguments explain the value of intelligence. When we want to formulate an intelligence, we must learn how to respect every intelligence owned by each human. Gardner provides requirements about something called an intelligence. To him, these requirements are essential in assuring that man's intelligence is original and useful for human civilisation. First, intelligence must be able to be symbolised (for example, mathematics has symbols, music has notation, kinaesthetics has symbols, etc.). Secondly, each intelligence has a progress history (unlike the IQ approach – which supposes that intelligence is absolutely fixed, has been fixed at birth and remains the same – MI holds that intelligence occurs in certain periods in childhood, has the potential to develop over the lifespan, and contains a unique pattern which slowly or rapidly reduces as people's ages increase. The earliest intelligence is music, followed by Logical-Mathematical Intelligence). Thirdly, each intelligence is vulnerable to defects due to damage or injury to certain parts of the brain. For example, someone with damage to the left of the frontal lobe will not be able to speak or write easily, but will not find problems in singing, drawing, or dancing. Someone whose right temporal lobes are damaged might find problems in music, but will find it easy to talk, read and write. Someone with a damaged right lobus oksipital might experience difficulties in recognising faces, and in imagining or observing visual details (Armstrong, 2000). Next, each intelligence is subject to circumstances based on cultural values; for example, the ability to ride a horse, or to practise archery.

This article discusses linguistic intelligences. As explained in the above paragraph, linguistic intelligence is closely related to the ability of a man to master language. The strongest reason for choosing linguistic intelligences in this discussion is that the intelligences formulated by Gardner are identical to aspects of development released by the Indonesian government in its content standard about achievement levels of children's development (Permendikbud, 2014). The government rules place these achievement levels in accordance with age and scope of development. There are six scopes of development, namely: (1) religious and moral values; (2) physical-motoric kinaesthetic; (3) cognitive; (4) language/linguistic; (5) social-emotional; and (6) art.

Experts vary on the question of linguistic intelligence. Gardner invokes "The ability to use language to convince other individuals of a course of action" (Gardner, 2011). Armstrong develops the theory about this MI in his book *Multiple Intelligences in the Classroom*, stating that intelligence is;

The capacity to use words effectively, whether orally (e.g., as a story-teller, orator, or politician) or in writing (e.g., as a poet, playwright, editor, or journalist). This intelligence includes the ability to manipulate the syntax or structure of language, the phonology or sounds of language, the semantics or meanings of language, and the pragmatic dimensions or practical uses of language (Armstrong, 2000)

According to Armstrong, linguistic intelligence is an expertise in applying vocabularies effectively and efficiently. This statement is strengthened by McCoog in his journal article "Integrated Instruction: Multiple Intelligences and Technology" (2007). He states that linguistic intelligence is "characterized as students with excellent written and oral skills. They excel in the humanities and focus on careers in journalism and politics". Similarly, Leila states that this intelligence "involves sensitivity to spoken and written language, the ability to learn languages, and the capacity to use language to accomplish certain goals" (Vakili, 2013). Linguistic intelligence, also known by the term "wordsmart" – is an intelligence which can be used by someone to master vocabularies, to play, and to sing. In childhood, this intelligence is closely related to abilities in obtaining and managing information into vocabularies which are sometimes very imaginative, for example listening to other people speaking in the mother tongue or other languages, and understanding two orders given at the same time.

The use of gadgets in educating children, in some literatures, is in line with the development of MI revealed by Gardner, as suggested in research on the use of mobile devices in teaching mathematics. This research proves the use of smartphones in stimulating mathematical logic intelligence (Zaranis, Kalogiannakis, & Papadakis, 2013). Another research project proves that using smartphones offers an alternative in stimulating natural intelligence, as Gardner revealed. The use of ICT in learning can offer an alternative in bringing children closer to the natural world by applying a QR code to gather information and complete tasks (Kalogiannakis; Papadakis, 2017). The researches mentioned above prove the effectiveness of smartphones in stimulating intelligence. This is the basis for selecting the theme of this chapter: discussing the implementation of the Duolingo application in stimulating linguistic intelligence, as stated in the introduction. This research is based on a qualitative approach to explore facts and to extract data for analysis.

Duolingo is an Android, IOS and Windows phone-based application launched in November 2011 by Luis von Ahn and Severin Hacker. It has a rating of 4.7 out of 5, meaning that more than 90% of users have given a rating of 5 and choose this application to teach languages for children. Indonesia has become one of the users of Duolingo. Approximately 30% of its 92 million users are children aged five to six. Indonesia has only one language that must be mastered, namely Bahasa Indonesia, and the application has teaching features in the Indonesian language. This gives the application additional value. Research conducted by the authors in the past year indicates that parents' shifting pattern of educating with the help of smartphones is one thing that cannot be avoided. Parents born as digital immigrants will certainly experience a few obstacles when teaching their children in the native digital generation category; the results of this research can therefore become reference material for parents in educating their children, as well as providing an intermediary line between technological development and raising children according to their age.

THE USE OF SMARTPHONES IN INDONESIA

Issues, Controversies, Problems Pemanfaatan Smartphone di Indonesia

In everyday life, we often hear about technology, but many might be confused in describing what technology is. Maybe we just imagine that technology is something that is sophisticated, futuristic, or cool? This is not totally incorrect, but the definition of technology is actually very broad. According to the Indonesian Language Dictionary (*Kamus Besar Bahasa Indonesia*/KBBI), technology is all means to provide goods needed for the continuity and comfort of a human's life. Poerbahawadja Harahap states that the use of the word technology basically refers to a science that investigates the procedure in the field of engineering. Technology is a form of process that increases added value. This process can use certain products or produce new ones, and these latter products are not separated from existing products. Furthermore, it is also mentioned that technology is among the integral elements contained within a particular system (Miarso, 2007). Technology is a means of solving the fundamental problems of every human civilisation. Without the use of technology, many problems will not be solved properly and perfectly (Sardar, 1987). Thus, it can be concluded that technology is a process of increasing the value of an item into equipment used with science, based on the fundamental problems of human life. Technology itself first originates from simple objects around us which are designed to be used as tools. The making of a tool itself exists intentionally or unintentionally. A tool is expected to be capable of easing human activities. However, humans from time to time gain more knowledge from these simple tools, and with the knowledge obtained, tools become more complex and their performance improves.

Education using technology should employ both elements hand in hand. However, this research indicates that many parents use technology without having an objective. For example, they allowed their children to watch YouTube in order to keep their children silent, without filtering out what information/content their children see. There are also parents busy with their jobs who neglect their responsibility to educate their children by allowing them to use gadgets all the time. This certainly has negative impacts on the development of their children, especially in social-emotional growth, as evidenced by research by Laili (2017) in Surabaya, Indonesia. This suggested that an excessive use of gadgets can have a negative impact on children's emotional-social development. These results are in line with research conducted by Suhana (2018), who states that the impact of uncontrolled use of gadgets can be to hamper children's emotional-social development. He provides advice on using gadgets, with good examples, on time duration, and on accompanying children playing with gadgets.

However, the results of other research showed positive effects in using gadgets. This is based on the rationale that it is not technology that prevents humans from surviving, but it is humans who hinder their lives adversely when using technology. Research in Kalimantan, Indonesia, by Frahasini (Frahasini, Marhaeni, Astuti, & Atmaja, 2018) explores the use of gadgets at various levels for children. The results showed that the use of this technology is intended for entertainment, aiming to reduce boredom by providing activities. The researchers point out the importance of parents' assistance in using gadgets, so that the direction and purpose of their use can be better controlled and monitored. In line with Frahasini, Laini (Alif & Lara, 2018) provides arguments about the results of research on the use of gadgets for children. Alif reveals that there is a positive relationship in the use of gadgets for positive purposes when parents accompany their children while they play with gadgets.

This research, using an expo facto approach, showed that, overall, the level of prosocial behaviour in low gadget usage (A2) is higher than in high gadget usage (A1); there is an interaction between the use of gadgets with parental involvement on child's prosocial behaviour; there is a difference in behavioural level prosocial use of low gadget (A2) parental involvement high (B1) higher than high prosocial use behaviour (A1) high parental involvement (B1), there is a difference in behaviour level prosocous low use gadget (A2) high parental involvement (B1) higher than prosocial behaviour of low gadget usage (A2) low parental involvement (B1) (Laini, 2018).

Problems are created with a solution. The explanation above shows the condition of the use of gadgets that occurs in Indonesia. This problem lies behind the authors' desire to conduct research on how parents use gadgets for positive purposes. The authors began conducting research in 2018, against the background of the problems the authors have described above. In accordance with the theme of the book, "Mobile Learning Application in Early Childhood Education", this chapter presents the results of the research that the author conducted into the use of mobile learning based applications in Indonesia. The authors chose the Duolingo application because it is a free application, widely used by parents in teaching language. Language here refers to the linguistic intelligence revealed by Gardner, such as the ability to understand, to recognise letters, to have vocabularies, and to communicate.

The Use of the Duolingo Application, Between Expectations and Reality

Duolingo is an application platform with a highly interactive display, and uses an owl icon as a symbol of knowledge. With this application, we can train our children to read, listen, and write by playing games (gamification). This application can be used in formal situations such as classroom learning, as well as in daily activities (non-formal) that are usually carried out by children. Duolingo has more than 120 million users worldwide, and more than one million users in Indonesia. There are more than 21 languages to learn, using the same features, from introductory games to tests. Applications that can be used on IOS, Android, and Windows phones are conceived so that users can continue to improve their language skills anytime and anywhere.

Duolingo learning is designed like a game, so it is interesting and easy to absorb. Duolingo won Google Play's Best of the Best 2013 award from Google because the gamification technique used is very effective in making many people acquire new knowledge. The gamification technique used by Duolingo makes it easy and fun to use. Even though Duolingo is designed for learning foreign languages, it does not have to be serious in doing so. Duolingo is included in the education application category, but the application is like playing games. We can use it casually, without pressure, and it has proved to be more effective in being accepted and absorbed by children because of its flexibility – the user can take four language courses at once without having to complete a single language chosen previously. In addition, level and reward systems in this application are very effective in creating motivation for learning a language. Users of Duolingo learn new languages through techniques including repetition, audibly speaking, and conversation; this is how this application teaches children to remember a language. First, the author will hear the voice of the Duolingo operator intoning a word in a previously chosen language, complete with the meaning of that word. After that, the user is trained to always remember what the mentioned word means through questions given. The initial appearance of the Duolingo application can be seen from the following Figure 1:

Figure 1 above shows the initial appearance when we have chosen to learn English for Indonesian people. The simple initial display makes this application very easy to use. This application can also

Figure 1. Display of Duolingo in Indonesian language

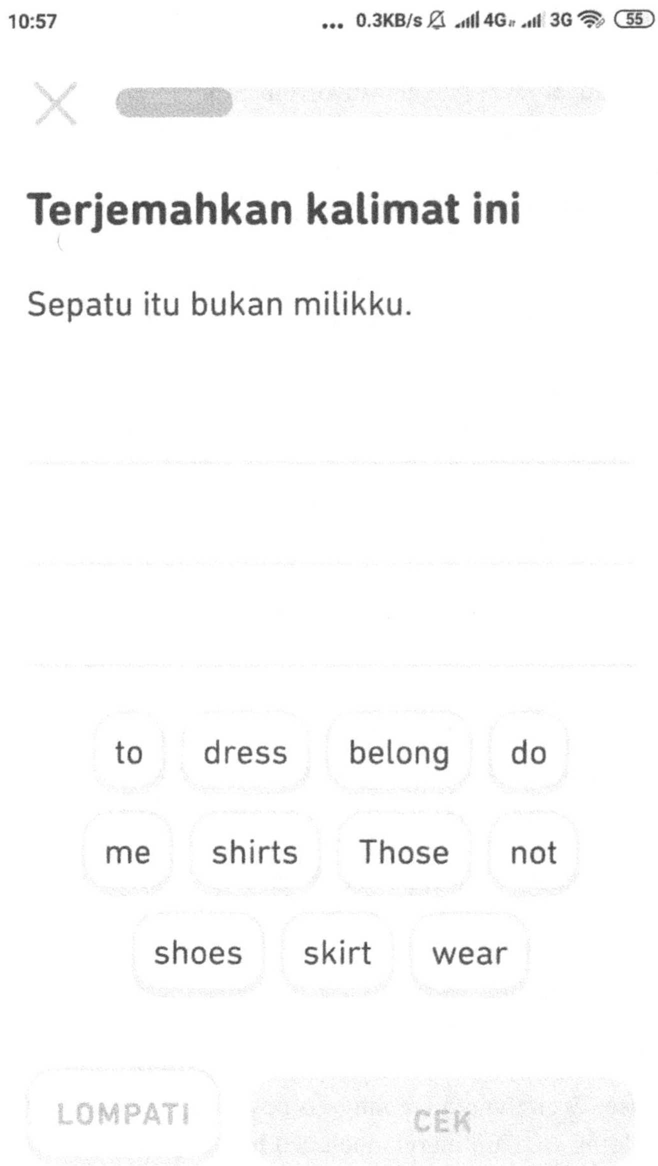

provide an initial assessment of our abilities, so in the following game, it can automatically measure our initial comprehension of language, and content later displayed will be in accordance with our initial understanding. This application also gives stages in mastering a language to make it easier for users to know and understand language sequentially; the aim is that knowledge acquired is not fragmented. Duolingo presents an interface that is very simple, as we can see in figure 2 below:

Figure 2 shows the activity after carrying out the initial understanding test. The image shows the initial abilities of the user in the Duolingo application. The application automatically detects the abilities of the user, and immediately assigns exercises accordingly. The Duolingo application is one of the favourite applications in language learning in Indonesia. It makes it easy for parents teaching children to recognise

Figure 2. Display after initial test

letters or words, to string together sentences and even to record children's linguistic activities, such as communicating using certain languages. Figure 2 illustrates the use of the Duolingo application, showing how a child uses it with the help of his parents. According to the results of interviews, the duration of using the application was established by agreement with parents; the child mentioned the use of the application with a limit of two hours per day. The authors also conducted interviews with the child's parents, ensuring that the child's answers corresponded with those of the parents. Discussions with parents also showed that the use of this application must be sanctioned by parents, with a maximum duration of use of two hours.

The results of a case study conducted by the authors on the use of applications for children showed that children find it easy to use this application, when they are accompanied by their parents. Secondly, the use of applications for learning, according to parents, should be maximised. The discussion with the parents showed that when they cooperate in their child's education, they can more easily control that child's growth. In this context, parents believe that using the Duolingo application can help them to teach correct vocabularies, especially when teaching foreign languages to their children.

SOLUTIONS AND RECOMMENDATIONS

This chapter discusses the use of the Duolingo application in practice in the context of linguistic intelligence through the gamification method. The large number of findings makes the authors aware of various ways to arrange solutions and recommendations. As already discussed, the use of technology has positive and negative sides. Our ability to rationalise this phenomenon should lead us to the best solution in the use of technology for our children. Readers are faced with the problems of combining the use of technology with the development and growth of their children. Important suggestions can offer a solution to this problem of providing applications for children. The authors suggest several problem-solving solutions in taking advantage of the gadget, referring to humanist and constructive approaches, as in the following:

1. Give good examples of how children use gadgets (duration of use, when to use, and with whom they can use),
2. Accompany your children when they are playing with gadgets,
3. Give a positive response when your children ask about something they are doing when playing with their gadget,
4. Limit the use of gadgets by turning on child-friendly modes (such as restrictions on YouTube content, application restrictions with freeze facilities),
5. Train your children to take charge of the gadgets they use, so that they know about charging when the battery runs out, cleaning their gadgets, and putting them in the proper place.

FUTURE RESEARCH DIRECTIONS

Background research studies have weaknesses, such as a limited number of respondents, and the population could be justified. However, this research could provide an in-depth picture of the phenomenon being studied. This chapter discusses thoroughly the basis for choosing a research approach, the results of research and recommendations on the use of the Duolingo application for children's linguistic intelligence. It is necessary to develop research on how to apply this application in formal classrooms; the effect is expected to be more generalisable to formal classes, making the research a contribution to science, and a reference to readers.

CONCLUSION

Duolingo is one of the favourite learning applications in Indonesia. Its massive use offers an alternative way of teaching children to master a language. This chapter starts from the background to the conclusion – that the authors coherently present data and facts about the use of mobile learning applications for children. Considerable debate has occurred about the use of gadgets for children. Those who disagree with it argue that too-early exposure to gadgets disrupts children's growth and development, whereas those who support mobile learning applications argue that gadgets help parents to teach their children. Gamification becomes one of the mediators of this debate. It offers a way of learning without reducing our interactions with others. Gamification methods can be used easily. Children learn while playing and parents can interact while they play. From the research into gamification performed by parents through the Duolingo application, it can become an adhesive tool for social-emotional relationships. Parents can be positively involved when children use their gadgets, and children feel comfortable because they use the Duolingo application with their parents' help.

ACKNOWLEDGMENT

This research received no specific grant from any funding agency in the public, commercial, or not-for-profit sectors.

REFERENCES

Ailwood, J. O. (2003). Governing early childhood. *Contemporary issues in early childhood*, *4*(3), 286–299. doi:10.2304/ciec.2003.4.3.5

Alif, L., & Lara, F. (2018). Influence of gadget usage and parent involvement to children's prosocial behavior. *Journal Imiah Pendidikan Dan Pembelajaran*, *2*(2), 174–179. doi:10.23887/jipp.v2i2.15366

Allen, K. E., & Marotz, L. R. (1989). *Developmental profiles: Birth to six*. Albany, NY: Delmar.

Armstrong, T. (2000). Multiple intelligences in the classroom. In *Intelligence* (2nd ed.). USA: Association for Supervision and Curriculum Development.

Bishop, J. (2014). Gamification for human factors integration: Social, education, and psychological issues. In *Gamification for human factors integration*. Social, Education, and Psychological Issues; doi:10.4018/978-1-4666-5071-8

BPS-Statistics. (2017). Telecommunication Statistics in Indonesia-2017 (Directorate of Communications and Information Technology). Jakarta: BPS-Statistics Indonesia.

Deterding, S., Khaled, R., Nacke, L., & Dixon, D. (2011). Gamification: toward a definition. *Chi 2011*, 12–15.

Fadhli, M. (2015). *Implementasi beyond center and circle time (BCCT) di TK Aisyiah Dagangan Madiun*. Ponorogo.

Fadhli, M. (2016). Pemikiran Howard Gardner dalam Pendidikan Anak Usia Dini. *Jurnal Indria, 1*(1), 80–87.

Findlay, J. (2016). Retrieved from https://trainingindustry.com/articles/learning-technologies/game-based-learning-vs-gamification-do-you-know-the-difference/

Frahasini, Marhaeni, T., Astuti, P., & Atmaja, H. T. (2018). *The impact of the use of gadgets in school of school age towards children' s social behavior in Semata Village, 7*(2), 161–168.

Gardner, H. (1988). *Frames of mind* (2nd ed.). doi:10.2307/3324261

Gardner, H. (2008). *Multiple intelligences: new horizons in theory and practice*. USA: Basic Group.

Gardner, H. (2011). *Frames of mind the theory of multiple intelligences* (2nd ed.). New York: Basic Books.

Guss, F. (2005). Reconceptualizing play: Aesthetic self-definitions. *Contemporary Issues in Early Childhood, 6*(3), 233–243. doi:10.2304/ciec.2005.6.3.4

Irmayanti, Y. (2018). *The rules of parents in assisting the use of gadget in preschool children.* (Thesis, Universitas Muhammadiyah Surakarta).

Kalogiannakis, M., & Papadakis, S. (2017). *Combining mobile technologies in environmental education : a Greek case study Michail Kalogiannakis * and Stamatios Papadakis, 11*(2), 108–130.

Kapp, K. M. (2014). *The gamification of learning and instruction fieldbook*. San Francisco, CA: Wiley.

Laili, U. (2017). Influence of gadget on social personal of early childhood. *Proceeding of Surabaya International Health Conference,* July 13-14, 2017. 273–277.

McCoog, I. (2007). Integrated instruction: Multiple intelligences and technology. *The Clearing House: A Journal of Educational Strategies, Issues and Ideas, 81*(November), 25–28. doi:10.3200/TCHS.81.1.25-28

Miarso, Y. (2007). Jurnal teknologi pendidikan. *Tekno-Pedagogi.*

Moon, K., & Reifel, S. (2008). Play and literacy learning in a diverse language pre-kindergarten classroom. *Contemporary Issues in Early Childhood, 9*(1), 49–65. doi:10.2304/ciec.2008.9.1.49

Neeli, B. K., Reiners, T., & Wood, L. C. (2015). Gamification in education and business. In Gamification in Education and Business. doi:10.1007/978-3-319-10208-5

Piaget, J. (n.d.). *The psychology of intelligence*. London, UK: Taylor & Francis e-Library.

Sardar. (1988). The touch of Midas - science, values and environment in Islam and the West. *Social Studies of Science.*

Sheridan, M., Howard, J., & Alderson, D. (2010). *Play in early childhood: from birth to six years*. Retrieved from https://books.google.com/books?hl=en&lr=&id=gvWsAgAAQBAJ&pgis=1

Suhana, M. (2018). *Influence of gadget usage on children's social-emotional development. 169*(Icece 2017), 224–227. doi:10.2991/icece-17.2018.58

Synodi, E. (2010). Play in the kindergarten: The case of Norway, Sweden, New Zealand, and Japan. *International Journal of Early Years Education, 18*(3), 185–200. doi:10.1080/09669760.2010.521299

UNICEF. (2018). *Learning through play: Strengthening learning through play in early childhood education programmes*. New York: UNICEF.

Vakili, L. (2013). *The relationship between linguistic intelligence and L2 learning strategies among EFL learners with intermediate level of proficiency, 1*, 89–93.

Zaranis, N., Kalogiannakis, M., & Papadakis, S. (2013). *Using mobile devices for teaching realistic mathematics in kindergarten education, 4*(7), 1–10. doi:10.4236/ce.2013.47A1001

ADDITIONAL READING

Fadhli, M. (2018). Gamification For Early Childhood Using 'Lingokids' application. *Jurnal Indria (Jurnal Ilmiah Pendidikan Prasekolah dan Sekolah Awal), 3*(1).

Kalogiannakis, M. (2017). *Mobile educational applications for children : what educators and parents need to know Stamatios Papadakis * and. 11*(3), 256–277.

Kalogiannakis, M., & Papadakis, S. (2017). Combining mobile technologies in environmental education: A Greek case study. *International Journal of Mobile Learning and Organisation, 11*(2), 108. doi:10.1504/IJMLO.2017.084272

Kalogiannakis, M., & Zaranis, N. (2016). *Comparing tablets and PCs in teaching mathematics : An attempt to improve mathematics competence in early childhood education. 4*(2), 241–253.

Papadakis, S., Teacher, S. E., Kalogiannakis, M., Orfanakis, V., Teacher, S. E., & Zaranis, N. (2017)... *The Appropriateness of Scratch and App Inventor as Educational Environments for Teaching Introductory Programming in Primary and Secondary Education., 12*(4), 58–77. doi:10.4018/IJWLTT.2017100106

Zaranis, N., Kalogiannakis, M., & Papadakis, S. (2013)... *Using Mobile Devices for Teaching Realistic Mathematics in Kindergarten Education., 4*(7), 1–10. doi:10.4236/ce.2013.47A1001

KEY TERMS AND DEFINITON

Duolingo: Duolingo is an application based on the Android platform which helps children to master other languages by practising speaking, reading, listening, and writing through play.

Gamification: Using game-based mechanics, aesthetics, and game-thinking to engage people, motivate action, promote learning, and solve problems.

Chapter 8
Mobile Application to Exercise Oral Language From a Systems Approach

Anabelem Soberanes-Martín
https://orcid.org/0000-0002-1101-8279
Universidad Autónoma del Estado de México, Mexico

Samuel Olmos Peña
https://orcid.org/0000-0001-7843-4771
Universidad Autónoma del Estado de México, Mexico

Magally Martínez Reyes
Universidad Autónoma del Estado de México, Mexico

ABSTRACT

The primary means of human communication is language; it requires support to develop the ability to speak and it is not the same for everyone. Intellectual Disability (ID) makes the child learn and develop more slowly, leading to the development of language software. The chapter develops an application for mobile devices called Gradual Interactive Language (GIL), created through five stages of Soft Systems Methodology (SSM) and the classic model of software development. The results of the pilot test indicated that it is an alternative support to move from pre-language to language, not limited to children with ID, the only thing that would depend is the moment in which it will use; in addition to showing improvement through gradual exercises that allow the child to advance, considers the roles involved in the process of learning language (teachers, educators or therapists, parents, children, and technology), finally, a technical virtue of GIL is to run on any device regardless of the operating system it uses (iOS or Android).

DOI: 10.4018/978-1-7998-1486-3.ch008

INTRODUCTION

The primary means of communication of human beings is language, with this reference is made to the set of sounds and signs that human beings have used since its creation until today to communicate with other individuals of the same species (RAE, 2019), to which it manifests both what it feels and what it thinks about any tangible or intangible social issue, natural or abstract; hence, as a species, human language is the most developed.

According to the theory of language from the perspective of linguistics, language composed of language (language, a general and constant model of a linguistic collectivity that integrates a community) and speech (momentary materialization of that resource, individual and voluntary action carried out through phonation and writing) (Saussere, 2016).

In technical terms three dimensions are recognized within language: form (phonology, morphology, and syntax), content (semantics) and use (pragmatics) and therefore different types of languages developed (cult, technical, colloquial, so forth). Thus, language is a gradual process that first goes through a pre-language stage and then develops its capacities extensively (Saussure, 2016). The pre-language is a rudimentary system of communication that becomes evident during the first years of life, and that implies capacities of neurophysiological and psychological characteristics, such as perception, motricity, imitation, and memory (Koerner, 1982). Therefore, this is an essential stage for children to be able to interact with their environment and begin to establish relationships with the individuals around them.

In this sense, when children have language problems it is crucial that they have support to develop their ability to speak through language, mainly because it is not the same for all children, for example when Intellectual Disability (ID) causes the child to learn and develop more slowly than children of the same age, which manifested in the fact that they may need more time to learn to speak, walk, dress or eat without the help of their parents, and of course they have learning problems in school.

The Mexican Academy of Pediatrics (2019), establishes some standards to determine reasonable progress or if you may need additional help. For example, children with speech disorders may have trouble producing sounds correctly; they may hesitate or stutter when speaking. While children with language disorders may have trouble understanding what others are saying or may have difficulty communicating their thoughts.

This need for help has led in recent years to a broadening of the scope from computing and software development to education at different stages, primarily in early education, and particularly concerning educational language support software. The traditional options that incorporate technology range from websites with webinars, online courses for parents and educators, mobile download applications (app), virtual didactic games, and others (API, 2019).

A recurrent modality presented as entertaining and eye-catching didactic games, which try to help children learn to read and write, develop cognitive and communication skills, acquire vocabulary, improve memory and concentration and awaken the imagination, among others (Saalem, 2019).

Although according to the Horizont Report (2017) there are successful developments and others with essential limitations in proposals that incorporate mobile devices, it is clear that it shares the same limitations of any technological application if care not taken in its design and a model not contemplated that allows to evaluate that each phase of the development of the application fulfills the requirements of the final user, in addition, a defined instructional design is necessary to reach a concrete educational objective, to support the development of language processes in children, with or without language problems.

This proposal is based on the need detected in children with language problems and proposes to develop an application that supports language transition from pre-language to oral language in children, through its use in mobile devices and a systems approach, proposes to apply a methodology that allows to reach a smooth system that through gradual exercises allows the child to move forward, defining the roles involved in the process of language learning (teachers, educators or therapists, parents, children, and technology) to measure the level of intervention of each and define their level of participation.

BACKGROUND

It begins with a review of the literature that presents language learning, learning with technology, and the systems approach.

Language Learning

Speaking, drawing, thinking and writing are products of what is called language, according to Vygotsky, in such a way that language is a means to solve problems and to relate to others. Those who access language have the opportunity to communicate and to introspect, and language becomes a social tool that makes the individual more efficient at solving problems. This approach linked to the socio-cultural gaze, and far from individualizing children's learning actions, it maintains that with the help of an expert, who is not exclusively the teacher, children will build increasingly elaborate learning and increase the complexity of their knowledge (Uresti & Cannon, 2014).

To speak requires specific biological capacities as expressed by Savater (2005), the human-animal differs from the rest of the animals by a specific feature: the conventional, intelligent and symbolic language; this condition is a necessary element, but not sufficient; the language does not arise spontaneously from this capacity, it requires to be taught, validated and interact with others so that it becomes a reality (Searle, 1994; Urgilés, 2016).

Mora (2009) mentions that children at an early age by listening and interacting with adults and other children become proficient in language and its components (sounds or phonemes, vocabulary, and grammar); by age two, the child is beginning to learn to speak and by age six has almost completed the sound system of the mother tongue (phonology), its grammar, and many of the meanings coded by language.

The fact of articulating or joining sounds and forming words, and articulating words to form concepts, is going to be evidenced in cognitive capacities, there is a directly proportional relationship between the formation of words and the formation of thought (Urgilés, 2016; Koerner, 1982). It generally recognized that early childhood is the critical period in which the foundations of achievement are laid in school and beyond. Early childhood care and education must, therefore, be central to both Education for All (EFA) and broader development programs (UNESCO, 2012).

Defining intelligence as a capability that Gardner makes a skill that can develop; intelligence implies the ability to solve problems or produce products or services for a community or cultural context; that skill must develop. It defines eight interrelated bits of intelligence: musical, inter and intrapersonal, naturalistic, body kinetic, spatial, visual, kinesthetic, mathematical logic and linguistics. The latter considers that if a student has more significant development of his linguistic ability it means that he thinks and expresses himself more frequently through the word, that he enjoys communicating orally or in writing using linguistic structures efficiently, it is where the individual forms a mental model of the world of letters and words (reading and writing), where he acquires the ability to write what he understands by reading and playing with words appropriately. In the case of children, Gardner recommends developing activities to augment this linguistic ability where they hear and see words, hear and tell stories, do crossword puzzles and frequently use a new word during the day (Ávila, 1999).

Gardner considers Piaget's cognitive theory in the development of linguistic intelligence, since Piaget assumes that children are not born with knowledge but that it is proper to their development to construct it from actions on space or hypotheses that attempt to give meaning to the world, in this sense both share the definition of intelligence as the capacity to solve problems. For Piaget, man socializes through communicative interaction and communication demands the active presence of the other, but if this communication is to be constructive, it must create conflict, defined as the contraposition of different points of view or different centralizations, so that language acquired with socialization. Reading and writing favors the formulation of general sentences and reflect linguistic intelligence and cognitive changes at the beginning of schooling, although it often ignores the particular richness of previous experiences (Ávila, 1999).

It is easy to learn to speak, but the same cannot be said for everyone, so it is necessary to incorporate a great diversity of tools to support oral language in the first place. The tools can be from the most traditional printed figures that associates sound to the most sophisticated technologies such as Augmented Reality (AR) or Virtual Reality (VR), but without forgetting the objective.

Learning With Technology

Recent research has addressed the use of technology in early childhood, Kalogiannakis et al. (2018a) introduced the concept of gravity and planets to 17 children aged five to seven using intelligent mobile devices (tablets), the ScratchJr programming environment, demonstrated how properly designed activities allow children to develop an interdisciplinary approach to science and technology and apply the skills to their daily lives, emphasizing that ICT serve better as a compliment than as a substitute for traditional teaching and learning tools.

For children in preschool, play, and learning go hand in hand; they gain meaningful experiences that allow them to learn. The same thing happens with digital and interactive games, in addition to the fact that children today are born with abilities that allow them to manipulate computers and the Internet (Santos & Osório, 2008). By making comparisons between the diagnosis found in the literature and that observed in children, it found that the child with technology becomes a user full of skills and able to know and adapt perfectly to any technological application (Rodríguez, 2006).

Papadakis et al. (2016) analyze the influence of the use of computers and tablets on the development of mathematical competence in early childhood education, the sample was 256 and among the results that teaching with tablets compared to teaching with computers contributes significantly to the development of children's mathematical ability.

Other features of small, lightweight, reliable and robust mobile devices with Internet connectivity, as well as an impressive number of easy-to-use software applications (apps), made exclusively for their use (Papadakis & Kalogiannakis, 2017a).

Papadakis et al. (2019) point out that the positive attitude of parents towards mobile learning limited by the lack of knowledge about the selection of mobile applications with real educational value; by the widespread boom in educational applications, parents and educators make their use more and more frequent; however, there are few well-designed applications to really be considered as such (Papadakis & Kalogiannakis, 2017b).

However, other authors assume that the implementation of technology in the classroom from an early age would only form hyperactive and mentally lethargic children. On the contrary, like any other educational tool, technology should be used as one of the classroom learning experiences, capable of promoting practical and life-saving exercises (Gutiérrez & Cannon, 2014).

Technology is an element that has incorporated into the world of children since early childhood. There are aspects that young children can learn through this medium in all its fields: cognitive, social, emotional, and linguistic; software currently designed for children up to nine months. This type of software is known as lap programs, children are placed on their parents' lap when they use it, and it designed for use by parents and children (Morrison, 2005). Pérez & Salmaron et al. (2006) indicate that educational software for children under the age of five is the one that spreads most rapidly.

Children rarely play to learn specific content; they play because they want to belong to a particular group and live meaningful experiences for them. ITC from infancy is inevitable, but this use can guide if adults assume this responsibility as part of education. Children, from birth to eight years of age, learn quickly using all their senses to access all the sensations and experiences around them. From this point of view, we believe that multimedia technologies can enrich experiences and promote new learning at the level of social and emotional development, linguistic, mathematical, physical-motor and universal culture (Santos & Osorio, 2008).

Mobile learning is a branch of ITC in education; however, since it employs technology that is more affordable and easier to acquire and use on its own than fixed computers, it requires a new conceptualization of traditional models of use and application of technologies. Mobile devices facilitate learning because they remove the boundaries between formal and non-formal education. With a mobile device, students can easily access complementary material to clarify doubts and develop skills (UNESCO, 2013).

Kokkalia & Drigas (2016) covering the role of mobile learning in special preschool education for a period from 2005 to 2015 show that initial skills (early literacy, early mathematics, cognitive) are a crucial factor in the development of mobile learning, are supported by mobile learning as visual material increases children's interest during the learning process, coupled with the fact that cell phones and tablets are conveniently accessible at home and easy to use by children, so performance improvements have been documented with these devices (Farhana *et al.*, 2013; Henderson & Yeow, 2012). Papadakis et al. (2018) emphasize that motor skills should consider as the coordination of children of this age, is still developing and that children generally cannot handle long periods of demanding work.

Also, the use of mobile technologies in children with ID, according to various studies, allows these children to tend to improve their behavior, communication, and relationships with their environment to improve their quality of life, learning, and daily life. (Liebermana *et al.*, 2009). According to Kärnä, *et al.*, (2007) the growing number of children in need of special education also requires an increasing amount of resources from teachers so mobile learning can be part of the solution; however, for the technological solution to work, technologies need to be designed and implemented in new ways. Concrete examples such as Picca for the development of learning skills for children with special educational needs (Fernandez & Lopez, 2013), and Cogmed (Shipstead *et al.*, 2012) that improve attention, concentration, control, social skills, and complex reasoning and memory capacity in children with Attention Disorder (AD).

In this sense, this proposal for a mobile application seeks methodological support for the development of the technological part and the educational part that enhances the benefits of mobile learning in children with AD.

System Approach

In order to approach the development of the application integrally, due to its complexity, the systems approach used. This interdisciplinary approach is based on the General Systems Theory (GST) which establishes that any real-life problem has the behavior of a system from the most general concept, that is, that it is a set of parts (subsystems) interrelated with a common objective, that contains inputs and outputs in the form of matter, energy or information, and is influenced by an environment, among others (Bajgoric & Moon, 2017). Systems thinking was first introduced and formalized in the 1950s. GST developed as a conceptual framework and mathematically expressed theory, most notably by Bertalanffy (Caws, 2015). In recent years important original academic work has been done in multiple disciplines to significantly increase the value and use of systemic thinking (Warren *et al.*, 2019).

Within the GST, there are different classifications of systems and methodologies to address the different types of problems; one of these classifications is "hard" systems and "soft" systems (Siriram, 2012). The first are those that are well defined, i.e., that do not change their values, e.g., a system of linear equations to optimize a given activity. Soft systems are those where the problem can vary depending on who observes it, that is, they are systems that depend on the perspective of the person involved, the simplest example of this can be politics, where a problem can be attacked depending on the political party to which it belongs (De Bruijn & Herder, 2009).

The problem addressed in this chapter is a flexible system that involves different visions of how it can be solved. Therefore, it was determined to use the Soft Systems Methodology (SSM) proposed by Checkland & Scholes (1994) to address this subjectivity. SSM provide a procedure for addressing complex real-world problems, starts from the premise that organizations are open systems that interact with their environment. Also, it recognizes that problem situations are not orderly or are a smooth system and involve people who have different views on the origins and nature of the problem, the ways to address it and the objectives to achieve. The methodology includes some activities: finding out the situation of a problem; formulating relevant activity models with a purpose; using the models to discuss the situation; identifying desirable and feasible changes that would improve the situation (Holland & Garfield, 2016)

SSM involves seven phases: (i) finding out about the unstructured problem situation; (ii) structuring the problem situation; (iii) constructing the basic definitions of the relevant human activity systems; (iv) developing conceptual models from the basic definitions; (v) comparing models with the problem; (vi) defining desirable and feasible changes; and (vii) carrying out actions in real life (problem situation). The result of applying the methodology is a comprehensive understanding process that researchers can use to generate systemic improvement recommendations for the problem under study. Together, it provides stakeholders with a detailed picture of the problem situation identified and the systems involved in which more information is needed. The result of the research could be a clearer picture of related systems, organized by stakeholders, organizational structures, activities, physical processes, and physical objects working in harmony with others, depending on how they are defined at the beginning of an investigation to address (Warren *et al.,* 2019). Therefore, the general objective of this research was to improve the oral language of children with different types of exercises through an application called Gradual Interactive Language (GIL), without being limited to small children with ID, the only thing that would depend is when it would use. The first five stages of SSM applied for the development of the application, in addition to the classic software development model.

MAIN FOCUS OF THE CHAPTER

Issues, Controversies, Problems

The use of IT in children with ID is an alternative to the complex problem of learning and development, as shown by studies by Drigas & Kokkalia (2014) on its incursion at the preschool level in a period from 2003 to 2013, likewise UNESCO (Kalas, 2010) points out that for this educational level language and cognitive and social skills are the basis for lifelong learning and suggests including IT as part of formal education to achieve this goal, especially to overcome learning barriers. Different studies (Aubrey, & Dahl, 2; DfES, 2007; Segers & Verhoeven, 2003; Korat, 2010) support the incorporation of IT to foster literature, language, early mathematics, cognitive, emotional, social and motor skills and to improve the creativity of preschool children. Thus, the use of IT in children with ID is oriented to favor the development of thinking structures in students with motor disabilities, to provide the support that allows a higher degree of autonomy in learning, which implies the need for careful design (Hinostroza *et al.*, 2013).

In order to address this problem, due to its complexity, the systems approach used. This interdisciplinary approach, based on the GST, establishes that any real-life problem behaves as a system, set of interrelated parts (subsystems) with a common goal, contains inputs and outputs in the form of information, and is influenced by an environment. In the case of language, the parties to be considered are language (the language of a community) and speech (individual and voluntary action through phonation and writing). So, these general components must be constituted in subsystems with information common to a context and interrelated through forms and sounds.

GST establishes as a premise totality of the system is more than the sum of its parts, and at the same time, the totality determines its behavior and structure of the parts. In the case of language, the set of sounds and signs that are used to communicate what it feels and what it thinks about any tangible or intangible issue is part of this structure. Form, content, and use are elements that any language support system must contemplate in order to develop it, but each part separately does not give meaning to what the child expresses without a context.

Learning Language and Technology

Children learn in different ways and approaches. For that reason, teachers need to employ various instructional techniques to match students with different learning styles (Zare & Othman, 2013). Ellis (2005) proposes ten principles in language learning. These general principles derive from the researcher's understanding based on theoretical perspectives, although she predominantly relies on the work of Lantolf (1996).

The principles proposed by Ellis (2005) is the use of technology as a means of support in language learning. The current advancement of technology, its accessibility, and ease of use of smartphones and tablets is a tool that presents itself as a feasible and exciting resource for language learning.

Several kinds of researches show that new touchscreen technologies make them attractive to younger and less skilled users. They provide immediate access to a world of learning, entertainment, and creativity for younger children, including those with special needs. Mobile applications offer an opportunity to learn, understand and use these tools by applying appropriate evidence-based methods that focus on the interaction that occurs when communication created (De Curtis & Ferrer, 2011; Khaddage *et al.*, 2011; Pierce & Cleary, 2016).

Implementation of applications as a teaching tool with young children 1-5 years of age has proven effective in developing speech, language, and social skills with a wide range of our clients' communication disorders, such as autism spectrum disorders, Down syndrome; specific language impairment, auditory processing disorders, and apraxia. Mobile technology, especially tablets, is best used as a way to improve early therapeutic intervention methods. (De Curtis & Ferrer, 2011).

Despite the theoretical/pedagogical advance in language learning and the advantages of using technology in this practice, there is no evidence in the literature on the development of applications for language learning from the systems approach. That is to say; they support the development of applications from a software engineering or pedagogical point of view without analyzing language learning from a multifactorial approach that takes into account all those environmental factors that can lead to the success of language learning.

RESEARCH

Method

A mixed methodology was employed that included qualitative data interviews and questionnaires to identify the requirements to include in the GIL application. The study explored the needs and opinions of parents and teachers about information technologies in the learning process, as well as exploring the possibility of implementation.

Participants described in the following section of SSM Phase 1. Participants recruited through letters sent describing aspects of the nature of the study, and all parents and teachers signed informed consent letters authorizing their participation in the research. All activities took place on the premises of the educational institutions.

Interviews digitally recorded for validation and data exchange between researchers. The research team met and discussed data analysis several times to ensure the internal validity of the process and agreement on interpretation.

For the GIL application development process, five phases of the SSM applied (detailed in the Results section). The measurement instruments that quantify the technological, pedagogical, content, and aesthetic elements described in phase 5 testing. The method of analysis of qualitative data was based on the proposal of Bogdan & Bicklen (2007) and consisted of coding and categorizing the responses and then computing them to allow quantitative comparisons.

SOLUTIONS AND RECOMMENDATIONS

The problem addressed in this chapter is a soft system that involves different visions of how a language learning problem can be solved, therefore, it was determined to use the SSM to address this subjectivity (figure 1), if not precisely in a scientific way, if at least in a way characterized by intellectual rigor. An essential software development model used together. Therefore, the general objective was to improve the oral language of children with different types and degrees of exercises through an application called GIL, without being limited to a specific language problem, since it only depends on the moment in which the

Figure 1. SSM

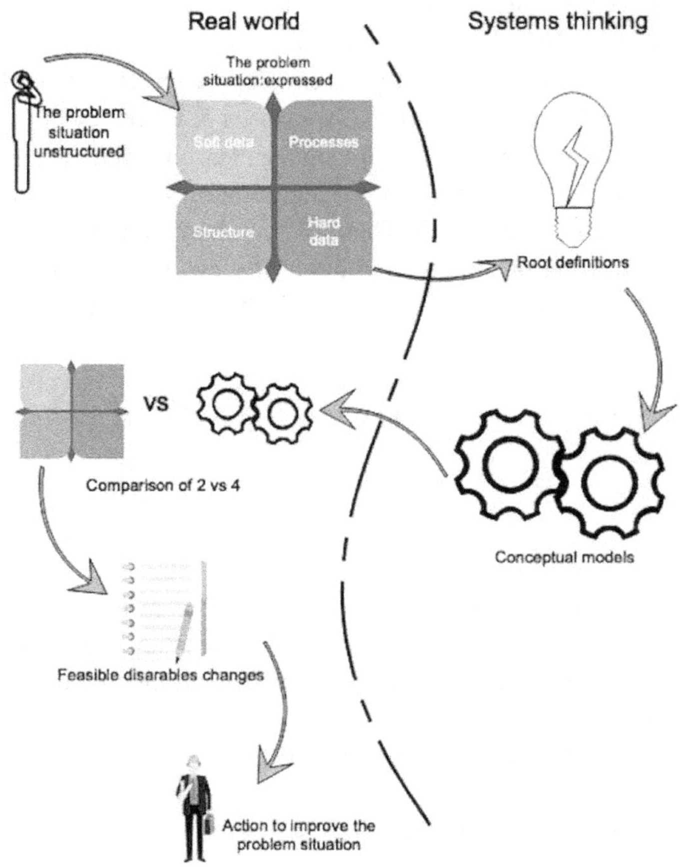

mobile application used. The first five stages of the SSM were carried out for the development of the application, together with the software life cycle. The preceding, because stages six and seven involve developing an implementation plan and executing the solution, which is outside of this chapter.

Use of SSM for GIL Development

The first five stages of the SSM were carried out for application development, and the software lifecycle. The steps six and seven involve an implementation plan and execution of the solution, which is outside of this chapter.

Stage 1. Unstructured situation.

Language problems are common in children and can be considered a public health problem in this sector of the population. In general, a sketch representing the problem was constructed based on the initial observation, with emphasis on the systems involved, such as teachers, children, parents or guardians, language therapists, infrastructure, technology, among others. Answers gave to questions such as: Who intervenes? How do they intervene? What processes should follow? How does information flow through different processes? How does technology influence children's learning? (real thinking). The construction of this stage, documentary research was carried out to describe in a general way what is related to children's communication, as well as the teaching methodologies that can be applied to stimulate learning. Also, seven institutions in the eastern part of the State of Mexico visited, and issues inherent to the problem identified. For example, as the parent or guardian who leaves the child in the institution (children's room) intervenes, the teacher, who has planned the session with the children, starts with games and makes use of cards (in very few cases with a projector) with images to associate them with the corresponding sound, in some cases there is a language therapist who supports the teacher, parents or guardians are asked to practice with the children to strengthen the activity, is when some parents make use of technological resources.

Stage 2. Structured situation.

The rich picture in Figure 2 puts language learning into context by showing an overview of the educational environment. It highlights the structure, process elements, soft and hard data, as well as the climate situation. This diagrammatic method helped to identify several problems that affected the success or failure of the issues addressed, such as low level of spoken language, poor infrastructure, inadequate training, and others. With this diagram, a bright, structured, and quick view of the problem obtained. The educational environment (figure 2) is typical in schools in Mexico. Education policy drove by government and industry. Schools react to the various pressures exerted on them by issues such as changes in legislation, funding, or the demand for applicants from each school. With the fieldwork carried out and with the support of experts in language therapy, it identified that exercise was required, and it was determined that an application at three levels should be carried out that would initially allow functions involved in the production of sounds for speech, such as humming and babbling, that is, exercise through sounds associated with illustrations that produce the sound (Soberanes *et al.*, 2018), with the categories: Animals, Onomatopoeia of the body, objects, transports, and vowels; in a second level the association with words and images (Soberanes et. al., 2017), the categories: animals, colors, food, body, family, fruits, numbers, transport, vowels, and others. Also, on the third level two- or three-words phrases associated with figures.

Stage 3. Root definitions.

In the case of this chapter, the first constructed root definition is proposed: teaching language skills to children using a mobile application that improves the comprehension and skills of the students and allows them to obtain better grades (figure 3).

CATWOE is applied:

C-> Students.
A-> Teachers.
T-> Achieve improvement in the success rate of language topics by providing students with additional support and feedback. To help students develop the language skills they need for proper language management in their daily activities.
W-> Through technology and with attractive applications for children, social inclusion, and satisfactory performance in the use of language in their academic trajectory can be guaranteed.

Figure 2. Rich picture

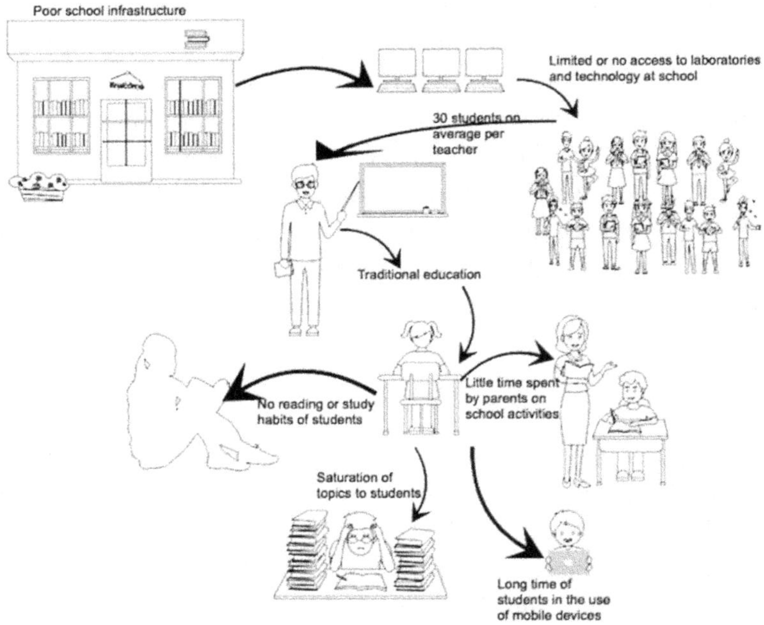

O-> School management.
E-> Educational environment; Faculty requirements, School infrastructure, access to technology.

The mnemonic CATWOE (Customers, Actors, Transformation, Weltanschauung, Owners, and Environment) the essence of the system was defined, and the relevant systems constructed, that is, the subsystems or components that give meaning to the system were determined. Once all the elements that shape the problem, their relationships, the type of input, and the output required is bright, the construction of the application begins. The result is a mobile application with three gradual levels to use with the support of parents or teachers on any platform, at any time and at the pace required to strengthen the initial oral language of children. Based on this root definition, the subsystems that make up the problem constructed. Once all the elements that give shape to the problem, its relationships, the type of entry and exit required are clear, the construction of the application begins. Briefly, the development of the computer application developed as follows way:

Figure 3. Root

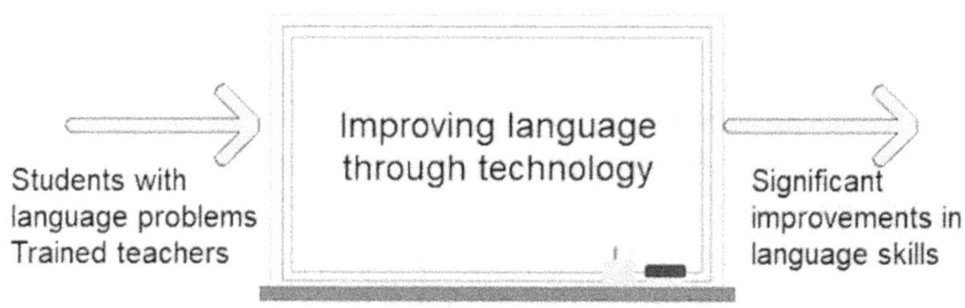

Stage 4. Application model for mobile devices for the ideal situation

This stage involves the design of the application model. Once the problem was understood in a structured way, and a first root definition obtained, the research methodology asks to develop a solution model or ideal system, which understudy. The length of the chapter and complexity of the problem studied, limited not to present all the perspectives involved in solving an interdisciplinary problem, only the perspective of the use of technology in language learning problems presented. Three aspects included:

a. ***Analysis of requirements*** are considered:
   i) The particular topic was identified, which based on the concepts that language therapists prioritize as fundamental and relied on content experts to determine the level of depth so that children could understand it. It was decided to divide the application into three stages, each one including seven to ten categories: 1) Sounds: animals, means of transport, objects, onomatopoeia of the body and vowels; 2) Name of the elements: animals, colors, food, family, fruits, means of transport, numbers, objects, and body; 3) Prayers (from three to five words): animals, family, food, body, and transport.
   ii) A document developed specifying the functional requirements of the supporting software. It should also include images that represent sound, which should be reproduced and should allow the children's progress to be verified, as well as the possibility of personalizing and increasing the elements of the application. Similarly, the application can run on any device regardless of the operating system its use (iOS or Android).

b. ***Design of the application*** need three activities:
   i) The categories to be considered in the application were determined, specifying the content of each one. Soberanes *et al.* (2018) show the description of the five categories of level 1; each category has initially from five to ten elements of the sounds that represent each object. Soberanes *et al.* (2017) present the elements of the second level is composed of ten categories, and each with the same number of elements originally. Finally, the third level of the application; it has ten categories with an equal number of sentences.
   ii) After determining the categories and elements of each stage, we proceeded to the selection of the images that would place in each case, which should be of free access or in their case to elaborate them. The images searched according to the words or sentences; 95% of the images made for the GIL application.
   iii) Different programming languages were analyzed to decide the most appropriate to the needs detected, and it was chosen to develop it in Xamarin language to generate a multiplatform application.

c. ***Construction***, the functional prototype of the application software generated.

GIL integrated into a package that requires the user to run for installation and generates an icon through which it entered then has a welcome screen and then presents a menu with the three levels of GIL (figure 4). When a level is selected, a screen of the application opened where each one of the five categories is shown using icons, which in turn have each one between 5 and 10 elements, for example, GIL level 2 (figure 5).

Once the category is selected, a screen like the one in figure 6 shown, for its description the elements were placed with letters, the A refers to the name of the category; B is a button of the application

Figure 4. Three level display

Figure 5. Menu categories Gil-2

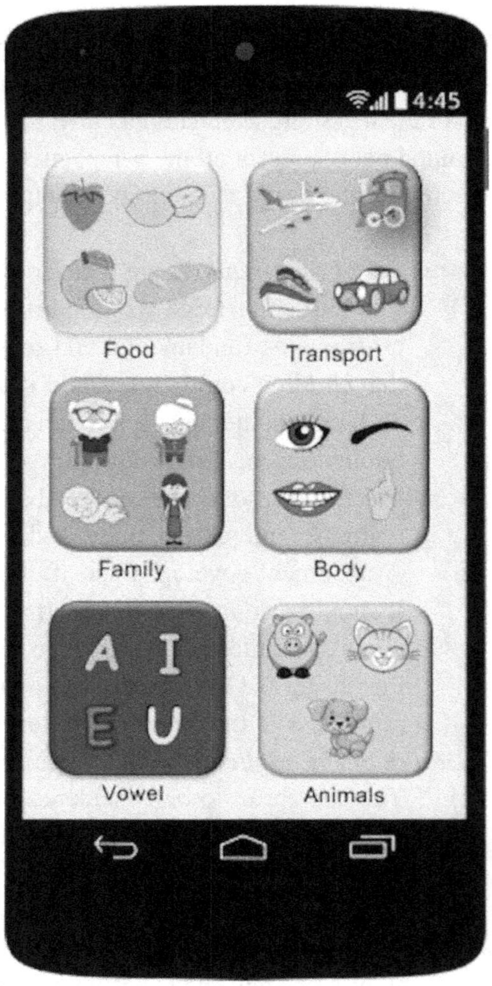

that allows you to return to the menu of the three levels; C indicates the level of the application (in this example GIL level 1), D and E indicate the image and the icon of listening, which when pressed allows you to hear the sound or word; F takes you to the next element of the category; G refers to the icon with which the microphone activated so that the child can repeat the word and if it one correctly it shows a window with a star and the sound of applause; otherwise it asks the child to repeat it.

In figure 7, three examples of GIL-1 presented, which are on an orange screen that characterizes this level.

In figure 8, three examples of GIL-2 presented, which are on a pink screen that characterizes this level, have the number two on the top right indicating the stage and change the icons of listening, speaking and following, in addition to having the name of the element below each image.

In figure 9, we present three examples of GIL-3, which are on a blue screen that characterizes this level, we have the number three in the upper right indicating the stage and change the icons of listening, speaking and following, also having a brief sentence on the element below each image.

Figure 6. Exercise GIL-1

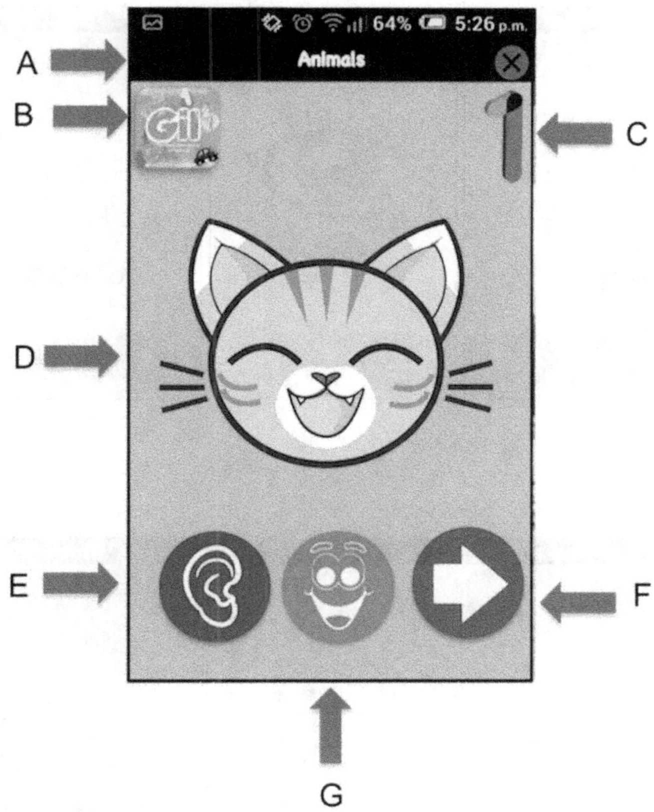

Figure 7. GIL-1

Figure 8. GIL-2

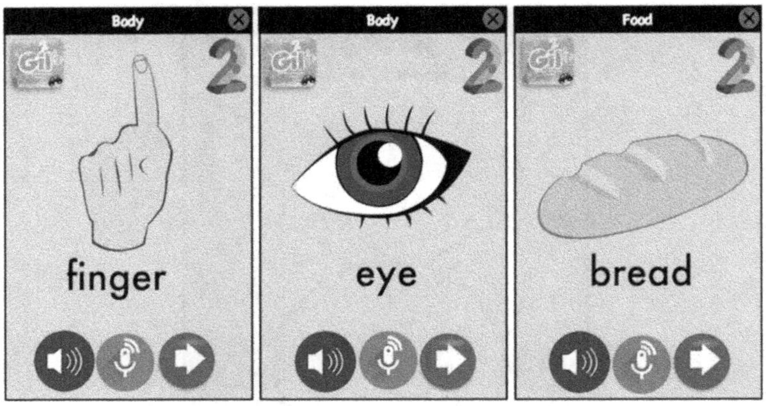

Figure 9. GIL-3

In figure 10, three examples of the GIL presented, in which a comparison is presented alluding to the same cat element, but in each of the three stages.

It allows the user to individualize the application by adding other elements, it requires the support of the teacher or tutor, since it is necessary to place the name of the object, then the categories are displayed, and it is necessary to select which one to associate, finally the image with which the sound is related is chosen.

Stage 5. Comparison of the structured problem situation against the solution model

The comparison consisted of carrying out a series of tests of the application for mobile devices. Technological, pedagogical, content, and aesthetic elements evaluated. Tests applied in two moments: (i) The review and pilot tests were carried out, identifying the elements that require adaptation, and (ii) the evaluation of the application, where measurement instruments are used that quantify the technological, pedagogical, content, and aesthetic elements.

Figure 10. Screens of the element cat in the three levels

Evaluation of GIL

GIL evaluated made by level, with parents of 48 children (20 from level 1, 15 from level 2 and 13 from GIL level 3) in a child stay, during three sessions and with two categories for each day (the survey was answered on the last day by the parent or tutor), the technological and aesthetic aspects considered were those indicated by Soberanes *et al.*(2017), an integrated rubric used for the aesthetic aspect of four questions and the technological aspect of five, and a type of survey to know the perception or opinion on the use of the application. The hypothesis test performed applying the t-student test of two queues for related tests to find out if there is a difference between the characteristics of the instrument with respect to each of the levels, a confidence level of 0.95 obtained, it determined that there is no significant difference for each of the levels, which implies that users agreed independently that each level is adequate with the characteristics of the software for language development.

In terms of aesthetics, 95% agreed on the colors used, 93% considered the design and distribution of the application to be acceptable, 100% of the images seemed adequate and 98% considered the font used in the application to be optimal (figure 11).

The technological aspect, 100% of the users stated that the installation of GIL was fast and straightforward, also 100% indicated that it is easy to use the application, in relation to the response time of the application 95% considers it fast, the same percentage was obtained with respect to using it in different mobile devices regardless of the platform (figure 12). The result t-student test a comparison of the sections of the instrument; the block was people and treatments for each aspect. The hypothesis tests were calculated to identify differences between two or more related fields, Cochran's Q test applied for related samples showing asymptotic meanings, with a significance level of 0.05, applied to the aspects (i) Aesthetic and (ii) Technological, then the results are presented:

(i) There is a direct relationship between the questions in the Aesthetics section with respect to color (Q=0.572), typeface (Q=0.174), images (Q=0.174), there is variation between the questions in the Aesthetics section with respect to design and distribution, it is recommended not to change the structure of the questions, but to change the order.

Figure 11. Graphics about esthetic

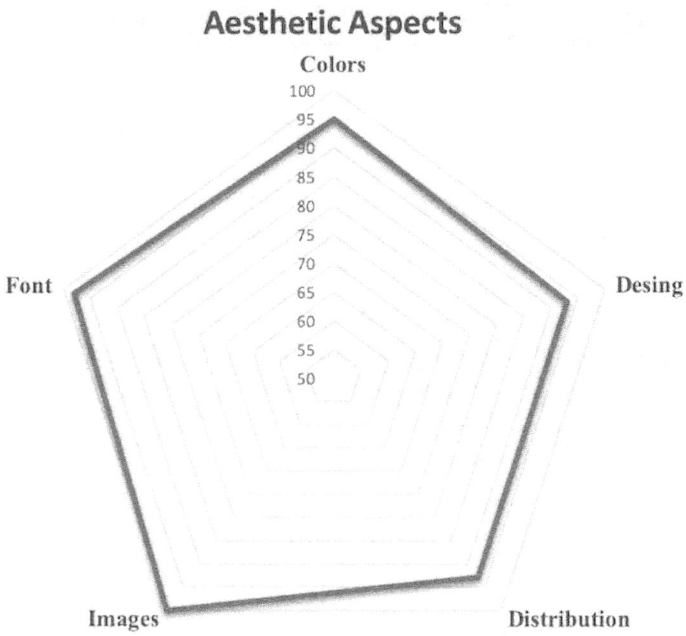

(ii) There is a direct relationship between each of the technology section questions concerning installation (Q=0.934), use (Q=0.368) and use in different devices (Q=0.457).

It can, therefore, concluded that each of the distributions indicates that if they are well designed, there is no difference between them and the objective of measuring what proposed in the instrument fulfilled.

Figure 12. Graphics about technology

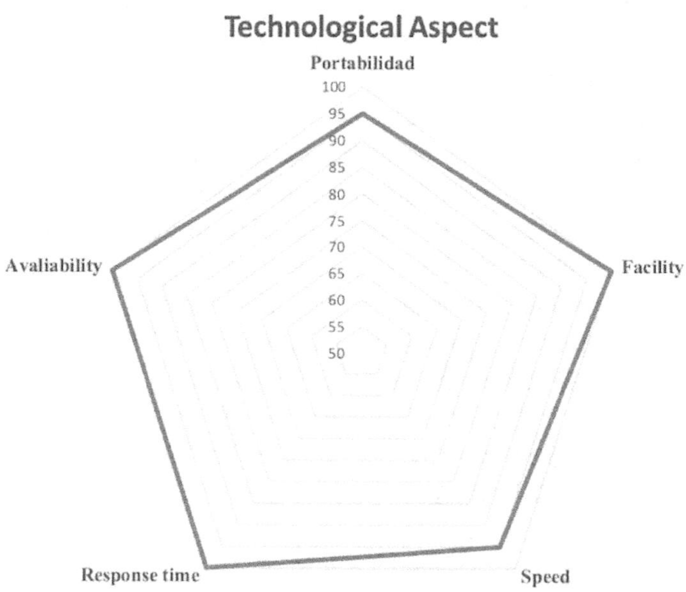

The evaluation of pedagogical and content issues was carried out by three experts. Regarding the pedagogical, the percentages obtained were: 100% consider that the application stimulates the language in the children, 100% indicated that the way of presenting the information was adequate, 87.2% consider that the way of identifying the child's progress in GIL should be improved, and 94% think that the application motivates the child to use it and the same percentage considers the interaction of the child with the application to be adequate. The results established by the validation of the application by experts have a rating of 9.33, SD=0.69, establishes uniformity among experts, obtaining a lower level of acceptance in the follow-up (figure 13).

FUTURE RESEARCH DIRECTIONS

The pertinence of incorporating mobile applications at an early age turns out to be a new paradigm in teaching, which when applied through the application of a methodology for its development gives it scientific rigor, which allows the results to be reproducible in similar conditions or, failing that, adapt the model to the particular conditions of the infant.

Currently, GIL contemplates a basic set of sounds and signs, which complies with the pre-language phase, so that children can communicate what they feel and what they think about any subject, it is necessary to incorporate more activities focused on this purpose and move towards the writing process.

The implementation of mobile applications can consider an emerging technique in the classroom. The experience of incorporating lap applications into a child's development will allow them to contribute to the diverse competencies required by children at an early age as future work is considered the development of other mobile applications having a control group to identify the impact of the application on the development of children.

Figure 13. Graphics by experts

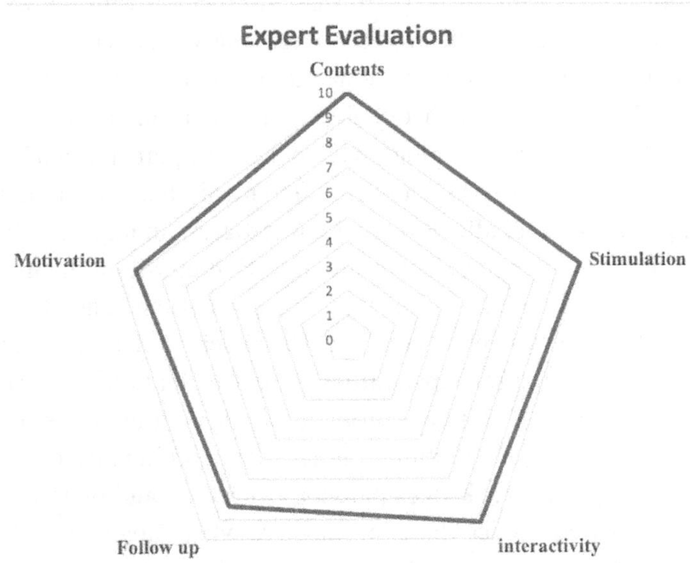

GIL is not limited to a specific linguistic problem, as it only depends on when it used. Therefore, as a future project, its functioning should be explored by sectors such as Attention Deficit Disorder, Hyperactivity Disorder, or Autism Spectrum Disorder, to confirm its effectiveness.

CONCLUSION

The results are encouraging as they coincide with what was stated by Papadakis et al. (2016), Farhana et al. (2013), and Henderson & Yeow (2012) that mobile learning through visual material increases children's interest during the learning process and that mobile devices are accessible and manageable for them. This finding coincides with the results of Flewitt et al. (2014) children using technologies at an early age to achieve motivation in their learning, thinking skills and ability to work independently at an early age; the latter aspects not identified with the instruments applied. young children explore and learn with mobile devices in ways that are natural to them (touch, repetition, trial, and error) (Papadakis & Kalogiannakis, 2017b).

What was mentioned by Ávila (1999) confirmed in order to develop activities that strengthen the linguistic capacity where words are heard and seen; however, other activities that allow them to listen to stories are lacking (Kalogiannakis et al., 2018b). Also identified is that expressed by Masek et al. (2012) educators could use mobile devices to allow students to work at different rhythms and styles; complemented with Kalogiannakis et al. (2018a) can be used without the need for separate devices (keyboard, mouse); emphasizing the responsibility of parents and educators to mitigate negative or aggressive incidents, such as inability to maintain attention, egocentric behaviors and inactive physical behaviors.

The SSM used to develop GIL allowed to address the subjectivity of the language problem in young children, identifying roles and actors (teacher, tutor, children, language therapist, and school infrastructure), as well as structuring the situation to improve children's oral language by integrating different types and degrees of exercises. The SSM identifies general components to become subsystems with information common to a context and interrelated through forms and sounds. Montoya & Montoya (2013) ratifies SSM contributes to make operative the strategy formation process for the researcher and to establish connections between the world he represented and the events that occur in the real world. Also, it allowed addressing the problem of learning and the development of language to encourage the development of thinking structures. According to linguistic theory, four components identified: perception, motor skills, imitation, and memory. GIL considers the imitation of sounds, the perception of identifying an object or form with sound, the motor skills necessary for the child to manipulate the application at the user level, and a basic level of memorization that allows the child to progress in regular activities. In the future, it remains to consolidate these four components with more activities within the mobile application.

GIL presents an appropriate design for SSM; the evaluation of the content reflects that it meets the objective of practicing the language of young children. The expert content needs analysis should include in the application that corresponds to user expectations; expert evaluation of information (9.33) provides confidence and security to the user. The initial problem of defining the roles involved and their function, the process they follow and the input/output information, and in particular the role of technology in children's language learning. Feedback from parents, educators, and therapists allowed the level of acceptance to adequately assessed. Hence the importance of what Papadakis & Kalogiannakis (2017a) expressed is the pedagogical practices of teachers which can facilitate through the use of ICTs, which could change education.

SSM was suitable for the development of a mobile application by the users who approved the application with 97%. The technical and aesthetic aspects make it possible to determine the level of access and interaction with the application, which, according to the study, is satisfactory. Besides, the tools that quantify technological, pedagogical, content, and aesthetic elements of GIL were adequate. One virtue of the application is to run on any mobile device regardless of the operating system it uses or the expiration date of the device. Thus, the use of GIL by children is alternative support to the complex problem of learning and language development, to move from pre-language to language, and aims to promote the development of thinking and communication structures at this age, especially if they also have an additional limitation by the ID. The development of an adaptable system such as GIL brings benefits as it helps to improve the learning process in different areas: cognitive, sensory and motor, for the advantages offered by tactile technology in mobile devices.

One of the elements that should highlight is that the diagrammatic method helped to identify the problems that affect the success of the issues addressed, especially the low level of spoken language in children, the poor infrastructure of educational institutions, inadequate training of educators and parents, among others. Despite this, the application represents a viable option for children to exercise language and move on to the next level of pre-language and writing. ID-centered mobile applications are scarce, an opportunity to continue the development of the application with tests to measure progress in children's learning.

This proposal is not new in its definition, it uses mobile technology to encourage the practice of language in children, it based on some previous work; but they do not contemplate ID nor do they consider educational achievements by transforming the educational environment into a classroom that they manipulate anywhere and at any time; GIL incorporates the possibility of trial and error, repetition on demand, practices focused on motivation and objects contained in real life of children. Besides, it has the support of the SSM and the appropriate pedagogical theories for the activities included in the application. Finally, it confirmed the use of ICT, and in particular mobile applications, supports learning processes, such as language learning in children, when a careful design and methodology is maintained that allow monitoring and evaluation of phases, functions, and activities.

ACKNOWLEDGMENT

To all participants in the research, especially children, parents, teachers, and experts, for supporting the work with children.

REFERENCES

Academia Mexicana de Pediatría. (2019). *El desarrollo del lenguaje*. Retrieved from https://www.healthychildren.org/spanish/ages-stages/baby/paginas/language-development-4-to-7-months.aspx

APNI. (2019). Centro de Adiestramiento e Información para Padres. https://www.parentcenterhub.org/

Aubrey, C., & Dahl, S. (2008). A review of the evidence on the use of ICT in the Early Years Foundation Stage [Ebook]. Retrieved from https://dera.ioe.ac.uk//1631/2/becta_2008_eyfsreview_report.pdf

Ávila, A. (1999). Inteligencias Múltiples: Una aproximación a la teoría de Howard Gardner. *Horizontes Pedagógicos, 1*, 19–27.

Bajgoric, N., & Moon, Y. B. (2017). Business model for digital economy era: A framework based on the Churchman's theory of design integrity. *International Journal of Business and Systems Research, 11*(3), 284–308. doi:10.1504/IJBSR.2017.085466

Bogdan, R., & Biklen, S. K. (2007). *Qualitative research for education: an introduction to theory and methods* (5th ed.). Boston, MA: Pearson.

Caws, P. (2015). General systems theory, its past and potential. *Systems Research and Behavioral Science, 32*(5), 514–521. doi:10.1002res.2353

Checkland, P., & Scholes, J. (1994). *La Metodología de los Sistemas Suaves en Acción*. México: Grupo Noriega.

De Bruijn, H., & Herder, P. (2009). System and actor perspectives on sociotechnical systems. *IEEE Transactions on systems, man, and cybernetics. Part a, systems and humans, 39*(5), 981–992. doi:10.1109/TSMCA.2009.2025452

De Curtis, L. L., & Ferrer, D. (2011). Toddlers and technology: Teaching the techniques. *ASHA Leader, 16*. Retrieved from http://leader.pubs.asha.org/article.aspx?articleid=2280052

DfES. (2007). *Department for Education and Skills, Special Educational Needs Code of Practice*. [Ebook]. London, UK. Retrieved from https://assets.publishing.service.gov.uk/government/uploads/system/uploads/attachment_data/file/228675/7092.pdf

Drigas, A., & Kokkalia, G. (2014). ICTs in kindergarten. *International Journal of Emerging Technologies in Learning, 9*(2), 52–58. doi:10.3991/ijet.v9i2.3278

Ellis, R. (2005). Principles of instructed language learning. *System*, 33.

Farhana, L., Borhan, N., & Yatim, M. (2013). A field study of understanding child's knowledge, skills and interaction towards capacitive touch technology (iPad). In *38th International Conference Information Technology in Asia (CITA)*. IEEE. 1-5.

Fernandez, J., & Lopez, M. (2013). Communication, destination brands and mobile applications. *Communicatio Socialis, 26*(2), 95–112.

Flewitt, R., Messer, D., & Kucirkova, N. (2014). New directions for early literacy in a digital age: The iPad. *Journal of Early Childhood Literacy, 0*(0), 1–22.

Fonseca, M. S. (2005). *Comunicación oral. Fundamentos y práctica estratégica*. México: Pearson.

Gutiérrez, A. D., & Cannon, B. Y. (2014). Impacto de los recursos educativos abiertos (REA) en el aprendizaje del inglés de alumnos de preescolar. In Y. Heredia & B. Y. Cannon (Eds.), Aplicaciones de la tecnología educativa en diversos ambientes de Aprendizaje, 27-52. Monterrey: ITESM.

Henderson, S., & Yeow, J. (2012). iPad in education: A case study of iPad adoption and use in a primary school. In *Proceedings 45th Hawaii International Conference on System Science (HICSS)*. IEEE. 78-87.

Hinostroza, J., Labbé, C., & Matamala, C. (2013). The use of computers in preschools in Chile: Lessons for practitioners and policy designers. *Computers & Education*, *68*, 96–104. doi:10.1016/j.compedu.2013.04.025

Holland, L., & Garfield, J. (2016). Linking research and teaching: an applied soft systems methodology case study. *International Journal of Information Technologies and Systems Approach*, *9*(2), 23–38. doi:10.4018/IJITSA.2016070102

Kalas, I. (2010). Recognizing the potential of ICT in early childhood education: analytical survey. Moscow: UNESCO; Retrieved from HTTPS://UNESDOC.UNESCO.ORG/ARK:/48223/PF0000190433

Kalogiannakis, M., Ampartzaki, M., Papadakis, S., & Skaraki, E. (2018). Teaching natural science concepts to young children with mobile devices and hands-on activities. a case study. *International Journal of Teaching and Case Studies*, *9*(2), 171–183. doi:10.1504/IJTCS.2018.090965

Kalogiannakis, M., Nirgianaki, G.-M., & Papadakis, S. (2018). Teaching magnetism to preschool children: The effectiveness of picture story reading. *Early Childhood Education Journal*, *46*(5), 535–546. doi:10.100710643-017-0884-4

Kärnä, E., Pihlainen, K., Sutinen, E. & Virnes, M. (2007). Technology in Finnish special education-toward inclusion and harmonized school days. *Informatics in Education*, *6*(1), 103-114.

Khaddage, F., Lattemann, C., & Bray, E. (2011). Mobile apps integration for teaching and learning, *Proceedings of Society for Information Technology & Teacher Education International Conference*, 2545-2552. Chesapeake, VA.

Koerner, J. (1982). *Ferdinand de Saussure. Génesis y evolución de su pensamiento en el marco de la lingüística occidental*. Madrid: Gredos.

Kokkalia, G., & Drigas, A. (2016). Mobile learning for special preschool education. International *Journal of Interactive Mobile Technologies*, *10*(1), 67–74. doi:10.3991/ijim.v10i1.5288

Korat, O. (2010). Reading electronic books as a support for vocabulary, story comprehension and word reading in kindergarten and first grade. *Computers & Education*, *55*(1), 24–31. doi:10.1016/j.compedu.2009.11.014

Lantolf, J. (1996). Second language theory building: Letting all the flowers bloom! *Language Learning*, *46*(4), 713–749. doi:10.1111/j.1467-1770.1996.tb01357.x

Liebermana, D., Batesa, C., & Soa, J. (2009). Young children's learning with digital media. *Computers in Schools*, *26*(4), 271–283. doi:10.1080/07380560903360194

Masek, M., Murcia, K. J., & Morrison, J. K. (2012). Getting serious with iPads: The intersection of game design and teaching principals. *Australian Educational Computing*, *27*(2), 34–38.

Montoya, I. A., & Montoya, L. A. (2013). La formación de estrategias deliberadas y emergentes: Una propuesta a partir de definiciones básicas de una metodología de sistemas suaves. *Investigación y Reflexión*, *26*(2), 67–96.

Mora, F. (2009). *Cómo funciona el cerebro*. Madrid: Alianza Editorial.

Morrison, G. (2005). Educación infantil. México. Pearson Education. 2-14.

Papadakis, S., & Kalogiannakis, M. (2017a). Combining mobile technologies in environmental education: A Greek case study. *International Journal of Mobile Learning and Organisation*, (Special Issue on Mobile Learning Applications and Strategies), 11(2), 108-128. doi:. doi:10.1504/IJMLO.2017.10003925

Papadakis, S., & Kalogiannakis, M. (2017b). Mobile educational applications for children. What educators and parents need to know. [Special Issue on Mobile Learning Applications and Strategies]. *International Journal of Mobile Learning and Organisation*, *11*(3), 256–277. doi:10.1504/IJMLO.2017.085338

Papadakis, S., Kalogiannakis, M., & Zaranis, N. (2016). Comparing tablets and PCs in teaching mathematics: An attempt to improve mathematics competence in early childhood education. *Preschool and Primary Education*, *4*(2), 241–253. doi:10.12681/ppej.8779

Papadakis, S., Kalogiannakis, M., & Zaranis, N. (2018). The effectiveness of computer and tablet assisted intervention in early childhood students' understanding of numbers. An empirical study conducted in Greece. *Education and Information Technologies*, *23*(5), 1849–1871. doi:10.100710639-018-9693-7

Papadakis, S., Kalogiannakis, M., & Zaranis, N. (2019). *Parental involvement and attitudes towards young Greek children's mobile usage*. International Journal of Child-Computer Interaction; doi:10.1016/j.ijcci.2019.100144

Pérez, P., & Salmerón, T. (2006). Desarrollo de la comunicación y del lenguaje: Indicadores de preocupación. *Revista Pediatría Atención Primaria*, *VIII*(32), 111–125.

Pierce, G. L., & Cleary, P. F. (2016). The K-12 educational technology value chain: Apps for kids, tools for teachers and levers for reform. *Education and Information Technologies*, *21*(4), 863–880. doi:10.100710639-014-9357-1

Ramos, J. R. (2015). Evaluación de un programa de maestría en Educación a través de la MSS. *Educación y Humanismo*, *17*(29), 236–255.

Real Academia Española. (2019). *Lenguaje*. Retrieved from https://dle.rae.es/?id=n7bnifo

Report, H. (2017). The NMC/CoSN Horizon Report: 2017 K–12. Retrieved from https://cdn.nmc.org/media/2017-nmc-cosn-horizon-report-k12-EN.pdf

Rodríguez, P. (2006). Redefiniendo el trabajo metodológico cualitativo con niños: El uso de la entrevista de grupo aplicada al estudio de la tecnología. *EMPIRIA. Metodología de Ciencias Sociales*, *12*(12), 65–88. doi:10.5944/empiria.12.2006.1135

Saalem, M., Mohamed, A., Gazala, Y., & Rajbunisa, A. (2019). Augmented reality based education for the improvement of sustainable learning ability in OMAN educational system. *International Advanced Trends in Computer Science and Engineering*, *8*(12), 110–115.

Santos, M., & Osório, J. A. (2008). Las TIC en la primera infancia: valorización e integración en la educación inicial a través del enlace @rcacomum. *Iberoamericana de Educación, (46), 9.*

Saussure, F. (2016). *Curso de lingüística general*. México: Fontamara.

Savater, F. (2005). *Las preguntas de la vida*. Barcelona: Ariel.

Searle, J. (1994). *Actos de habla, ensayo de filosofía del lenguaje*. Barcelona: Planeta-Agostini.

Segers, E., & Verhoeven, L. (2003). Effects of vocabulary training by computer in kindergarten. *Computer Assisted Learning*, *19*(4), 557–566. doi:10.1046/j.0266-4909.2003.00058.x

Shipstead, Z., Hicks, K., & Engle, R. (2012). Cogmed working memory training: Does the evidence support the claims? *Applied Research in Memory and Cognition*, *1*(3), 185–193. doi:10.1016/j.jarmac.2012.06.003

Siriram, R. (2012). A soft and hard systems approach to business process management. *Systems Research and Behavioral Science*, *29*(1), 87–100. doi:10.1002res.1095

Soberanes-Martín, A., Castillo, J. L., & Peña-Martín, A. (2018). GIL-1 aplicación para la ejercitación inicial del lenguaje. In M. Prieto-Méndez, S. J. Pech-Campos, & A. Francesa-Alfaro (Eds.), Tecnologías y aprendizaje: Investigación y práctica. 283-291. Ciudad Real, España: Editorial CIATA.org-UCML.

Soberanes-Martín, A., Peña-Martín, A., Castillo-Mendoza, J. L., & Martínez-Reyes, M. (2017). Aplicación para la ejercitación del lenguaje de niños con discapacidad intelectual grado de leve a moderado. *Sistemas y Gestión Educativa*, *4*(11), 14–23.

Swaab, D. (2014). *Somos nuestro cerebro*. Barcelona: Plataforma. The iPad.

UNESCO. (2012). Los *Jóvenes y las competencias: Trabajar con la educación*. Paris: Organización de las Naciones Unidas para la Educación, la Ciencia y la Cultura.

UNESCO. (2013). *Directrices de la UNESCO para las políticas de aprendizaje móvil*. Paris: ONU.

Uresti, A. S., & Cannon, B. Y. (2014). Uso de Recursos Educativos Abiertos (REA) y su impacto en la adquisición de la lectura convencional en niños de tercer grado de preescolar. In Y. Heredia & B. Y. Cannon (Eds.), Aplicaciones de la tecnología educativa en diversos ambientes de Aprendizaje 53-90. Monterrey: ITESM.

Urgilés, G. (2016). Aula, lenguaje y educación. *Sophia, Colección de Filosofía de la Educación*, (20), 221-244.

Warren, S., Sauser, B., & Nowicki, D. (2019). A bibliographic and visual exploration of the historic impact of soft systems methodology on academic research and theory. *Systems*, *7*(10), 1–15.

Zare, P., & Othman, M. (2013). Classroom debate as a systematic teaching/learning approach. *World Applied Sciences Journal*, *28*(11), 1506–1513.

ADDITIONAL READING

Bouck, E. C., Satsangi, R., & Flanagan, S. (2016). Focus on inclusive education: evaluating apps for students with disabilities: supporting academic access and success. *Childhood Education*, *92*(4), 324–328. doi:10.1080/00094056.2016.1208014

Crescenzi-Lanna, L., & Grané-Oró, M. (2016). An analysis of the interaction design of the best educational apps for children aged zero to eight. *Comunicar*, *24*(46), 77–85. doi:10.3916/C46-2016-08

Evans, M. A., Nowak, S., Burek, B., & Willoughby, D. (2017). The effect of alphabet eBooks and paper books on preschoolers' behavior: An analysis over repeated readings. *Early Childhood Research Quarterly*, *40*, 1–12. doi:10.1016/j.ecresq.2017.02.002

Kerckaert, S., Vanderlinde, R., & van Braak, J. (2015). The role of ICT in early childhood education: Scale development and research on ICT use and influencing factors. *European Early Childhood Education Research Journal*, *23*(2), 183–199. doi:10.1080/1350293X.2015.1016804

Lovato, S. B., & Waxman, S. R. (2016). Young children learning from touch screens: Taking a wider view. *Frontiers in Psychology*, *7*, 1078. doi:10.3389/fpsyg.2016.01078 PMID:27486421

Múzquiz, N. Y. (2018). *Entorno virtual de aprendizaje ubicuo en el proceso de lecto-escritura para niños con déficit de atención* (Unpublished master dissertation). Universidad Autónoma del Estado de México.

Papadakis, S., Kalogiannakis, M., & Zaranis, N. (2017). Designing and creating an educational app rubric for preschool teachers. *Education and Information Technologies*, 1–19.

Papadakis, S., Kalogiannakis, M., & Zaranis, N. (2018). Educational apps from the Android Google Play for Greek preschoolers: A systematic review. *Computer Education*, *116*, 139–160. doi:10.1016/j.compedu.2017.09.007

KEY TERMS AND DEFINITIONS

Educational Technology: Information and communication technology incorporated into education to help improve activities in teaching and learning processes.

General Systems Theory (GST): Establishes that any real-life problem has the behavior of a general system, that is, that it is a set of interrelated parts (subsystems) with a common objective, that contains inputs and outputs in the form of matter, energy or information, and influenced by an environment.

Gradual Interactive Language (GIL): Mobile application to practice the oral language of children with intellectual disabilities.

Intellectual Disability: When a person learns and develops more slowly than other people of the same age, which manifested in the fact that they may need more time to learn to talk, walk, dress or eat without help and of course they have learning problems in school.

Language: Set of sounds and signs that human beings have used since its creation until today to communicate with other individuals of the same species.

Mobile Learning: It is a branch of Information and Communication Technologies in education to make it more affordable through technological means.

Soft Systems Methodology (SSM): Provides a procedure to address complex real-world problems, through the premise that organizations are open systems that interact with their environment and provides a way of thinking and reflecting on the situation of the problem.

Software Engineering: Set of methods, techniques, and tools for the development of computer programs (software).

Section 3
Attitudes to Mobile Learning

Chapter 9
Comparing the Attitudes of Greek Kindergarten Teachers Against Primary School Teachers Towards ICT

Nicholas Zaranis
University of Crete, Greece

Vassilios Oikonomidis
University of Crete, Greece

Michalis Linardakis
University of Crete, Greece

ABSTRACT

This chapter examines the main factors of Greek kindergarten against primary school teachers' attitudes towards ICT in class. The participants of this research were drawn from teachers of public schools in the area of Athens and Crete (Greece). A 5-point Likert scale questionnaire with 30 questions designed to assess attitudes of teachers toward computers and their computer knowledge in primary schools learning environment. Varimax with Kaiser Normalization was applied to extract the main five factors of kindergarten teachers' attitudes. Also, a three 2-way MANOVA was performed between the kindergarten and primary school teachers to see the interactions of these main factors on: a) the ICT knowledge of teachers, b) the frequency of using ICT at home by the teachers and, c) the ICT use in the teaching process. Finally, the changes in the main factors of kindergarten and primary school teachers were discussed in the hope that they may provide useful information on the school curriculum.

DOI: 10.4018/978-1-7998-1486-3.ch009

INTRODUCTION

Nowadays, children accept the use of computers in learning as simply as any other learning activity (Palaiologou, 2016). Computers have taken a permanent place in all levels of education, but their appearing in the classroom has not been without some concern; the future is not without challenges for both kindergarten teachers and students (Chen & Chang, 2006; Fisher, et al., 2012; Livingstone, 2012; Pelgrum, 2001; Shade & Davis, 1997).

Greek kindergarten and primary school teachers come from varied educational backgrounds which may include two year studies in a public academy named "Paidagogiki Akadimia" or four year studies in a public University of Department of Education. Some graduates of two years studies may have gone through two years of additional in-service education at a University (Zaranis, & Oikonomidis, 2009). For the past ten years in Greece, a new Cross-Thematic Curriculum Framework for nursery, primary and middle school education (Hellenic Ministry of National Education - Pedagogical Institute, HMNE-PI, 2003) has been in place and is being tried out. According to this, individual subjects are maintained within the Cross-Thematic Curriculum, while at the same time the horizontal and vertical linking of subject matter content are promoted (Tselfes & Paroussi, 2008). The ICT portion of this program introduces the use of computers as a teaching aid, a cognitive tool and as a means of communication. Simultaneously, the training programs, hitherto, primarily had cognitive orientation. It seems that in the near future it will focus on enhancing teaching skills through the use of ICT (Jimoyiannis & Komis, 2007).

Although the fact that the primary level learning environment has many unresolved issues regarding the use of computers, the fact remains that the first educational contact for many students with ICT begins at primary school level. As a result, it is important to examine this environment in terms of how kindergarten and primary school teachers and students learn about and use computers. These educators are an important group to study because they are among the first to have an influence on how students learn in the classroom.

THEORETICAL BACKGROUND

School Teachers' Attitudes and ICT

It is important to understand the attitudes of kindergarten and primary school teachers on ICT because these traits form their profiles (Scherer, Siddiq & Teo, 2015). Although a lot of research has been done wide word about the attitudes of kindergarten and primary school teachers toward ICT (Sanchez, A.B., Marcos, Gonzalez, &Guanlin, 2012; Slechtova, 2014;Umar, & Hassan, 2015), only a small amount of research has been done in Greece (Gialamas, &Nikolopoulou, 2010; Liua, Toki, &Pange, 2014; Zaranis, & Oikonomidis, 2014).

Researchers explained that although software has become increasingly easy to use, it is much more difficult to convince teachers to use various computer applications than a student (Angeli & Valanides, 2005). Teachers may not have a positive attitude toward the computer without some type of prerequisite training. One of the hurdles teachers have to overcome is when a computer is placed in their classroom and they are given a broad overview of how to use it having little, if any, previous training (Raman et al., 2015). Although many educators have computers, there has not been enough training on how to actively use them in imaginative ways in order to stimulate interest among students and as a result, the computers are used exclusively for secretarial support by the teachers (Tsitouridou & Vryzas, 2004).

A key problem related to school educators' attitudes toward the use of computers in schools is the limited support, which present for integrating this unfamiliar technology into instruction. Also, teachers had some concerns about their training or lack of training with computers. Teachers worried that their lack of knowledge might result in damage to the computers. They also lacked confidence in their ability to help children learn how to use and care for the computer. Included within this concern were the issues surrounding the selection and the use of proper software in the classroom (Chen & Chang, 2006). The researchers were also concerned with fully integrating computers into the curriculum so that computers became more than just an add-on (Kalogiannakis, 2010).

Research (Mumtaz, 2000) showed that a main variable influencing the beginning of teachers' uptake of computers is the amount and adequacy of their pre-service training. Also, a gap exists between what pre-service teachers are taught in their ICT courses and what they are expected to do with ICT in a real classroom (Pope, Hare, & Howard, 2002). There is no doubt that basic computing skills constitute the foundation of ICT literacy, but they are not enough for adequately preparing pre-service teachers to teach with ICT (Selinger, 2001; Wetzel, Wilhelm, & Williams, 2004). Understandings of how pre-service and in-service teachers are trained and prepared for working with new technologies are critical issues for the field of teacher education and technology (Mertala, 2018). These factors are related to instructors' general attitudes and how their attitudes and computer use are affected and improved through computer courses (Angeli, 2004; Chen & Chang, 2006; Kumar, & Kumar, 2003; Yildirim, 2000). Evidently, the training of pre-service teachers will be pivotal in determining the future role of ICT in education (Kalogiannakis, 2010; Wedman, & Diggs, 2001; Wheeler, 2001).

Moreover the majority of the teachers surveyed had positive attitudes towards computer use in early childhood and primary education (Jimoyiannis & Komis, 2007). There did appear to be a relationship between teachers' attitudes towards computers and number of years of teaching experience or level of their education (Nikolopoulou & Gialamas, 2015). In addition, there did seem to be a relationship between the attitudes of teachers towards computers and the types of computer training they had received (Nasrin & Islahi, 2019). The teachers who had received computer training from a college or university appeared to have a more positive attitude than those who had received in-service training or were self-taught (Chen & Chang, 2006).

In addition, another research conducted with elementary school children indicates that ICT can have a positive impact on student learning only when teachers know how to use ICT to promote student thinking, expression, and knowledge building (Loveless, & Dore, 2002).

A positive attitude toward computers played an important role in whether teachers use technology in the classroom, according to Yildirim (2000) and Pelgrum (2001). The researchers stated that teachers teach the way they have been taught and it is unlikely that computer skills will be communicated to students and promoted by teachers unless the teachers have positive attitudes toward computer use. However, students graduating from universities have a better understanding of computers. They view computers as just another part of their world and everyday life assuming they can be in full control of it (Ma, Anderson, & Streith, 2005).

Research Questions

Based on various exposures to computer training, the viewpoint of kindergarten and primary school teachers regarding ICT use may be more easily influenced by their attitude profile and relative amount of knowledge. As a result, this study seeks to examine the following research questions:

(1) What are the main factors of kindergarten and primary school teachers' attitudes towards ICT?
(2) What is the sequencing of kindergarten teachers' factors during 2012-2013?
(3) What is the sequencing of primary school teachers' factors during 2014-2015?
(4) What factors were affected by the knowledge of the teachers on various software applications?
(5) What factors were affected by computer use of the teachers in their home during the week?
(6) What factors were influenced by the use of computers in the classroom during the teaching process?

The above six research questions can be classified in two groups. The former group includes the first three research questions and examines the changes of the factors towards ICT. The latter group involves the latest three research questions and investigates the main reasons for the change of attitude of kindergarten and primary school teachers towards ICT.

METHODOLOGY

The present survey was conducted in two phases. This survey was carried out in 2012 - 2013 and 2014 – 2015 school years and used kindergarten and primary school teachers as participants respectively. The purpose of the two phases of the survey was to point out differences and make comparisons between the sample from 2012-2013and the sample from 2014 – 2015 school years.

Participants

In the first phase, the survey was conducted with 295 kindergarten teachers from the areas of Athens and Crete (Greece) and their attitudes towards ICT were examined during the 2012 – 2013 year. The vast majority of kindergarten teachers in the sample were women. About 77% of them had a second degree or in-service training in educational methods. In addition, 43.1% had less than ten years of teaching experience, 29.2% had between 11 and 20 years, and about 27.8% had over 21 years of teaching experience. Also, more than 66% of them graduated from public universities with 4-year studies and less than 34% received their first degree with 2-year studies. Over 60% of them used the computer at home more than three hours a week. Less than 80% of them had a European Computer Driving License or formal training in computers.

In the second phase, a similar survey was conducted with 320 primary school teachers that served in public primary schools during the 2014 – 2015 years from the same areas (Athens and Crete) and the last 3 years had taught in one of the first three grades (A, B, C) of primary school. About 72% of primary school teachers in the sample were women. Over 85% of them had a second degree or in-service training in educational methods. In addition, 36.3% had less than ten years of teaching experience, 29.7% had between 11 and 20 years, and about 34% had over 21 years of teaching experience. Also, less than 64% of them graduated from public universities with 4-year studies and nearly 36% received their first degree through 2-year studies. Over 62% of them used the computer at home more than three hours a week. Less than 85% of them had a European Computer Driving License or formal training in computers.

We selected the participants of our study from kindergarten and primary school teachers as a single sample for the following reasons: a) Both kindergarten teachers and primary school teachers are teachers of the first cycle of compulsory education in Greece, b) None of these two kinds of teachers are ICT experts, c) We wanted to search a developmental impact of the views of teachers about ICT on the of children in their early years of compulsory education (kindergarten, elementary school. For this reason, we conducted Part 2 of the survey (2014-2015) two years after the Part 1 (2012-2013), so that all children who had teachers from our kindergarten teachers' sample (2012-2013) would attend primary schools in 2014-2015. d) The common characteristics of kindergarten and primary school teachers are more than their different characteristics. Studies show that this two kinds of teachers share common attitudes and opinions on a variety of educational issues, so we can face them as a single population (Ashton, 2015). (e) The objectives for ICT, as set out in the curricula for the kindergarten and for the first grades of the primary school, have a lot of similarities between them (HMNE-PI, 2003).

Survey Tools

During the first (2012-2013) and the second phase (2014-2015) of the survey an anonymous questionnaire was generated to collect teachers' attitudes towards ICT in the classroom. The questionnaire consisted of thirty questions (Table 1) derived from the principal axes of the Cross Thematic Integrated Curriculum for Kindergarten (Tselfes, & Paroussi, 2008) and based on a previous research about the main factors of the attitudes of Greek kindergarten teachers towards ICT (Zaranis & Oikonomidis, 2014). The questions were measured with a 5-point Likert type scale ranging from 1 (*extremely disagree*) to 5 (*extremely agree*). Demographic data of kindergarten and primary school teachers such as educational background, training, gender and ICT knowledge were collected through the same questionnaire; as well as their willingness to accept and to use ICT in education as an instructional alternative.

The questionnaire was answered by 615 kindergarten and primary school teachers in total. The initial Cronbach's alpha was 0.859 for the 30-item perception survey, which demonstrates a high reliability. The questionnaire took about fifteen minutes to complete.

Survey Design

The present two-phase survey attempts to discover and greater understand the main factors of the attitudes of Greek kindergarten and primary school teachers towards computers. It was performed through a two-phase survey carried out in 2012-2013 and 2014-2015. The questionnaire was answered by kindergarten and primary school teachers who served in public kindergartens or and primary schools located in the areas of Athens and Crete (Greece). For this survey, Athens has been chosen as a metropolitan city, as it is the capital of Greece, and Crete as a representative case of a geographic area with urban cities.

In both phases a total of 812 questionnaires were sent randomly to kindergarten and primary school teachers in Athens and Crete. In total 653 of them were returned to us. After discarding 38 incomplete questionnaires, 615 questionnaires were entered into our analysis yielding a final response rate of 75.7%. However the case to variable ratio (615 cases versus 30 variables) in our study is higher than the recommended general rule of thumb of at least 5:1 for an adequate factor analysis as suggested in Hair et al.

Table 1. Group statistics of the total pre test

No	Question description
2.1 The computer may be used in kindergarten or primary school for:	
1	preparation and planning of teaching
2	creation of supervisory and teaching materials
3	performing administrative tasks
4	recording data on the assessment of each student's development
2.2 The use of computer in kindergarten or primary school affects the all-around growth of students:	
5	undermining their social/emotional development
6	reduces psychomotor development
2.3 The use of computer in teaching:	
7	facilitates the conduct of teaching and the achievement of its objectives
8	renews the teaching process
9	can be applied to all areas of the curriculum
2.4 The use of computers by students in kindergarten or primary school:	
10	enhances their imagination and creativity
11	provide possibilities for their own initiatives
12	respects their personal learning rhythm
13	promotes their active participation in learning
14	does not allow students' cooperation
2.6 55 The educational use of computers in kindergarten or primary school provides students with opportunities to improve their:	
15	linguistic capabilities
16	concepts of space
2.7 66 The following objectives of the program Child and ICT of the CROSS THEMATIC CURRICULUM FRAMEWORK (C.T.C.F.) may be achieved in the kindergarten or primary school:	
17	computer is a means of communication, work and entertainment
18	recognition of computer's units
19	the use of keyboard and mouse
20	creation of drawings and paintings
21	the use of CD-ROMs as a teaching material
22	the use of exploration games and problem solving
23	recognition of correct body position and rules when working with computer
2.9 77 The introduction of computer in kindergarten or primary school:	
24	removes the dominant role from kindergarten or primary school teachers during teaching process
25	reduces communication between other students and teachers
26	reduces the prestige of kindergarten or primary school teachers towards students
2.10 88 The introduction of computer in kindergarten or primary school depends on the following factors:	
27	adequacy of the schools' area
28	appropriateness of technological equipment
29	interest of kindergarten or primary school teachers
30	training of kindergarten or primary school teachers

(1998). Specifically, this survey preformed in 2012-2013 and 2014-2015 used 295 and 320 participants respectively.

Moreover, the designed questionnaire was anonymous and the ethical principles relating to basic individual protection requirements were met with regard to information, informed consent, confidentiality and the use of data.

RESULTS

The results are briefly described in the order of kindergarten and primary school teachers' profile, factor analysis and cluster analysis.

Factor Analysis

The factor structure of the questionnaire that had been previously found on kindergarten teachers (Zaranis, & Oikonomidis, 2015) was tested on the current data set through confirmatory factor analysis. The indices were not satisfactory even after the addition of significant covariances in the model, that were suggested by the modification indices (CFI=0.755, RMSEA=0.09). Thus, the model was not adequate for the primary school teachers' responses. Instead, we used an exploratory factor analysis for the current data set.

On the sample of the first phase the Kaiser–Meyer–Olkin (KMO) test for sampling adequacy and Bartlett's test for sphericity were used to first examine the suitability of selected variables for the factor analysis (Bryman, 1989). The KMO test resulted in a 0.908 value that was greater than the suggested minimum value of 0.5 for adequacy, and Bartlett's test also demonstrated a very good sphericity ($v2$ = 5924.18, df= 435, p< 0.001).

Factor analysis was used to extract the principal components of kindergarten teachers' attitudes on the 30-item questionnaire items of the previous research sample (Zaranis, & Oikonomidis, 2015). Only those extracted factors with eigenvalues bigger than one were selected. In the end, a total of five factors emerged, which together explained the 59.75% of the total variance.

Varimax with Kaiser Normalization was applied to ensure the extracted factors were distinct from one another. Table 2 shows the result of the factor analysis. Each factor was named to reflect a common higher level group constructed of those questions that related to the same factor.

Factor one (F1) included 12 question items regarding how a computer helps the child universally in the teaching process, with a Cronbach's alpha of 0.895. Seven question items, in relation to the use of computers by the pupils as cognitive tool, made up the second factor (F2), which resulted in a high Cronbach's alpha of 0.908. The third factor (F3) included three question items concerning the changes caused by the computer on the role of the kindergarten teacher in the classroom, with a Cronbach's alpha of 0.844. Four questions examining that kindergarten teachers use the computer only to organize their administrative work constituted the fourth factor (F4), with a Cronbach's alpha of 0.754. The fifth factor (F5) had a Cronbach's alpha of 0.737 and was composed of four question items regarding the conditions under which the computer may integrate into the kindergarten. Cronbach's alpha of the overall 30-item questionnaire remained high at 0.916

On the sample of the second phase the Kaiser–Meyer–Olkin (KMO) test for sampling adequacy and Bartlett's test for sphericity were used to first examine the suitability of selected variables for the factor

Table 2. Factor analysis results of the kindergarten teachers' attitudes

No	Factor	Question number	Cronbach's alpha
F1	how computer helps the child universally in the teaching process	5, 6, 7, 8, 9, 10, 11, 12, 13, 14, 15, 16	0.895
F2	the use of computers by the pupils as cognitive tool	17, 18, 19, 20, 21, 22, 23	0.908
F3	the changes caused by the computer on the role of the kindergarten teacher in the classroom	24,25,26.	0.844
F4	kindergarten teachers use the computer only to organize their administrative work	1, 2, 3, 4	0.754
F5	the conditions under which the computer may integrate into the kindergarten	27, 28, 29, 30	0.737
	Overall		0.916

analysis (Bryman, 1989). The KMO test resulted in a 0.847 value that was greater than the suggested minimum value of 0.5 for adequacy, and Bartlett's test also demonstrated the existence of latent factors (v2 = 4730.684, df = 435, p < 0.001).

Factor analysis was used to extract the principal components of kindergarten teachers' perception on the 30-item questionnaire items of the previous research sample (Zaranis, & Oikonomidis, 2015). Only those extracted factors with eigenvalues bigger than one were selected. In the end, a total of five factors emerged, which together explained the 57.06% of the total variance.

Varimax with Kaiser Normalization was applied to ensure the extracted factors were distinct from one another. Question items 2 and 6 were dropped from the analysis due to their low factor loadings on every extracted factor. Cut off value of item loadings is kept less than 0.5 and as a result the number of variables was reduced to 28. Table 3 shows the result of the factor analysis. Each factor was named to reflect a common higher level group constructed of those questions that related to the same factor.

Factor one (F'1) included 9 question items regarding how computer helps the child universally in the teaching process, with a Cronbach's alpha of 0.888. Seven question items, in relation to the use of computers by the pupils as cognitive tool, made up the second factor (F'2), which resulted in a high Cronbach's alpha of 0.779. The third factor (F'3) included eight question items concerning the changes caused by the computer on the role of the preschool teacher in the classroom, with a Cronbach's alpha of 0.835. Two questions examining that primary school teachers use the computer only to organize their administrative work constituted the fourth factor (F'4), with a Cronbach's alpha of 0.789. The fifth factor (F'5) had a Cronbach's alpha of 0.571 and was composed of two question items regarding the conditions under which the computer may integrate into the primary school. Cronbach's alpha of the overall 28-item questionnaire remained high at 0.908

Comparing the Attitudes of Kindergarten Against Primary School Teachers

The attitudes collected from the two phases of this survey during the academic years 2012-2013 and 2014-2015 were classified into five factors. Descriptive statistics were calculated and the results are displayed in Table 4. In the first phase (2012-2013), the first factor has the lowest value and the fifth factor follows with the second lowest value. Next the third factor follows and the second factor performed higher than all the previous factors. Then the second factor has a higher value than the fourth. Finally,

Table 3. Factor analysis results of the primary school teachers' attitudes

No	Factor	Question number	Cronbach's alpha
F'1	how computer helps the student universally in the teaching process	5, 7, 8, 10, 11, 12, 13, 15, 16.	0.888
F'2 F2K F3D	the use of computers by the students as cognitive tool	17, 18, 19, 20, 21, 22, 23.	0.779
F'3 F3K F2D	the changes caused by the computer on the role of the primary school teacher in the classroom	1, 3, 4, 9, 14, 24, 25, 26.	0.835
F'4	the conditions under which the computer can be integrated into the primary school according to the teachers' capabilities on ICT	29, 30.	0.789
F'5	the conditions under which the computer can be integrated into the primary school in accordance with the appropriate technological equipment	27, 28.	0.571
	Overall		0.908

the fourth factor has the highest value of all. Overall, the values of the factors of the first phase are in the following order: ICT in educational process (F1), the terms of ICT integration in kindergarten (F5), ICT and the role of the kindergarten teacher (F3), ICT as cognitive tool (F2), and kindergarten teachers' administrative work with ICT (F4), (F1, F5, F3, F2 and F4). In the second phase (2014-2015), the factors have the same accenting order but the fifth factor (the terms of ICT integration in kindergarten regarding the technological equipment) was changed from the second place to the fourth place (F'1, F'3, F'2, F'5 and F'4).

Then paired-samples t-tests were conducted after Bonferroni adjustment to compare the values between of the lower and the higher factors according to the ascending order of the kindergarten teachers' factors during the academic year 2012 - 2013. The modified alpha level was adjusted according to the comparisons across the five scores of the factors of the attitudes of kindergarten teachers towards computers (the alpha level had to be 0.01). It was pointed that there was a significant difference between the values of the first factor (M=3.68, SD=.50) and the values of the fifth factor (M=4.05, SD=.58), $t(294)=-7.99$, $p < .001$. Moreover, there was not a significant difference between the values of the fifth

Table 4. Descriptive statistics of the main factors of the attitudes of teachers for the two phases

	Factor	First phase 2012-2013 kindergarten teachers (n=295)		Second phase 2014-2015 primary school teachers (n=320)	
	Factor	M	SD	M	SD
F1, F'1:	ICT and the teaching process	3.68	.50	3.62	.65
F2, F'2:	ICT as cognitive tool	4.17	.50	3.93	.50
F3, F'3:	ICT and the role of the kindergarten teacher / primary school teacher	4.08	.67	3.73	.64
F4, F'4:	Administrative work and ICT of Kindergarten teachers/ primary school teachers' ICT capabilities	4.34	.57	4.22	.62
F5, F'5:	The terms of ICT integration in kindergarten	4.05	.58	4.05	.68

factor (M=4.05, SD=.58) and the values of the third factor (M=4.08, SD=.67), t(294)=-.494, p = .622. On the contrary, there was a significant difference between the values of the third factor (M=4.08, SD=.67), and the values of the second factor (M=4.17, SD=.50), t(294)=-2.62, p= .009. Also, there was significant difference between the values of the second factor (M=4.17, SD=.50), and the values of the fourth factor (M=4.34, SD=.57), t(294)=-4.90, p<.001.

Also, paired-samples t-tests for each factors was performed after Bonferroni adjustment during the year 2014-2015. The modified alpha level was adjusted according to the comparisons across the five scores of the factors and had to be 0.01. There was a significant difference between the values of the first factor (M=3.62, SD=.65) and the values of the third factor (M=3.73, SD=.64), t(319)=-3.76, p < 0.001. Moreover, there was a significant difference between the values of the third factor (M=3.73, SD=.64) and the values of the second factor (M=3.93, SD=.50), t(319)=-5.88, p<.001. Also, there was a significant difference between the values of the second factor (M=3.93, SD=.50) and the values of the fifth factor (M=4.05, SD=.68), t(319)=-2.71, p=.007. On the contrary, there was no significant difference between the values of the fifth (M=4.05, SD=.68) and the values of the fourth factor (M=4.22, SD=.62), t(319)=-3.64, p<.001.

Performing t-test controls among kindergarten and primary school teachers, we found that the first factor had not increased significantly in the kindergarten teachers of academic year 2012-2013 (M= 3.68, SD= .50) than did those in the primary school teachers of academic year 2014-2015 (M= 3.62, SD= .36), t(593.502) = 1.27, p=.204. On the contrary, the third factor had increased significantly in the kindergarten teachers of academic year 2012-2013 (M= 4.08, SD= .67) than did those in the primary school teachers of academic year 2014-2015 (M= 3.73, SD= .03), t(613) = 6.55, p<.001. The second factor of kindergarten teachers attitudes had significant difference between the kindergarten teachers of academic year 2012-2013 (M= 4.17, SD= .50) and the primary school teachers of academic year 2014-2015 (M= 3.93, SD= .50), t(613) = 5.935, p<.001. Similarly, the fourth factor had significant difference between the kindergarten teachers of academic year 2012-2013 (M= 4.34, SD= .57) and the primary school teachers of academic year 2014-2015 (M= 4.22, SD= .62), t(613) = 2.257, p=.024.On the other hand, the fifth factor had not decreased significantly in the kindergarten teachers of academic year 2012-2013 (M= 4.05, SD= .58) than did those in the primary school teachers of academic year 2014-2015 (M= 4.05, SD= .68), t(613) = -.042, p=.967.

The Change of Attitudes Among the Primary School and Kindergarten Teachers

As mentioned above, the attitudes of kindergarten and primary school teachers during the two phases of study in the years of 2012-2013 and 2014-2015 were classified into five factors which are influenced by a number of different variables. These variables may be: a) the knowledge of kindergarten teachers have on various software applications, b) the number of hours that kindergarten teachers use computers at home during the week and c) the use of computers in the classroom during the teaching process.

The Influence of the Software Knowledge on Teachers' Attitudes Towards ICT

In order to examine whether the software knowledge level of kindergarten and primary school teachers influences their attitudes during the years 2012-2013 and 2014-2015 respectively, we initially performed a 2-way MANOVA of the five factors. The 2-way MANOVA, with 'school' (kindergarten versus primary) and 'knowledge of software' (beginner versus intermediate versus advanced) used as between-factor variables and the five attitude factors used as dependent variables, showed a significant main effect of 'school', *Pillai's trace* =0.068, $F(5, 605) = 8.831$, $p<0.001$, *partial* $\eta 2 = 0.68$. Additionally, there was a significant main effect of 'knowledge of software', *Pillai's trace* =0.141, $F(10, 1212) = 9.166$, $p<0.001$, *partial* $\eta 2 = 0.07$.The interaction of 'school' and 'knowledge of software' was statistically significant,*Pillai's trace* =0.056, $F(10, 1212) = 3.501$, $p<0.001$, *partial* $\eta 2 = 0.028$.111++

For this reason, a post hoc analysis for each factor was performed after Bonferroni adjustment. The modified alpha level was adjusted according to the comparisons across the five scores of the factors of kindergarten and primary school teachers' attitudes towards computers (the alpha level had to be 0.01). Five two-way fixed-effects analyses of variance (ANOVAs) were computed to compare differences between the level of knowledge in various software applications (beginner versus intermediate versus advanced) and the school of measurement (kindergarten versus primary) on the main five factors of the attitudes of various teachers towards ICT. Both 'knowledge of software' and 'school' were treated as fixed-effects- factors. Composite scores for the five factors served as dependent variables. In all cases, the analyses of variance showed that there were not significant interactions between the effects of 'school' and 'knowledge of software' to all five factors of kindergarten teachers' attitudes. On the contrary, there was a significant effect of 'school' to the third and fifth factors as depicted in Table 5. Also, there was a significant effect of 'knowledge of software' to the first, second and fourth factors of the attitudes of teachers towards ICT (Table 6).

The Effect of The Hours of Computer Use by Teachers at Home During the Week on Their Attitudes Towards ICT

In order to examine how the hours of computer use by teachers at home during the week influences the attitudes collected during the years of 2012-2013 and 2014-2015, we initially performed a 2-way MANOVA of the five factors. The 2-way MANOVA with 'year' and 'the use of computers at home' (none versus 1 to 2 hours versus more than 3 hours) used as a between-factor variables and the five attitude factors used as dependent variables showed a significant main effect of 'school',*Pillai's trace* =0.09, $F(5, 668) = 13.44$, $p<0.001$, *partial* $\eta 2 = 0.09$. Additionally, there was a significant main effect

Table 5. Analyses of variance of 'school' on the five factors the attitudes of teachers towards ICT for 'knowledge of software'

	2012-2013 Kindergarten		2014-2015 Primary school		ANOVA		
Factors	**M**	**SD**	**M**	**SD**	**F**		**Partial Eta Squared**
F1: ICT in educational process	3.51	.62	3.68	.50	1.34		.002
F2: ICT as cognitive tool	4.14	.56	4.17	.50	.20		.000
F3: ICT and the role of the kindergarten teacher	3.70	.75	4.08	.67	17.85	**	.026
F4: Kindergarten teachers' administrative work and ICT	4.30	.56	4.34	.57	.25		.000
F5: The terms of ICT integration in kindergarten	4.32	.58	4.05	.80	27.45	**	.039

Note: $**p<.001$

Table 6. Analyses of variance of 'knowledge of software' on the five factors of the attitudes of teachers towards ICT

	beginner		intermediate		advanced		ANOVA		
Factors	**M**	**SD**	**M**	**SD**	**M**	**SD**	**F**		**Partial Eta Squared**
F1: ICT in educational process	3.36	.61	3.56	.55	3.85	.51	10.23	**	.030
F2: ICT as cognitive tool	4.00	.55	4.15	.51	4.30	.55	4.77	**	.014
F3: ICT and the role of the kindergarten teacher	3.50	.87	3.87	.67	4.13	.71	2.90		.009
F4: Kindergarten teachers' administrative work and ICT	4.20	.59	4.28	.55	4.53	.52	10.87	**	.031
F5: The terms of ICT integration in kindergarten	4.32	.59	4.20	.59	4.14	.62	1.70		.005

Note: **p<.001

of'the use of computers at home',*Pillai's trace* =0.06, $F(10, 1338) = 4.40$, $p<0.001$, *partial* $\eta 2 = 0.03$. The interaction of 'school' and 'the use of computers at home' was statistically significanct,*Pillai's trace* =0.04, $F(10, 1338) = 2.51$, $p=0.005$, *partial* $\eta 2 = 0.02$.

For this reason, after the Bonferroni adjustment, the modified alpha level was adjusted to be 0.01. Five two-way fixed-effects analyses of variance (ANOVAs) were computed to compare differences between 'the use of computers at home' (none versus 1 to 2 hours versus more than 3 hours) and the school of measurement (kindergarten versus primary) on the main five factors of teachers' attitudes towards ICT. Both 'the use of computers at home' and 'school' were treated as fixed-effects- factors. The scores for the five factors served as dependent variables. The analyses of variance showed that there were only significant interactions between the effects of 'school' and 'the use of computers at home' to the third factor $F(2, 672) = 5.15$, $p = .006$, and the fifth factor of teachers' attitudes $F(2, 672) = 4.79$, $p = .009$, as displayed in Table 7. Also, there was a significant effect of 'the use of computers at home' to the first, second, third and fourth factors of the attitudes of teachers towards ICT (Table 8).

Table 7. Analyses of variance of 'school' on the five factors the attitudes of teachers towards ICT for 'the use of computers at home'

	2012-2013 Kindergarten		2014-2015 Primary school		ANOVA		
Factors	**M**	**SD**	**M**	**SD**	**F**		**Partial Eta Squared**
F1	3.51	.62	3.68	.50	.497		.001
F2	4.14	.56	4.17	.50	2.34		.003
F3	3.69	.75	4.08	.67	23.47	**	.034
F4	4.30	.55	4.34	.57	2.27		.003
F5	4.32	.58	4.05	.58	38.01	**	.054

Note: **p<.001

Table 8. Analyses of variance of 'the use of computers at home' on the five factors of the attitudes of teachers towards ICT

	none		1 to 2 hours		more than 3 hours		ANOVA		
Factors	M	SD	M	SD	M	SD	F		Partial Eta Squared
F1	3.37	.63	3.58	.49	3.76	.574	11.59	**	.033
F2	4.02	.56	4.16	.50	4.25	.53	7.77	**	.023
F3	3.57	.86	3.83	.67	4.11	.63	5.13	*	.015
F4	4.21	.64	4.27	.54	4.46	.49	12.11	**	.035
F5	4.29	.60	4.20	.57	4.15	.61	3.12		.009

Note: *p<.01, **p<.001

The Effect of Computer Use in The Classroom During the Educational Process on the Attitudes of Teachers Towards ICT

In order to examine how computer use in the classroom during the educational process influences teachers' attitudes during the years of 2012-2013 and 2014-2015, we initially performed a 2-way MANOVA of the five factors. The 2-way MANOVA with 'school' (kindergarten versus primary) and 'the use of computers in the teaching process' (no computer use versus the use of computers) used as between-factor variables and the five attitude factors as dependent variables showed a significant main effect of 'school',*Pillai's trace* =0.04, $F(5, 669) = 5.87$, $p<0.001$, *partial* $\eta 2 = 0.42$. Additionally, there was a significant main effect of 'the use of computers in the teaching process',*Pillai's trace* =0.06, $F(5, 669) = 8.83$, $p<0.001$, *partial* $\eta 2 = 0.06$.The interaction of 'school' and 'the use of computers in the teaching process' was not a statistical significance,*Pillai's trace* =0.01, $F(5, 669) = 1.66$, $p=0.14$, *partial* $\eta 2 = 0.01$.

For this reason, after the Bonferroni adjustment, the modified alpha level was adjusted to be 0.01. Five two-way fixed-effects analyses of variance (ANOVAs) were computed to compare differences between 'the use of computers in the teaching process' (no computer use versus the use of computers) and 'school' of measurement (2012-2013 versus 2014-2015) on the main five factors. Both 'the use of computers in the teaching process' and 'school' were treated as fixed-effects- factors. The scores for the five factors served as dependent variables. In all cases, the analyses of variance showed that there were not significant interactions between the effects of 'school' and 'the use of computers in the teaching process' to all five factors. On the contrary, there was a significant effect of 'the use of computers in the teaching process' to the first, second, third and fifth factors of the attitudes of kindergarten teachers towards ICT (Table 10). Also, there was a significant effect of 'year' to the third and fifth factors as depicted in Table 9.

DISCUSSION

The results of the factor analysis produced the following five factors of the attitudes of kindergarten teachers towards ICT: a) How ICT helps children universally in the teaching process (F1), b) The use of ICT by the pupils as cognitive tool (F2), c) The changes caused by ICT on the role of the teacher in the classroom (F3), d) Kindergarten and primary school teachers use of ICT only to organize their adminis-

Table 9. Analyses of variance of 'school' on the five factors the attitudes of teachers towards ICT for 'the use of computers in the teaching process'

	2012-2013 Kindergarten		2014-2015 Primary school		ANOVA		
Factors	**M**	**SD**	**M**	**SD**	**F**		**Partial Eta Squared**
F1	3.51	.62	3.68	.50	2.92		.944
F2	4.14	.56	4.17	.50	.81		.229
F3	3.69	.75	4.08	.67	14.41	**	7.236
F4	4.30	.55	4.34	.57	.35		.109
F5	4.32	.58	4.05	.58	10.80	*	3.498

Note: *p<.01, **p<.001

Table 10. Analyses of variance of 'the use of computers in the teaching process' on the five factors of the attitudes of teachers towards ICT

	no		yes		ANOVA		
Factors	**M**	**SD**	**M**	**SD**	**F**		**Partial Eta Squared**
F1	3.53	.60	3.72	.51	7.64	*	2.470
F2	4.11	.56	4.26	.47	11.02	*	3.102
F3	3.74	.75	4.12	.66	14.64	**	7.352
F4	4.29	.58	4.38	.52	1.86		.579
F5	4.31	.57	3.97	.57	21.45	**	6.946

Note: *p<.01, **p<.001

trative work (F4), e) The conditions under which ICT may integrate into the classroom (F5). The above factors have characteristics similar to our initial conceptualization from the literature review (Yildirim 2000; Kumar and Kumar 2003; Tsitouridou and Vryzas 2003, 2004; Angeli 2004; Chen and Chang, 2006; Tondeur et al, 2007; Tao and Rosa Yeh 2008). These findings support the first research question.

Moreover, we can classify the factors into categories. The first category includes the first and the third factor. This category is characterized by the fact that computers can be used by teachers and students across the spectrum of activities in school and affects the teaching process and the role of teacher in the classroom. These factors are strongly believed and support the use of ICT in education. Also, the second category consists of the second and the fourth factor and considers that the integration of computers in the educational process may be successful only as a teaching aid (Slutsky, Kragh-Müller, Rentzou et al., 2019) and can also help the bureaucratic procedures (Sabanci & Omeroglu, 2015). In other words, these factors use the computer mainly as a cognitive tool. Finally, the third category includes only the fifth factor, which expresses high support on the conditions under which the computer may integrate into the classroom (Slutsky, Kragh-Müller, Rentzou et al., 2019). These conditions are used as an excuse in order to avoid the implementation of ICT in kindergarten.

Then, continued comparisons among the main factors of the year 2012-2013 classified into an ascending order, as it is described below. On the lowest level is the first factor. The fifth factor has a significantly greater value than the first. Then, the third factor follows with a significant difference from the previous factors. Also, the second factor follows with a significant difference. On the contrary, the fourth factor has the greatest value of all. Overall, the consecutive order of the main factors of the year 2012-2013 on the attitudes of kindergarten teachers towards ICT is as follows: F1, F5, F3, F2 and F4. The above findings addressed the second research question.

As we can see from the ascending order of the factors during the year 2012-2013, second category follows with larger values than all the others. Finally the first and the third category consists of the lowest values. This is probably because the greater majority they did not have the background to fully accept the incorporation of ICT in teaching process. Moreover, it appears from the above results that kindergarten teachers in 2012-2013 school year supported the use of computers as a cognitive tool more than the use of computers as a teacher's aid during the teaching process (Nikolopoulou & Gialamas, 2015).

Furthermore, the results of continued comparisons among the main factors of the kindergarten teachers classified them into a different ascending order than in the primary school teachers. An important change is that the fifth factor of kindergarten teachers' attitudes towards ICT, which demonstrates the conditions under which the ICT may integrate into the kindergarten, has been displaced from the lower position it held in the year 2012-2013 to the place between the second and fourth factor of the year 2014-2015. As a result, the fifth factor occupies the fourth position in the ascending order of the factors in the year 2014-2015. Thus the ordering of the factors has changed in: F'1, F'3, F'2, F'5 and F'4, and the third research question has been addressed. These results agree with other research indicating that nowadays that kindergarten teachers are more positive about the changes caused by ICT on the quality of learning for the children and the role of the kindergarten teachers themselves (Pelgrum 2001; Loveless and Dore 2002; Hu, Clark & Ma, 2003; Angeli & Valanides 2005; Zaranis & Oikonomidis 2009; Kalogiannakis 2010).

Also, the MANOVA analysis showed that the knowledge of kindergarten teachers on various software applications significantly influences the first, second and the fourth factor of their attitudes towards ICT (Nikolopoulou & Gialamas, 2015; Nasrin & Islahi, 2019). We observed that the more 'knowledge of software' from teachers influenced their attitudes on the use of ICT as: a) a cognitive tool, b) the organizing of the bureaucratic procedures in kindergarten and c) the use of computers in the educational process (Raman et al., 2015). Also, the analysis of variable 'school', showed that it significantly affects the third and the fifth factors of kindergarten and primary school teachers' attitudes. As a result the fourth research question has been addressed.

In addition, we have a similar situation with the effect of the hours of computer use by teachers at home during the week on their attitudes towards ICT (Balisane, 2015). We found that the hours of computer use by teachers at home during the week significantly influences the first, second, third and the fourth factor of their attitudes towards ICT. As a result the fifth research question has been addressed.

Finally, the MANOVA analysis showed that the computer use in the classroom during the educational process influences the first, second and the fifth factor of their attitudes towards ICT. As a result the fifth research question has been addressed.

CONCLUSION

The above discussion should be referenced in light of some limitations of this study. The first limitation of this study is that data was collected only from the cities of Athens and the island of Crete (Greece). The second limitation is the generalizability of this study which was limited specifically to participants attending public schools. Therefore, the results from this survey can be generalized only to similar groups of kindergarten and primary school teachers. The results may not adequately describe teachers from other regions of Greece.

The present survey finds that the factors of teachers' attitudes towards ICT have been changed among the kindergarten and the primary school teachers. The most important finding is that the fifth factor, which demonstrates the conditions under which ICT may integrate into school, changes its place in the ascending order during the period 2012-2013 of kindergarten sample, having moved to a lower position in the year 2014-2015 of primary school teachers. This may be due to the ICT support of elementary school while kindergarten schools do not.

We believe offering more ICT courses to educators may increase the percentage of teachers who support ICT in education (Namodi et al., 2015). We propose that the courses for ICT offered by universities to the teachers should have the greatest approach to educational practice and should not be consumed only in theoretical in the classroom (Raman et al., 2015). Thus, the emphasis should be on teaching with technology, instead of teaching about technology (Tondeur, van Braak, Siddiq, & Scherer, 2016). Moreover, it is necessary that the continuing in-service training of teachers should be oriented to the specific nature of their professional specialty.

The integration of computers in education is a direction for policy makers and individual educators to focus on. Thus, recent calls for educational reform in teacher education and the preeminent implementation of ICT, which is becoming more than passing trend, place the proper and formal training of pre-service teachers in effective ways of using ICT use as a top priority.

REFERENCES

Angeli, C. (2004). The effects of case-based learning on early childhood pre-service teachers' beliefs about the pedagogical uses of ICT. *Journal of Educational Media*, *29*(2), 139–151. doi:10.1080/1358165042000253302

Angeli, C., & Valanides, N. (2005). Preservice teachers as ICT designers: an instructional design model based on an expanded view of pedagogical content knowledge. *Journal of Computer Assisted Learning*, *21*(4), 292–302. doi:10.1111/j.1365-2729.2005.00135.x

Ashton, P. (2015). Historical overview and theoretical perspectives of research on teachers' beliefs. In H. Fives & M. G. Gills (Eds.), *International handbook of research on teachers' beliefs* (pp. 31–47). London: Routledge.

Balisane, H. (2015). Investigation of teachers' attitudes towards computer use in primary education in Kurdistan. *Journal of Education and Practice*, *6*(14), 37–49.

Bryman, A. (1989). *Research methods and organizational studies*. London: Unwin Hyman.

Chen, J., & Chang, C. (2006). Using computers in early childhood classrooms: Teachers' attitudes, skills and practices. *Journal of Early Childhood Research*, *4*(2), 169–188. doi:10.1177/1476718X06063535

Fisher, T., Denning, T., Higgins, C., & Loveless, A. (2012). Teachers' knowing how to use technology: Exploring a conceptual framework for purposeful learning activity. *Curriculum Journal*, *23*(3), 307–325. doi:10.1080/09585176.2012.703492

Gialamas, V., & Nikolopoulou, K. (2010). In-service and pre-service early childhood teachers' views and intentions about ICT use in early childhood settings: A comparative study. *Computers & Education*, *55*(1), 333–341. doi:10.1016/j.compedu.2010.01.019

Hair, J. F. Jr, Anderson, R. E., Tatham, R. L., & Black, W. G. (1998). *Multivariate data analysis* (5th ed.). London, UK: Prentice-Hall.

Hellenic Ministry of Greek National Education & Pedagogical Institute (HMNE-PI). (2003). *A cross thematic curriculum framework for compulsory education (Diathematikon Programma)*. Retrieved from http://www.pischools.gr/programs/depps/

Hu, P., Clark, T., & Ma, W. (2003). Examining technology acceptance by school teachers: A longitudinal study. *Information & Management*, *41*(2), 227–241. doi:10.1016/S0378-7206(03)00050-8

Jimoyiannis, A., & Komis, V. (2007). Examining teachers' beliefs about ICT in education: Implications of a teacher preparation program. *Teacher Development*, *11*(2), 149–173. doi:10.1080/13664530701414779

Kalogiannakis, M. (2010). Training with ICT for ICT from the trainer's perspective. A Greek case study. *Education and Information Technologies*, *15*(1), 3–17. doi:10.100710639-008-9079-3

Kumar, P., & Kumar, A. (2003). Effect of a web-based project on preservice and inservice teachers' attitude toward computers and their technology skills. *Journal of Computing in Teacher Education*, *19*(3), 87–92.

Liua, X., Toki, E. I., & Pange, J. (2014). The use of ICT in preschool education in Greece and China: A comparative study. *Procedia: Social and Behavioral Sciences*, *112*, 1167–1176. doi:10.1016/j.sbspro.2014.01.1281

Livingstone, S. (2012). Critical reflections on the benefits of ICT in education. *Oxford Review of Education*, *38*(1), 9–24. doi:10.1080/03054985.2011.577938

Loveless, A., & Dore, B. (Eds.). (2002). *ICT in the primary school*. Buckingham, UK: Open University Press.

Ma, W., Anderson, R., & Streith, K. (2005). Examining user acceptance of computer technology: An empirical study of student teachers. *Journal of Computer Assisted Learning*, *21*(6), 387–395. doi:10.1111/j.1365-2729.2005.00145.x

Mertala, P. (2018). *Two worlds collide? Mapping the third space of ICT integration in early childhood education*. (Ph.D. Dissertation, University of Oulu).

Ministry of Education & University of Ioannina. (2008). *Program for academic & professional upgrade of primary school teachers*. Retrieved from http://protagoras.edu.uoi.gr/eksomoiosi/ypoerg.htm

Mumtaz, S. (2000). Factors affecting teachers' use of information and communications technology: A review of the literature. *Journal of Information Technology for Teacher Education, 9*(3), 319–341. doi:10.1080/14759390000200096

Namodi, M. O., Nyaoga, O., Sirma, J., Obegi, F., & Moriasi, J. (2015). The relationship between teacher characteristics and Information and Communication Technology (ICT) integration in Nasrin & Islahi, F. (2019). Exploring teacher attitude towards information technology with a gender perspective. *Contemporary Educational Technology, 10*(1), 37–54.

Nikolopoulou, K., & Gialamas, V. (2015). ICT and play in preschool: Early childhood teachers' beliefs and confidence. *International Journal of Early Years Education, 23*(4), 409–425. doi:10.1080/09669760.2015.1078727

Palaiologou, I. (2016). Children under five and digital technologies: Implications for early years pedagogy. *European Early Childhood Education Research Journal, 24*(1), 5–24. doi:10.1080/1350293X.2014.929876

Pelgrum, W. (2001). Obstacles to the integration of ICT in education: Results from a worldwide educational assessment. *Computers & Education, 37*(2), 163–178. doi:10.1016/S0360-1315(01)00045-8

Pope, M., Hare, D., & Howard, E. (2002). ICT integration: Closing the gap between what preservice teachers are taught to do and what they can do. *Journal of Technology and Teacher Education, 10*, 191–203.

Namodi, M. O., Nyaoga, R., Sirma, J., Obegi, F., & Moriasi, J. (2015). The relationship between teacher characteristics and information and communication technology (ICT) integration in public secondary schools in Nakuru Town Sub-County, Kenya. *European Journal of Business and Management, 7*(30), 200–216.

Raman, A., Malik, B., Dzahir Kasa, M., & Sofian,, O. F. M., & Hussin, F. (2015). Teachers' attitude toward computer use in classroom practice. *Scholars Journal of Arts, Humanities, and Social Sciences, 3*(3A), 646–659.

Sabanci, A., & Omeroglu, M. (2015). Preschool teachers' views and experiences about ICT use in instruction: a case study. *International Journal of Academic Research in Business and Social Science, 5*(6), 170–183.

Sanchez, A. B., Marcos, J. J. M., Gonzalez, M., & Guanlin, H. (2012). In service teachers' attitudes towards the use of ICT in the classroom. *Procedia: Social and Behavioral Sciences, 46*, 1358–1364. doi:10.1016/j.sbspro.2012.05.302

Scherer, R., Siddiq, F., & Teo, T. (2015). Becoming more specific: Measuring and modeling teachers' perceived usefulness of ICT in the context of teaching and learning. *Computers & Education, 88*, 202–214. doi:10.1016/j.compedu.2015.05.005

Selinger, M. (2001). Learning information and communications technology skills and the subject context of the learning. *Journal of Information Technology for Teacher Education, 10*(1-2), 143–154. doi:10.1080/14759390100200108

Shade, D., & Davis, B. (1997). The role of computer technology in early childhood education. In J. Isenberg & M. Jalongo (Eds.), *Major trends and issues in early childhood education* (pp. 90–103). New York: Teachers College Press.

Slutsky, R., Kragh-Müller, G., Rentzou, K., Tuul, M., Gol Guven, M., Foerch, D., & Paz-Albo, J. (2019). A cross-cultural study on technology use in preschool classrooms: Early childhood teacher's preferences, time-use, impact and association with children's play. *Early Child Development and Care*, 1–13. doi:10.1080/03004430.2019.1645135

Tao, Y.-H., & Rosa Yeh, C.-C. (2008). Typology of teacher perception toward distance education issues – A study of college information department teachers in Taiwan. *Computers & Education, 50*(1), 23–36. doi:10.1016/j.compedu.2006.03.005

Tondeur, J., van Braak, J., Siddiq, F., & Scherer, R. (2016). Time for a new approach to prepare future teachers for educational technology use: Its meaning and measurement. *Computers & Education, 94*, 134–150. doi:10.1016/j.compedu.2015.11.009

Tondeur, J., van Braak, J., & Valcke, M. (2007). Towards a typology of computer use in primary education. *Journal of Computer Assisted Learning, 23*(3), 197–206. doi:10.1111/j.1365-2729.2006.00205.x

Tselfes, V., & Paroussi, A. (2008). Science and theatre education: a cross-disciplinary approach of scientific ideas addressed to student teachers of early childhood education. *Science & Education, 18*(9), 1115–1134. doi:10.100711191-008-9158-2

Tsitouridou, M., & Vryzas, K. (2003). Early childhood teachers' attitudes towards computer and information technology: the case of Greece. *Information Technology in Childhood Education*, 187-207.

Tsitouridou, M., & Vryzas, K. (2004). The prospect of integrating ICT into the education of young children: The views of Greek early childhood teachers. *European Journal of Teacher Education, 27*(1), 29–45. doi:10.1080/0261976042000211838

Umar, I. N., & Hassan, A. S. A. (2015). Malaysian teachers' levels of ICT integration and its perceived impact on teaching and learning. *Procedia: Social and Behavioral Sciences, 197*, 2015–2021. doi:10.1016/j.sbspro.2015.07.586

Wedman, J., & Diggs, L. (2001). Identifying barriers to technology-enhanced learning environments in teacher education. *Computers in Human Behavior, 17*(4), 421–430. doi:10.1016/S0747-5632(01)00012-7

Wetzel, K., Wilhelm, L., & Williams, M. K. (2004). The introductory technology course: A tool for technology integration. *Contemporary Issues in Technology & Teacher Education, 3*, 453–465.

Wheeler, S. (2001). Information and communication technologies and the changing role of the teacher. *Journal of Educational Media, 26*(1), 7–17. doi:10.1080/1358165001200069292

Yildirim, S. (2000). Effects of an educational computing course on preservice and inservice teachers: A discussion and analysis of attitudes and use. *Journal of Research on Computing in Education, 32*(4), 479–496. doi:10.1080/08886504.2000.10782293

Zaranis, N., & Oikonomidis, V. (2009) ICT in preschool education. Athens: Grigoris Publications (Text in Greek).

Zaranis, N., & Oikonomidis, V. (2014). The main factors of the attitudes of Greek kindergarten teachers towards information and communication technology. *European Early Childhood Education Research Journal*, 1–18. doi:10.1080/1350293X.2014.970853

Zaranis, N., & Oikonomidis, V. (2015). Profiling the attitudes of Greek kindergarten teachers towards computers. *Education and Information Technologies*, *20*(1), 201–215. doi:10.100710639-013-9296-2

ADDITIONAL READING

Avramidis, E., & Kalyva, E. (2007). The influence of teaching experience and professional development on Greek teachers' attitudes towards inclusion. *European Journal of Special Needs Education*, *22*(4), 367–389. doi:10.1080/08856250701649989

Jimoyiannis, A., & Komis, V. (2007). Examining Teacher's beliefs about ICT in Education. *Teacher Development*, *11*(2), 149–173. doi:10.1080/13664530701414779

Ma, W. W. K., Andersson, R., & Streith, K. O. (2005). Examining user acceptance of computer technology: An empirical study of student teachers. *Journal of Computer Assisted Learning*, *21*(6), 387–395. doi:10.1111/j.1365-2729.2005.00145.x

Papadakis, St., & Kalogiannakis, M. (2010). eTwinning in the early childhood as starting line of innovative practices for the didactic of natural sciences. In M. Kalogiannakis, D. Stavrou, & P.G. Michaelidis (Eds.) *Proceedings of the 7th International Conference on Hands-on Science, Bridging the Science and Society Gap Rethymno*, 25-31 July 2010, 235-240.

Papadakis, St., & Kalogiannakis, M. (2019). Evaluating a Course for Teaching Advanced Programming Concepts with Scratch to Preservice Kindergarten Teachers: A Case Study in Greece. In D. Farland-Smith (Ed.), *Early Childhood Education,* 1-19, UK-London: IntechOpen Limited https://www.intechopen.com/online-first/evaluating-a-course-for-teaching-advanced-programming-concepts-with-scratch-to-preservice-kindergart/

Papadakis, S., Kalogiannakis, M., & Zaranis, N. (2019). Parental involvement and attitudes towards young Greek children's mobile usage. *International Journal of Child-Computer Interaction*. doi: . doi:10.1016/j.ijcci.2019.100144

Tondeur, J., Van Braak, J., & Valcke, M. (2007). Towards a typology of computer use in primary education. *Journal of Computer Assisted Learning*, *23*(3), 197–206. doi:10.1111/j.1365-2729.2006.00205.x

KEY TERMS AND DEFINITIONS

Early Childhood Education: Is a broad term used to describe any type of educational program that serves children in their preschool years, before they are old enough to enter kindergarten.

ICT: Information and communications technology (ICT) is an extensional term for information technology (IT).

Preschool Education: Is education that focuses on educating children from the ages of infancy until six years old.

Preservice Teachers: Students enrolled in an initial educator preparation program.

Chapter 10
Investigating the Factors Influencing Pre-Service Teachers' Acceptance to Use Mobile Devices for Learning:
The Case of a Greek University

Georgios K. Zacharis
https://orcid.org/0000-0003-1158-9175
Aristotle University of Thessaloniki, Greece

ABSTRACT

This chapter determines the factors that significantly influence pre-service teachers' acceptance to use mobile devices as resources for learning in a university context. Based on the methodological framework of the UTAUT, a modified contextualized model of evaluation was created. A data collection instrument was designed, validated contextually, and optimized for mobile learning in higher education. A total of 320 Greek university students from a Faculty of Education participated in the study. Results demonstrated that the instrument constructed showed a notable internal consistency, with a high validity for data collection in 8 of its 9 factors. Results indicated Performance Expectancy, Effort Expectancy, Social Influence, Facilitating Conditions, Hedonic Motivation, and Empowerment in Learning as factors which affected participants' Behavioral Intention to use mobile technology for learning. Behavioral Intention, Social Influence and Empowerment in Learning affected university students' behavior to use mobile devices for learning.

INTRODUCTION

Mobile devices such as smartphones, tablets and laptops are gaining popularity as they are directly connected with our everyday life. Ownership of mobile devices is on the rise, with the majority of adults having more than one mobile device. It is worth mentioning that in the year 2012, the total number of

DOI: 10.4018/978-1-7998-1486-3.ch010

mobile devices exceeded the human population. Most mobile device users are between 18-29 years old, which is the standard age of University students (Crompton & Burke, 2018). New versions of mobile devices bring innovative features that makes them more convenient and affordable.

Mobile devices constitute one of the most popular emerging technologies, as they are light and convenient to carry and give access to multiple applications such as those related to health, banking and mobile learning (Alalwan, Dwivedi, & Rana, 2017; Briz-Ponce, Pereira, Carvalho, Juanes-Méndez, & García-Peñalvo, 2017; Crompton and Burke, 2018; Muñoz-Leiva, Climent-Climent, & Liébana-Cabanillas, 2017; Nikou and Economides, 2017). As a result, they might influence how students learn. Mobile technology for educational purposes, often referred to as mobile learning, has attracted researchers' attention for their pedagogical implications (Crompton and Burke, 2018; Hamidi & Chavoshi, 2018; Hamidi & Jahanshaheefard, 2019; Heflin, Shewmaker, & Nguyen, 2017; Kim et al., 2017; Nikou & Economides, 2017; Wu et al., 2012). The aim of mobile learning is to facilitate and extend teaching and learning anytime and anywhere to boost motivation, build knowledge, collect and exchange information, promote collaborative, as well as independent and lifelong learning (Sánchez-Prieto, Hernández-García, García-Peñalvo, Chaparro-Peláez, & Olmos-Migueláñez, 2019; Mojarro, Duarte, Guzmán, & Aguaded, 2019).

Undoubtedly, attitudes play an important role in our lives. On a daily basis, the way we accept people and situations determines our attitude and behavior towards them. The relationship between attitude and behavior not only matters in Interpersonal relationships but also in school environments (OECD, 2016). Over recent years, there has been an information and communication explosion in education. Therefore, stakeholders of the educational system - students and educators – must be prepared to use new technologies. This has brought studies on teachers' attitudes towards technology to the forefront as a key factor in the acceptance and adoption of technologies in education and generally in learning.

For teachers, the use of mobile technology in the classroom setting provides an opportunity to redefine teaching and learning (Heflin, Shewmaker, & Nguyen, 2017; Wu et al., 2012). Any effort to integrate technology into the educational process depends largely on whether teachers themselves support this effort (Teo, 2011). It is reasonable to assume that if teachers themselves do not believe that the use of technology will satisfy their own teaching and learning needs, they are likely to avoid using it. In contrast with other technologies, and despite the use of mobile phones by teachers in their daily lives as well as the high rate of use of mobile devices among teachers, this does not seem to automatically affect their teaching practice, or methodology (Sánchez-Prieto et al., 2019).

An important factor that influences teachers' intention to use technology is their attitude towards technology where research confirms that positive attitudes towards Information and Communication Technology (ICT) are directly related to their increased acceptance and use in the classroom (Scherer, Tondeur, Siddiq, & Baran, 2018). Students from the Faculty of Education (pre-school and primary education) of a Greek University were selected for the current study due to their position as current students and future vocation as teachers in addition to their active use of mobile devices in their everyday lives. Pre-service teachers as future teachers are the gatekeepers to technology integration in the classroom, including the inevitable use of mobile technologies in education and the development of mobile learning. In this context, the aim of the current study was to develop and evaluate an instrument based on Venkatesh, Morris, Davis & Davis's (2003) Unified Theory of Acceptance and Use of Technology (UTAUT) variables in order to identify the factors affecting pre-service teachers' Behavioral Intention and Use Behavior of mobile technologies for learning. Thus, the study at hand had a two-fold purpose: (a) to examine the factors of the modified UTAUT model that influence pre-service teachers' acceptance to use mobile technology for mobile learning; and (b) to examine the factors of the modified UTAUT model that influence pre-service teachers' actual use of mobile technology for mobile learning.

THEORETICAL BACKGROUND

Mobile Learning

Mobile learning is a relatively new term in ICT for education (Traxler & Vosloo, 2014) and its definition is constantly changing. It usually refers to the advantages provided by mobile devices and their applications in an educational context. According to West & Voslooo (2013) "mobile learning involves the use of mobile technology, either alone or in combination with other information and communication technology (ICT), to enable learning anytime and anywhere" (p. 6). The portability of digital technologies creates new opportunities for learning and a more flexible learning model that gives both teachers and students access to multiple sources of information, supporting the transition from a classic learning structure based on individual learning to a structure based on the concept of coexistence (Zydney & Warner, 2016). It empowers teachers to teach everywhere and anywhere, allowing continuous learning outside the classroom where learning is genuine (Huang, Lin, & Cheng, 2010).

Mobile technology provides students with different ways of associating their physical experience with scientific knowledge. The use of mobile technology applications can provide students with the conditions to become researchers, it generates autonomous learning (Gerger, 2014), encourages connectivity, interactivity and instant feedback, in a context of cooperative learning that promotes peer collaboration and collaborative feedback during the learning process (Jahnke & Kumar, 2014). In general, mobile technology helps students increase their engagement, technological awareness, and manages their self-directed learning to find answers to their questions and facilitate teamwork and socializing, in order to enable knowledge sharing, and therefore to successfully apply their learning outcomes (Al-emran, Elsherif, & Shaalan, 2016; Badia & Marta, 2016).

Mobile learning as a new research trend attracts many researchers to explore the use of mobile technologies, to study their impact on students and teachers and to develop the required infrastructure. Nowadays, there is an endeavor to incorporate mobile devices into higher education institutions as part of an educational context (Alemran, Elsherif, & Shaalan, 2016). Recent empirical evidence suggests that mobile learning can be used to support student learning in tertiary education environments (Crompton & Burke, 2018). It is also a known fact that many students bring their own mobile devices to the university, such as smartphones and tablets, and expect to access academic resources using them (Han & Sug, 2016). However, simple possession of mobile devices does not guarantee their effective use in the educational process by both students and educators. The attitude towards any educational technology could assess the extent to which students and educators have the intention to use technology as well as their level of acceptance and readiness for mobile learning (Ardies, De Maeyer, Gijbels, & van Keulen, 2014).

Models of Acceptance

The user acceptance of a new technology is considered an essential factor that determines the success or failure of the technology (Davis, 1993). Measuring teachers' acceptance of technology and intentions to use new technologies in their educational practice is a way of determining their willingness to

Figure 1. Technology acceptance model (Davis, 1989)

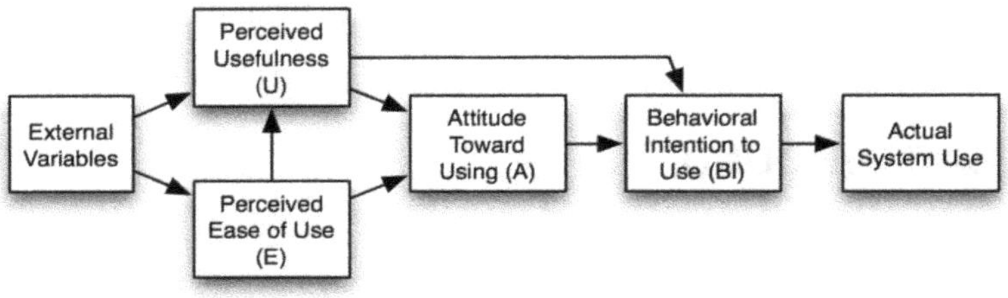

apply the latter (Scherer, Siddiq, & Tondeur, 2019). For Higher Institutions, the use of mobile devices into educational practices is crucial when considering students' needs and technology acceptance. In this context, several models of technology acceptance have been proposed to describe the mechanism behind the factors affecting technology adoption, such as the Technology Acceptance Model (TAM), the Unified Theory of Acceptance and Use of Technology (UTAUT) and the extended Unified Theory of Acceptance and Use of Technology (UTAUT2).

The Technology Acceptance Model (TAM) (Figure 1) proposed by Davis (1985), was developed from the Theory of Reasoned Action (TRA). The model claims that the actual use of a technology depends on the person's intention to use it and proposes Perceived Usefulness and Perceived Ease of Use as the two main predictors of acceptance and use of a new information system or technology. Attitudes toward technology is another core variable. The outcome variables of the model are Behavioral Intention and Actual System Use. The external variables of the model such as Subjective Norms, Self-Efficacy, and Facilitating Conditions explain variations in Perceived Usefulness and Perceived Ease of Use (Sánchez-Prieto et al., 2019; Scherer, Siddiq, & Tondeur, 2019).

The TAM was frequently used as a research tool to investigate the acceptance of technological innovations by users. However, the TAM can only explain 40% of the variance in technology acceptance. This is because the model considers factors regarding personal beliefs about the use of technology but does not include theoretical constructs spanning social and contextual influence processes (Venkatesh & Davis, 2000; Venkatesh, Morris, Davis, & Davis, 2003).

In 2003, Venkatesh, Morris, Davis, & Davis, designed the Unified Theory of Acceptance and Use of Technology (UTAUT), which incorporates eight original models and theories of individual acceptance and can explain 70% of the variance in user intention to use technology thus suggesting that the UTAUT model is the most effective model for analyzing technology acceptance. This model includes: The Theory of Reasoned Action (TRA), the Technology Acceptance Model (TAM), the Motivational Model (MM), the Theory of Planned Behavior (TPB), the combined TAM and TPB (C-TAM-TPB), the Model of PC Utilization (MPCU), the Innovation Diffusion Theory (IDT) and Social Cognitive Theory (SCT). The UTAUT model consists of four critical factors influencing Behavioral Intention and Use Behavior to use technology and are namely Performance Expectations, Effort Expectancy, Social Influence and Facilitating Conditions while moderators of the four factors are Gender, Age, Experience and Voluntariness of Use (Figure 2).

To overcome this limit and to better explain individuals' technology acceptance, Venkatesh, Thong, & Xu (2012) extended the UTAUT model (Figure 3) with three new factors and as such presented the extended UTAUT (UATUT2). These new factors are Hedonic Motivation, Price and Habit. The researchers

Figure 2. Unified theory of acceptance and use of technology (UTAUT) (Venkatesh et al., 2003)

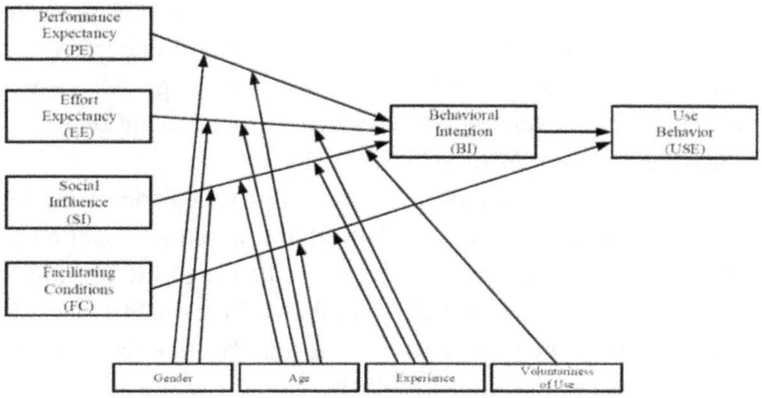

claimed that the suggested additions in UTAUT2 exhibit significant changes in the variance explained in Behavioral Intention and Use of Technology.

LITERATURE REVIEW

The TAM, UTAUT and extended versions of these have been widely used to investigate the factors affecting university students' Behavioral Intention (BI) and Use Behavior (UB) of mobile technologies for learning. Phuangthong & Malisawan (2005) reported that TAM was helpful to understand factors affecting mobile learning adoption with 3G technology while Chang, Yan, & Tseng (2012) reported an extended TAM framework as effective at predicting and explaining the acceptance of English mobile learning from college students in Taiwan. Teo & Noyes (2011) extended the TAM framework by adding

Figure 3. Extended unified theory of acceptance and use of technology (UTAUT2) (Venkatesh et al., 2012)

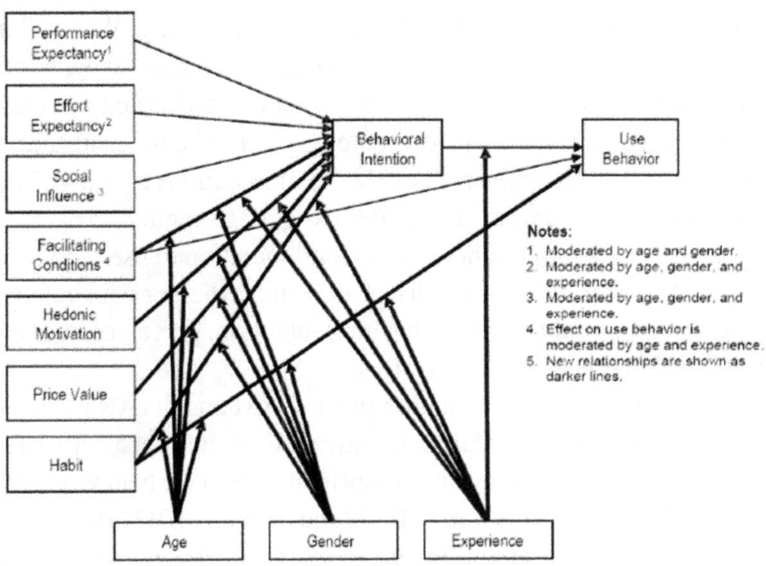

subjective norm, facilitating conditions, and technological complexity as external variables in order to examine pre-service teachers' attitudes toward the use of computers. Results show that that the proposed model explained 48.7% of the attitude for computer use. Teo (2012) integrated the TAM and TPB to examine pre-service teachers' self-reported intention to use technology. The results indicated that the TAM constructs were significant in explaining pre-service teachers' intention to use technology, which is in agreement with the study of Park, Nam, & Cha (2012).

Yang (2013) studied undergraduate students' mobile learning acceptance in a consumer context based on an extended UTAUT model in China. The results from the structural equation analysis showed that Hedonic Motivation, Performance Expectancy and Social Influence positively affected students' mobile learning adoption. This is in contrast with the studies of Iqbal & Qureshi (2012) and Jackman (2014) who reported that Social Influence had a negative impact or no significant effect in undergraduate students' acceptance of mobile learning technologies respectively.

Uğur, Koç & Koç (2016) conducted an analysis on mobile learning acceptance among college students in Turkey. The regression analysis revealed that the four UTAUT predictors affected students' behavioral intention to use mobile learning in agreement with the findings of similar studies (Nassuora, 2013; Thomas, Singh, & Gaffar, 2013; Zainol, Yahaya, Yahaya, & Zain, 2017). Onaolapo & Oyewole (2018) applied the UTAUT model to examine the effect of Performance Expectancy, Effort Expectancy and Facilitating Conditions on the use of smart phones for mobile learning by postgraduate students in the University of Ibadan, Nigeria. Performance Expectancy was found to be the strongest predictor of smart phone use for mobile learning in accordance with the studies of Kim-Soon, Ibrahim, Ahmad & Sirisa (2015), Mtebe & Raisamo (2014) and Jawad & Hassan (2015). In addition, Jackman (2014) determined the critical factors influencing undergraduate students' acceptance of mobile learning technologies at the University of the West Indies in Barbados. The results from the multiple linear regression analysis revealed that Performance Expectancy, Effort Expectancy, and Facilitating Conditions were significant positive determinants of students' intentions to adopt mobile technologies for learning, whereas Social Influence had no significant effect in this context. Tey & Moses (2018) explored the influence of UTAUT factors on undergraduate students' behavioral intention to use technology. The results from the regression analysis indicated that the undergraduates' behavioral intention to use technology was significantly influenced by Effort Expectancy, Performance Expectancy and Social Influence in agreement with the studies of Jairak, Praneetpolgrang, & Mekhabunchakij (2009) and Wang, Wu, & Wang (2009).

Using the UTAUT model as a theoretical framework, Darren Pullen, Swabey, Abadooz, & Ranjit Sing (2015) examined Malaysian pre-services teachers' use of mobile technology in education. The results showed that performance expectancy, effort expectancy, social influence, attitude toward technology and self-efficacy were significant determinants of behavioral intentions to use mobile devices for learning. Sánchez-Prieto, Olmos-Migueláñez, & García-Peñalvo (2016) applied an extended TAM to explore Spanish pre-service primary education teachers' behavioral intention to use mobile devices within their future teaching practice. According to the results of the study, the stronger relationships were those established between perceived usefulness and behavioral intention, perceived ease of use and perceived usefulness, and self-efficacy and perceived ease of use.

Baydas & Yilmaz (2018) applied a combination of the TAM and UTAUT model to determine pre-service teachers' intentions to adopt mobile learning from a motivational perspective. Pre-service teachers' attitudes and cognitive needs were found to have an influence on their behavioral intention while their affective and social needs did not. Adel Ali & Rafie Mohd Arshad (2018) developed a research model through extending the UTAUT by incorporating two additional factors namely; learners' autonomy and content quality design. The results highlighted that learners' autonomy, performance expectancy, facilitating conditions, and social influence are significant in relation to behavioral intention to use mobile learning while effort expectancy (EE) did not have any impact on the intention to use mobile learning.

Drawing on the Technology Acceptance Model (TAM), Çakıroğlu & Öztürk, (2017) examined Turkish pre-service teachers' adoption of mobile technologies through factors such as current use, instructional use and future use in their teaching practices. Results indicated that current use and instructional use had a strong positive correlation. In addition, there was a similar correlation with the factor of future use and current use. Relationships among current, instructional and future use of mobile technologies were explained within the context of the Perceived Usefulness, Ease of Use and Behavioral Intention constructs of the TAM. Mojarro, Dcuarte, Guzmán & Aguaded (2019) aimed to determine the factors that significantly influence the acceptance and use of smartphones and tablets as resources for learning in a university context. Results from statistical analyses are in agreement with the study of Chaka & Govender (2017) and reveal Performance Expectancy, as well as Effort Expectancy to direct and significantly influence students' intention to use and accept mobile devices for learning.

METHODOLOGY

Students form the Faculty of Education (pre-service teachers) together with in-service teachers, are the stakeholders to integrate mobile learning into classrooms as young children tend to use mobile technology in their lives. In this context, the main purpose of this research was to inquire into the factors that affect the technological acceptance and intent to use mobile devices for learning among university students. Pre-service teachers were selected for the current study due to their position as current students and future vocation as teachers. Students from the Faculty of Education (kindergarten and primary education) as future teachers are the gatekeepers to technology integration in the classroom along with the inevitable use of mobile technologies in education and the development of mobile learning.

A Google online questionnaire was developed based on the survey recommended by Venkatesh et al. (2003) and administered between April and May 2019. Empirical data was collected by means of the online questionnaire containing 37 items. Email addresses were obtained from faculty departments and the link to the questionnaire was emailed to all students (undergraduates and postgraduates) enrolled in both Schools of Faculty of Education. Respondents were kindly asked to forward the link to the questionnaire. Two weeks after the first email was sent requesting participation, a reminder was sent to the same list. After four weeks, the data collection process ended. Overall 320 pre-service teachers completed the online questionnaire.

The research instrument was divided into two parts. The first part included demographic statements about students' gender, age, year of studies, use of mobile devices in daily life, ownership of an internet connection and mobile device. The second included items used to measure the modified UTAUT constructs.

Demographic Statements

Participants in the current research were 320 undergraduate students from the Faculty of Education (School of Primary Education and School of Early Childhood Education), of the Aristotle University of Thessaloniki, in Greece. Greek university students seem to be familiar with mobile devices, as they own at least one mobile device. They also use mobile devices to communicate or search for information.

As shown in Table 1, the demographical characteristics revealed that 299 (93.4%) of the study participants were women whereas 21 participants were men (6.6%), which is common for a Faculty of Education in Greece. The mean age was 22.07 years (STD = 5.98). Concerning the faculty of the participants, there appears to be a slightly higher participation rate (162 students, 50.6%) from students at the School of Early Childhood Education than from the School of Primary Education (158 students, 49.4%).

Results in Table 2 show that university students access the internet via a smartphone (219 students, 68.4%) or laptop (85 students, 26.6%) while 286 students (89.4%) own a fixed broadband connection. As for the frequency of use of mobile devices, 309 students (96.6%) stated "always" which is significantly high. Finally, all university students own a smartphone (100.0%) while the majority of them also own a laptop (204 students, 63.7%).

Research Framework and Hypotheses

In this study, the UTAUT model was chosen as a ground theory to develop the research model for investigating pre-service teachers' perceptions on using mobile devices for learning because of its novelty in technology acceptance research, especially in the mobile learning technology field. Increasing the number of external variables can enhance a model's ability to predict the acceptance of information technology (Chao, 2019). The UTAUT model was adapted to a context of use of mobile devices for learning from pre-service teachers and modified by incorporating new constructs. Hedonic Motivation and Habit (Venkatesh et al., 2012) respectively are the extent to which university students experience enjoyment and pleasure from using mobile devices to learn and the extent to which they believe that using mobile devices is an automatic response. Given that university students use mobile devices to access educational material and information, another factor, Empowerment in Learning (EiL) (Blackwell,

Table 1. Profile of university students

			Percentage
Gender	Female	299	93.4%
	Male	21	6.6%
	Total	393	100%
Age	<25	283	88.4%
	25-30	9	2.8%
	>30	28	8.8%
Field	School of Primary Education	162	50.6%
	School of Early Childhood Education	158	49.4%
	Total	320	100%
Year of studies	A	94	29.4%
	B	72	22.5%
	C	78	24.4%
	D	46	14.4%
	Graduate	30	9.4%
	Total	320	100%

Table 2. Access end use of technology by the participants

			Percentage
To access the Internet I mainly use a	Smartphone	219	68.4%
	Tablet	4	1.3%
	Laptop	85	26.6%
	Desktop	12	3.8%
	Total	320	100%
I access the internet via a fixed broadband connection	Yes	286	89.4%
	No	34	10.6%
	Total	320	100%
I use the internet	Always	309	96.6%
	Often (2-3 times a week)	11	3.4%
	Sometimes (about once week)	0	0%
	Rarely	0	0%
	Total	320	100%
Mobile devices I own	Smartphone	320	100.0%
	Laptop	204	63.7%
	Tablet	66	20.6%

Lauricella, Wartella, Robb, & Schomburg, 2013) was added to the questionnaire. EiL is defined as the extent to which university students believe that using mobile devices empowers their learning.

The dependent variables as shown in Table 3 were Behavioral Intention (BI) and Use Behavior (UB) of mobile devices in educational practices. In addition, this study did not investigate the effect of Gender, Age or Experience on one's Behavioral Intention to use mobile devices for learning. The majority of students in Higher Education are of almost the same age and with small variations in technological experience as shown in Table 1, similar to findings obtained in other studies around the globe (Chaka & Govender, 2017; Jairak, Praneetpolgrang, & Mekhabunchakij, 2009; Nassuora, 2013). Price Value was excluded seeing that it was irrelevant to the purpose of the study. The independent variables of the questionnaire as shown in Table 3 were Performance Expectancy (PE), Effort Expectancy (EE), Social Influence (SI), Facilitating Conditions (FC), Hedonic Motivation (HM), Habit (H) and Empowerment in Learning (EiL).

All items were measured using a five-point Likert scale, with the anchors being "strongly disagree" and "strongly agree." The online questionnaire was first conducted in English and reviewed for content validity by a group of experts. As the questionnaire was administered in Greece, the English questionnaire was translated to Greek and then back to English to ensure translation equivalence. A professional translator and two research assistants independently translated the original items from English into Greek. They analyzed the independently translated Greek versions of the items and came to an agreement on the final version for the questionnaire. Another professional translator translated the questionnaire to confirm translation equivalence. To reduce measurement error, the development of an online questionnaire involved the following steps. First, two experts from the field of ICT in Education were asked to take part in the pre-test of the questionnaire. Experts were asked to check the content of the items being

Table 3. Independent and Dependent Variables of the research model

Type	Variable codes
Independent Variables	Performance Expectancy (PE) Effort Expectancy (EE) Social Influence (SI) Facilitating Conditions (FC) Hedonic Motivation (HM) Habit (H) Empowerment in Learning (EiL)
Dependent Variables	Behavioral Intention (BI) Use Behavior (UB)

measured and to examine the meaningfulness, relevance and clarity of the questionnaire and then return the questionnaire together with their observations. With respect to the feedback, we refined the wording of the questionnaire. The questionnaire is attached as Appendix A.

Research Model and Hypotheses

Figure 4 presents the proposed research model that measures pre-service students' acceptance and intention to use mobile devices for learning and the hypothesized relationships between the variables.

Based on the online questionnaire described above and the scales of the acceptance and intention of use of mobile devices from pre-service teachers, nine hypotheses were put forward.

Figure 4. Conceptualized modified UTAUT model for measuring pre-service students' acceptance and intention to use mobile devices for learning

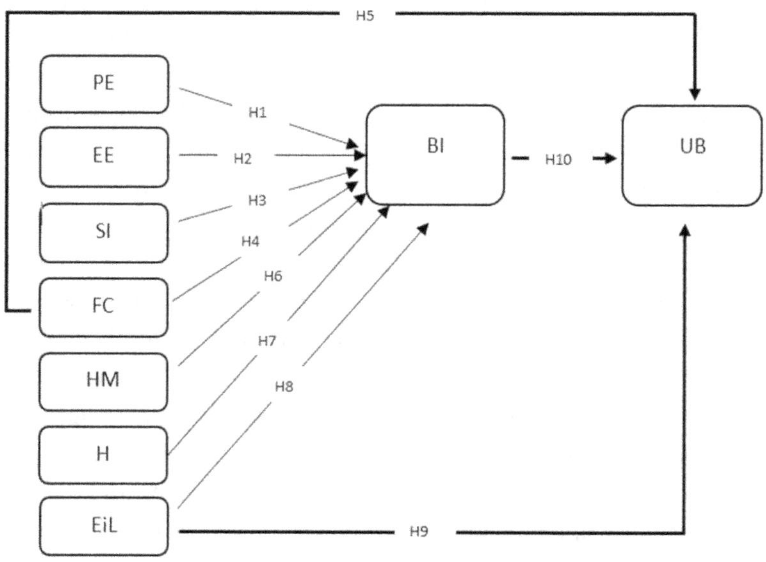

Performance Expectancy

Venkatesh et al. (2003, p. 447) defined Performance Expectancy (PE) as "*the degree to which an individual believes that using the system will help him or her to attain gains in job performance*". PE is the strongest predictor of intention to use a technology (Venkatesh et al., 2003; Venkatesh et al., 2012). In the current study, PE stands for university students' belief that the use of mobile devices will contribute to his or her teaching performance. Accordingly, the following hypothesis was proposed:

Hypothesis One: Performance Expectancy will have a significant influence on university students' Behavioral Intention to use mobile devices to learn.

Effort Expectancy

Venkatesh et al. (2003, p. 450) defined Effort Expectancy (EE) as "*the degree of ease associated with the use of the system*". When a system is perceived to be easier to use then the perceived ease of use is more likely to induce perception BI. In the current study, EE represents university students' beliefs about the ease of using mobile devices to learn. According to the base UTAUT model, the following hypothesis was put forward:

Hypothesis Two: Effort Expectancy will have a significant influence on university students' Behavioral Intention to use mobile devices to learn.

Social Influence

Venkatesh et al. (2003, p. 451) defined Social Influence (SI) as "*the degree to which an individual perceives how important others believe he or she should use the new system*". In the context of mobile learning, vital others are the teachers, parents, peers, etc. In the current study, SI stands for university students' perceptions about how important other people believe they should use mobile devices. The researchers stated that SI was a direct determinant of BI and UB. Accordingly, the following hypotheses were proposed:

Hypothesis Three: Social Influence will have a significant influence on university students' Behavioral Intention to use mobile devices to learn.

Hypothesis Four: Social Influence will have a significant influence on university students' Use Behavior to use mobile devices to learn.

Facilitating Conditions

Venkatesh et al. (2003, p. 453) defined Facilitating Conditions (FC) as "*the degree to which an individual believes that an organizational and technical infrastructure exists to support use of the system*". In this study, the FC represents university students' belief about the existence of support that is necessary to enable the use of mobile devices in a university context. According to the base UTAUT model, the following hypothesis was put forward:

Hypothesis Five: Facilitating Conditions will have a significant influence on the pre-service students' Behavioral Intention to use mobile devices to learn.

Hedonic Motivation

Venkatesh et al. (2012, p. 161) defined Hedonic Motivation (HM) as "*the fun or pleasure derived from using a technology which has been shown to play an important role in determining technology acceptance and use*". In the proposed research context, HM is the extent of university students' experience of enjoyment and pleasure from using mobile devices to learn. HM will positively influence university students' intention to accept and use mobile learning. According to the base UTAUT2 model, the following hypothesis was put forward:

Hypothesis Six: Hedonic Motivation will have a significant influence on university students' Behavioral Intention to use devices to learn.

Habit

Habit was differentiated in two distinct ways according to Venkatesh et al. (2012). As prior automatic behavior because of learning and as the extent to which an individual believes the behavior to be automatic. In this study, we added Habit as a predictor of university students' automatic behavioral intention to use mobile devices. According to the extended UTAUT model, the following hypothesis was put forward:

Hypothesis Seven: Habit will have a significant influence on university students' Behavioral Intention to use mobile devices to learn.

Empowerment in Learning

Students perceived Empowerment in Learning from the use of mobile devices (Blackwell et al., 2013). In the proposed research context, this is the extent to which university students believe that using mobile devices empower their learning. EiL is a predictor of pre-service students' BI and UB to use mobile devices for learning. Therefore, the following hypotheses were proposed:

Hypothesis Eight: Empowerment in Learning will have a significant influence on university students' Behavioral Intention to use mobile devices to learn.

Hypothesis Nine: Empowerment in Learning will have a significant influence on university students' Use Behavior of mobile devices to learn.

Behavioral Intention

Behavioral Intention (BI) is a significant determinant behind the actual use of technology in different intention models (Venkatesh et al., 2003; Venkatesh et al., 2012). In the proposed research context, BI is the extent to which pre-service students are willing to use and continue to use mobile devices to learn. Therefore, the following hypothesis was proposed:

Hypothesis Ten: Behavioral Intention will have a significant influence on university students' Use Behavior of mobile devices to learn.

Use Behavior

In the proposed research context, Use Behavior (UB) is the extent to which pre-service students use mobile devices for educational purposes in the context of higher education.

Table 4 summarizes the hypotheses of the current study.

RESULTS

Analysis Validity and Reliability

The measurement model was assessed by examining internal reliability and convergent validity. Convergent validity (CV) is the correlation between the multiple indicators of an aspect. In this study, in addition to Cronbach's alpha, composite reliability (CR) and the average variance extracted (AVE) were examined for all constructs. Measures of the average variance extracted (AVE) was assessed for the evaluation of CV. The analysis of the results was conducted with the SPSS v26 software.

The reliability and consistency of the instrument were verified, obtaining a high Cronbach's alpha (0.874, Number of items = 29) well above DeVellis's (2016) recommendations (> .70). Their validity was also measured and included in the measurement of each of the constructs that were proposed, and correlations between them were calculated for the ad hoc observation of their relationships with the BI and UB dependent variables.

Construct reliability indicates how well a construct is measured by its items, and can be assessed based on Cronbach's alpha and CR. The internal consistency reliability and construct validity of each variable was assessed by computing Cronbach's alpha coefficients with a range from 0.648 for FC to 0.942 for BI (Table 5). Cronbach's alpha or ordinal alpha higher than 0.6 was considered sufficient, higher than

Table 4. Hypothesis of the research

	Hypothesis
H1	Performance Expectancy will have a significant influence on Behavioral Intention to use mobile devices to learn.
H2	Effort Expectancy will have a significant influence on Behavioral Intention to use mobile devices to learn
H3	Social Influence will have a significant influence on Behavioral Intention to use mobile learning.
H4	Facilitating Conditions will have a significant influence on Behavioral Intention to use mobile devices to learn
H5	Facilitating Conditions will have a significant influence on the Use Behavior of mobile devices to learn.
H6	Hedonic Motivation will have a significant influence on Behavioral Intention to use mobile devices to learn.
H7	Habit will have a significant influence on Behavioral Intention to use mobile devices to learn.
H8	Empowerment in Learning will have a significant influence on Behavioral Intention to use mobile devices to learn.
H9	Empowerment in Learning will have a significant influence on the Use Behavior of mobile devices to learn.
H10	Behavioral Intention will have a significant influence on the Use Behavior of mobile devices to learn.

0.7 was acceptable, and higher than 0.8 was good. A lower value of Cronbach's alpha suggests an insufficient number of questions or poor inter-relatedness between the questions and factors tested. As can be observed, those related to Habit (H) did not exceed the degrees of confidence needed (0.206<<0.7) and were excluded from the research. With eight elements left, Cronbach's alpha was recalculated and found 0.891 (Number of items = 26).

As shown in Table 6, CR values ranged from 0.81 for FC to 0.97 for BI and exceeded the recommended cutoff of 0.7 suggesting high internal reliability (Hair, Black, Babin, & Anderson, 2010). In addition, AVE ranged from 0.59 (FC) to 0.94 (BI) indicating CV as it was greater than 0.5 for each construct (Hair, et al., 2010).

Level of Acceptance of Mobile Learning

In order to provide an understanding of the level of perception and acceptance of mobile learning a statistical analysis using a mean value and standard deviation was conducted and shown in Table 7.

The results show that the majority of university students in a Faculty of Education of a Greek university have a high level of acceptance for mobile learning, which indicates a high level of acceptance of mobile learning from students as demonstrated with the mean value of the PE indicator $\left(\overline{x} = 3.84\right)$. University students tend to strongly agree that mobile devices are easy to use and easy to facilitate learning with them, as evidenced by the high EE mean value ($\overline{x} = 4.13$). Also, the results suggest that university students may be influenced by others who think they should use mobile devices as the mean value of SI was high ($\overline{x} = 3.26$). Descriptive statistics support students' beliefs that they have the necessary resources, knowledge and support to use mobile learning at a moderate level (mean value of FCs, $\overline{x} = 4.10$). The HM mean value ($\overline{x} = 3.82$) indicates university students' feeling of enjoyment or pleasure using mobile technology. University students' perceived Empowerment in Learning due to the use of mobile devices as EiL has a high mean value ($\overline{x} = 4.29$) and the result suggests a strong use of mobile learning in terms of behavioral intention as the mean value of BI was 3.50.

Table 5. Reliability of each factor of scale used

Variables	No items	Cronbach's Alpha
PE	4	0.844
EE	4	0.828
SI	4	0.796
FC	3	0.648
HM	2	0.838
H	3	0.189
EiL	3	0.775
BI	2	0.942
UB	4	0.780

Table 6. Construct reliability results

Variable	No. of Items	AVE	CR
PE	4	0.68	0.90
EE	4	0.66	0.89
SI	4	0.62	0.87
FC	3	0.59	0.81
HM	2	0.86	0.93
EiL	3	0.69	0.87
BI	2	0.94	0.97
UB	4	0.61	0.86

Table 7. Descriptive data regarding the factors

Factor	Mean	S.D.
PE	3.84	0.75
EE	4.13	0.72
SI	3.26	0.78
FC	4.10	0.66
HM	3.82	0.86
EiL	4.29	0.64
BI	4.48	0.77
UB	3.84	0.75

Regression Analysis

The Kaiser–Meyer–Olkin (KMO) test for sampling adequacy and Bartlett's test for sphericity were used to first examine the suitability of selected variables for the factor analysis (Bryman, 1989). The KMO test resulted in a 0.862 value that was greater than the suggested minimum value of 0.5 for adequacy, and Bartlett's test also demonstrated a very good sphericity (v2 = 4165.742, df = 325, $p < 0.001$).

A regression analysis was conducted to assess the relationship between six main factors (PE, EE, SI, FC, HM, EiL) and BI to use mobile learning and the relationship between three main factors (FC, EiL, BI) and UB. Table 8 illustrates the graphical presentation of the β-value for each of the factors.

According to the results from the regression analysis, nine of ten hypotheses were supported (exclusion of the Habit factor from the research automatically excluded hypothesis 7), as shown in Table 9.

Specifically, according to Table 9 all six factors (PE, EE, SI, FC, HM and EiL) were significantly correlated with the Behavioral Intention of students to use mobile devices to learn, and in turn supported Hypothesis 1, Hypothesis 2, Hypothesis 3, Hypothesis 4 and Hypothesis 6. Facilitating conditions, Empowerment in Learning and Behavioral Intention of using mobile devices also significantly predicted Use Behavior and supported Hypothesis 5, Hypothesis 9 and Hypothesis 10.

Table 8. Inter-Item correlation matrix

	PE	EE	SI	FC	HM	BI	EiL	UB
PE	1							
EE	0.203**	1						
SI	0.445**	0.081	1					
FC	0.162**	0.597**	0.071	1				
HM	0.436**	0.415**	0.279**	0.339**	1			
BI	0.322**	0.285**	0.210**	0.352**	0.381**	1		
EiL	0.559**	0.307**	0.299**	0.321**	0.474**	0.496**	1	
UB	0.258**	0.240**	0.108*	0.316**	0.339**	0.413**	0.513**	1

**. Correlation is significant at the 0.01 level (2-tailed).

*. Correlation is significant at the 0.05 level (2-tailed).

Table 9. Hypotheses conclusions

Hypothesis	Result	Conclusion
H1. PE will have a significant influence on BI.	Yes: Significant (Beta = 0.322, $p < 0.001$)	Supported
H2. EE will have a significant influence on BI.	Yes: Significant (Beta = 0.285, $p < 0.001$)	Supported
H3. SI will have a significant influence on BI.	Yes: Significant (Beta = 0.210, $p < 0.001$)	Supported
H4. FC will have a significant influence on BI.	Yes: Significant (Beta = 0.352, $p < 0.001$)	Supported
H5. FC will have a significant influence on UB	Yes: Significant (Beta = 0.316, $p < 0.001$)	Supported
H6. HM will have a significant influence on BI.	Yes: Significant (Beta = 0.322, $p < 0.001$)	Supported
H7. H will have a significant influence on BI.	Excluded	Not Supported
H8. EiL will have a significant influence on BI.	Yes: Significant (Beta = 0.496, $p < 0.001$)	Supported
H9. EiL will have a significant influence on UB.	Yes: Significant (Beta = 0.513, $p < 0.001$)	Supported
H10. BI will have a significant influence on UB.	Yes: Significant (Beta = 0.413, $p < 0.001$)	Supported

Figure 5 presents the research models' explanatory power. The model explains a substantial portion of the variance in all endogenous variables: PE (32.2%), EE (28.5%), EiL (49.6%) and HM (38.1%). The model explains a 29.8% and 34.8% of the BI and UB variance respectively of pre-service teachers' intention to use mobile devices and technologies for learning.

Discussion

The current study, based on a modified model of the UTAUT, examined the factors affecting university students' Behavioral Intention and Actual Use of mobile learning. The results of the regression analysis highlighted interesting findings. The research model explained 29.8% of the variance in Behavioral Inten-

Figure 5. Results of the research model

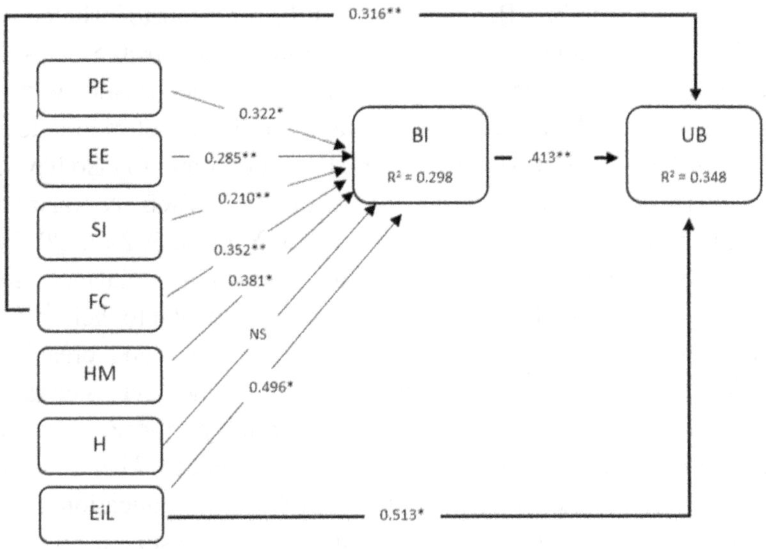

* p<0.05, ** p<0.01, NS = Not Supported (factor excluded)

tion suggesting that adding external factors to the proposed model of the study will possibly increase the model's ability to explain pre-service students' behavioral intention to use mobile devices for learning. According to the proposed model, all six factors studied in this research had a positive impact on university students' intention to use mobile devices for their learning. These findings are consistent with previous research in which most, or all of the factors examined, are positive contributors to one's behavioral intention to adopt mobile learning (Chaka & Govender, 2017; Gaffar, 2013; Nassuora, 2013; Tey & Moses, 2018; Thomas, Singh, & Yang, 2013; Uğur, Koç, & Koç, 2016; Wang, Wu, & Wang, 2009). Such results confirm the value of the UTAUT as a tool to predict behavioral intention to use mobile technologies. The three factors most affecting BI to adopt mobile learning were EiL ($\beta = 0.496$, $p<0.01$), HM ($\beta = 0.322$, $p<0.01$) and PE ($\beta = 0.322$, $p<0.01$). All these factors seem to be person-related, in contrast to context-related factors like FC ($\beta = 0.316$, $p<0.01$) and SI ($\beta = 0.210$, $p<0.01$), that are less significant positive predictors. EiL, as a determinant of BI to adopt mobile learning has not been adequately studied, although it seems to be the most influential factor of the modified UTAUT model. The next two most influential factors, FC and EE, have also been found to be significant positive indicators of BI to adopt mobile learning (Chaka & Govender, 2017; Jackman, 2014; Jairak, Praneetpolgrang, & Mekhabunchij, 2009; Mtebe & Raisamo, 2014; Nassuora, 2013; Onaolapo & Oyewole, 2018; Thomas, Singh, & Gaffar, 2013; Uğur, Koç, & Koç, 2016; Venkatesh et al., 2003; Wang, Wu, & Wang, 2009). Although SI was found to be the less significant positive factor for BI this may indicate that despite university students being familiar with mobile devices given that they own and use them daily, important third parties (like family, teachers and friends) can influence their intention to use mobile devices for learning. This result is in line with existing studies (Chaka & Govender, 2017; Jairak, Praneetpolgrang, & Mekhabunchakij, 2009; Tey & Moses, 2018; Thomas, Singh, & Gaffar, 2013; Uğur, Koç, & Koç, 2016). However it is in contrast with studies that reveal that SI had no or negative effect on Behavioral Intention (Iqbal & Qureshi, 2012; Jackman, 2014; Kim-Soon, Ibrahim, Ahmad, & Sirisa, 2015), which may be due to a different cultural context of the studies.

The research model also explained 34.8% of the variance in Use Behavior. UB of mobile devices for learning was examined in relation to EiL, BI and FC, as these factors seem to influence university students' use of mobile technologies. According to the results of the study, EiL ($\beta=0.513$, $p<0.01$) affected more the UB of university students for mobile learning. This result is in accordance with (Cacciamani et al., 2018) which also indicate EiL as the most significant factor in the use of mobile learning in secondary students. BI ($\beta=0.413$, $p<0.01$) and FC ($\beta=0.316$, $p<0.01$) were found to also have a significant effect on the UB of mobile devices for learning (Jawad & Hassan, 2015; Mtebe & Raisamo, 2014; Onaolapo & Oyewole, 2018; Uğur, Koç, & Koç, 2016; Zainol, Yahaya, Yahaya, & Zain, 2017).

According to the results of the study, Empowerment in Learning is a crucial factor in order to determine the acceptance of mobile learning and future studies should implement it to their research models. Based on the findings of the current study, using mobile devices for learning is an increasingly crucial learning method for students. When university students find the use of mobile devices engaging and easy to use for learning and consider them essential to improve their learning performance and effectiveness, their Behavioral Intention and Use Behavior toward using these are enhanced. Therefore, pre-service teachers as future educators will use mobile devices for learning in their own educational practice. In addition, the results of this study urge educational providers to understand how students use mobile technologies as a key component of their university studies.

FUTURE RESEARCH DIRECTIONS

This study presents several limitations that could be addressed in future studies. First, this study was cross-sectional in nature and conducted within a short period. Students' perceptions of the eight factors used in this study (PE, EE, FC, SI HM, EiL, BI, and UB) toward mobile learning can change over time as new knowledge and experiences are accumulated. Therefore, future research employing a longitudinal design will obtain findings that are more accurate from a specific group. Second, we used data collected only from university students enrolled in a Faculty of Education. Further studies are also needed to investigate university students' intention and use of mobile devices for learning from different departments and a wider variety of educational levels (primary and secondary school) conducted with the same model. Future studies should also investigate academic staff's intention and use of mobile devices for learning and determine possible differences by comparing the faculty's perceptions to students' perception. In addition, future research is needed to examine moderators such as, Gender, Age, and Experience that may influence university faculty and students' acceptance on mobile learning. Finally, it is crucial to investigate the perceptions of students who have been exposed to mobile learning into their courses and compare them with those who have not been exposed.

CONCLUSION

In this study, a modified UTAUT model was chosen to investigate the factors affecting the acceptance of mobile devices for learning from pre-service students as potential educators in a Greek university. Based on the results obtained, and the analysis conducted on the instrument designed, it was verified that it had a high consistency and validity for gathering information related to six of the seven factors that composed it (Performance Expectancy, Effort Expectancy, Social Influence, Facilitating Conditions, Hedonic Motivation and Empowerment in Learning). The results revealed that all hypotheses were supported in this research. Therefore, this concludes that the results indicate support for the modified UTAUT model for university students' acceptance of mobile learning. In agreement with what the UTAUT postulates, PE, EE, SI and FC were found to have a significant influence on students' intention to use mobile devices for learning. HM and EiL have the strongest impact on students' mobile learning adoption than that of PE and highlights the importance of those two factors in university students' decision making regarding the adoption of mobile devices for learning. In addition, we found that SI had the lowest significant effect on students' intention to adopt mobile devices for learning. Therefore, Greek university students are aware of the important force of social influences as they incorporate them to their everyday communication needs. However, the results may indicate that there are additional factors beyond the scope of this study that could be incorporated to improve the power of the proposed model.

ACKNOWLEDGMENT

The current research was supported by the State Scholarships Foundation (IKY) funded by the "Strengthening Postdoctoral Researchers" Initiative and resources from the OP "Human Resource Development, Education and Lifelong Learning" with Priority Axes 6, 8, 9 and co-funded by the European Commission Social Fund - ECSF and the Greek State.

REFERENCES

Adel Ali, R., & Rafie Mohd Arshad, M. (2018). Empirical analysis on factors impacting on intention to use M-learning in basic education in Egypt. *International Review of Research in Open and Distance Learning*, *19*(2), 253–270. doi:10.19173/irrodl.v19i2.3510

Al-emran, M., Elsherif, H. M., & Shaalan, K. (2016). Investigating attitudes towards the use of mobile learning in higher education. *Computers in Human Behavior*, *56*, 93–102. doi:10.1016/j.chb.2015.11.033

Alalwan, A. A., Dwivedi, Y. K., & Rana, N. P. (2017). Factors influencing adoption of mobile banking by Jordanian bank customers: Extending UTAUT2 with trust. *International Journal of Information Management*, *37*(3), 99–110. doi:10.1016/j.ijinfomgt.2017.01.002

Ardies, J., De Maeyer, S., Gijbels, D., & van Keulen, H. (2014). Students' attitudes towards technology. *International Journal of Technology and Design Education*, *25*(1), 43–65. doi:10.100710798-014-9268-x

Baydas, O., & Yilmaz, R. M. (2018). Pre-service teachers' intention to adopt mobile learning: A motivational model. *British Journal of Educational Technology*, *49*(1), 137–152. doi:10.1111/bjet.12521

Blackwell, C. K., Lauricella, A. R., Wartella, E., Robb, M., & Schomburg, R. (2013). Adoption and use of technology in early education: The interplay of extrinsic barriers and teacher attitudes. *Computers & Education*, *69*, 310–319. doi:10.1016/j.compedu.2013.07.024

Briz-Ponce, L., Pereira, A., Carvalho, L., Juanes-Méndez, J. A., & García-Peñalvo, F. J. (2017). Learning with mobile technologies – Students' behavior. *Computers in Human Behavior*, *72*, 612–620. doi:10.1016/j.chb.2016.05.027

Cacciamani, S., Villani, D., Bonanomi, A., Carissoli, C., Olivari, M. G., Morganti, L., & Confalonieri, E. (2018). Factors affecting students' acceptance of tablet PCs: A study in Italian high schools. *Journal of Research on Technology in Education, 50*(2), 120–133. doi:10.1080/15391523.2017.1409672

Çakıroğlu, Ü., & Öztürk, M. (2017). Pre-service computer teachers' tendencies towards the use of mobile technologies: A technology acceptance. *European Journal of Open Distance and E-Learning, 20*(1), 175–190.

Chaka, J. G., & Govender, I. (2017a). Students' perceptions and readiness towards mobile learning in colleges of education: A Nigerian perspective. *South African Journal of Education, 37*(1), 1–12. doi:10.15700aje.v37n1a1282

Chaka, J. G., & Govender, I. (2017b). Students' perceptions and readiness towards mobile learning in colleges of education. A Nigerian perspective. *South African Journal of Education, 37*(1), 1–12.

Chang, C. C., Yan, C. F., & Tseng, J. S. (2012). Perceived convenience in an extended technology acceptance model. *Mobile Technology and English Learning for College Students, 28*(5), 809–826.

Chao, C. M. (2019). Factors determining the behavioral intention to use mobile learning: an application and extension of the UTAUT model. *Frontiers in Psychology, 10*(July), 1–14. PMID:31379679

Crompton, H., & Burke, D. (2018). The use of mobile learning in higher education: A systematic review. *Computers & Education, 123*, 53–64. doi:10.1016/j.compedu.2018.04.007

Davis, F. D. (1993). User acceptance of information technology: System characteristics, user perceptions and behavioral impacts. *International Journal of Man-Machine Studies, 38*(3), 475–487. doi:10.1006/imms.1993.1022

Domingo, M. G., & Garganté, A. B. (2016). Exploring the use of educational technology in primary education: Teachers' perception of mobile technology learning impacts and applications' use in the classroom. *Computers in Human Behavior, 56,* 21-28.

Hamidi, H., & Chavoshi, A. (2018). Telematics and informatics analysis of the essential factors for the adoption of mobile learning in higher education: A case study of students of the University of Technology. *Telematics and Informatics, 35*(4), 1053–1070. doi:10.1016/j.tele.2017.09.016

Hamidi, H., & Jahanshaheefard, M. (2019). Essential factors for the application of education information system using mobile learning: A case study of students of the university of technology. *Telematics and Informatics, 38*(December 2017), 207–224.

Han, I., & Sug, W. (2016). The use of a mobile learning management system and academic achievement of online students. *Computers & Education, 102*, 79–89. doi:10.1016/j.compedu.2016.07.003

Heflin, H., Shewmaker, J., & Nguyen, J. (2017). Impact of mobile technology on student attitudes, engagement, and learning. *Computers & Education, 107*, 91–99. doi:10.1016/j.compedu.2017.01.006

Iqbal, S., & Qureshi, I. A. (2012). M-learning adoption: A perspective from a developing country. *International Review of Research in Open and Distance Learning, 13*(3), 147–164. doi:10.19173/irrodl.v13i3.1152

Jackman, G.-A. (2014). Investigating the factors influencing students' acceptance of mobile learning: The Cave Hill Campus experience. *Caribbean Educational Research Journal. The University of the West Indies, 2*(2), 14–32. Retrieved from http://www.cavehill.uwi.edu/fhe/hum/publications/EducationCERJ.htm

Jahnke, I., & Kumar, S. (2014). Digital didactical designs: teachers' integration of iPads for learning-centered processes. *Journal of Digital Learning in Teacher Education, 30*(3), 81–88. doi:10.1080/21532974.2014.891876

Jairak, K., Praneetpolgrang, P., & Mekhabunchakij, K. (2009). An acceptance of mobile learning for higher education students in Thailand. In *The Sixth International Conference on eLearning for Knowledge-Based Society*. Retrieved from http://ijcim.th.org/SpecialEditions/v17nSP3/36_Full_Kallaya Jairak_Online.pdf

Jawad, H. H., & Hassan, Z. (2015). Applying Utaut to evaluate the acceptance of mobile learning in higher education in Iraq. *International Journal of Science and Research, 4*(5), 2319–7064. Retrieved from www.ijsr.net

Kim-Soon, N., Ibrahim, M. A., Ahmad, A. R., & Sirisa, N. M. X. (2015). Factors influencing intention to use mobile technologies for learning among technical universities students. *Innovation and Sustainable Economic Competitive Advantage: From Regional Development to Global Growth*, (2010), 1–12.

Mojarro, A., Duarte, A., Guzmán, M., & Aguaded, I. (2019). Mobile learning in university contexts based on the unified theory of acceptance and use of technology (UTAUT). *Journal of New Approaches in Educational Research, 8*(1), 7–17. doi:10.7821/naer.2019.1.317

Mtebe, J., & Raisamo, R. (2014). Investigating students ' behavioural intention to adopt and use mobile learning in higher education in East Africa. *International Journal of Education and Development Using Information and Communication Technology, 10*(3), 4–20.

Muñoz-Leiva, F., Climent-Climent, S., & Liébana-Cabanillas, F. (2017). Determinants of intention to use the mobile banking apps: An extension of the classic TAM model. *Spanish Journal of Marketing - ESIC, 21*(1), 25–38.

Nassuora, A. (2013). Students acceptance of mobile learning for higher education in Saudi Arabia. *International Journal of Learning Management Systems, 1*(1), 1–9. doi:10.12785/ijlms/010101

Nikou, S. A., & Economides, A. A. (2017). Mobile-based assessment: Integrating acceptance and motivational factors into a combined model of self-determination theory and technology acceptance. *Computers in Human Behavior, 68*, 83–95. doi:10.1016/j.chb.2016.11.020

Onaolapo, S., & Oyewole, O. (2018). Performance expectancy, effort expectancy, and facilitating conditions as factors influencing smartphones use for mobile learning by postgraduate students of the University of Ibadan, Nigeria. *Interdisciplinary Journal of E-Skills and Lifelong Learning, 14*(June), 95–115.

Park, S. Y., Nam, M. W., & Cha, S. B. (2012). University students' behavioral intention to use mobile learning: Evaluating the technology acceptance model. *British Journal of Educational Technology, 43*(4), 592–605. doi:10.1111/j.1467-8535.2011.01229.x

Pullen, D., Swabey, K., Abadooz, M., & Sing, T. K. R. (2015). Pre-service teachers' acceptance and use of mobile learning in Malaysia. *Australian Educational Computing, 30*(1).

Sánchez-Prieto, J. C., Hernández-García, Á., García-Peñalvo, F. J., Chaparro-Peláez, J., & Olmos-Migueláñez, S. (2019). Break the walls! Second-order barriers and the acceptance of mLearning by first-year pre-service teachers. *Computers in Human Behavior, 95*(December 2018), 158–167.

Sánchez-Prieto, J. C., Olmos-Migueláñez, S., & García-Peñalvo, F. J. (2016). MLearning and pre-service teachers: An assessment of the behavioral intention using an expanded TAM model. *Computers in Human Behavior*, *72*, 644–654. doi:10.1016/j.chb.2016.09.061

Scherer, R., Siddiq, F., & Tondeur, J. (2019). The technology acceptance model (TAM): A meta-analytic structural equation modeling approach to explaining teachers' adoption of digital technology in education. *Computers & Education*, *128*(0317), 13–35.

Scherer, R., Tondeur, J., Siddiq, F., & Baran, E. (2018). The importance of attitudes toward technology for pre-service teachers' technological, pedagogical, and content knowledge: Comparing structural equation modeling approaches. *Computers in Human Behavior*, *80*, 67–80. doi:10.1016/j.chb.2017.11.003

Teo, T. (2012). Examining the intention to use technology among pre-service teachers: An integration of the technology acceptance model and theory of planned behavior. *Interactive Learning Environments*, *20*(1), 3–18. doi:10.1080/10494821003714632

Tey, T. C. Y., & Moses, P. (2018). UTAUT: Integrating achievement goals and learning styles for undergraduates' behavioural intention to use technology. *EAI Endorsed Transactions on E-Learning, 5*(17), 155573.

Thomas, T. D., Singh, L., & Gaffar, K. (2013). The utility of the UTAUT model in explaining mobile learning adoption in higher education in Guyana. *Journal of Education and Education and Development Using Information and Communication Technology*, *9*(3), 71–85.

Traxler, J., & Vosloo, S. (2014). Introduction: The prospects for mobile learning. *Prospects*, *44*(1), 13–28. doi:10.100711125-014-9296-z

Uğur, N. G., Koç, T., & Koç, M. (2016). An analysis of mobile learning acceptance by college students. *Journal of Educational and Instructional Studies in the World*, *6*(2), 2146–7463.

Venkatesh, V., Morris, M. G., Davis, G. B., & Davis, F. D. (2003). User acceptance of information technology: toward a unified view. *Management Information Systems Quarterly*, *27*(3), 425–478. doi:10.2307/30036540

Venkatesh, V., Thong, & Xu. (2012). Consumer acceptance and use of information technology: extending the unified theory of acceptance and use of technology. *Management Information Systems Quarterly*, *36*(1), 157–178. doi:10.2307/41410412

Wang, Y. S., Wu, M. C., & Wang, H. Y. (2009). Investigating the determinants and age and gender differences in the acceptance of mobile learning. *British Journal of Educational Technology*, *40*(1), 92–118. doi:10.1111/j.1467-8535.2007.00809.x

West, M., & Voslooo, S. (2013). *Policy guidelines for mobile learning. Policy guidelines for mobile learning*. 978-92-3-001143-7. France

Wu, W. H., Jim Wu, Y. C., Chen, C. Y., Kao, H. Y., Lin, C. H., & Huang, S. H. (2012). Review of trends from mobile learning studies: A meta-analysis. *Computers & Education*, *59*(2), 817–827. doi:10.1016/j.compedu.2012.03.016

Yang, S. (2013). Understanding undergraduate students' adoption of mobile learning model: a perspective of the extended UTAUT2. *Journal of Convergence Information Technology*, *8*(10), 969–979. doi:10.4156/jcit.vol8.issue10.118

Zainol, Z., Yahaya, N., Yahaya, N. M., & Zain, N. N. B. M. (2017). Factors influencing mobile learning among higher education students in Malaysia. *International Journal of Advanced Scientific Research and Management*, *2*(8), 86–91.

Zydney, J. M., & Warner, Z. (2016). Computers & education mobile apps for science learning: Review of research. *Computers & Education*, *94*, 1–17. doi:10.1016/j.compedu.2015.11.001

ADDITIONAL READING

Christensen, R., & Knezek, G. (2017). Readiness for integrating mobile learning in the classroom : Challenges, preferences and possibilities. *Computers in Human Behavior*, *76*, 112–121. doi:10.1016/j.chb.2017.07.014

Gikas, J., & Grant, M. M. (2013). Mobile computing devices in higher education: Student perspectives on learning with cellphones, smartphones & social media. *Internet and Higher Education*, *19*, 18–26. doi:10.1016/j.iheduc.2013.06.002

Papadakis, S., Kalogiannakis, M., & Zaranis, N. (2018). The effectiveness of computer and tablet assisted intervention in early childhood students' understanding of numbers. An empirical study conducted in Greece. *Education and Information Technologies*, *23*(5), 1849–1871. doi:10.100710639-018-9693-7

Papadakis, S., Kalogiannakis, M., & Zaranis, N. (2019). Parental involvement and attitudes towards young Greek children's mobile usage. *International Journal of Child-Computer Interaction*. doi: 10.1016/j.ijcci.2019.100144

Teo, T., & Huang, F. (2019). Investigating the influence of individually espoused cultural values on teachers' intentions to use educational technologies in Chinese universities. *Interactive Learning Environments*, *27*(5–6), 813–829. doi:10.1080/10494820.2018.1489856

Testa, N., & Tawfik, A. (2017). Mobile, but Are We Better? Understanding Teacher's Perception of a Mobile Technology Integration Using the Unified Theory of Acceptance and Use of Technology (UTAUT) Framework. *Journal of Formative Design in Learning*, *1*(2), 73–83. doi:10.100741686-017-0010-4

KEY TERMS AND DEFINITIONS

Behavioral Intension: In the context of mobile learning, is the extent to which a student is willing to use and continue to use, mobile devices to learn.

Early Childhood Education: Formal and informal educational programs that serves children in their preschool years, before enter to primary education.

Empowerment in Learning: Is the process of creating intrinsic task motivation to students by providing an educational environment and tasks, which increase one's sense of self-efficacy during education process.

Mobile Device: A computing device with a touch-screen interface, small enough to hold and operate with one hand.

Mobile Learning: Learning conducted by means of mobile devices such as smartphones or tablet computers.

Pre-service Teachers: Students trained from higher education institutions to become professional teachers.

Primary Education: Is typically the first stage of formal education, between early childhood and secondary education.

Use Behavior: Refers to the way students use mobile devices for educational purposes in the context of higher education.

APPENDIX: SURVEY ITEMS

Performance Expectancy

PE1. I would find mobile devices useful for teaching.
PE2. Using mobile devices would enable me to accomplish teaching tasks more quickly.
PE3. Using mobile devices would increase my productivity while teaching.
PE4. Using mobile devices would increase the effect of my teaching.

Effort Expectancy

EE1. Learning to operate mobile devices will be easy for me.
EE2. My interaction with mobile devices will be clear and understandable.
EE3. I will find mobile devices easy to use.
EE4. It would be easy for me to become skillful at using mobile devices.

Social Influence

SI1. People who influence my behavior think that I should use mobile devices.
SI2. People who are important to me will think that I should use mobile devices.
SI3. In general, the university climate would be supportive towards the use of mobile devices.

Facilitating Conditions

FC1. I would have the necessary mobile devices to use.
FC2. I would have the knowledge necessary to use mobile devices.
FC3. I can get help from others when I have difficulties participating in mobile devices.

Hedonic Motivation

HM1. Using mobile devices is fun.
HM2. Using mobile devices is enjoyable.
HM3. Using mobile devices is entertaining.

Habit

H1. The use of mobile devices has become a habit for me.
H2. I am addicted to using mobile devices.
H3. I must use mobile devices.

Behavioral Intension

BI1. I intend to use mobile devices in the next 12 months (or next semester)
BI2. I predict I will use mobile devices in the next 12 months.

Empowerment in learning

EiL1. I can personalize my study with mobile devices, building my knowledge up the more interesting topics
EiL2. Mobile devices enhances some of my skills.
EiL3. Thanks to mobile devices, I can cooperate more with my fellow students.

Use Behavior

UB1. I use mobile devices as a research tool.
UB2. I use mobile devices within group works
UB3. I use mobile devices for making assigned homework.
UB4. I use mobile devices to communicate with my fellow students about the courses.

Chapter 11
Preschool Children's Use of Tablet at Home and Parents' Views

Kleopatra Nikolopoulou
https://orcid.org/0000-0002-2175-1765
University of Athens, Greece

ABSTRACT

This chapter investigated young children's use of tablets at home and parents' views on tablet benefits and their concerns. A questionnaire was completed by the parents of 100 children aged 4-6. Young children engage in a range of activities such as playing games (76%), watching cartoons (75%), listening to music (65%), watching videos (60%), and using educational apps (54%). Fewer children look at pictures/photos/books or take pictures, while visiting websites and using email are never carried out by most of the children. Gender and age had an occasional isolated impact on children's tablet activities. 4-5 year olds tend to do tablet activities with an adult, while 5.5-6.5 year olds with siblings or alone. Most parents "agree and strongly agree" that tablets "teach basic technology skills" (85%), "teach foreign languages", and "can make learning fun". The parents' main concerns included dependence, reduction of communication, and inappropriate content. Implications regarding links between home and kindergarten are discussed.

INTRODUCTION

Digital technologies are increasingly evident in children's play worlds (Arnott et al., 2018) and home experiences (Hadlington et al., 2019), with the home environment to be often the first context where children access and experience these technologies (Kervin et al., 2018). Mobile digital touch screen devices (such as tablets, tablet devices of the type of iPad and smartphones) are, with increasing tendency, very popular among young children compared to the desktop or laptop computer and this popularity is attributed to the following features/functions (Kucirkova, 2014): portability, light weight, intuitive interface of the touch screen, no need for separate input devices (such as the keyboard or mouse), flexible

DOI: 10.4018/978-1-7998-1486-3.ch011

design for easy installation (and running of applications), manageability for even very young children and autonomy. Software features including multiple representations of information, such as pictures, video and animation, varying levels of task difficulty, learner control, task feedback and repetition, serve to create an individualized learning environment, placing the children in active control of their learning (Outhwaite et al., 2017). Tablets appear to be the most preferred/popular portable device (used daily) for children under 7 years of age (Ebbeck et al., 2016), and also for children aged 2-9 years; due to portability, ease of use and convenience (Zrim, 2015). As many preschool children have not adequately developed their fine motor skills required to handle peripheral devices of conventional computers, tablets are an attractive means of carrying out digital educational activities (Papadakis & Kalogiannakis, 2017), since they only require the use of the finger.

The popularity of tablets in the homes of young children is rapidly increasing (Marsh et al., 2018) and in many parts of the world touch-screen tablet technologies have become part of a shared activity between adults and young children in the home environment (Kirkonian & Pempek, 2013). Mobile applications present a significant opportunity for out-of-school, informal learning when designed in educationally appropriate ways (Hirsh-Pasek et al., 2015). Neumann (2018) found that parents played a key role in supporting young children's interactions with tablets, while various research studies examined the impact of touch screen tablets on young children's learning (e.g., Neumann & Neumann, 2014, Herodotou, 2017). In general, there has been a positive impact on motivation (MacCallum & Bell, 2019), literacy skills (Neumann, 2014; Gray et al., 2017; Clarke & Abbott, 2016), mathematical skills (Gray et al., 2017; Papadakis et al., 2018a; Outhwaite et al., 2017), science, problem-solving, and self-efficacy (Herodotou, 2017). There is an agreement that a major role for parents (and teachers) is to curate and mediate the content children receive via technology and to personalize their reading experiences (Kucirkova & Flewitt, 2018). The views of parents on the benefits and disadvantages of tablets are important, because parents' perspectives, engagement and support influence children's learning experiences at home. Although research on tablets with preschool children has emerged, in Greece there is little empirical evidence (e.g., Papadakis et al., 2018a; 2018b; Kalogiannakis & Papadakis, 2019; Kalogiannakis et al., 2018), at kindergarten or at home environments.

The purpose of this chapter is to examine preschool children's use of tablet at home and their parents' views on educational benefits and concerns of using tablets. For the purpose of this chapter, the terms "early childhood", "preschool" and "kindergarten" are used as synonymous, indicating the formal educational settings that attend, in most countries, children aged between 3 and 6 years old. The specific research questions of this chapter are:

1. What are the activities children do on tablet and with whom they do these activities?
2. Is there an impact of children's gender or age on their activities?
3. What are parents' views on benefits and their concerns regarding their children's use of tablets?

Next section presents the literature review, followed by the research method (procedure, sample and research instrument) and the results. The last section discusses the findings and concludes the chapter (this section also includes limitations, implications and suggestions for future research).

LITERATURE REVIEW

Tablet Use By Young Children At Home

As digital table technologies become more available, accessible and affordable, children from very young age are using these technologies in their homes (Kervin et al., 2018). A study involving six European countries (Chaudron, 2015) reported that many young children have daily contact with a wide range of mobile digital devices/tools. For example, in the United Kingdom, in a survey of 100 parents, 73% of under-5 year old children were using a tablet or a laptop computer, compared to just 23% in 2012. Livingstone (2016), in the United Kingdom, found that 25% of children aged 0 to 2 years have their own tablet, this figure rising to 36% for children aged 3 to 5 years, while the time devoted to tablet use was found to be one hour a day. The recently published Ofcom report (Ofcom, 2019) indicated that 42% of 5-7 year old children have their own tablet, young children watch TV on other devices –mostly tablets (30% for ages 3-4 and 44% for ages 5-7), while over 60% of 3-7 years old use a tablet to go online.

Children use their mobile devices primarily for playing games (63%), followed by watching videos (43%) and far fewer (30%) are using the devices for educational content, such as reading (Rideout, 2013). In a study with 6 year old children in Jordan (Oliemat et al., 2018), almost 50% of the sample reported they spend on their tablet one to two hours a day and they use them with their parents or older siblings (particularly when they are guided to how a game works or when apps are downloaded); touch-screen tablets were used for different purposes with 'playing games' being the most common purpose, followed by watching videos/movies on YouTube. Neumann (2018), in Australia, found that the majority of 3,5 years old children used a tablet at home to play entertainment app games (84%) and literacy apps (82%); 40% of children played entertainment app games on a daily basis, around a third of children typed/wrote on tablets by themselves, while reading e-books on tablets occurred rarely. In another study (Kumi-Yeboah & Campell, 2015) tablets were an advanced tool for drawing as compared to the traditional media tools for children, such as computers.

Edwards et al. (2016) in their research examining internet cognition in 4-5 years old children, noted that children of preschool age have only ever experienced a world where internet accessibility through touch screen devices is a common, everyday resource. Arnott et al. (2019) investigated practices with internet-connected toys across homes and early childhood settings, across Scotland, England, N.Ireland and Greece. Their findings indicated play-based internet of toy use at home (versus structured adult directed teaching in early childhood practice); play and interaction with the toys (including tablets) in the home was reminiscent of older studies of technologies where children just 'pick it up'. It has also been reported that tablets are an incentive/motivation for children (Neumann, 2014; Flewitt et al., 2015; Clarke & Abbott, 2016).

Parents' Views on Benefits and Concerns –Regarding Children's Tablet Use

Since touch screen tablets are widely available and increasingly used by young children in homes, it is important to consider parents' views on this issue. The role of parents in making decisions about access to technology for young children is a relatively new research focus (Valcke et al., 2010). For example, Stephen et al. (2013) revealed that parents' perspectives about "the value of play with technology", "their understanding about how learning should be supported" and "their typical ways of interacting with their children" influence young children's use of digital technologies at home. When children are encouraged to use tablets in doing homework, this reflects the fact that parents believe in the role of tablets as learning tools (Wood et al., 2016). Thus, the investigation of parents' views on their children's use of tablets is significant; so far there is little empirical research evidence on this issue (Nikolopoulou, 2019).

Grunwald Associates (2013) in the USA explored parents' (with children aged 3-18 years) perceptions on mobile learning and young children. Parents of children 3-5 years old, generally had positive attitudes, and seemed to prefer tablets as the main mobile device for their children (in comparison to mobile phones). They believed that mobile learning opens up new learning opportunities for their children, for example, it can make learning more fun, teach basic technological skills and promote curiosity. Mobile devices and applications were considered to have some educational content, to provide help in different cognitive areas and especially in mathematics. Another research study in Canada (Archer, 2017) found that parents of children aged 1-2 years give their children portable devices at increasingly younger ages. While parents expressed different opinions and concerns regarding the use of mobile technology by young children, this did not discourage them from using technology with their children (in order to spend time with them, to take away the children's boredom and to provide them with some educational value). At the same time, the study revealed that while children were initially interested in the new mobile device, their interest was not necessarily maintained, and verbal interactions between parents and children were reduced during the use of mobile devices.

Gray's et al. (2017) project in N. Ireland involved five nursery schools and indicated that a large majority of parents (92%) said their child had access to a range of digital devices at home with approximately half using them on a daily basis. The most common educational benefits of using portable devices identified by parents included making learning easier, exposure to a wider range of learning resources and greater experience with technology. Concerns about the use of portable devices included the potentially negative impact they might have on a child's attitude to work, concentration and awareness of the world around them (several also mentioned the need for vigilance to ensure children's safety). In a study in Turkey, parents' concerns when their preschool children played games on tablets had to do with the lack of outdoor games (Yılmaz Genç & Fidan, 2017). Neumann (2014) examined the views of 109 parents of Australian pre-schoolers aged 3–5 years old and found that only 35% of parents believed young children should have access to tablets at home (while 53% of parents believed that tablets should be available at preschool). Neumann et al. (2018) reported on an interview study which explored the perspectives of four parents of children aged 20–36 months in one early learning centre in England. Parents had a positive perspective on young children's use of tablets and viewed them as educationally valuable. They said that they did not wish children to be left behind in learning about new technologies and they also expressed concerns over the potential overuse and misuse of tablets at home (suggesting more guidance for parents is needed to support young children's experiences with tablets).

Palaigeorgiou et al. (2018) interviewed 54 Greek parents regarding their attitudes towards the educational usage of tablets both at school and at home. Parents' responses revealed that they had conflicting views on the educational value of tablets, they were mis/un-informed and they had several concerns regarding excessive usage, access to unsupervised content and less physical activity. Parents reported using restrictive mediation practices and, they felt outsmarted by their children. Ihmeideh & Shawareb (2014), in Jordan, examined parents' perspectives on their children's internet use and internet parenting styles; children's internet use typically involved playing games online and visiting websites, while the authoritative parenting style was the most commonly used (the neglectful style being used the least). Chen et al. (2019), in Singapore found that parents use digital devices (including tablets) with their 5 year old children; parents viewed the appropriate use of digital devices, with their rich educational content, flexibility, and interactivity, as basic in providing further opportunities for learning. It is noteworthy to mention that it is not only parents who influence young children's experiences with digital technologies, but also other family members such as older siblings (Aliagas et al., 2017).

METHOD

Procedure and Sample

The collection of data took place in March and April 2019. A questionnaire exploring children's home experiences of tablets and parents' views was distributed –via the early childhood teachers- to parents whose children were attending eight state (not private) kindergartens in Athens and North Greece. The anonymous questionnaire was followed by an informative consent letter, including the purpose of the research. 217 questionnaires were distributed of which 100 were returned without missing values (the response rate was 46%). This low rate of returned questionnaires was attributed to that not all children own tablets and to the low level of communication between parents and kindergartens (although early childhood teachers distributed the questionnaires with great willingness, many parents did not respond). Official permission was obtained from the Greek Ministry of Education since this study constituted part of a larger project carried out by the author-researcher. The sample of the study consisted of 100 parents with children aged 4-6,5 years old (56 children were 4-5 years old, and 44 children 5.5-6.5 years old). Out of the 100 children, 48 were boys and 52 were girls. Table 1 shows the demographic characteristics of the children. Regarding the frequency of tablet use by the child, 23% of the parents reported "daily" use, 39% "weekly (2-4 times per week)", 17% "monthly (2-4 times per month)", 19% "about once a month" and 2% "less than once a month".

The Questionnaire

Data were collected by the use of a questionnaire which included two parts (see Appendix). The first part was adapted from an earlier study that explored young children's use of ICT at home (Nikolopoulou et al., 2010); this part included questions regarding children's gender, age, frequency of tablet use at home, as well as different activities that children potentially do on tablet (play games, watch video, take pictures etc.) and whether they do these activities on their own or with other persons (Table 8, in Appendix). The second part included 12 statements/items aiming to investigate parents' views on benefits

Table 1. Demographic characteristics of the children

			Age of child		Total
			4-5 years old	**5.5-6.5 years old**	
Gender	boy	Count within age %	38 67.9%	10 22.7%	48 48%
	girl		18 32.1%	34 77.3%	52 52%
Total			56 100%	44 100%	100 100%

regarding their child's use of tablets (Table 9, in Appendix). These statements were taken-adopted from the Grunwald Associates LLC (2013) study; the statements were not modified since they regard universal early childhood skills or processes and could fit to the Greek context as well. Parents were asked to rate their views on a five-point Likert type scale: 1 (strongly disagree), 2 (disagree), 3 (I am not sure), 4 (agree) and 5 (strongly agree). At the end two open ended questions asked parents to shortly write/ note their views on the educational benefits and their concerns regarding the use of tablet by the child.

RESULTS

Activities Children Do on Tablet (And With Whom)

Table 2 displays the percentages of children (n=100) and their tablet activities at home: the activities are displayed in descending order according to the column "alone" (the child usually does this alone). Overall, children engage in a range of activities such as playing games (76%), watching cartoons (75%), listening to music (65%), watching videos/ You Tube (60%) and using educational apps (54%). Fewer children look at pictures/photos (44%), look at (listen to) books (44% for each activity) or take pictures (31%). The activities children usually do on their own are, watching cartoons (26%), watching videos/ You Tube (20%) and playing games (17%); these activities are also carried out with siblings or with an adult. Looking at pictures/photos, listening to music and using educational apps are activities usually carried out with an adult. Visiting websites and sending/receiving email messages are never carried out by the majority the children (83% and 100% respectively).

The Impact of Children's Gender or Age on Their Tablet Activities

Regarding the impact of gender on children's tablet activities, there were statistically significant differences (occasional isolated effect) with regard to the following activities: "watches cartoons" (Table 3),

Table 2. Percentages of children (n=100) and their tablet activities at home: the activities are displayed in descending order according to the column "alone" (the child usually does this alone)

	alone	needs help	with siblings	with adult	never
Watches cartoons	26	15	10	24	25
Watches videos/ YouTube	20	18	9	13	40
Plays games	17	18	23	18	24
Takes pictures	11	2	2	16	69
Looks at pictures/photos	9	2	7	26	56
Listens to music	6	9	16	34	35
Visits websites	4	4	7	2	83
Looks at (Listens to) books	4	3	2	35	56
Uses educational apps	2	13	0	39	46
Sends and receives email messages	0	0	0	0	100

Table 3. Relationship between gender and the activity "Watches cartoons"

			Gender		Total
			1 boy	**2 girl**	
Watches cartoons	alone	Count % within gender	8_a	18_b	26
			16.7%	34.6%	26.0%
	needs help		4_a	11_a	15
			8.3%	21.2%	15.0%
	with siblings		10_a	0_b	10
			20.8%	0.0%	10.0%
	with adult		14_a	10_a	24
			29.2%	19.2%	24.0%
	never		12_a	13_a	25
			25.0%	25.0%	25.0%
Total			48	52	100
			100.0%	100.0%	100.0%

Each subscript letter denotes a subset of gender categories whose column proportions do not differ significantly from each other at the .05 level

$\chi^2(4)=17.7$, p=0.001

"uses educational apps" (Table 4) and "looks at (listens to) books" (Appendix, Table 10). Regarding the activity "watches cartoons" the significant differences regarded the rows "alone" and "with siblings". Overall, 34.6% of the girls versus 16.7% of the boys do this activity "alone" (on their own), while 20.8% of the boys versus 0% of the girls do this activity with siblings (Table 3). Regarding the activity "uses educational apps", the significant differences regarded the rows "needs help" and "with adult". Overall, 21.2% of the girls versus 4.2% of the boys need help for this activity, while 56.2% of the boys versus 23.1% of the girls do this activity with an adult (Table 4). Regarding the activity "looks at (listens to) books" the significant differences mainly regarded the rows "with adult" and "never". Overall, 50% of the boys versus 21.2% of the girls do this activity with an adult, while 45.8% of the boys versus 65.4% of the girls never do this activity (Table 10 in Appendix).

Regarding the impact of age on children's tablet activities, there were statistically significant differences (at specific points) with regard to most of the activities: "plays games", "watches videos/ YouTube", "looks at pictures/photos", "takes pictures", "listens to music", and "uses educational apps". An emerging pattern is that younger children (4-5 years old) tend to do these tablet activities with adult, while older children (5.5-6.5 years old) with siblings or alone. For example, for the activity "plays games", 32.1% of the younger children versus 0% of the older children do this activity with an adult, while 36.4% of the older children versus 12.5% of the younger ones do this activity with siblings (Table 5). Regarding the activity "uses educational apps", 58.9% of the younger children versus 13.6% of the older children do this activity with an adult (Table 6). The Tables for the remaining activities with statistical differences are shown in Appendix (Tables 11, 12, 13 and 14).

Table 4. Relationship between gender and the activity "Uses educational apps"

			Gender		Total
			boy	girl	
Uses educational apps	alone	Count % within gender	0_a	2_a	2
			0.0%	3.8%	2.0%
	needs help		2_a	11_b	13
			4.2%	21.2%	13.0%
	with adult		27_a	12_b	39
			56.2%	23.1%	39.0%
	never		19_a	27_a	46
			39.6%	51.9%	46.0%
Total			48	52	100
			100.0%	100.0%	100.0%

Each subscript letter denotes a subset of gender categories whose column proportions do not differ significantly from each other at the .05 level

$\chi^2(3)=15.3$, $p=0.002$

Parents' Views on Benefits and Concerns

Regarding parents' views on benefits of using tablets, a descriptive analysis was performed. Table 7 shows parents' response percentage frequencies on the 12 items of the questionnaire, S1-S12 (n=100

Table 5. Relationship between age and the activity "Plays games"

			Age		Total
			4-5 years	5.5-6.5 years	
Plays games	alone	Count % within age	9_a	8_a	17
			16.1%	18.2%	17.0%
	needs help		12_a	6_a	18
			21.4%	13.6%	18.0%
	with siblings		7_a	16_b	23
			12.5%	36.4%	23.0%
	with adult		18_a	0_b	18
			32.1%	0.0%	18.0%
	never		10_a	14_a	24
			17.9%	31.8%	24.0%
Total			56	44	100
			100.0%	100.0%	100.0%

Each subscript letter denotes a subset of age categories whose column proportions do not differ significantly from each other at the .05 level

$\chi^2(4)=23.4$, $p<0.001$

Table 6. Relationship between age and the activity "Uses educational apps"

			Age		Total
			4-5 years	5.5 -6.5 years	
Uses educational apps	alone	Count % within age	0_a	2_a	2
			0.0%	4.5%	2.0%
	needs help		4_a	9_b	13
			7.1%	20.5%	13.0%
	with adult		33_a	6_b	39
			58.9%	13.6%	39.0%
	never		19_a	27_b	46
			33.9%	61.4%	46.0%
Total			56	44	100
			100.0%	100.0%	100.0%

Each subscript letter denotes a subset of age categories whose column proportions do not differ significantly from each other at the .05 level

$\chi^2(3)=22.9$, $p<0.001$

parents). The last column of the Table has added together the percentages of those who "agree" and "strongly agree". More specifically, over 50% of the sample, "agree and strongly agree" with the items S2, S11, S1, and S9. Parents reported as benefits, "teach basic technology skills" (agreement for S2: 85%), "teach foreign languages" (for S11: 73%), "can make learning fun" (for S1: 67%), and "teach math" (for S9: 51%).

Regarding parents' answers in the open questions regarding benefits and concerns (at the end of the questionnaire), only some parents answered them. Regarding tablets' advantages they mainly mentioned

Table 7. Parents' response percentage frequencies on the 12 items, S1-S12 (n=100 parents)

Tablets	Strongly disagree	Disagree	Not sure	Agree	Agree & Strongly agree
S2. Teach basic technology skills	2	2	11	67	85
S11. Teach foreign languages	9	3	15	57	73
S1. Can make learning fun	0	2	31	49	67
S9. Teach math	19	17	13	41	51
S7. Allow my child to relax	6	16	29	49	49
S8. Teach reading	19	21	18	42	42
S6. Teach problem solving	19	31	15	31	35
S3. Promote curiosity	7	7	52	32	34
S10. Teach science	18	14	35	33	33
S12. Teach writing	19	36	32	13	13
S4. Foster creativity	3	24	29	9	9
S5. Allow my child to express himself/herself	25	17	52	6	6

familiarization with the technology (28 references), means of entertainment (9 references), learning foreign languages (6 references) and other things. Indicative excerpts were: "Familiarizes herself with the use of technology/develops technological skills", "learns foreign languages…she has learned many English words from children's programs on YouTube", "learns letters, numbers, songs, various stories, mythology", "learns for places, things, animals that he may never see in his real life". A few parents emphasized the role of parents/adults (4 references). For example, in their words: "under an adult's supervision, the tablet can be used as a supportive tool", "if its use is made under the guidance and control of parents, then it can be a means of fun but to also learn things…such as enhancing the use of language and their socialization", and "the use should be made under the supervision/mentoring of the parent/teacher so as to reward the advantages for learning".

Parents also expressed concerns about young children using tablets at home, and these mainly regarded dependence, addiction from inconsiderate/daily use (29 references), reduction of social communication and interaction (18 references) and inappropriate content (12 references). A few parents raised the issue of loss of time from outdoor play (5 references). Indicative excerpts were: "It affects negatively, after a long time of use, affecting concentration (distraction), causing nervousness", "avoids socialization, communication (e.g., with other children), prefers to stay alone", "via the web he can find websites with inappropriate content", "it (tablet) does not promote creativity, curiosity, play, distances herself from other activities", "I think the use of the tablet removes children from reading, from love for books and writing. It reduces children's ability to spell, isolates them, it causes introversion, depression. It affects thinking through guided information". Isolated references regarded advertisements and impact on health. For example, "advertisements: because at a time that he will be unsupervised, he can see video games that I would not like. It is also difficult to disable it (for the child to accept that the time of tablet use is over)…increasing the risk of addiction", "exposure to tablet cumulatively with other monitors (e.g., of TV)" and "damage to the eyes".

DISCUSSION AND CONCLUSION

This study investigated preschool children's use of tablet at home and their parents' views on benefits and concerns of using tablets. The study adds to the body of empirical evidence, particularly in the Greek context. This investigation is important because children's use of technology at home impacts on their learning/skills, while parents' views have an important role in providing children with opportunities to access/use digital tools and tablets at home, which in turn influences their learning (Kervin et al., 2018; Neumann et al., 2018).

With regard the first research question (what are the activities children do on tablet and with whom they do these activities?), it was shown that children engage in a range of activities such as playing games (76%), watching cartoons (75%), listening to music (65%), watching videos/ You Tube (60%) and using educational apps (54%). Fewer children look at pictures/photos (44%), look at (listen to) books (44%) or take pictures (31%), while visiting websites and sending/receiving email messages are never carried out by most of the children (83% and 100% respectively). There is some agreement with earlier research which indicated that playing games (Neumann, 2018; Oliemat et al., 2018; Rideout, 2013), and watching videos (Oliemat et al., 2018; Rideout, 2013) were the most popular activities carried out by young children. In this study, the activities children usually do on their own are watching cartoons (26%), watching videos/ You Tube (20%) and playing games (17%); these activities are also carried out with siblings or with an adult. Looking at pictures/photos, listening to music and using educational apps are activities usually carried out with an adult. Other researchers (e.g., Ihmeideh & Shawareb, 2014) also indicated that children's use of tablets was carried out with the parent's guidance, and this raises the issue of the role of adults (and more knowledgeable people, such as siblings, friends) at home, in supporting children's use-learning via applications.

With regard the second research question (is there an impact of children's gender or age on their activities?) it was shown that both gender and age had an occasional isolated impact on children's tablet activities. The variable gender had an impact/influence only on the activities "watches cartoons", "uses educational apps" and "looks at (listens to) books". Although watching cartoons is an activity carried out by the same percentage of boys and girls (75%), the significant difference derives from that the girls do this activity on their own, while the boys do it with siblings. In using educational applications and looking at (listening to) books, main gender differences regard that boys tend to do these activities with an adult (and more often than girls). This finding may be partially linked to the synthesis of the sample (Table 1) which included more young boys (4-5 years old) and older girls (5.5-6.5 years old); as children grow up they tend to carry out some activities on their own, rather than with the help of another person. The variable age had an impact/influence on the majority of the activities carried out by the children. Younger children (4-5 years old) tend to do tablet activities with adult, while older children (5.5-6.5 years old) with siblings or alone. Also, watching videos is an activity more frequently carried out by older rather than younger children. More research is needed on the influence of gender or age on children's tablet activities at home.

With regard the third research question (what are parents' views on benefits and concerns of tablet use by young children?), over half of the sample "agree and strongly agree" with the benefits that tablets "teach basic technology skills" (agreement 85%), "teach foreign languages" (73%), "can make learning fun" (67%) and "teach maths" (51%). There is an agreement with earlier research which indicated parents' views on tablet benefits; tablet can make learning more fun (Grunwald Associates, 2013) and provides experiences with technology (Grunwald Associates, 2013; Gray's et al., 2017). In this study, parents' main concerns included dependence, reduction of social communication and interaction, inappropriate content and loss of time from outdoor play. This finding is also in agreement with earlier research; on the issues of potential overuse/misuse (Neumann et al., 2018; Palaigeorgiou et al., 2018) and on access to unsupervised content and less physical activity (Palaigeorgiou et al., 2018). Parents' views reveal the essential role of parents on different issues such as the time young children spend on tablets at home and the guidance on children's tablet use.

The results of this study cannot be generalized because of the small size and the synthesis of the sample (more young boys and older girls). Additionally, the questionnaire did not include statements regarding parents' concerns on tablet use (only a few parents answered the open question on concerns), and the impact of parents' characteristics (age, gender, income, educational background, prior experience with technology, motivation etc.) on their views was not investigated. For example, research indicated that parents' motivation facilitated scaffolding their children's literacy skills development (See et al., 2019) when preschool children used tablets with multilingual eBooks. It was also found that parental mediation is linked with parents' gender, education, cultural/socioeconomic background and computer skills (Kumpulainen & Gillen, 2017). Thus, modifications for future research are suggested to investigate the above mentioned issues. Also, a number of interviews could be conducted with some parents in order to complement the quantitative data.

The findings of this study have implications for closer links and increased communication-collaboration between home and kindergarten. For example, formal early childhood education settings could inform parents about their tablet use initiatives, while kindergarten teachers could raise parents' awareness of children's tablet use at home (providing, for example, information about age appropriate educational apps). Gray et al. (2017) suggested for parental training in child protection and safeguarding, in order to ensure the wellbeing of young children who have free access to tablets, while Edwards et al. (2018) stressed the need for early childhood cyber-safety education (since many children are online, often independently of adult supervision). Parent involvement may have an impact on kindergarten technology decisions. The introduction/implementation of tablets in Greek kindergarten classes is a new issue, and there is limited empirical evidence (e.g., Hatzigianni et al., 2018; Papadakis et al., 2018a). It is suggested for teachers to be aware of children's tablet experiences at home.

In the 21st century, tablets have become one of the leading instructional technology devices in K-12 schools; the emerging use of tablets in K-12 schools include various applications such as Google Earth, Google Apps, etc. (Kumi-Yeboah & Campbell, 2019). Tablet technology can provide individualized effective support for early mathematics development, when software is age appropriate (Outhwaite et al., 2017). Marsh et al. (2018) indicated that children under 5 year old were using a variety of apps some of which were not aimed at their age range; the design features of such apps can lead to the support/inhibition of play and creativity. Papadakis & Kalogiannakis (2017) suggested the need for a comprehensive and up-to-date guide for evaluating mobile applications, an easy-to-use guide for parents, educators and software designers. Particular attention should be given to the design and the content of educational mobile applications (Papadakis et al., 2018b), in order to exploit the potential of tablets. It is note-worthy that the features of tablets are not sufficient conditions for positive learning effects. There is a need to further investigate family contexts to understand how tablets can enrich (rather than prevent) young children's learning experiences. For example, internet-connected tablets are being increasingly used by young children at home. Future research could investigate social interactions with family members (during engagement with tablets), as well as children's perceptions/feelings when they use technology at home.

ACKNOWLEDGMENT

The author would like to thank prof. Vasilis Gialamas for his help in statistical analyses, the teachers who distributed the questionnaires and the parents who completed them.

COMPETING INTERESTS

The author declares that she has no competing interests.

FUNDING

The author received no financial support for the research, authorship, and/or publication of this article.

REFERENCES

Aliagas, C., Matsumoto, M., Morgade, M., Correro, C., Galera, N., & Poveda, D. (2017). *Young children (0-8) and digital technology 0-8. What changes in one year?* National report SPAIN. *Papers Infancia, 20*, 1–65.

Archer, K. (2017). *Infants, toddlers and mobile technology: examining parental choices and the impact of early technology introduction on cognitive and motor development. (Theses and Dissertations (Comprehensive)*, Wilfrid Laurier University).

Arnott, L., Palaiologou, I., & Gray, C. (2018, September). Digital devices, internet-enabled toys and digital games: The changing nature of young children's learning ecologies, experiences and pedagogies. *British Journal of Educational Technology, Special Issue*, *49*(5), 803–806. doi:10.1111/bjet.12676

Arnott, L., Palaiologou, I., & Gray, C. (2019). Internet of toys across home and early childhood education: Understanding the ecology of the child's social world. *Technology, Pedagogy and Education*, 1–12. doi:10.1080/1475939X.2019.1656667

Chaudron, S. (2015). Young children (0-8) and digital technology: a qualitative exploratory study across seven countries. Luxembourg: Publications Office of the European Union.

Chen, W., Teo, M. H., & Nguyen, D. (2019). Singapore parents' use of digital devices with young children: Motivations and uses. *The Asia-Pacific Education Researcher*, *28*(2), 1–12. doi:10.100740299-019-00432-w

Clarke, L., & Abbott, L. (2016). Young pupils', their teachers' and classroom assistants' experiences of iPads in a Northern Ireland school: Four and five years old, who would have thought they could do that? *British Journal of Educational Technology*, *47*(6), 1051–1064. doi:10.1111/bjet.12266

Ebbeck, M., Yim, H., Chan, Y., & Goh, M. (2016). Singaporean parents' views of their young children's access and use of technological devices. *Early Childhood Education Journal*, *44*(2), 127–134. doi:10.100710643-015-0695-4

Edwards, S., Nolan, A., Henderson, M., Mantilla, A., Plowman, L., & Skouteris, H. (2018). Young children's everyday concepts of the internet: A platform for cyber-safety education in the early years. *British Journal of Educational Technology*, *49*(1), 45–55. doi:10.1111/bjet.12529

Edwards, S., Nolan, A., Henderson, M., Skouteris, H., Mantilla, P., & Bird, J. (2016). Developing a measure to understand young children's internet cognition and cybersafety awareness: A pilot test. *Early Years*, *36*(3), 322–335. doi:10.1080/09575146.2016.1193723

Flewitt, R., Messer, D., & Kucirkova, N. (2015). New directions for early literacy in a digital age: the iPad. *Journal of Early Childhood Literacy*, *15*(3), 289–310. doi:10.1177/1468798414533560

Gray, C., Dunn, J., Moffett, P., & Mitchell, D. (2017). *Mobile devices in early learning: Evaluating the use of portable devices to support young children's learning*. Report commissioned by Irish Education Authority.

Grunwald Associates LLC. (2013). *Living and learning with mobile devices: What parents think about mobile devices for early childhood and K-12 learning*. USA: Learning First Alliance.

Hadlington, L., White, H., & Curtis, S. (2019). "I cannot live without my [tablet]": Children's experiences of using tablet technology within the home. *Computers in Human Behavior*, *94*, 19–24. doi:10.1016/j.chb.2018.12.043

Hatzigianni, M., Gregoriadis, A., Karagiorgou, I., & Chatzigeorgiadou, S. (2018). Using tablets in free play: The implementation of the digital play framework in Greece. *British Journal of Educational Technology*, *49*(5), 928–942. doi:10.1111/bjet.12620

Herodotou, C. (2017). Young children and tablets: A systematic review of effects on learning and development. *Journal of Computer Assisted Learning*, *34*(1), 1–9. doi:10.1111/jcal.12220

Hirsh-Pasek, K., Zosh, J. M., Golinkoff, R. M., Gray, J. H., Robb, M. B., & Kaufman, J. (2015). Putting education in "educational" apps lessons from the science of learning. *Psychological Science in the Public Interest*, *16*(1), 3–34. doi:10.1177/1529100615569721 PMID:25985468

Ihmeideh, F., & Shawareb, A. (2014). The association between internet parenting styles and children's use of the internet at home. *Journal of Research in Childhood Education*, *28*(4), 411–425. doi:10.1080/02568543.2014.944723

Kalogiannakis, M., Ampartzaki, M., Papadakis, S., & Skaraki, E. (2018). Teaching natural science concepts to young children with mobile devices and hands-on activities. A case study. *International Journal of Teaching and Case Studies*, *9*(2), 171–183. doi:10.1504/IJTCS.2018.090965

Kalogiannakis, M., & Papadakis, S. (2019). Evaluating preservice kindergarten teachers' intention to adopt and use tablets into teaching practice for natural sciences. *International Journal of Mobile Learning & Organization*, *13*(1), 113–127. doi:10.1504/IJMLO.2019.096479

Kervin, L., Verenikina, I., & Rivera, C. (2018). Digital play and learning in the home: Families' perspective. In S. J. Danby, M. Fleer, C. Davidson, & M. Hatzigianni (Eds.). Digital childhoods: Technologies and children's everyday lives (pp. 117-130). Springer: International Perspectives on Early Childhood Education and Development, 22.

Kirkorian, H., & Pempek, T. (2013). Toddlers and touch screens: Potential for early learning? *Zero to Three*, *33*(4), 32–35.

Kucirkova, N. (2014). iPads in early education: Separating assumptions and evidence. *Frontiers in Psychology*, *5*(715). Retrieved from https://www.frontiersin.org/articles/10.3389/ PMID:25071678

Kucirkova, N., & Flewitt, R. (2018). The future-gazing potential of digital personalization in young children's reading: Views from education professionals and app designers. *Early Child Development and Care*, 1–15. doi:10.1080/03004430.2018.1458718

Kumi-Yeboah, A., & Campbell, K. S. (2019). Emerging use of tablets in K-12 environments: Issues and implications in K-12 Schools. In *Early childhood development: concepts, methodologies, tools, and applications* (pp. 857–876). Hershey, PA: IGI Global. doi:10.4018/978-1-5225-7507-8.ch042

Kumi-Yeboah, A., & Campell, K. S. (2015). Emerging use of tablets in K-12 environments: Issues and implications in K-12 schools. In Tablets in K-12 education: Integrated experiences and implications (pp. 46-63). Hershey, PA: IGI Global. doi:10.4018/978-1-4666-6300-8.ch004

Kumpulainen, K., & Gillen, J. (2017) Young children's digital literacy practices in the home: A review of the literature. *COST ACTION ISI1410 DigiLitEY*. Retrieved from http://www.digilitey.eu

Livingstone, S. (2016). What are preschoolers doing with tablets and is it good for them? Retrieved from http://eprints.lse.ac.uk/

MacCallum, K., & Bell, H. R. (2019). Improving teaching practice in early childhood supported by mobile technology. In *Early childhood development: concepts, methodologies, tools, and applications* (pp. 1066–1082). Hershey, PA: IGI Global. doi:10.4018/978-1-5225-7507-8.ch053

Marsh, J., Plowman, L., Yamada-Rice, D., Bishop, J., Lahmar, J., & Scott, F. (2018, September). Play and creativity in young children's use of apps. *British Journal of Educational Technology, Special Issue, 49*(5), 870–882. doi:10.1111/bjet.12622

Neumann, M. (2014). An examination of touch screen tablets and emergent literacy in Australian preschool children. *Australian Journal of Education, 58*(2), 109–122. doi:10.1177/0004944114523368

Neumann, M. (2018). Parent scaffolding of young children's use of touch screen tablets. *Early Child Development and Care, 188*(12), 1654–1664. doi:10.1080/03004430.2016.1278215

Neumann, M., & Neumann, D. (2014). Touch screen tablets and emergent literacy. *Early Childhood Education Journal, 42*(4), 231–239. doi:10.100710643-013-0608-3

Neumann, M. M., Merchant, G., & Burnett, C. (2018). Young children and tablets: The views of parents and teachers. *Early Child Development and Care*. doi:10.1080/03004430.2018.1550083

Nikolopoulou, K. (2019, June). Mobile technologies and early childhood education. In M. Tsitouridou, J. A. Diniz, & T. Mikropoulos (Eds.), *Technology and innovation in learning, teaching and education. TECH-EDU 2018, CCIS 993* (pp. 444–457). Springer Nature Switzerland AG; doi:10.1007/978-3-030-20954-4_33

Nikolopoulou, K., Gialamas, V., & Batsouta, M. (2010). Young children's access to and use of ICT at home. *Review of Science, Mathematics, and ICT Education, 4*(1), 25–40.

Ofcom. (2019). *Children and parents: Media use and attitudes report 2018*. Making sense of Media.

Oliemat, E., Ihmeideh, F., & Alkhawaldeh, M. (2018). The use of touch-screen tablets in early childhood: Children's knowledge, skills, and attitudes towards tablet technology. *Children and Youth Services Review, 88*, 591–597. doi:10.1016/j.childyouth.2018.03.028

Outhwaite, L. A., Gulliford, A., & Pitchford, N. J. (2017). Closing the gap: Efficacy of a tablet intervention to support the development of early mathematical skills in UK primary school children. *Computers & Education, 108*, 43–58. doi:10.1016/j.compedu.2017.01.011

Palaigeorgiou, G., Kamarina, K., Bratitsis, T., & Xefteris, S. (2018). Parental mediation of tablet educational use at home and at school: Facilitators or preventers? *International Conference on Interactive Mobile Communication, Technologies and Learning* (IMCL 2017), 339-350. 10.1007/978-3-319-75175-7_90

Papadakis, S., & Kalogiannakis, M. (2017). Mobile educational applications for children. What educators and parents need to know. *International Journal of Mobile Learning and Organization, 11*(3), 256–277. doi:10.1504/IJMLO.2017.085338

Papadakis, S., Kalogiannakis, M., & Zaranis, N. (2018a). The effectiveness of computer and tablet assisted intervention in early childhood students' understanding of numbers. An empirical study conducted in Greece. *Education and Information Technologies, 23*(5), 1849–1871. doi:10.100710639-018-9693-7

Papadakis, S., Kalogiannakis, M., & Zaranis, N. (2018b). Educational apps from the Android Google Play for Greek preschoolers: A systematic review. *Computers & Education, 116*, 139–160. doi:10.1016/j.compedu.2017.09.007

Rideout, V. (2013). *Zero to eight: children's media use in America 2013.* Commonsense Media.

See, K. T., Madhubala, B. H., & Koo, A. C. (2019). Motivation of parents towards reading multilingual eBooks to pre-school children. *International Journal of Interactive Mobile Technologies, 13*(1), 20–36. doi:10.3991/ijim.v13i01.9060

Stephen, C., Stevenson, O., & Adey, C. (2013). Young children engaging with technologies at home: The influence of family context. *Journal of Early Childhood Research, 11*(2), 149–164. doi:10.1177/1476718X12466215

Valcke, M., Bonte, S., Wever, B., & Rots, I. (2010). Internet parenting styles and the impact on internet use of primary school children. *Computers & Education, 55*(2), 454–464. doi:10.1016/j.compedu.2010.02.009

Wood, E., Petkovski, M., De Pasquale, D., Gottardo, A., Evans, M. A., & Savage, R. S. (2016). Parent scaffolding of young children when engaged with mobile technology. *Frontiers in Psychology, 7*, 1222. doi:10.3389/fpsyg.2016.00690 PMID:27242603

Yılmaz Genç, M. M., & Fidan, A. (2017). Children, parents and tablets: Preschool children's tablet use. *Egitim ve Ögretim, 7*(3), 367–398. doi:10.14527/pegegog.2017.014

Zrim, L. (2015). Child's play: Connected Aussie kids spend up to equivalent of three school days online every week. *Nielsen Australia e-Generation Report.* Retrieved from http://www.nielsen.com/au/en/insights/news/2015/childs-play-connected-aussie-kidsspend-up-to-equivalent-of-three-school-days-online.html

ADDITIONAL READING

Gillen, J., Aliagas, C., Bar, L., Flewitt, R., Jorge, A., Kumpulainen, K., . . . Tomé, V. (2019). A day in the digital lives of children aged 0-3: the tablet and TV:Working Group 1 symposium. In Reconceptualising early childhood literacies: an international conference, 2019-03-072019-03-08, Manchester Conference Cenre.

Holloway, D., Haddon, L., Green, L., & Stevenson, K. (2019). The parent-child-app learning assemblage: Scaffolding early childhood learning through app use in the family home. In N. Kucirkova, J. Rowsell, & G. Falloon (Eds.), *The Routledge International Handbook of Learning with Technology in Early Childhood* (pp. 210–218). Routledge. doi:10.4324/9781315143040-15

Kılıç, A. O., Sari, E., Yucel, H., Oğuz, M., Polat, E., Acoglu, E., & Senel, S. (2019). Exposure to and use of mobile devices in children aged 1–60 months. *European Journal of Pediatrics, 178*(2), 221–227. doi:10.100700431-018-3284-x PMID:30402717

Pablo Hidelbrando, S. S., Nunes Santos, J., Ribeiro Leite, H., Hadders-Algra, M., Conceição Guedes, S., Nogueira Pontes Nobre, J., ... Luzia de Souza Morais, R. (2019). Tablet Use in Young Children is Associated with Advanced Fine Motor Skills. *Journal of Motor Behavior*. doi:10.1080/00222895.2019.1602505 PMID:31007146

Zaranis, N., & Valla, V. (2019). Tablets in Learning Mathematics for Kindergarten Students. In L. Daniela (Ed.), *Didactics of Smart Pedagogy* (pp. 267–284). Cham: Springer. doi:10.1007/978-3-030-01551-0_14

KEY TERMS AND DEFINITIONS

Digital Technologies: synonym for "Information and Communication Technologies" (ICT); include desktop computers, as well as mobile technologies (tablets, etc.).

Early Childhood (settings): synonym to "preschool" and "kindergarten" settings; indicating the formal educational settings that attend, in most countries, children aged between 3 and 6 years old.

ICT (Information and Communication Technologies): synonym to the terms "computers" and "digital technology". Historically the focus has tended to be on computers, but this has extended to include interactive whiteboards and tablets; apart from computer software, a number of products that incorporate some aspects of ICT are available to young children (such as electronic musical keyboards, programmable interactive toys and digital cameras), while lately mobile technologies and, in particular, the tablets are very popular among young children.

Mobile Applications: synonym for "mobile apps" or "apps".

Mobile Devices/Technologies: tablets, smartphones, e-readers, MP3 players etc.; these technologies continue to expand and evolve with tablets, Chromebooks, and even laptops viewed as "mobile" devices.

Mobile Learning: The learning mode that employs mobile technology/devices to facilitate or support learning; can be defined as facilitating and enhancing the learning process via mobile devices anytime and anywhere.

Mobile Phones: synonym to smartphones; lately, most are equipped with Wi-Fi, Bluetooth, near field communication capabilities, as well as Global Positioning System (GPS) capabilities.

Mobile/Portable Touch Screen Devices: Tablets, tablet devices of the type of iPad and smartphones.

Tablet: Synonym for "touch-screen tablet" or "mobile touch-screen tablet device". For example, iPads, Blackberry Playbook, Android tablets etc.

APPENDIX

Gender: boy girl
Age: 4-5 5.5-6.5

FREQUENCY OF USAGE

How often does the child use the tablet at home?
Every day weekly (2-4 times per week) monthly (2-4 times per month)
About once a month Less than once a month

Table 8. Which activities does your child do on the tablet? Who does the child do them with? (Please tick one box only for each activity)

	The child...				
	...usually does it alone	**...usually does it alone, but needs help from time to time**	**...usually does this with siblings**	**...usually does this with an adult**	**...never does this**
Plays games					
Watches videos/ YouTube					
Watches cartoons					
Looks at pictures/photos					
Takes pictures					
Visits websites					
Listens to music					
Uses educational apps					
Looks at (Listens to) books					
Sends and receives email messages					
Other. Please say what: 1…………..					
2……………					

Table 9. Parents' views regarding benefits of using tablets

Tablets	strongly disagree	disagree	not sure	agree	strongly agree
S1. Can make learning fun	1	2	3	4	5
S2. Teach basic technology skills	1	2	3	4	5
S3. Promote curiosity	1	2	3	4	5
S4. Foster creativity	1	2	3	4	5
S5. Allow my child to express himself/herself	1	2	3	4	5
S6. Teach problem solving	1	2	3	4	5
S7. Allow my child to relax	1	2	3	4	5
S8. Teach reading	1	2	3	4	5
S9. Teach math	1	2	3	4	5
S10. Teach science	1	2	3	4	5
S11. Teach foreign languages	1	2	3	4	5
S12. Teach writing	1	2	3	4	5

What do you want to say about the benefits of tablet's usage by the child?
Which are your concerns of the tablet's usage by the child?

Table 10. Relationship between gender and the activity "Looks at (Listens to) books"

			Gender		Total
			boy	girl	
Looks at (Listens to) books	alone	Count % within gender	0_a	4_b	4
			0.0%	7.7%	4.0%
	needs help		0_a	3_a	3
			0.0%	5.8%	3.0%
	with siblings		2_a	0_a	2
			4.2%	0.0%	2.0%
	with adult		24_a	11_b	35
			50.0%	21.2%	35.0%
	never		22_a	34_b	56
			45.8%	65.4%	56.0%
Total			48	52	100
			100.0%	100.0%	100.0%

Each subscript letter denotes a subset of gender categories whose column proportions do not differ significantly from each other at the .05 level

$\chi^2(4)=16.3$, $p=0.003$

Table 11. Relationship between age and the activity "Watches videos/ YouTube"

			Age		Total
			4-5 years	**5.5 -6.5 years**	
Watches videos/ YouTube	alone	Count % within age	11_a	9_a	20
			19.6%	20.5%	20.0%
	needs help		3_a	15_b	18
			5.4%	34.1%	18.0%
	with siblings		2_a	7_b	9
			3.6%	15.9%	9.0%
	with adult		7_a	6_a	13
			12.5%	13.6%	13.0%
	never		33_a	7_b	40
			58.9%	15.9%	40.0%
Total			56	44	100
			100.0%	100.0%	100.0%

Each subscript letter denotes a subset of age categories whose column proportions do not differ significantly from each other at the .05 level

$\chi^2(4)=26.9, p<0.001$

Table 12. Relationship between age and the activity "Looks at pictures/photos"

			Age		Total
			4-5 years	**5.5 -6.5 years**	
Looks at pictures/photos	alone	Count % within age	2_a	7_b	9
			3.6%	15.9%	9.0%
	needs help		0_a	2_a	2
			0.0%	4.5%	2.0%
	with siblings		2_a	5_a	7
			3.6%	11.4%	7.0%
	with adult		23_a	3_b	26
			41.1%	6.8%	26.0%
	never		29_a	27_a	56
			51.8%	61.4%	56.0%
Total			56	44	100
			100.0%	100.0%	100.0%

Each subscript letter denotes a subset of age categories whose column proportions do not differ significantly from each other at the .05 level

$\chi^2(4)=20.3, p<0.001$

Table 13. Relationship between age and the activity "Takes pictures"

			Age		Total
			4-5 years	**5.5 -6.5 years**	
Takes pictures	alone	Count % within age	0_a 0.0%	11_b 25.0%	11 11.0%
	needs help		2_a 3.6%	0_a 0.0%	2 2.0%
	with siblings		2_a 3.6%	0_a 0.0%	2 2.0%
	with adult		10_a 17.9%	6_a 13.6%	16 16.0%
	never		42_a 75.0%	27_a 61.4%	69 69.0%
Total			56 100.0%	44 100.0%	100 100.0%

Each subscript letter denotes a subset of age categories whose column proportions do not differ significantly from each other at the .05 level

$\chi^2(4)=18.1$, p=0.001

Table 14. Relationship between age and the activity "Listens to music"

			Age		Total
			4-5 years	**5.5 -6.5 years**	
Listens to music	alone	Count % within age	2_a 3.6%	4_a 9.1%	6 6.0%
	needs help		9_a 16.1%	0_b 0.0%	9 9.0%
	with siblings		6_a 10.7%	10_a 22.7%	16 16.0%
	with adult		23_a 41.1%	11_a 25.0%	34 34.0%
	never		16_a 28.6%	19_a 43.2%	35 35.0%
Total			56 100.0%	44 100.0%	100 100.0%

Each subscript letter denotes a subset of age categories whose column proportions do not differ significantly from each other at the .05 level.

$\chi^2(4)=13.9$, p=0.008

Chapter 12
Primary School Students' Perceptions About the Use of Mobile Games in the Classroom

George Koutromanos
National and Kapodistrian University of Athens, Greece

ABSTRACT

This chapter investigates primary school students' perceptions regarding the use of games on mobile devices (i.e., smartphone, tablet) in classroom environment for teaching purposes. Data was collected from 10,381 students (Grades 4, 5, and 6) using a survey questionnaire. A series of semi-structured interviews were used to more deeply understand student perceptions. Those perceptions tended to range from neutral to positive regarding the use of games and their content. Students prefer games that enhance their knowledge and develop their thinking skills. They also believe lessons will be more enjoyable and interesting through the use of games, and learning will be easier and more effective. The results showed differences in students' perceptions in terms of their gender and grade level. Implications of this study and further research are discussed.

INTRODUCTION

Nowadays, tablets and smartphones have become the central computing and communication devices in people's lives and activities (Courage, 2019; OFCOM, 2019). These devices differ from others (e.g. mobile phones, computers, laptops) due to their comprehensive and relatively advanced features, such as geo-positioning systems and web browsing capabilities, touch screen displays, Wi-Fi connectivity, and sophisticated built-in applications (Lee et al., 2019). These features have given rise to new communication capabilities (e.g. WhatsApp, Viber, Skype, and Snapchat) and forms of leisure time entertainment for young people (Ericsson, 2015, 2016), who are often referred to as "digital natives" (Barak, 2018; Prensky, 2001a, 2001b; Thompson, 2013), "the net generation" (Barak, 2018; Tapscott, 1998; Tapscott, 2008), or "the gamer generation" (Beck & Wade, 2004). According to national surveys conducted in different countries, the number of young people who use smartphones and tablets is increasing year by

DOI: 10.4018/978-1-7998-1486-3.ch012

year (Eurostat, 2015; Ericsson, 2016; Lenhart, 2015; OFCOM, 2019; Pew Research Center, 2015). It has been argued that playing games on tablets and smartphones has become the primary leisure activity for the new generation. These games are called mobile games (Huizenga, Admiraal, ten Dam, & Voogt, 2019; Koutromanos & Avraamidou, 2014) and are usually included in either the currently existing game categories (e.g. simulation games, puzzles, adventure games) (Kirriemuir & McFarlane, 2004) or in new categories, such as location-based games (Reinhardt & Heinig, 2016), augmented reality games (Alha, Koskinen, Paavilainen, & Hamari, 2019; Koutromanos, Sofos & Avraamidou, 2015), or pervasive games (Arango-López, Valdivieso, Collazos, Vela, & Moreira, 2018; Kasapakis & Gavalas, 2015).

The growing popularity of games on tablets and smartphones as a leisure activity among young people (The Asian Parent, 2014; ESA, 2015; OFCOM, 2019) has led researchers to conclude that these games should also be used in formal learning environments to promote student engagement (Koutromanos & Avraamidou, 2014; Koutromanos et al., 2015). Previous review articles support that the games being played on mobile devices can have a significant impact on learning and the development of various skills among students in both formal and informal learning environments (Koutromanos & Avraamidou, 2014; Koutromanos et al., 2015). Furthermore, there are researchers who claim that games on mobile devices could overcome both the technological and curricular problems that arise in schools the moment students move to the computer laboratory to play digital games (Brom, Levcic, Buchtova, & Klement, 2015, p. 683).

Nonetheless, to successfully utilize educational games on tablets/smartphones in formal learning environments, it would be interesting to first examine the perceptions of the already engaged as gamers students, who already possess experience in playing mobile games as a recreational activity and they possibly have formed positive perceptions regarding the particular games. The perceptions of primary education students, in particular, must be examined. Primary education constitutes the first stage of compulsory education for students, where the foundations for their future performance in secondary school and higher education is laid (Hainey, Connolly, Boyle, Wilson, & Razak, 2016).

However, until now no studies seem to have been conducted on the perceptions of primary school students regarding the use of mobile games in the classroom (see Acquah & Katz, 2020; Boyle, Connolly, Hainey, & Boyle, 2012; Connolly, Boyle, MacArthur, Hainey, & Boyle, 2012; Hainey et al., 2016; Hung et al., 2018; Xu et al., 2019). The recent study by Özdener and Demirci (2019) focused on students' perceptions about using mobile games for homework. In addition, previous studies have focused on students' perceptions regarding games that are not played specifically on tablets and smartphones, but rather, on other devices (e.g. computer, portable consoles, television consoles, and mp3 players) (Bourgonjon, Valcke, Soetaert, & Schellens, 2010) or older generation mobile phones (Ha, Yoon, & Choi, 2007; Liu & Li, 2011). Playing games on tablets and smartphones may yield different experiences for students compared to their experiences on other devices and platforms. On the other hand, previous studies in the area of digital games have focused on older student populations, such as secondary school students and university students (e.g. Bourgonjon et al., 2010; Hsiao & Tang, 2016). The mentality of primary school students is different from that of adolescents and adults, and their perceptions of games played on tablets/smartphones may likely differ from those characterizing older generations.

In conclusion, unanswered questions remain regarding primary school students' perceptions towards educational games on tablets/smartphones, as well as the factors that will encourage users to use these games at school. This study focused on primary school students between the ages of 9 to 12, who had experience playing games on tablets and smartphones in their leisure time. The aim of this study is to answer the following research questions:

1. What is the students' profile regarding the use of educational games on tablets/smartphones in their leisure time?
2. Do students prefer to use educational games on tablets/smartphones in the classroom?
3. For which subject do students mostly prefer to utilize educational games on tablets/smartphones?
4. What are students' perceptions regarding the content that they would like the educational games on tablets/smartphones to feature?
5. For which reasons do students prefer to utilize educational games on tablets/smartphones in the classroom?
6. What are students' perceptions regarding the factors contributing to the use of educational games on tablets/smartphones at their school?
   Additionally, this study examine the following questions:
7. Do the preferences of students regarding the use of educational games on tablets/smartphones in their learning differ by gender and grade?
8. Do students' perceptions regarding the learning content that they would like educational games on tablets/smartphones to feature differ by gender and grade?

This chapter is organized as follows. Section 2 provides a review of previous research regarding students' perceptions towards the use of digital games in education. Section 3 describes the research methodology of the empirical study. Section 4 reports the main results. Finally, in Section 5, implications are discussed, while important limitations are recognized, along with future research directions.

STUDENTS' PERCEPTIONS ABOUT DIGITAL GAMES IN EDUCATION

As no previous studies have been conducted concerning primary school students' perceptions regarding the use of mobile games in classroom, it might be of interest to explore other studies which focus on the use of digital games that can be played on computers. These studies can generally fall into two categories.

The first category of studies relates to the examination of students' attitudes regarding the use of digital games in lessons. One such study was conducted by Bourgonjon et al. (2010). In their study, used variables from the Technology Acceptance Model in order to examine which factors influenced students' acceptance of the use of video games in their classroom. 858 secondary school students participated in this study. Of these, 12.8% reported not playing video games at all. The findings of this study showed that students' preference for using video games in the classroom was significantly influenced by their perceptions regarding the usefulness, ease of use, learning opportunities, and personal experience with video games. An analysis of the data regarding differences in the averages among the variables of this study and students' gender indicated that female students do not prefer video games in the classroom, whereas male students are more supportive of the idea. Despite this, Bourgonjon et al. (2010) found that both male and female students tend to believe that video games offer learning opportunities.

Variables of the Technology Acceptance Model were also utilized in the study conducted by Hwang, Hong, Cheng, Peng, & Wu (2013). They examined the effect of gameplay anxiety, cognitive load, perceived ease of use, perceived usefulness, and attitude towards play in students' intention to play the "Radical Assembly Game." 235 students from a primary school located in Taipei City, China participated in this study. Hwang et al. (2013) found that students showed a positive attitude and intention to play the game in question. In a more recent study, Camilleri and Camilleri (2019) utilized the variables of perceived ease of use, perceived usefulness, perceived enjoyment, and normative pressures to investigate primary school students' intention to play educational games at home and at school. This study was conducted on 148 third-grade students (ages 8-9). The results of the studies showed that students have a favorable intention to play games at school.

The second category includes studies in which students' preferences and attitudes regarding the use of games in their learning did not serve as the main purpose of the study, but rather, one of its objectives. For example, Barendregt and Bekker (2011) examined the in situ use of and opinions of students regarding an educational computer game for learning English introduced in three primary schools offering a varying degree of freedom to choose school activities. Specifically, one of the aims of their study was to examine the degree to which students wanted to play a game in the formal learning setting (at school) or in the informal learning setting (at home). The first school offered only free-choice learning activities, while the second school offered "free-choice activities in the sense that students can choose from a range of activities and have to argue for their choice" (p. 83). The third school did not offer any free-choice learning activities. The results of this study showed that students at the schools offering only limited-choice or no free-choice learning activities were the most interested in playing the computer game at school. The students from the school offering only free-choice learning activities were less interested in playing the game at school.

Okanlawon, Fakokunde, Yusuf, Abanikannda, and Oyelade (2017) examined 360 second year Junior Secondary School students' attitude towards instructional games (board and computer games) on peace education. The results of their study showed that students had positive attitudes towards the instructional games. In addition, the results revealed that there was a significant relationship between students' attitudes towards the instructional games and their achievement in peace education.

In a more recent study, Özdener and Demirci (2019) examined 28 third grade primary school students' opinions regarding the use of an educational mobile application which is supported by sensors as a homework tool. One of the objectives of their study was to examine students' perceptions about using mobile games for homework. The results showed that almost all of the students stated that they would like to use mobile games for homework.

Corresponding examples of studies from the second category are present in higher education (Bekebrede, Warmelink, & Mayer, 2011; Gumusgul, 2019; Hainey et al., 2013; Kron, Gjerde, Sen, & Fetters, 2011). For example, Hainey et al. (2013) used a sample of 887 students from 13 different higher education institutes in Scotland and the Netherlands for the purpose of examining students' traits with regards to their game playing habits, gaming preferences, as well as their thoughts and perceptions of the use of digital games in education. In terms of students' perceptions, they found that a large number of them believed that games could be used in higher education setting for educational purposes. In another study, Bekebrede et al. (2011) examined the perceptions of college students regarding the use of digital games in learning. The sample was composed of 1432 respondents from eight Dutch institutes of higher education. Bekebrede et al. (2011) found that students deemed digital games to be both of value and added value to their own education. More specifically, the students deemed digital games to be more of an added rather than a general value.

METHODOLOGY

The current study used both quantitative and qualitative data collection and analysis approach. In order to collect data regarding students' perceptions on the use of educational games on tablets/smartphones in the teaching process, the survey method was used. In addition, in order to more deeply understand the students' perceptions regarding the reasons for the use of these games in classroom, as well as their perceptions regarding the factors that influence the use of these games in classroom, a series of semi-structured interviews were used.

Sample

The participants who took part in the survey were 10.381 primary - aged students from 360 schools in Greece. These schools came from regions all across Greece. Of the 10.381 students in the research, 5.339 (51.4%) were boys and 5.042 (48.6%) were girls. The students were aged 9 – 12 years from Grades 4 (N=2.558, 24.6%), 5 (N=3.585, 34.5%) and 6 (N=4.238, 40.8%). Eighteen students from three primary schools participated in interviews. Two students from the three highest grades in each school participated. 12 (67%) were girls and 6 (33%) were boys.

Instruments

Quantitative and qualitative data collection tools were used in the research. In order to investigate students' perceptions regarding games on tablets/smartphones in classroom, an online questionnaire was used. The survey questionnaire was divided into two parts. The first part was composed of two sections. The first section was composed of questions about the demographic characteristics of the students (i.e. gender, grade). The second section of the questionnaire elicited the following data regarding students' use of games on smartphones and tablets in leisure time: frequency of use, years of game playing, ownership of the mobile device, game genre, times of day when the students play the games, their location when playing the games, and who taught them to play the games. The question about the use frequency of the games was adapted from a study by Lu, Zhou, and Wang (2009). The items for this question were presented on a four-point Likert scale (ranging from "Very Rarely" to "Very Frequently"). The questions regarding the years that students played the games (ranging from "1 year" to "3 years"), who taught (e.g. family member, friends) them to play the games (see Table 2) (ranging from "Strongly disagree" to "Strongly agree"), the ownership of the device (i.e. smartphone, tablet) and the time of the day when students play games (e.g. morning, afternoon, evening) (ranging from "I have never played" to "Very often") were formulated after a pilot study. The question about how long students play games during each sitting (1=Less than 1 hour, 2=1 to 2 hours, 3= More than 2 hours) was adapted from a study by Lowrie and Jorgensen (2011). The question regarding the genres of games (see Table 2) was adapted by Hong, Cheng, Hwang, Lee, & Chang (2009). The items for this question were presented on a five-point Likert scale (ranging from "I have never played" to "Very often").

The second part of the questionnaire was aimed at measuring students' preference for mobile games use in teaching as well as their perceptions of the content of the relevant games. The preference for mobile games consisted of three items (i.e. *If I had the choice, I would choose to follow courses in which educational games on tablets/smartphones are used, If I had to vote, I would vote in favor of using educational games on tablets/smartphones in the classroom, I am enthusiastic about using educational games on tablets/smartphones in the classroom*). This section adapted from Bourgonjon et al. (2010). The section regarding the content of mobile games consisted of seven categories of perceptions and 30 items. Table 4 shows these items. The seven categories are: mentality change, emotional fulfillment, knowledge enhancement, thinking skill development, spatial ability development, bodily coordination and interpersonal skills. The items of these categories were adapted from Hong, et al. (2009). All items of the second part of the questionnaire were based on a five-point Likert scale, ranging from 1 (''Strongly disagree") to 5 (''Strongly agree"). In addition, this part of the questionnaire asked students to identify which subject they would most want these games to be used in during school hours. The subjects were listed and students were asked to choose only one.

The questionnaire was presented to 15 experts in ICT in education and six teachers of primary education to improve face and content validity. In addition, an earlier version of the questionnaire was trialled with 120 primary education students in order to investigate the clarity and appropriateness of the wording of the items. These students were excluded from the subsequent study. Based on experts and teachers' remarks, and students' feedback, some of the items were removed from the questionnaire, while other were rephrased.

Additionally, a semi-structured interview - based study was conducted with 18 students. Its purpose was to collect qualitative data regarding i) students' perceptions of the reasons that they desire to use mobile games during lessons, as well as ii) their perceptions of the factors that will facilitate the use of these games.

Data Collection

The study was conducted in collaboration with the Greek Ministry of Education, Research, and Religions' Directorate of Studies, Programs, and Organization for Primary Education. The ministry sent an e-mail to all the primary schools, informing their head teachers and technology teachers about the purpose of the study. The pre-requisite for students' participation in this study was that they play digital games in their leisure time on tablets or smartphones. Participation was voluntary and anonymous. The ICT teachers assisted students on how to fill out the online survey designed on Google Forms. Students filled out the questionnaire during the lesson. Not more than 15 minutes were spent on the questionnaire.

The qualitative data from this study, which was gathered from the interviews, was collected from the students of three primary schools in the Athens area, who had also participated in the questionnaire. The researcher of this study initially informed the head teacher and teachers of the three schools regarding the purpose of the study, and subsequently requested permission from parents to allow their children to participate in the interview. The interviews were conducted in a classroom at the end of the school day. Each interview was recorded by the researcher and lasted approximately 10-15 minutes.

Data Analysis

All statistical analyses for the questionnaire were performed using Statistical Package for the Social Sciences (SPSS, version 21). Data analysis consisted of the following four methods: First, a descriptive analysis of the items was performed to calculate the mean and standard deviation of the variables. Second, Cronbach's alpha was calculated regarding preference for mobile games and perceptions on the content of mobile games for teaching variables. As shown in Table 1, all variables exhibited an α-value greater than 0.7 (see Nunnally 1978). Third, an independent samples t-test was conducted, to examine

Table 1. Cronbach alpha for variables of the study

	No of items	Cronbach's α
Preference for mobile games	3	.834
Mentality change	2	.750
Emotional fulfillment	4	.869
Knowledge enhancement	10	.935
Thinking skill development	6	.917
Spatial ability development	1	-
Bodily coordination	2	.826
Interpersonal skills	5	.924

if there were any significant differences between students' age and their preferences for mobile games and perceptions regarding the content of mobile games for teaching using mean scores. In addition, to examine any differences between the responses of the three student grades (i.e. 4th, 5th, 6th grade) and their preferences and perceptions, a one – way - ANOVA was conducted. Post hoc analysis using a Bonferroni test was performed on those results found to be significant at $p<0.05$ level.

The data from students' interviews was analyzed thematically, organized, and interpreted according to the reasons and factors that influence the acceptance of ICT in education.

RESULTS

Use of Games in Leisure Time

The students stated that they play games relatively frequently on smartphones (*M*=2.00, *SD*=1.096) and more frequently on tablets (*M*=2.31, *SD*=1.116). Furthermore, they have been playing over a year and a half on both smartphones (*M*=1.65, *SD*=1.063) and tablets (*M*=1.64, *SD*=.984). 52.2% (N=5.419) of students stated that when they play games on a smartphone, it belongs to a family member (i.e., parents, older siblings), 27.9% (2.895) stated that it belongs to relatives (i.e., uncles/aunts, cousins), 19.3% (N=2.002) stated that it belongs to them, and 0.6% (N=65) stated that it belongs to a friend. On the other hand, 61.1% (N=6.342) of students stated that when they play games on a tablet, it belongs to them, 21.2% (N=2.198) stated it belongs to a member of their family, 16.2% (N=1.683) stated it belongs to a relative, and 1.5% (N=158) stated that it belongs to a friend.

51.7% (N=5371) of students stated that they play less than an hour during each sitting, 35.5% (N=3684) play between 1 to 2 hours, while 12.8% (N=1326) play more than 2 hours (*M*=1.61, *SD*=.702). The time of day when students play games the most is the afternoon (*M*=3.29, *SD*=1.230) and evening (*M*=2.74, *SD*=1.346). They play less in the morning (when they don't have school or on weekends) (*M*=2.72, *SD*=1.334) and at noon (*M*=2.61, *SD*=1.222). The students stated that they download the games they play to their device (*M*=3.57, *SD*=1.404), that these games are preinstalled on their devices (*M*=3.28, *SD*=1.410), or that they play them online (*M*=2.91, *SD*=1.423).

The genres of games that students play are presented in Table 2. Action and adventure games (*M*=2.97, *SD*=1.338), sports game (*M*=2.67, *SD*=1.509) and combat games (*M*=2.59, *SD*=1.548) were played most

Table 2. Genres of games played by students in their leisure time

	M	SD
Action and adventure games	2.97	1.338
Sports games	2.67	1.509
Combat fighting games	2.59	1.548
Role-playing games	2.34	1.380
Strategy games	2.31	1.362
Racing games	2.21	1.227
Simulation games	2.11	1.252
Puzzle games	2.11	1.216

frequently, while role-playing games (*M*=2.34, *SD*=1.380) and strategy games (*M*=2.31, *SD*=1.362) were played least frequently. Racing games (*M*=2.21, *SD*=1.227), simulation games (*M*=2.11, *SD*=1.252) and puzzle games (*M*=2.11, *SD*=1.216) were played not very frequently as well. The location where students play these games is at their home (*M*=4.07, *SD*=1.123), or the homes of friends and relatives (*M*=2.61, *SD*=1.273).

Regarding who taught them to play the games listed in Table 2, most students responded that they learned "by themselves" (*M*=3.77, *SD*=1.298) or that they "read a magazine or article on online" (*M*=3.77, *SD*=1.298). Fewer students stated that they were taught by a family member (*M*=2.78, *SD*=1.443) or friends (*M*=2.50, *SD*=1.316).

Students' Preference for Mobile Games in Classroom

The mean for the preference for mobile games scale (3 items) was calculated as 3.05, with a standard deviation of 1.124. The mean score for the boys was 3.14 (*SD*=1.151) and for girls was 2.95 (*SD*=1.087). An independent-samples t-test indicated that mean score was significantly higher for boys than for girls, $t(10378.99)=8.673, p<.000$. In addition, there was a statistically significant difference between the three groups of students' grade as determined by one-way ANOVA ($F(2, 10378)=12.879, p=.000$). A Bonferoni post hoc test revealed that the mean score for the preference scale of the students in the 6th grade group (*M*=3.11, *SD*=1.110) was statistically significantly higher than in the 5th (*M*=3.02, *SD*=1.121) and 4th (*M*=2.98, *SD*=1.147) grade group ($p<.005$).

Subjects in Which Students Want to Play Games in Classrooms

Table 3 shows the subjects regarding which students stated a desire to play educational games on smart phones/tablets during school instruction. According to the results displayed in the table, students' interest in playing games is highest for the subjects of Physics, Mathematics, and Modern Greek, followed by History, ICT, English, and their second foreign language. The subjects generating the least interest from students were Religion, Social and Civic Education, Music, Physical Education, Theater, and Art.

Table 3. Subjects in which students want to play games in classroom

Subject	N	%
Physics	2485	23,9
Mathematics	2217	21,4
Language Arts	2158	20,8
History	806	7,8
ICT	528	5,1
English	456	4,4
2nd Foreign Language	443	4,3
Religion	289	2,8
Social and Civic Education	284	2,7
Music	271	2,6
Physical Education	160	1,5
Theater	155	1,5
Art	129	1,2

Students' Perceptions Regarding the Content That They Would Like to See in Games

Table 4 shows the descriptive summary of the results with the mean scores of perceptions about the content of mobile games and the standard deviation of the two gender groups. In addition, this table shows the results of t-test regarding the differences in students' perceptions related to their gender. As we can see in this table, the overall mean score of the seven categories of perceptions varied from 3.14 (Mentality change) to 3.38 (Thinking skill development).

The categories of "Thinking skill development", "Bodily coordination", "Knowledge enhancement", "Emotional fulfillment", "Interpersonal skills" had the highest mean scores and "Spatial ability development", "Mentality change" the lowest. In addition, the overall mean score of the 30 items of perceptions varied from 3.07 (see item 1) to 3.53 (see item 18). These mean scores indicate that students tend to have higher perceptions regarding the specific content of mobile games which are presented in Table 4.

Inspection of the overall means and the standard deviations per category of perceptions indicate that the boys had higher means than the girls. In addition, the boys had higher means than the girls in the majority of the 30 items. The three most important boys' perceptions of the content of mobile games were "To inspire my imagination" (M=3.59, SD=1.225), "To develop my memories capacity" (M=3.51, SD=1.257) and "To develop my ability to construct and apply technology in life" (M=3.49, SD=1.237). On the other hand, "To inspire my imagination" (M=3.47, SD=1.198), "To develop my memories capacity" (M=3.39, SD=1.239) and "To develop my vocabulary" (M=3.36, SD=1.253) were the most important perceptions for girls. The t-test results showed that boys' perceptions were significantly more positive than girls' perceptions. Table 4 also shows that girls had higher mean scores than boys in the following perception: "To cultivate my musical abilities, such as rhythm and tone" (M=3.27, SD=1.222). However, this difference was not statistically significant.

Table 4. The Mean (M), Standard deviation (SD), and t-test results for gender differences

	Overall		Boys		Girls		*t*
	M	SD	M	SD	M	SD	
Mentality change	3.14	1.146	3,22	1.187	3.05	1.093	8.004*
1. To cultivate adventure.	3.07	1.329	3.19	1.378	2.93	1.261	10.153*
2. To cultivate my sense of responsibility.	3.21	1.232	3.25	1.263	3.16	1.196	3.948*
Emotional fulfillment	3.29	1.069	3.34	1.085	3.23	1.050	5.059*
3. To encourage cooperation.	3.39	1.219	3.43	1.241	3.34	1.194	3.759*
4. To enhance my concentration.	3.25	1.243	3.33	1.266	3.17	1.213	6.448*
5. To cultivate fairness and a sense of justice.	3.21	1.257	3.28	1.276	3.13	1.233	6.028*
6. To decrease aggressive behavior.	3.30	1.326	3.31	1.336	3.28	1.315	1.110
Knowledge enhancement	3.30	1.003	3.34	1.021	3.26	.983	4.237*
7. To develop my vocabularies.	3.38	1.270	3.41	1.285	3.36	1.253	1.910
8. To develop my literature reading.	3.34	1.248	3.40	1.266	3.29	1.226	4.527*
9. To develop my writing skills.	3.30	1.302	3.34	1.323	3.27	1.279	2.747*
10. To develop my ability to solve mathematical questions.	3.31	1.317	3.35	1.323	3.28	1.310	2.465*
11. To cultivate my musical abilities, such as rhythm and tone.	3.26	1.263	3.25	1.300	3.27	1.222	-.932
12. To develop my ability to construct and apply natural science concepts.	3.24	1.250	3.29	1.280	3.18	1.215	4.393*
13. To develop my ability to construct and apply technology in life.	3.36	1.227	3.49	1.237	3.23	1.202	10.736*
14. To develop my ability to construct and apply knowledge in history and geography.	3.31	1.258	3.37	1.276	3.25	1.235	5.082*
15. To develop my ability to appreciate and express art.	3.27	1.228	3.29	1.266	3.24	1.185	2.237*
16. To develop my ability to create art.	3.22	1.265	3.23	1.300	3.21	1.227	.822*
Thinking skill development	3.38	1.047	3.44	1.052	3.31	1.038	6.684*
17. To cultivate my ability to develop various modes of cognition.	3.36	1.246	3.41	1.258	3.30	1.232	4.357*
18. To inspire my imagination.	3.53	1.213	3.59	1.225	3.47	1.198	5.308*
19. To cultivate my critical thinking.	3.31	1.224	3.38	1.240	3.23	1.201	6.384*
20. To cultivate my ability to analyze and solve problems in life.	3.27	1.302	3.35	1.308	3.19	1.291	6.264*
21. To cultivate my powers of observation.	3.34	1.243	3.42	1.264	3.26	1.214	6.459*
22. To develop my memories capacity.	3.46	1.250	3.51	1.257	3.39	1.239	4.922*
Spatial ability development							
23. To develop my spatial navigation skills.	3.26	1.247	3.35	1.268	3.16	1.216	7.914*
Bodily coordination	3.28	1.175	3.40	1.184	3.15	1.152	10.664*
24. To develop my hand-eye coordination.	3.27	1.268	3.37	1.282	3.16	1.243	8.568*
25. To improve my reflexes.	3.29	1.278	3.42	1.289	3.15	1.250	11.097*
Interpersonal skills	3.29	1.114	3.34	1.119	3.23	1.106	4.719*
26. To develop my communicative expressive ability.	3.30	1.274	3.35	1.290	3.25	1.257	3.699*
27. To develop one's care for others and the society.	3.23	1.266	3.28	1.281	3.17	1.247	4.234*
28. To inspire one's respect for others and the society.	3.26	1.259	3.31	1.273	3.21	1.242	3.989*
29. To emphasize the importance of helping others.	3.33	1.267	3.37	1.279	3.28	1.253	3.631*
30. To emphasize the importance of trusting others.	3.32	1.298	3.38	1.312	3.25	1.280	5.088*

* $p<0.05$.

Table 5 shows the results of one way ANOVA regarding the differences in students' perceptions related to their grade. As we can see from this table the mean score of the three grade groups varied from 3.00 (see item 1) to 3.43 (see item 18) for 4th grade, 3.09 (see item 1) to 3.54 (see item 18) for 5th grade and 3.09 (see item 1) to 3.58 (see item 18) for 6th grade. As is shown in Table 5 there were statistically significant differences in the overall means of the seven categories of perceptions as well as in the means of the majority of the 30 items, among the three grade groups. The Bonferroni post hoc test showed (see last column in Table 5) that in the 6th grade the perceptions were significantly more positive than in the 5th and 4th grade groups. In addition, the majority of perceptions of students in 5th grade were significantly more positive than from students in 4th grade.

Reasons Why Students Want to Play Games at School

The analysis of the students' answers during their interviews generated two categories of reasons regarding why students would like educational games to be utilized in school on tablets/smartphones. The first reason is that it would make the lesson more pleasant and interesting. The majority of students mentioned that their lessons are often boring – especially when their teachers use only the textbook during instruction. The response of a sixth-grader is indicative of this sentiment:

I don't doubt that our teacher is very good and loves us. I know this, and I realize it. However, all day long we are taught our lessons from the textbooks and board. Once in a while, she takes us to the computer lab to watch a video, but we don't like this... and it creates a commotion. If we had tablets in the classroom and she told us to play a game related to Modern Greek... I think – I'd like to do this very much – that the lesson would be better, we'd have fun, and we wouldn't be bored...

The second reason is that games make instruction easier and more effective for students. In particular, students emphasized that games could be used for subjects that they find challenging, such as Physics and Mathematics. The answers that follow are indicative of the prevailing sentiment:

At home, my dad showed me a game about fractions and I play it on the tablet. The game is in English and I play it all the time. I understand the operations with uncommon denominators better... Why couldn't we play these sort of games at school during Mathematics lessons? (Sixth-grade student)

I have difficulty with Physics. I don't understand a lot of things about electrons, as well as the current topic we're studying – thermal energy. A game with experiments and movement would help me learn faster and more effectively. (Fifth-grade student)

It should be noted that two out of the 18 students who participated in the interview cited entertainment as the sole reason for their desire to play games. They were both boys from Grade 4, approximately nine years old. Based on further analysis of the data, it appears that they viewed digital games purely as a means of entertainment and not as a means that could aid them in learning.

Table 5. The Mean (M), Standard deviation (SD), and one way ANOVA results for the grade differences

	Overall		4th Grade		5th Grade		6th Grade			Grades**
	M	SD	M	SD	M	SD	M	SD	F	
Mentality change	3.14	1.146	3.07	1.191	3.14	1.137	3.18	1.124	8.052*	6→4
1. To cultivate adventure.	3.07	1.329	3.00	1.356	3.09	1.322	3.09	1.318	4.379*	5→4 6→5
2. To cultivate my sense of responsibility.	3.21	1.232	3.14	1.252	3.18	1.237	3.27	1.212	11.342*	6→5 6→4
Emotional fulfillment	3.29	1.069	3.20	1.089	3.26	1.064	3.36	1.057	19.344	6→5 6→4
3. To encourage cooperation.	3.39	1.219	3.30	1.239	3.38	1.220	3.45	1.203	11.965*	5→4 6→5 6→4
4. To enhance my concentration.	3.25	1.243	3.17	1.264	3.22	1.246	3.33	1.223	15.940*	6→5 6→4
5. To cultivate fairness and a sense of justice.	3.21	1.257	3.13	1.280	3.17	1.261	3.28	1.236	14.275*	6→5 6→4
6. To decrease aggressive behavior	3.30	1.326	3.21	1.320	3.26	1.347	3.38	1.307	14.264*	6→5 6→4
Knowledge enhancement	3.30	1.003	3.23	1.039	3.30	.995	3.34	.987	9.945*	5→4 6→4
7. To develop my vocabularies.	3.38	1.270	3.26	1.289	3.39	1.276	3.45	1.248	16.950*	5→4 6→4
8. To develop my literature reading.	3.34	1.248	3.24	1.278	3.35	1.254	3.40	1.221	12.535*	5→4 6→4
9. To develop my writing skills.	3.30	1.302	3.20	1.328	3.31	1.310	3.36	1.276	11.392*	5→4 6→4
10. To develop my ability to solve mathematical questions.	3.31	1.317	3.24	1.314	3.31	1.332	3.36	1.305	6.292*	6→4
11. To cultivate my musical abilities, such as rhythm and tone.	3.26	1.263	3.24	1.271	3.25	1.280	3.28	1.242	1.058	ns
12. To develop my ability to construct and apply natural science concepts.	3.24	1.250	3.17	1.267	3.25	1.255	3.27	1.234	5.563*	6→4
13. To develop my ability to construct and apply technology in life.	3.36	1.227	3.24	1.261	3.35	1.225	3.45	1.200	24.912*	5→4 6→5 6→4
14. To develop my ability to construct and apply knowledge in history and geography.	3.31	1.258	3.26	1.272	3.31	1.277	3.35	1.231	4.300*	6→4
15. To develop my ability to appreciate and express art.	3.27	1.228	3.23	1.256	3.26	1.236	3.29	1.202	2.131	ns
16. To develop my ability to create art.	3.22	1.265	3.21	1.283	3.24	1.273	3.20	1.248	.704	ns
Thinking skill development	3.38	1.047	3.28	1.078	3.37	1.048	3.44	1.024	17.215*	5→4 6→5 6→4
17. To cultivate my ability to develop various modes of cognition.	3.36	1.246	3.25	1.269	3.35	1.261	3.43	1.215	16.869*	5→4 6→5 6→4
18. To inspire my imagination.	3.53	1.213	3.43	1.242	3.54	1.222	3.58	1.184	12.014*	5→4 6→4

continued on following page

Table 5. Continued

	Overall		4th Grade		5th Grade		6th Grade			Grades**
	M	SD	M	SD	M	SD	M	SD	F	
19. To cultivate my critical thinking.	3.31	1.224	3.24	1.255	3.31	1.218	3.35	1.207	7.398*	6→4
20. To cultivate my ability to analyze and solve problems in life.	3.27	1.302	3.20	1.299	3.26	1.323	3.32	1.283	7.482*	6→4
21. To cultivate my powers of observation.	3.34	1.243	3.25	1.269	3.33	1.257	3.41	1.210	12.216*	5→4 6→5 6→4
22. To develop my memories capacity.	3.46	1.250	3.33	1.284	3.46	1.254	3.53	1.220	20.407*	5→4 6→5 6→4
Spatial ability development										
23. To develop my spatial navigation skills.	3.26	1.247	3.22	1.262	3.24	1.259	3.30	1.226	3.898*	6→4
Bodily coordination	3.28	1.175	3.20	1.176	3.26	1.190	3.34	1.158	12.978*	6→5 6→4
24. To develop my hand-eye coordination.	3.27	1.268	3.19	1.279	3.27	1.287	3.32	1.241	9.591*	5→4 6→4
25. To improve my reflexes.	3.29	1.278	3.21	1.286	3.26	1.291	3.36	1.258	13.094*	6→5 6→4
Interpersonal skills	3.29	1.114	3.22	1.114	3.29	1.129	3.33	1.099	7.112*	5→4 6→4
26. To develop my communicative expressive ability.	3.30	1.274	3.21	1.279	3.31	1.291	3.34	1.255	8.442*	5→4 6→4
27. To develop one's care for others and the society.	3.23	1.266	3.17	1.264	3.23	1.280	3.26	1.254	3.993*	6→4
28. To inspire one's respect for others and the society.	3.26	1.259	3.20	1.259	3.25	1.281	3.31	1.238	6.595*	6→4
29. To emphasize the importance of helping others.	3.33	1.267	3.27	1.260	3.32	1.276	3.37	1.262	5.105*	6→4
30. To emphasize the importance of trusting others.	3.32	1.298	3.25	1.298	3.33	1.307	3.34	1.289	4.478*	5→4 6→4

* $p<0.05$. ** This column presents the statistically significant differences among students' grades.

Factors That Influence the Use of Games on Mobile Devices in School

The interviews with the students yielded three primary factors that facilitate playing games in school. The first factor relates to the existence of tablets in school. Students in every school where an interview was conducted stated that there were no tablets on site and that their teachers did not bring any tablets to the classroom either. Several students (11 out of the 18) mentioned that if they were allowed to play games in school, they would bring their own tablets from home. The second factor relates to Wi-Fi availability. Seven students mentioned that their classroom does not have a wireless or wired internet connection.

One day, our teacher brought her laptop to show us a simulation in geography regarding earthquakes. We didn't have a connection in the classroom... both our teacher and we were aware of this, but our teacher connected to the internet through her mobile phone network... (Fifth-grade student)

The third factor has to do with classroom teachers' desire for students to use games. All of the students who participated in the interview reported that their teachers never allowed them to play a digital game, nor spoke to them about digital games. A sixth-grade student noted the following:

I think that my teacher doesn't want us to play games... [...] this is the sense I have gotten... she would have mentioned something after all this time. Even when we go to the computer lab to look at some webpages and videos, she never told us to play a game. We want to... My parents have me play at home on programs designed by the Ministry... They aren't so great, but I don't have any other games to play...

DISCUSSION AND CONCLUSION

This study focused on primary school students who had experience playing games on tablets and smartphones in their leisure time. The first research question was related to these students' profiles regarding the use of educational games on tablets/smartphones in their leisure time. The empirical results of this study demonstrated that students in the last three grades of primary school play games less frequently on smartphones and more frequently on tablets. Their frequency of use of these games is relatively moderate and can be termed reasonable for students at this age group. This result contradicts the results of prior studies (e.g. Dindar, 2018; Mysirlaki & Paraskeva, 2007), which found that older students play games very frequently in their leisure time. What is interesting to note is that a large percentage of students (61.1%) stated that they play games on their own tablets. A notable percentage 19.3% stated that they have their own smartphone. These results indicate that today's generation is fairly familiar with playing these types of games on mobile devices from a young age.

In terms of the second research question, the empirical results of this study indicated that primary school students had a neutral mean score regarding their preference for games on smartphones/tablets in their learning. This finding is similar to that of a previous study on digital games (Bourgonjon et al., 2010), which indicated that the preference for using video games in the classroom among secondary school students was relatively neutral. One possible explanation for this relatively neutral preference may be the fact that in their leisure time, students have grown accustomed to viewing digital games more as a means of entertainment than as a means of learning.

In terms of the third research question, the empirical results of this study showed that students prefer to play games in association with lessons related to Physics, Mathematics, and Language Arts. These are the main subjects in most curriculums used by countries around the world and are usually considered to be the most challenging by a majority of students. This result has implications for game designers. Game designers should design digital games to align with the needs of these three specific subjects.

The fourth research question examined students' perceptions regarding the content that they wanted to see in games used for teaching purposes. The results indicated that students' perceptions tended to range from relatively moderate to positive for most of categories in Table 4. The more positive perceptions were related to the categories "Thinking skill development" and "Knowledge enhancement". The findings of this study suggest that these two categories may play an important role regarding student engagement in games on smartphones/tablets in their learning. Therefore, in order to promote the use of educational games on smartphones/tablets in schools, game designers will have to include activities that enhance students' knowledge and develop their thinking skills.

The results related to the fifth research question showed that students would prefer that games be played on mobile devices at their schools to make the lessons more pleasant and interesting. Additionally, another reason for their preference includes making their learning easier and more effective. In terms of the sixth research question, the results from the interviews associated with this study indicated that in order for games to be utilized at school, issues relating to the availability of tablets and internet service must be resolved. Also, teachers must be favorably disposed to the use of such games in teaching and learning. A recent study of the factors for success in mobile learning at a rural primary school by Al-Huneini, Walker, and Badger (2020) supports these findings.

Finally, the seventh and eighth research questions had to do with the differences between the preferences and perceptions of students based on their gender and grade level. The findings revealed that boys' preference in playing games on mobile devices for learning is higher in comparison to girls. In addition, the boys were found to have more positive perceptions of the content that these games should have in comparison to the girls. This result is congruent with the results of other studies that found that boys hold more positive perceptions of digital games compared to girls (e.g. Lowrie & Jorgensen, 2011) or boys spent more time on video games and have higher skills in video game play than girls (Desai, Krishnan-Sarin, Cavallo, & Potenza, 2010; Dindar, 2018; Ekinci, Yalcin & Ayhan, 2019; Homer, Plass, Raffaele, Ober, & Ali, 2018). In this study, perhaps boys find games in mobile devices much more attractive than girls. A recent study has shown that the differences between males and females might be due to their game preferences (Dindar, 2018).

Finally, it was found that fifth and sixth grade primary school students have a greater preference for these games than fourth graders. The same results regarding these grade levels were found to hold true in terms of students' perceptions of the content that they want these games to have. This means that younger students likely view these kind of games more as entertainment than a learning tool. This was also confirmed by the results of the interviews with fourth graders, which were presented in a previous section.

This paper provides an important contribution regarding primary school students' perceptions of mobile games. These students had experience playing games on tablets and smartphones in their leisure time. As mentioned earlier, previous studies focused on samples from other grade levels and adults. Another important contribution has to do with the devices being studied. No studies had previously addressed primary school students' (ages 9-12) use of games on tablets and smartphones. The present study focused on more modern devices (i.e., tablets and smartphones) from the standpoint of technology and its capabilities, and thus enriches research literature regarding students' perceptions of using these devices compared to others (i.e., computers, laptops). Finally, by understanding younger students' perceptions on the use of mobile games in daily life and school, we may better understand how to design games for learning and how to incorporate them in formal learning environments starting from primary school.

In this study, various quantitative data were collected (e.g., frequency of use, mean hours of playtime, game genre) regarding the use of games by 10.381 students on smartphones/tablets in their leisure time. Future studies could focus on collecting qualitative data through interviews about the content and quality of games that students play, as well as what they believe that they learn from them. Moreover, the qualitative data of this study was collected from 18 students through interviews. It is noteworthy that the findings of these interviews cannot be generalized to other samples without additional research. There is also a need to conduct further research to examine the differences in students' perceptions based on their academic performance and profile as gamers. Furthermore, future studies could examine if there is a correlation in students' perceptions and their attitudes towards specific subjects in school. In addition, future research is recommended to investigate the reasons for the gender differences which this study found. Finally, the students of this study were gamers on smartphones/tablets in their leisure time. Further work must be undertaken with a non-gamer sample to provide additional evidence. It would be interesting to compare a sample of gamers and non-gamers to see if there are differences regarding their perceptions on the use of mobile games in their learning.

REFERENCES

Acquah, E. O., & Katz, H. T. (2020). Digital game-based L2 learning outcomes for primary through high-school students: A systematic literature review. *Computers & Education*, *143*, 103667. doi:10.1016/j.compedu.2019.103667

Al-Huneini, H., Walker, S. A., & Badger, R. (2020). Introducing tablet computers to a rural primary school: An Activity Theory case study. *Computers & Education*, *143*, 103648. doi:10.1016/j.compedu.2019.103648

Alha, K., Koskinen, E., Paavilainen, J., & Hamari, J. (2019). Why do people play location-based augmented reality games: A study on Pokémon GO. *Computers in Human Behavior*, *93*, 114–122. doi:10.1016/j.chb.2018.12.008

Arango-López, J., Valdivieso, C. C. C., Collazos, C. A., Vela, F. L. G., & Moreira, F. (2018). CREANDO: Tool for creating pervasive games to increase the learning motivation in higher education students. *Telematics and Informatics*, *38*, 62–73. doi:10.1016/j.tele.2018.08.005

Barak, M. (2018). Are digital natives open to change? Examining flexible thinking and resistance to change. *Computers & Education*, *121*, 115–123. doi:10.1016/j.compedu.2018.01.016

Barendregt, W., & Bekker, T. M. (2011). The influence of the level of free - choice learning activities on the use of an educational computer game. *Computers & Education*, *56*(1), 80–90. doi:10.1016/j.compedu.2010.08.018

Beck, J. C., & Wade, M. (2004). *Got game: How the gamer generation is reshaping business forever*. Boston, MA: Harvard Business School Press.

Bekebrede, G., Warmelink, H. J. G., & Mayer, I. S. (2011). Reviewing the need for gaming in education to accommodate the net generation. *Computers & Education*, *57*(2), 1521–1529. doi:10.1016/j.compedu.2011.02.010

Bourgonjon, J., Valcke, M., Soetaert, R., & Schellens, T. (2010). Students' perceptions about the use of video games in the classroom. *Computers & Education*, *54*(4), 1145–1156. doi:10.1016/j.compedu.2009.10.022

Boyle, E. A., Connolly, T. M., Hainey, T., & Boyle, J. M. (2012). Engagement in digital entertainment games: A systematic review. *Computers in Human Behavior*, *28*(3), 771–780. doi:10.1016/j.chb.2011.11.020

Brom, C., Levcic, D., Buchtova, M., & Klement, D. (2015). Playing educational micro-games at high schools: Individually or collectively? *Computers in Human Behavior*, *48*, 682–694. doi:10.1016/j.chb.2015.02.025

Camilleri, A. C., & Camilleri, M. A. (2019). The students' perceived use, ease of use and enjoyment of educational games at home and at school. *13th Annual International Technology, Education and Development Conference*. Valencia, Spain (March, 2019). International Academy of Technology, Education, and Development (IATED). Retrieved from https://papers.ssrn.com/sol3/papers.cfm?abstract_id=3339163

Connolly, T. M., Boyle, E. A., MacArthur, E., Hainey, T., & Boyle, J. M. (2012). A systematic literature review of the empirical evidence on computer games and serious games. *Computers & Education, 59*(2), 661–686. doi:10.1016/j.compedu.2012.03.004

Courage, M. L. (2019). *Interactive technologies: tablets, smartphones, and learning. Reference module in neuroscience and biobehavioral psychology*. Elsevier; doi:10.1016/B978-0-12-809324-5.23561-X

Desai, R. A., Krishnan-Sarin, S., Cavallo, D., & Potenza, M. N. (2010). Video-gaming among high school students: Health correlates, gender differences, and problematic gaming. *Pediatrics, 126*(6), 1414–1424. doi:10.1542/peds.2009-2706 PMID:21078729

Dindar, M. (2018). An empirical study on gender, video game play, academic success and complex problem solving skills. *Computers & Education, 125*, 39–52. doi:10.1016/j.compedu.2018.05.018

Ekinci, N. E., Yalcin, I., & Ayhan, C. (2019). Analysis of loneliness levels and digital game addiction of middle school students according to various variables. *World Journal of Education, 9*(1), 20–27. doi:10.5430/wje.v9n1p20

Ericsson. (2015, June). *Ericsson Mobility Report, On the pulse of the network society*. Retrieved from https://www.ericsson.com/4980fe/assets/local/news/2015/6/ericsson-mobility-report-june-20152.pdf

Ericsson. (2016, June). *Ericsson Mobility Report, On the pulse of the network society*. Retrieved from https://www.abc.es/gestordocumental/uploads/internacional/EMR_June_2016_D5%201.pdf

ESA. (2015). *Essential facts about the computer and video game industry*. Retrieved from http://www.theesa.com/wp-content/uploads/2015/04/ESA-Essential-Facts-2015.pdf

Eurostat, Statistical books. (2015). *Being young in Europe today*. Retrieved from http://ec.europa.eu/eurostat/documents/3217494/6776245/KS-05-14-031-EN-N.pdf/18bee6f0-c181-457d-ba82-d77b314456b9

Gumusgul, O. (2019). Investigation of University students' attitudes to play educational games and games consisting of physical activity. *World Journal of Education, 9*(2), 31–37. doi:10.5430/wje.v9n2p31

Ha, I., Yoon, Y., & Choi, M. (2007). Determinants of adoption of mobile games under mobile broadband wireless access environment. *Information & Management, 44*(3), 276–286. doi:10.1016/j.im.2007.01.001

Hainey, T., Connolly, T. M., Boyle, E. A., Wilson, A., & Razak, A. (2016). A systematic literature review of games-based learning empirical evidence in primary education. *Computers & Education, 102*, 202–223. doi:10.1016/j.compedu.2016.09.001

Hainey, T., Westera, W., Connolly, T. M., Boyle, L., Baxter, G., Beeby, R. B., & Soflano, M. (2013). Students' attitudes toward playing games and using games in education: Comparing Scotland and the Netherlands. *Computers & Education, 69*, 474–484. doi:10.1016/j.compedu.2013.07.023

Homer, B. D., Plass, J. L., Raffaele, C., Ober, T. M., & Ali, A. (2018). Improving high school students' executive functions through digital game play. *Computers & Education, 117*, 50–58. doi:10.1016/j.compedu.2017.09.011

Hong, J. C., Cheng, C. L., Hwang, M. Y., Lee, C. K., & Chang, H. Y. (2009). Assessing the educational values of digital games. *Journal of Computer Assisted Learning, 25*(5), 423–437. doi:10.1111/j.1365-2729.2009.00319.x

Hsiao, C. H., & Tang, K. Y. (2016). On the post-acceptance of mobile movie-themed games. *Electronic Commerce Research and Applications, 18*, 48–57. doi:10.1016/j.elerap.2016.06.003

Huizenga, J., Admiraal, W., ten Dam, G., & Voogt, J. (2019). Mobile game-based learning in secondary education: Students' immersion, game activities, team performance and learning outcomes. *Computers in Human Behavior, 99*, 137–143. doi:10.1016/j.chb.2019.05.020

Hung, H. T., Yang, J. C., Hwang, G. J., Chu, H. C., & Wang, C. C. (2018). A scoping review of research on digital game-based language learning. *Computers & Education, 126*, 89–104. doi:10.1016/j.compedu.2018.07.001

Hwang, M. Y., Hong, J. C., Cheng, H. Y., Peng, Y. C., & Wu, N. C. (2013). Gender differences in cognitive load and competition anxiety affect 6th grade students' attitude toward playing and intention to play at a sequential or synchronous game. *Computers & Education, 60*(1), 254–263. doi:10.1016/j.compedu.2012.06.014

Kasapakis, V., & Gavalas, D. (2015). Pervasive gaming: Status, trends and design principles. *Journal of Network and Computer Applications, 55*, 213–236. doi:10.1016/j.jnca.2015.05.009

Kirriemuir, J., & Mcfarlane, A. (2004). *Literature review in games and learning*. A NESTA Futurelab Research report - report 8. Retrieved from http://www.coulthard.com/library/Files/kirriemuir%2Dfuturelabs%5F2004%5Fgamesreview%2Epdf

Koutromanos, G., & Avraamidou, L. (2014). The use of mobile games in formal and informal learning environments: A review of the literature. *Educational Media International Journal, 51*(1), 49–65. doi:10.1080/09523987.2014.889409

Koutromanos, G., Sofos, A., & Avraamidou, L. (2015). The use of augmented reality games in education: A review of the literature. *Educational Media International Journal, 52*(4), 253–271. doi:10.1080/09523987.2015.1125988

Kron, F. W., Gjerde, C. L., Sen, A., & Fetters, M. D. (2010). Medical student attitudes toward video games and related new media technologies in medical education. *BMC Medical Education, 10*(1), 50. doi:10.1186/1472-6920-10-50 PMID:20576125

Lee, U., Han, K., Cho, H., Chung, K. M., Hong, H., Lee, S. J., ... Carroll, J. M. (2019). Intelligent positive computing with mobile, wearable, and IoT devices: Literature review and research directions. *Ad Hoc Networks, 83*, 8–24. doi:10.1016/j.adhoc.2018.08.021

Lenhart, A. (2015, April). *Teen, social media, and technology overview 2015*. Pew Research Center. Retrieved from http://www.pewinternet.org/2015/04/09/teens-social-media-technology-2015/

Liu, Y., & Li, H. (2011). Exploring the impact of use context on mobile hedonic services adoption: An empirical study on mobile gaming in China. *Computers in Human Behavior, 27*(2), 890–898. doi:10.1016/j.chb.2010.11.014

Lowrie, T., & Jorgensen, R. (2011). Gender differences in students' mathematics game playing. *Computers & Education, 57*(4), 2244–2248. doi:10.1016/j.compedu.2011.06.010

Lu, Y., Zhou, T., & Wang, B. (2009). Exploring Chinese users' acceptance of instant messaging using the theory of planned behavior, the technology acceptance model, and the flow theory. *Computers in Human Behavior, 25*(1), 29–39. doi:10.1016/j.chb.2008.06.002

Mysirlaki, S., & Paraskeva, F. (2007). Digital games: Developing the issues of socio-cognitive learning theory in an attempt to shift an entertainment gadget to an educational tool. In *Proceedings of the first IEEE international workshop on digital game and intelligent toy enhanced learning [DIGITEL2007]* (pp. 147–151). Jhongli, Taiwan.

Nunnally, J. C. (1978). *Psychometric theory* (2nd ed.). New York: McGraw-Hill.

OFCOM. (2019). *Children and parents: Media use and attitudes report, 2018*. London, UK: OFCOM. Retrieved from https://www.ofcom.org.uk/__data/assets/pdf_file/0024/134907/Children-and-Parents-Media-Use-and-Attitudes-2018.pdf

Okanlawon, A. E., Fakokunde, J. B., Yusuf, F. A., Abanikannda, M. O., & Oyelade, A. A. (2017). Attitudes towards instructional games on Peace Education among second year students in junior secondary schools in South-west Nigeria. *International Journal of Education and Development Using Information and Communication Technology, 13*(3), 98–108. Retrieved from https://files.eric.ed.gov/fulltext/EJ1166610.pdf

Özdener, N., & Demirci, F. (2019). Determining students' views about an educational game-based mobile application supported with sensors. *Technology Knowledge and Learning, 24*(1), 143–159. doi:10.100710758-018-9368-x

Pew Research Center. (2015, April). *The smartphone difference*. Retrieved from http://www.pewinternet.org/2015/04/01/us-smartphone-use-in-2015/

Prensky, M. (2001a). Digital natives, digital immigrants part 1. *On the Horizon, 9*(5), 1–6. doi:10.1108/10748120110424816

Prensky, M. (2001b). Digital natives, digital immigrants, part II: Do they really think differently? *On the Horizon, 9*(6), 2–6. doi:10.1108/10748120110424843

Reinhardt, D., & Heinig, C. (2016). Survey-based exploration of attitudes to participatory sensing tasks in location-based gaming communities. *Pervasive and Mobile Computing, 27*, 27–36. doi:10.1016/j.pmcj.2016.01.001

Tapscott, D. (1998). *Growing up digital: The rise of the Net generation*. New York: McGraw-Hill.

Tapscott, D. (2008). *Grown up digital: How the net generation is changing your world*. New York: McGraw-Hill.

The Asian Parent. (2014). *Mobile device usage among young kids, a Southeast Asia study*. Retrieved from https://s3-ap-southeast-1.amazonaws.com/tap-sg-media/theAsianparent+Insights+Device+Usage+A+Southeast+Asia+Study+November+2014.pdf

Thompson, P. (2013). The digital natives as learners: Technology use patterns and approaches to learning. *Computers & Education*, *6*, 512–533. doi:10.1016/j.compedu.2012.12.022

Xu, Z., Chen, Z., Eutsler, L., Geng, Z., & Kogut, A. (2019). A scoping review of digital game-based technology on English language learning. *Educational Technology Research and Development*, 1–28. doi:10.100711423-019-09702-2

ADDITIONAL READING

An, Y. J., & Cao, L. (2017). The Effects of Game Design Experience on Teachers' Attitudes and Perceptions regarding the Use of Digital Games in the Classroom. *TechTrends*, *61*(2), 162–170. doi:10.100711528-016-0122-8

Becker, K. (2017). *Choosing and Using Digital Games in the Classroom, A Practical Guide*. Switzerland: Springer; doi:10.1007/978-3-319-12223-6

Bokolas, V., Amanatidis, N., & Koutromanos, G. (2015). Students as Digital Games' Evaluators. Enhancing Media Literacy and Learning through Game Playing and Evaluation Methods. In R. Munkvold, & L. Kolås (Eds.), *Games Based Learning, ECGBL 2015*. 9th European Conference, Nord-Trondelag University College Steinkjer, Norway, 8-9 October (pp. 75–81).

Cheng, M. T., Chen, J. H., Chu, S. J., & Chen, S. Y. (2015). The use of serious games in science education: A review of selected empirical research from 2002 to 2013. *Journal of Computers in Education*, *2*(3), 353–375. doi:10.100740692-015-0039-9

Koutromanos, G., & Styliaras, G. (2015). The Buildings speak about our city: A location based augmented reality game, in *Information, Intelligence, Systems and Applications (IISA)*, 2015 6th International Conference (pp. 1–6), 6-8 July 2015, 10.1109/IISA.2015.7388031

Koutromanos, G., Tzortzoglou, F., & Sofos, A. (2018). Evaluation of an augmented reality game for environmental education: "Save Elli, Save the environment". In: Mikropoulos T. (eds) Research on e-Learning and ICT in Education (pp. 231–241). Springer, Cham. Springer. doi:10.1007/978-3-319-95059-4_14

Riemer, V., & Schrader, C. (2015). Learning with quizzes, simulations, and adventures: Students' attitudes, perceptions and intentions to learn with different types of serious games. *Computers & Education*, *88*, 160–168. doi:10.1016/j.compedu.2015.05.003

Touati, A., & Baek, Y. (2018). What Leads to Player's Enjoyment and Achievement in a Mobile Learning Game? *Journal of Educational Computing Research*, *56*(3), 344–368. doi:10.1177/0735633117713022

KEY TERMS AND DEFINITIONS

Computer Game: A computer game is any game that is played on a personal computer.

Digital Game: A digital game is any game that is played with the assistance of a computer or other electronic device.

Mobile Game: A mobile game is a game that is played on mobile devices, such as a smartphone, tablet, PDA or portable media player.

Smartphone: A smartphone is a mobile phone with highly advanced features such as wireless capability, global positioning system (GPS), camera/video, speech recognition, various sensors and applications.

Tablet: A tablet is defined as a mobile device with a touchscreen interface, WiFi connectivity, and advanced operating systems.

Section 4
Mobile Learning in General

Chapter 13
Mobile Learning in the Era of IoT:
Is Ubiquitous Learning the Future of Learning?

Koralia Papadokostaki
Hellenic Mediterranean University, Greece

Spyros Panagiotakis
Hellenic Mediterranean University, Greece

Athanasios Malamos
Hellenic Mediterranean University, Greece

Kostas Vassilakis
Hellenic Mediterranean University, Greece

ABSTRACT

Teaching is always affected by the advent of technology. Nowadays, mobile devices can offer an air of innovation in classrooms and multiple benefits in learning. On the other hand, IoT is expanding rapidly and promises to provide education with new dynamics: sensors and beacons may contribute to pervasive provision of educational content to students, whereas wearables can track the students' interaction with educational objects. As a result, learning is changing and may happen anywhere, anytime, and with any means. This evolution, described under the term Ubiquitous learning, promises to be the future of education for all ages and needs. This chapter presents the transformation of learning from traditional to e-learning, mobile learning, and Ubiquitous learning, and discusses the features and applications of the latter. Furthermore, authors describe the Experience API specification and investigate how it can be used to implement adaptive learning applications and make Ubiquitous learning a reality not only in typical but also in Early Childhood learning.

DOI: 10.4018/978-1-7998-1486-3.ch013

INTRODUCTION

Education is reformed. Following the accelerated progress in technology, high-tech breakthrough could not be missed in classrooms. Already, schools across the world have introduced tablets in classrooms or adopted the BYOD (Bring Your Own Device) movement (Ballagas, Rohs, Sheridan, & Borchers, 2004) and 3D printers are installed in many school labs. The use of Virtual Reality (VR) and Augmented Reality (AR) in classrooms seems more and more promising (Horváth, 2019) and innovations such as holograms ('7 Ways Holographic Technology Will Make Learning More Fun', n.d.; 'Potential and Applications of Holograms To Engage Learners', n.d.) or biometrics (Dafoulas, Maia, Clarke, Ali, & Augusto, 2018) are the near future of education. Moreover, Learning Management Systems (LMSs) are indispensable in large educational organizations, but with the spread of social media, social platforms have been used or developed (such as the educational social media Edmodo) ('Edmodo', n.d.) to enhance communication between teachers and students ('The Evolution of Technology in the Classroom | Purdue University Online', n.d.). The importance of involving video streaming platforms, such as YouTube into education has also been identified (Fleck, Beckman, Sterns, & Hussey, 2014), whereas Massive Online Open Courses (MOOCs) become more and more popular and competitive. Moreover, the use of technology in almost every aspect of our everyday life reinforces learning that happens outside the classroom, the so-called informal learning. Current technology offers a vast variety of means to support education and contributes significantly to pluralism in educational resources.

Children of very young age are becoming more and more familiar with new technologies and the internet. As the use of smartphones and tablets rises dramatically and mobile devices dominate our lives, their use by preschoolers and even toddlers seems inevitable (Kalogiannakis, Ampartzaki, Papadakis, & Skaraki, 2018; Papadakis & Kalogiannakis, 2017). At the same time, older children often own their own mobile device (Papadakis, Kalogiannakis, & Zaranis, 2016), while devices and applications specially designed for kids are constantly rising in demand (Papadakis & Kalogiannakis, 2017) . Children nowadays spend a lot of time using mobile devices and have evolved from "digital natives" (Papadakis et al., 2016), without exaggeration, to "iLearners" (Zaranis, Kalogiannakis, & Papadakis, 2013).

The advent of such a breakthrough, the Internet of Things (IoT) (Papadokostaki, Mastorakis, et al., 2017), may change education insurmountably. The vision of smart School (Abdel-Basset, Manogaran, Mohamed, & Rushdy, 2019) is not far: Wi-Fi locks and sensors will provide registered access to the school's space, sensors may be placed on students for security or healthcare reasons, whereas their wearable's may be equipped with sensors for automatic athletics tracking or monitoring. Sensors, beacons and physical computing devices along with fast connectivity all over the school facilities will improve the learning procedure and learning outcome (e.g. interactive dashboards connected to LMSs, personalized content etc.). Cyber-physical systems may simplify the management of the building (Abdel-Basset et al., 2019) (e.g. temperature/ lighting control, automatic cleaning, preventive maintenance) and take better use of resources (Papadokostaki et al., 2017). The applications of IoT in education are literally limitless (Mcrae, Ellis, & Kent, 2018) and the benefits for everyone will be grand.

Using technology in classroom not only captivates young children or teenagers, who are technology experts, but also exhibits a plethora of sources to care for all educational needs. Information may come in the format of text or hypertext, audio, video, image, graph, simulation or experiment. The learner may choose the format he prefers or the one that matches his learning profile; for it is nowadays clear that not all people learn in the same way. Therefore, as an instructor adapts his lesson to match his own students, adaptivity in technology-enhanced learning should be pursued for optimal results. Similarly, as a teacher gets feedback from his students concerning the learning procedure, so should learning -with the means of technology- make use of Learning Analytics.

The rest of this work is structured as below: First, the authors describe how learning transformed from traditional modus into electronic (e-learning) and then mobile learning (m-learning). Then they demonstrate the need for a more agile form of learning that considers all forms and sources of learning and summarize the term of Ubiquitous learning (U-learning) and its features. Additionally, they shortly present the need for adaptivity in learning and its main categories. In the next section the specification Experience API and the perception behind it are discussed, along with its basic architecture, interactions and relation to Learning Analytics. Later the authors present their research upon Ubiquitous learning with the use of Experience API, to demonstrate whether the specification may serve Ubiquitous learning and support adaptivity at the same time. They outline their novel implementation that records learning activities from various sources, including autonomous courses and standalone YouTube videos and interactive content and adapts its content analogously. In addition, they illustrate the architecture of their implementation and show the valuable graphs and statistics in the form of Learning Analytics their application provides. In the following, they discuss their contribution and report their future work, and lastly they develop their conclusions regarding their research in Experience API and Ubiquitous learning.

THE REVOLUTION OF LEARNING AND ITS EVOLUTION INTO UBIQUITOUS LEARNING

The Revolution of E- Learning

Teachers all over the world have always made use of advances of technology in order to enhance educational procedure, improve learning outcomes and make teaching more attractive. According to researchers (Sharples, Taylor, & Vavoula, 2007), the main technological achievements of an era influence the educational theories and practices of that era. More specifically, current rapid technological advances have radically changed education and learning: traditional learning, where the presence of teacher and learner were necessary, has evolved to electronic learning (e-learning), where distance learning is achieved with the use of computers and Internet (Ally & Prieto-Blâzquez, 2014). E-learning has opened new horizons to learning as it offers a wide range of possibilities to learners and instructors; it allows the upload of educational resources in electronic platforms on behalf of the instructor, the distant access of the material on 24h base, the electronic cooperation of the learners and the asynchronous interaction between the instructor and the learner (Papadokostaki Koralia, 2018). With e-learning, distance is nullified, whereas great learning opportunities are given to disabled people or people living in inaccessible places (Ally & Prieto-Blâzquez, 2014; Zhang, Zhao, Zhou, & Nunamaker, 2004). Moreover, e-learning allows each learner to learn at his own pace and potential, as the material is available for him to repeat or browse/consult it various times.

The Era of Mobile Learning

We live in the era of ‘personal and technical mobility’ (Balacheff et al., 2008), where mobile devices, such as phones, tablets, MP3 players are carried everywhere (Balacheff et al., 2008). Due to their small size, light weight and always increasing features, which is combined with the availability of connection to the Internet through Wi-Fi and 4G (Godwin-jones, 2011) they become an integral part of our everyday lives and an important tool in our hands.

As mobile devices seem to gain ground in education, we should consider their advantages for the learners. They are small and lightweight and can be used literally everywhere (Papadakis & Kalogiannakis, 2017). Their interface is based on the "What You See is What You Get" concept (WYSIWYG) and the touch screens are very attractive and easy to use (Kalogiannakis et al., 2018). They implement virtual keyboards and have very easy input; they are simple to maintain and may be connected to the Internet through Wi-Fi. Additionally, they are more affordable than computers, whereas they have built-in additional devices, i.e. microphones, speakers, and camera (Kalogiannakis et al., 2018; Wong, 2010). Finally, with such a plethora of - easy to download - applications for various uses, most of which are free or at very low cost they become increasingly popular (Kalogiannakis et al., 2018). With these advantages and the constant rise in users of mobile devices, it is no surprise that mobile technologies and learning practices converge towards what we call mobile learning (m-learning) (Balacheff et al., 2008; Sharples, 2013).

O'Malley et al. described m-learning as "Any sort of learning that happens when the learner is not at a fixed, predetermined location, or learning that happens when the learner takes advantage of the learning opportunities offered by mobile technologies" (Sharples, 2013). Deriving from this definition, mobile learning can be seen from various perspectives; the use of mobile equipment or learning in mobile circumstances itself (Sharples, 2013), or it may have extra dimensions, as described in (Balacheff et al., 2008):

- the literal meaning of the word, where the user moves from place to place
- the capability to synchronize content through multiple devices
- the mobility in conceptual space: which means versatile spaces and learning objects to wander through
- The use through different social contexts
- The learning through diverse formal and informal contexts, as learning is a 'cumulative process'

As smartphones and tablets are rated among the top six emerging technologies that have a serious impact on teaching, learning and research in (primary) education (Zaranis et al., 2013), it is important to look inside m-learning and how it affects education.

Research shows, that use of mobile devices as a part of the learning procedure increases the motivation of the learners and urges them to engage more in the learning procedure (Ciampa, 2014; Sharples et al., 2007). It is considered a pleasant experience, especially when combined with multimedia applications or games and seems to offer significant enhancement in student work. Furthermore, the use of differentiated teaching methods is thought to stimulate the brains of the learners (Zaranis et al., 2013) and makes learning more enjoyable. Moreover, m-learning supports constructivism, a promising learning theory (Ciampa, 2014); learners discover knowledge through simulations or experiments and are not pathetic receivers. To continue with, m-learning can contribute to self-paced and self-directed learning, as learners can learn at their own step and make their own choices. Last but not least, mobile devices can be used as collaborative tools and collaboration plays an important role in the process of learning, according to contemporary learning theories, such as social constructivism (Hung, 2001; Sharples, 2013) .

Case studies of successful implementation of m-learning include (Sharples, 2013):

- Use of mobile equipment in the classroom for interaction with educational context, educational (serious games) simulations, group-based activities and assessments.
- Use of mobile equipment outside the classroom, e.g. in home or in museums. Studies show that use of mobile equipment acts in connecting learning outside and inside classroom, while supports inquiry learning.
- Use of mobile equipment on visits and field trips, such as archaeological sites, environmental fields or outdoor excursions,
- Use of mobile devices as tools that support transitions between episodes of learning in formal and informal context
- Use of mobile devices in mixed reality learning, which includes games that are played in a physical environment augmented with virtual artifacts.
- Use of mobile devices to achieve seamless learning. Seamless learning implies "that students can learn whenever they are curious, using personal devices and embedded learning technology to store, share and recall contextualized knowledge, creating an experience of continuity and the ability to switch rapidly from one learning project to another" (Sharples, 2013).

The Rise of Ubiquitous Era

In the rapidly changing era we live in, electronic educational content has evolved from online courses and Learning Management Systems (LMSs) to social networks, online streaming platforms, such as YouTube or online resources, like Khan Academy. Mobile apps and MOOCs (Mobile Open Online Courses) are also powerful learning resources (Yuan & Powell, 2014).

The mode of learning has gradually shifted, from online courses and LMSs to learning through social media, such as Facebook and Twitter and through online resources or streaming platforms, such as Khan Academy and YouTube respectively (Papadokostaki Koralia, 2018). Additionally, learning is no longer achieved in the spatial restrictions of a classroom but may take place in educational visits at theme parks, museums or zoos. In the latter cases, learning is augmented through mobile devices (Christopoulou & Ringas, 2013) and is not only formal but informal as well (Murray, Berking, Haag, & Hruska, 2012).

In our mobile ages, microprocessors are part of various devices and products of our everyday life (Christopoulou & Ringas, 2013), e.g. smart cars and smart traffic lights are already in use, whereas clothes, glasses and jewelry will soon carry sensors. The omnipresence of computers in our everyday life is called Ubiquitous computing and is already happening (Hwang, 2006). Extending the concept of Ubiquitous computing in education raises the concept of Ubiquitous learning. With the omnipresence of computers and smart digital devices in our lives, the transformation of e-learning, to new modus of learning seems inevitable, as IoT devices, like sensors and beacons may as well play the role of educational material if properly used. Ubiquitous learning (u-Learning) describes a paradigm where "all students have access to a variety of digital devices and services, including computers connected to the Internet and mobile computing devices, whenever and wherever they need them" (Park, 2011).

The evolution of traditional learning to e-learning, m-learning and finally u-learning is depicted in Figure 1.

To summarize and efficiently describe Ubiquitous learning five characteristics have been proposed (Yahya, Ahmad, & Jalil, 2010):

- **Permanency:** The content is permanent, unless the provider removes it.

Figure 1. Learning has evolved from its traditional form, to electronic, mobile and nowadays ubiquitous

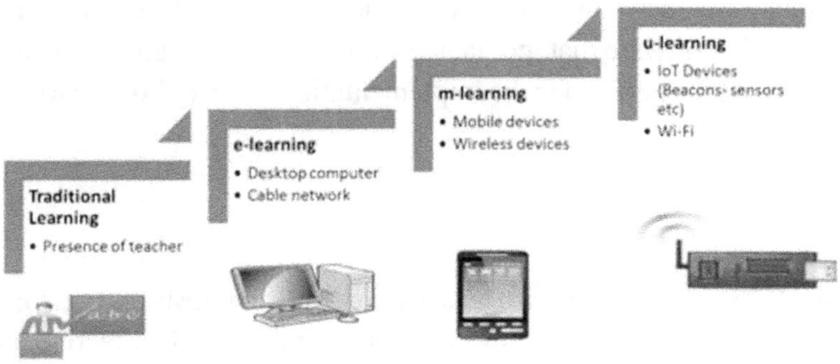

- **Accessibility:** The content is always available to the users.
- **Immediacy:** The content can be immediately retrieved by the users.
- **Interactivity:** The learners may interact with peers, teachers, and experts efficiently and effectively.
- **Context-awareness:** The environment may adapt to the learners' real situation to provide adequate content for the learners (e.g. adapt the content whether viewed on a smartphone or on a smart television).

Ubiquitous learning, described above, is a new vision not only for learning, but for our society and future. It is an educational paradigm that will change learning as we know it, as it will (Cope & Kalantzis, 2008):

- **Overcome traditional institutional, spatial and temporal boundaries** of education, and enable 'lifelong and life wide education'. It implements informal, semi-formal learning and 'over-the-shoulder-learning' from friends and colleagues.
- **Change balances**. Shifts the role of the student from a passive recipient of transmitted knowledge to a collaborative co-designer of knowledge, through the ability to select, modify and alter the content.
- **Identify learner differences**. Respects the learner's differences, interests and needs. Collaborative and teamwork is also enabled.
- **Include a vast range of representational modes** and change conceptualizing capacities. Ubiquitous learning involves oral, written, visual and audio mode of content and knowledge upon describing it and categorizing it should be developed.
- **Connect one's own thinking into the social mind** of distributed cognition. Now that knowledge is freely and widely available, knowledge should focus in deep understanding and not on simple memorization.
- **Create collaborative knowledge cultures**. Ubiquitous can build 'communities of practice' to support learning, regardless of gender, race or origin.

Ubiquitous learning is a promising paradigm in all ages of learning. It can be combined with collaborative learning and may exploit experiential learning through its coordination with sensors. Incorporating Ubiquitous learning in Early Childhood is a challenge and may offer serious advantages in children's learning. In a relative research (Chen, Hung, & Wei, 2010); the authors developed a ubiquitous learning system based upon theories of experiential learning, constructivism learning and joyful learning. It makes use of learning robot, sensing input device, mobile computing device, mobile output device, wireless local network and a software that includes instruction, collaborative learning and self-learning. This was an interesting and novel effort for the implementation of ubiquitous learning in actual learning circumstances.

Adaptive Learning

Students in a traditional classroom come with a variety of traits, interests and background. Taking these into consideration and adapting teaching to meet the students' needs is effective and desirable (Shute & Towle, 2003), but not always feasible. In e-learning, personalized learning (i.e., tailored-learning according to the user's needs and capabilities) cannot be accomplished, at a massive scale, using traditional approaches (Paramythis & Loidl-reisinger, 2003). Therefore, the importance of adaptivity in e-learning has been identified and integrating adaptation in e-learning has been one of the goals of e-learning for a long time (Chrysoulas, 2017; Paramythis & Loidl-reisinger, 2003; Shute & Towle, 2003).

In order to improve learning and performance and to provide with a more individualized learning, the characteristics of a particular learner need to be correctly identified. Such characteristics include (Chrysoulas, 2017; Shute & Towle, 2003; Zapata-rivera & Shute, 2012):

- **incoming knowledge**, i.e. the prior knowledge each learner has
- **cognitive abilities**, e.g. memory or spatial conception
- **learning styles**, e.g. if one is more visual or audio learner
- **skills**, e.g. if one is more theoretical or practical
- **personality traits**, e.g. if one is social or lonely
- **special needs/ disabilities**, e.g. if one is hearing impaired
- **demographic and sociocultural differences**, e.g. the age of the user

Additionally, interests and needs of the learner should be analyzed in order to provide him with the most useful content. For e-learning to be most effective, we should capitalize on these characteristics when delivering content.

Embedded assessments can also improve e-learning, as they may be the tool for diagnosis of problems and for follow-up, e.g. to remediate on the topic, if the learner failed the assessment or supply him with extra material if he excelled in it and the content is among his interests (Shute & Towle, 2003).

With Adaptive Learning we should modify any individual student's learning experience (Monova-zheleva, 2005), in order to deliver (Shute & Towle, 2003):

- the right content
- to the right person,
- at the proper time,
- in the most appropriate way

Adaptivity in learning environments may be categorized as below (Paramythis & Loidl-reisinger, 2003):

- **Adaptive Interaction,** where adaptation focuses on the user's interaction with the learning system through its interface, without affecting the content. Changing the size or colors of the learning system may be an example of such adaptation
- **Adaptive Course Delivery** is the most common type of adaptivity in e-learning and entails changing the content in order to meet the user's requirements and characteristics. This category includes: dynamic course structuring, adaptive navigation support and adaptive selection of content so that the learning result is maximized, with the minimum time and interactions.
- **Content Discovery and Assembly** uses adaptive techniques to track and assemble learning content from distributed sources / repositories.
- **Adaptive Collaboration Support** aims to implement adaptive support in learning processes that are collaborative and involve social interaction

Integrating characteristics from the aforementioned categories in a learning system should be achieved in order to meet the diverse requirements of several users. In an Early Childhood Learning Environment, adaptivity is also crucial as children may come from different backgrounds and therefore have different levels of knowledge. Additionally, children need to interact with learning environment that attract their interest and challenges them to interact with it. It is positive that children would have better learning results with a learning environment that combines their personal likes; e.g. their favorite cartoon hero, preferred colors and music and so on.

Dynamic Sequencing

SCORM (Sharable Content Object Reference Model) was designed to overcome the major problems of interoperability and reusability of learning content. Before SCORM was proposed, the process of tracking the learner's progress was tailor-made for each platform; if the company or foundation changed its Learning Management System (LMS), the tracking process had to be redesigned and re-implemented. With the use of the SCORM model, the learning content is packaged into a format which can be transferred through various LMSs (Lim, 2015; Moisa, 2013; 'SCORM Overview', n.d.) accomplishing thus not only interoperability, but also reusability, traceability and longer lifecycle. SCORM Sequencing defines the behaviors and data model used by the SCORM run-time environment to determine how a content package will be delivered as a learning experience (Mellon, 2003). It defines the functionality that a SCORM-conformant LMS must implement to process sequencing information based on learner interaction with content objects. Sequencing refers to the behaviors that an LMS or run-time system follows to deliver a specific learning experience as intended by an author or content developer. This learning experience may be free choice, in which the learner can choose any item in the content organization in any order, or it may be guided by flow through the structure of the content organization. SCORM sequencing implies that the navigation through the content is not conducted in the learning modules but is done by the LMS or the learning system. It determines the navigational controls and options that are available to the users ('SCORM Sequencing: Learn more about SCORM Sequencing', n.d.). Further, it orchestrates the flow and status of the course. This allows the reusability of the modules and their independence from the navigation decided by the content developer. Sequencing can be adopted not only for SCORM objects, but in general for e-learning applications and LMSs ('SCORM Sequencing', n.d.).

THE EXPERIENCE API (XAPI) SPECIFICATION

The Need for Change

As described above, learning occurs also besides common educational platforms, e.g. LMSs and lately takes place through smartphones, tablets, laptops, smart TVs, gaming consoles and other mobile devices. Therefore an urge for inclusion of both, informal and ubiquitous education as well (Murray et al., 2012) has arisen, which is accurately expressed with the phrase "Learning is happening everywhere" ('Experience API', n.d.).

Although several specifications were proposed, the one that prevailed (Papadokostaki, Panagiotakis, & Vassilakis, 2018) was proposed by ADL in 2010 and focused on standardized experience tracking capabilities. It evolved as an innovatory project called: the "Experience API" and offers capturing learning activities from various and versatile resources; it may track activities from LMSs, autonomous xAPI-enabled sites or apps, simulations or serious games, even IoT devices, such as beacons and sensors. Its flexibility and vast range of uses makes it a pioneer in tracking of learning activities.

A Dive into xAPI

The promising technical specification that was born to track activities from several sources is called Experience API (shortly xAPI and previously Tin Can API) and was formally introduced by ADL in 2012, but has been constantly updated to add extra functionality and clarify identified issues('Experience API, Appendix A: Revision History', n.d.). The xAPI is an open source, interoperability specification on learning technologies that defines precisely how tracking learner activities should be stored and transferred between technologies ('xAPI-Spec', n.d.). Licensed under the Apache License, Version 2.0, it enjoys the warm support of the community which often updates it ('xAPI-Spec', n.d.). Recently a new specification called cmi5 ('CMI-5 and the Experience API - Experience API', n.d.) was introduced to facilitate communication between the xAPI protocol and LMSs ('CMI-5 and the Experience API - Experience API', n.d.). This specification is out of the scope of the present work, as our research does not use LMSs.

Experience API tracks Activity Streams, i.e. Statements indicating the actions a user made. XAPI describes how the statements are to be transferred and stored in a repository, called Learning Record Store (LRS). It also provides with tools for accessing and aggregating the statements. The applications generating Activities are called Activity Providers, whereas Activity Consumers are the applications consuming the Activities.

The statements are stored and transferred in JSON (JavaScript Object Notation) format ('Anatomy of a Experience API Statement', n.d.), generally follow the syntax <Actor> <verb> <Activity> and are identified by a unique UUID (Universally Unique Identifier). The format is inspired by social media, in order to track experiences, e.g. Mary liked a video. The first component of the triplet, the Actor, indicates who is the person that 'acts'. The Actor may be an Agent or a Group. An Agent, that is a user or a learner, is easily identified by his email and there the pairing between property *mbox* and an email address can be used. The second and vital part of the statement, the Verb, describes the action being taken, i.e. the experience being recorded. The action is described with the property id which points to a URI (or IRI), as in the example:

{id:http://adlnet.gov/expapi/verbs/completed}

As the xAPI specification evolves and covers various needs, it supports not only traditional learning activities (attempted, passed, failed, completed) but social media actions (shared, like, tweeted, followed) as well. The list of statements is extended occasionally and may involve actions defined by ADL or other providers, which demonstrates the versatility and wide horizons of the specification.

The third part of the xAPI statement, the Activity, describes the "thing" or the object. The activity is what is to be tracked and depends upon the requirements of the implementation. It may be a quiz, a course, a test or even a video, a button or an event, expanding learning activities outside the classroom and overcoming formal education.

The triplet <Actor> <verb> <Activity> may be further extended to support a variety of objects. This is achieved by augmenting the statement with extra data, like the timestamp of the activity, score or result of a quiz (Glahn, 2013; 'xAPI-Spec', n.d.).

A wide range of learning technologies from a social learning platform to an interactive whiteboard or from an e-learning course to an event attendance application may play the role of an activity Provider in the xAPI Infrastructure (Moisa, 2013). This support of multiple and versatile Activity Providers makes xAPI an ideal solution for the realization of the vision "Learning Everywhere" and a promising technology to incarnate Ubiquitous learning.

The 'heart' of the specification, the LRS, need not coexist or communicate with a Learning Management System (LMS), but it may exist autonomously. At the same time, it can seamlessly exchange data with other LRSs that may serve different purposes. This expands the potential of xAPI significantly. Due to the fact that an LRS collects data from diverse learning environments it is ideal for data aggregation and analytics (Moisa, 2013; Murray et al., 2012). In addition to providing reports and analytics, LRSs may play a significant role on personalization of teaching material, as statements that are stored in the triplet <Actor> <verb> <Activity> are ideal for easy processing.

XAPI into Action

xAPI becomes more and more popular in many educational applications, as it mainly supports learning activities from various resources. It has been adopted by popular LMSs, such as Moodle, Blackboard Docebo, Talent LMS, SAKAI ('Experience API Adopters', 2018), in two ways: either by support of xAPI content or by offering a built-in LRS. To be more specific, the courses are hosted in LMSs and the LMSs either produce xAPI statements that are stored in standalone LRSs or store their xAPI statements in their own LRS. XAPI-enabled content may be created with authoring tools that export xAPI-enabled content, e.g. Articulate Storyline και 360, H5P and Adobe Captivate ('Experience API Adopters', n.d.). A step further, companies have developed software such as Simulations, Assessment Systems, Games and Social Learning Systems to support this novel specification and transfer activities from these modules to LRSs. Finally, its potential is recognized in various fields, as it has been adopted in training, games and Defense (Hruska, Amburn, Long, & Poeppelman, 2014).

XAPI and Learning Analytics

As learning becomes a 'ubiquitous' procedure and students may deploy learning in various places, formats and contexts, assessment of learning is even more difficult and important. Educators and instructors worldwide not only need to assess their students, but also need to evaluate upon their content, courses and effectiveness of their teaching. This assessment leads to the improvement of the teaching process and the maximization of the educational results. To this direction, Learning Analytics (LA) has emerged to improve learning and education. Learning Analytics is a field that is closely related to business intelligence (BI), web analytics, academic analytics, educational data mining, and action analytics (Elias, 2011). It aims to the statistical evaluation of data and the identification of patterns, which are used to predict future behavior and support decisions aimed at improving outcomes and detect problems in student performance. Extracting Analytics means production of knowledge from collected data; this knowledge can be useful to students for self-assessment; to teachers for self-evaluation and revelation of possible problems in their classroom or as an incentive for improvement. Finally, knowledge may serve to school organizations for decision- making, long-term planning and feature of Business Intelligence (Del Blanco, Serrano, Freire, Martinez-Ortiz, & Fernandez-Manjon, 2013; Elias, 2011; Siemens & Long, 2011).

The xAPI specification may contribute as the source for data aggregation and analytics and can be the 'storehouse' for production of precious information (Moisa, 2013; Murray et al., 2012). The LRS not only stores the streams of learning activities, but also provides access to the statements, allowing for the retrieval of these statements by external applications. It may host activities from various Activity Providers: if the statement derives from a serious game, comes from a mobile application or is created by an xAPI-enabled webpage it can be saved in the same format in the LRS (Figure 2). It is obvious that most, if not all, of these Activity Providers may play the role of learning resources for very young children, as children use mobile electronic devices more and more and may also interact with related software in games or simulators.

At the same time, an LRS can seamlessly swap data with different LRSs that may serve different purposes. It is now clear that an LRS may gather statements from various sources and store them under the same format ; it is therefore a precious tool for data aggregation and analytics (Moisa, 2013; Murray et al., 2012). Except from the features of reporting and analytics, LRSs provide significant amounts of data for personalized approaches, as statements deriving from diverse sources can be easily aggregated or processed.

Interaction with Other Modules

As stated earlier, xAPI statements may derive from varied sources; such as e-learning courses, LMSs, Games, Simulations, Mobile Apps or even Business Systems. Moreover, it may interact with other

Figure 2. LRS may host xAPI statements from assorted sources. All arrows indicate xAPI statements

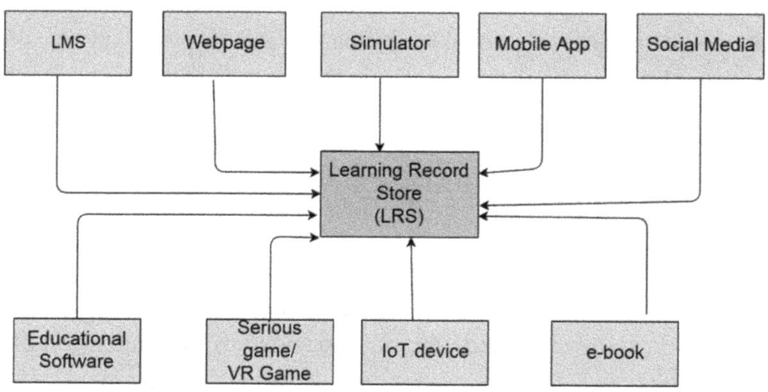

modules, such as Learning Analytics, Open Badges and Dashboards, in order to provide spherical and valuable information.

An LRS may exist autonomously or may be built-in an LMS. Whatever the case, it should be able to interact with activity providers outside the LMS, e.g. serious games or mobile Apps. Moreover, it may 'draw' activities from operational business systems to record job tasks. Additionally, an LRS is meant to communicate with another LRS. In some cases, organizations maintain two LRSs: an LRS embedded in the LMS to gather data from it and an external LRS to collect data from external Activity providers. The two LRSs may exchange data and provide deeper analytics. The wide spectrum of interactions in an xAPI ecosystem is given in Figure 3; only some Activity Providers are mentioned for economy of space.

AUTHORS' CONTRIBUTION TO UBIQUITOUS AND ADAPTIVE LEARNING

Our Related Work

The researchers conducted a thorough bibliographical research concerning xAPI applications without the use of an LMS but with the features of Learning Analytics. They have come across several interesting projects, such as Oregon Trail Game (Lim, 2015), the LIME project (Corbi & Burgos, 2014) and the Mobler Cards App (Glahn, 2013)- none of which adapts its content upon the user's history (Panagiotakis, Papadokostaki, Vassilakis, & Malamos, 2019). The AubiLearn (Zhao & Okamoto, 2011) provides adaptive multimedia content but no learning analytics. Concluding, in deep bibliographical research, the researchers have not discovered an LMS- independent application that engages adaptation based on a learners' earlier activities and at the same time provides LA and they have themselves implemented an adaptive xAPI enabled application that is totally autonomous, does not require the use of an LMS and uses the learner's email address to access a tailored course (Panagiotakis et al., 2019; Papadokostaki, Panagiotakis, Vassilakis, & Malamos, 2017). Later, they augmented the course with extra tools, such as video-activity tracker and the H5P xAPI enabled content to add extra value to their implementation and learning Analytics Dashboard. This extension of their implementation will be presented in the following section.

Figure 3. Interaction of LRS with other modules

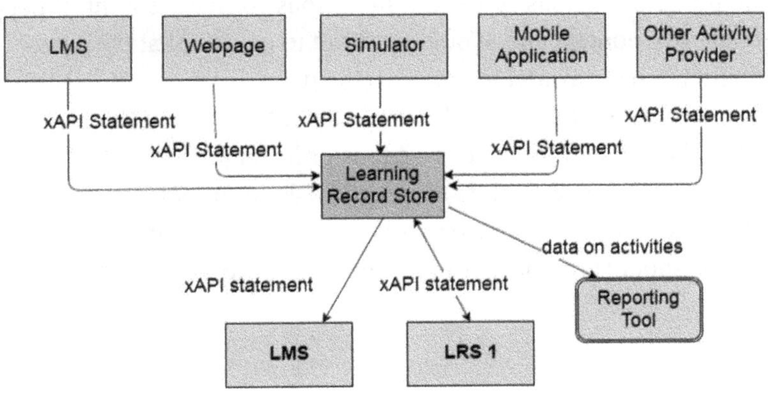

Our Vision

The researchers' vision was to make use of the tracking activities in order to create an adaptive learning system (Moisa, 2013; Paramythis & Loidl-reisinger, 2003) that complies with the Learner Model in Adaptive Learning Environments (ALE) (Paramythis & Loidl-reisinger, 2003). The content will be modified according to a user's history, as it 'also maintains a "live" account of the user's actions within the system.' (Paramythis & Loidl-reisinger, 2003). The learning activities should be gathered from various sources, to simulate the vision of ubiquitous learning. More specifically, there will be a simulation of a Ubiquitous Learning Environment (ULE) (Jones & Jo, 2004) that will make Ubiquitous Learning a reality, though without the use of sensors. According to the features of Ubiquitous learning above mentioned, the content should also:

- Be permanent, until the owner removes it
- Be accessible by the users (learners)
- Be immediately retrieved by the users.
- Be highly interactive
- Be Context-aware: the course adapts according to the user's environment (e.g. adjust to a user's smartphone screen) (Panagiotakis, 2012).

Additionally, the ULE has to match the following criteria (Hwang, 2006):

- be context- aware; that is, the learner's situation or the situation of the real-world environment in which the learner is located can be sensed.
- actively provides personalized support to the learner, fitting his learning portfolio
- enables seamless learning
- adapts its content to meet needs of various mobile devices

Finally, as the usability of LA has been manifested, LA will be embedded to the implementation. In summary, the following features of the implementation are considered to be critical:

1. The implementation has to be LMS-free, so that the content can be hosted literally 'everywhere'. The authors' vision is to make a Learning System that would be part of 'Ubiquitous Learning', and therefore no restriction upon the content should be allowed
2. The content should be of various types and in various formats, e.g. html page, YouTube video, autonomous interactive content and should not limit to educational purposes.
3. The content should be made available online to the users
4. The content should be accessible through mobile devices
5. The content should be adapted to the user's needs or history
6. The adaptation of the content should be accomplished seamlessly
7. The tracking of activities should be made without any intervention.
8. The implementation should provide Learning Analytics to the instructor, without the use of external tools.

For their implementation, the researchers followed the ADDIE (Analysis, Design, Development, Implementation, Evaluation) model, as proposed by ADL ('ADDIE', n.d.). More specifically they:

1. Analyzed the specifications of the project, the level of the learners and the content to be presented (Analysis Phase)
2. Designed the experiences to be tracked with xAPI and chose the most suitable tools and programming languages to be used (Design Phase)
3. Made use of available code libraries and xAPI tools (Development Phase)
4. Uploaded the content to the cloud (Implementation Phase)
5. Evaluated the implementation (Evaluation Phase)

As described above, the researchers wanted their implementation to seamlessly gather statements from dispersed autonomous content; sequentially store the statements in an infrastructure and provide access to them, without the use of an LMS. Activities' consumption is mostly used for aggregating data and extracting conclusions. The application is innovative and original in the sense that decision making is based upon activities produced by various sources. Furthermore, it provides a complete Dashboard to the tutor, granting him with graphs and reports for his content and courses. The statements from various activity providers are directly delivered towards the LRS and the learning system is immediately accessing them, via the LRS, in order to adapt its content. The exchange of statements between the implementation and the LRS is bi-directional and immediate.

The described implementation adopts the principles of Adaptive Course Delivery (Paramythis & Loidl-reisinger, 2003): the content a user accesses changes dynamically, so he is not repeating already watched content. This way the time he spends on the course is minimized, the user's interest is gained, and he makes the maximum out of the course. Although, it was in the researchers' plans to adopt Adaptive Content Delivery as well, that was not achieved at this stage of development.

System Architecture

The architecture of the implementation is illustrated in Figure 4 and consists of three main modules:

- The Adaptivity Support Module which hosts the learning content, enables tracking of user activities via xAPI and implements content adaptation
- The LRS which stores the xAPI statements
- The Dashboard which provides aggregated and visualized information to the teacher.

The LRS hosts the statements and plays the role of the server in our implementation. The user may interact with the content through a computer or a mobile device. The Adaptivity Support Module apart from interacting with the user, seamlessly tracks specific activities of the user and sends them to the LRS. Furthermore, it adapts the learning content according to the activities recorded in the LRS regarding the specific user. The instructor, on the other hand, operates visualizing and monitoring tools provided from the Dashboard to scan the student's activities/scores. For the reports to be produced and the statement viewer to provide access to the statements, a Learning Analytics Engine has been developed and the xAPI statement viewer module ('xapi-statement-viewer', n.d.) has been adapted to the research needs,

Figure 4. The architecture of the presented implementation

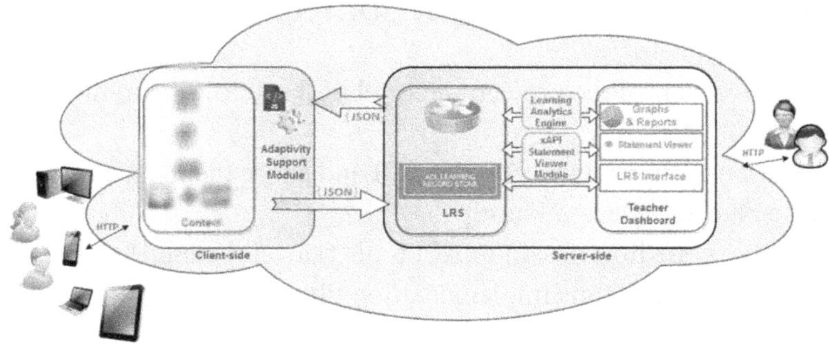

respectively. For the implementation, the researchers installed an instance of ADL's Open Source Learning Record Store (LRS) ('ADL LRS', n.d.).

The learning content of the implementation consists of four independent parts that can be hosted in different hosts and can be accessed via web-browsers.

- **The LMS- independent Adaptive System** makes use of the JavaScript Library tincan.js ('TinCan Javascript Library', n.d.) and is developed in Articulate Storyline 2. When specific actions are fired (e.g. the learner starts a part of a course, ends a quiz, or presses a button), a JavaScript script is triggered that sends an xAPI statement to the LRS. Each xAPI statement includes the email of the user and the action that has been fired. The implementation includes a module in JavaScript that checks whether related xAPI statements have been sent to the LRS regarding the specific user and adapts the content accordingly. The recommended system may record user activities that were part of a different course and adapt its content to match the user's background and history. A full description of the Adaptive System is given in (Papadokostaki, Panagiotakis, et al., 2017). The implementation uses independent courses, but the content could be of different sources, e.g. a smartphone app and an online course would interact through the LRS and adapt their content accordingly. An example of the online course is given in Figure 5.
- **An autonomous course implementation** that is developed without the use of an authoring tool and implements sequencing. It is based on bootstrap ('BootStrap', n.d.) and making use of the JavaScript Library tincan.js sends statements to the LRS and navigates the user through the content based upon his previous activities. It makes use of sequencing which was presented earlier in this chapter. Access to the second part of the content is delivered only if the user has completed the first part of it. This info is mined from the LRS and the sequencing is implemented by the course implementation. An example of the course is given in Figure 6.
- **YouTube xAPI-enabled content** that makes use of xAPI-youtube module and bootstrap to provide a list of videos that enable the monitoring of the user's actions (e.g. played, stopped, seeked a video). This module manifests the use of xAPI in popular streaming applications, such as YouTube, proving this way the vast applications of xAPI. The interface of the xAPI YouTube Video Tracking of the implementation is given in Figure 7.

Figure 5. The online course is adapted based on the history of the user

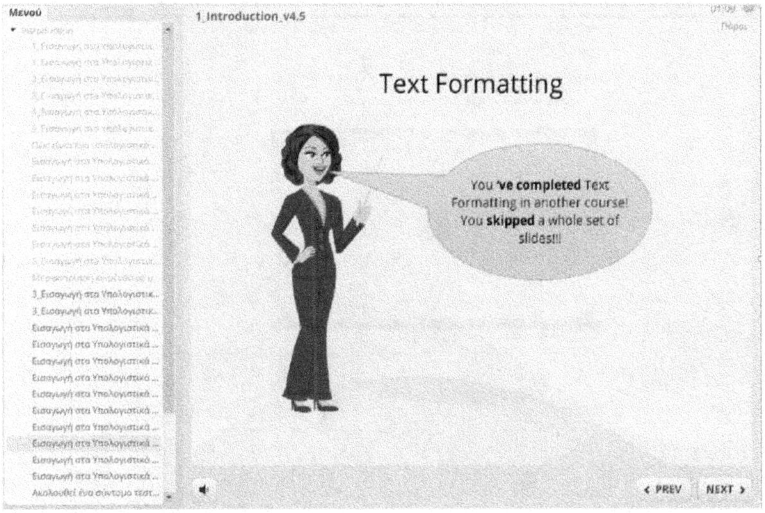

Figure 6. An example of the autonomous course. The rights to the content depend on the user's history and the implementation is based on sequencing

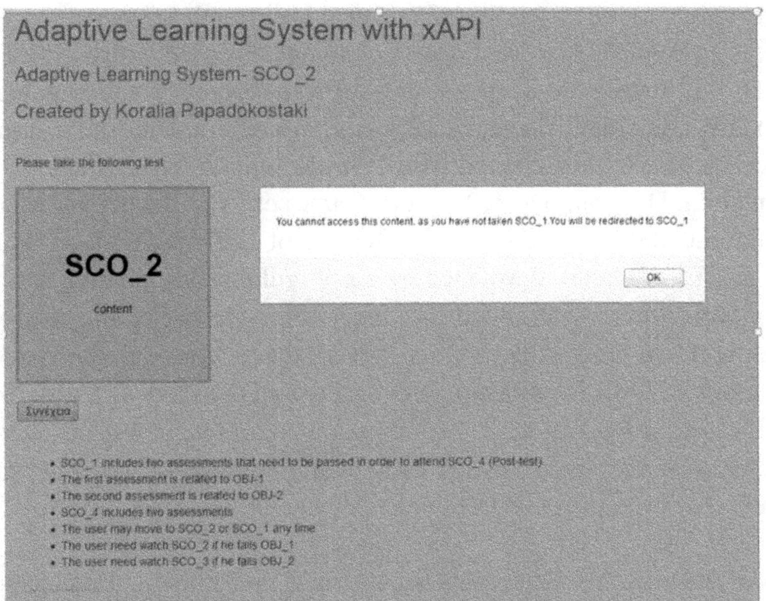

Figure 7. YouTube-xAPI enabled video of the implementation

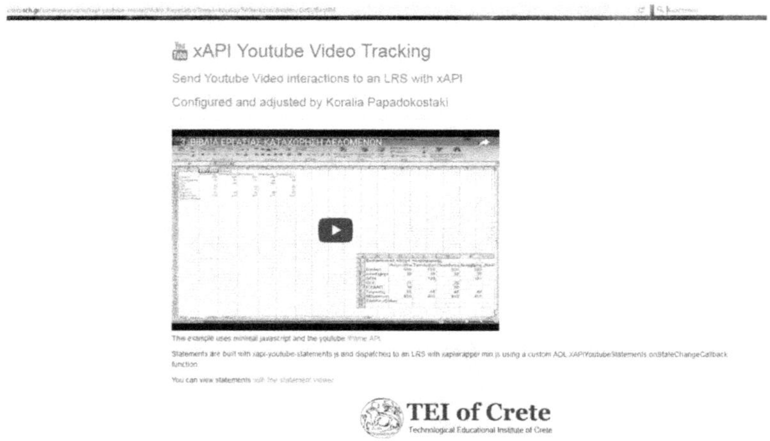

- **Interactive video created in H5P** ('H5P', n.d.), where monitoring of user's activities is enabled via plugins. Interactive videos, where information is provided and then the users answer related questions are popular in education. H5P is an open source authoring tool that creates interactive learning content, such as videos, presentations, quizzes and games. It creates HTML5 content that is rich, responsive and mobile friendly. H5P content is currently supported by WordPress, Moodle and Drupal. It is a completely free and open technology, licensed with the MIT license ('H5P and xAPI', n.d.). H5P is among the adopters of xAPI and supports the communication between the content and the LRS, in order to send the activities and achievements of the user ('H5P and xAPI', n.d.). Specific interactive tools created with H5P do support xAPI, such as Interactive Videos, Course Presentation, Drag and Drop, Memory Games etc. For the deployment of the H5P-based learning content the authors used the Blogs Service of Hellenic Mediterranean University (formerly known as TEI of Crete), developed over a WordPress installation, and for this reason the H5P plugin for WordPress was installed. The latter allows the use of an already existing H5P content by uploading it to a blog or the creation of H5P content through the WordPress plugin. H5P also provides minimal LMS functionality and may record the score of registered users of our H5P content. In this case, the wp-h5p-xapi ('wp-h5p-xapi', n.d.) plugin was installed in our WordPress instance, so xAPI activities from the H5P content are sent directly to the designated LRS. Figure 8 shows a screenshot from the H5P xAPI enabled content.

All the modules of the learning content that use the tincan.js library communicate with the LRS via PUT/POST and GET HTTP methods (Papadokostaki, Panagiotakis, et al., 2017), whereas the YouTube xAPI-enabled video makes use of the HTPP PUT/POST and GET functions implemented by xAPI Wrapper. The plugin for the H5P xAPI-enabled video uses HTTP POST and GET functions.

The Dashboard provides the teacher with four powerful tools for monitoring and visualization of the user's activities. As earlier mentioned, these tools may be used in order to provide learning analytics. More specifically they can be used to measure user's activity or performance, to track the impact of the content upon the learner, or even to investigate the difficulty of several parts of the content. As the content may come from various sources the frequency of each medium may as well be measured. The

Figure 8. The settings for the interaction between H5P content and xAPI specification

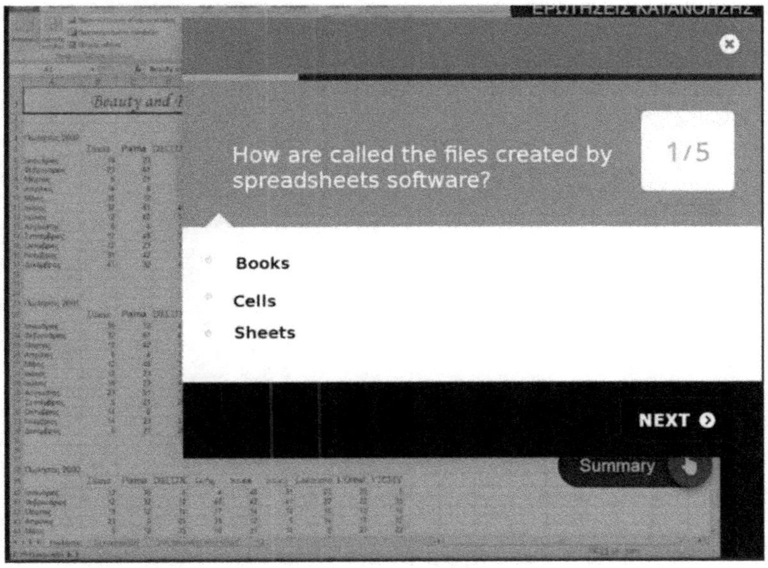

Teacher's Dashboard integrates, expands and configures existing modules into a complete monitoring tool for the instructor of the course. In order to provide a thorough mechanism for the monitoring of users' activities and performance the researchers have assembled four modules that provide an exhaustive review of the users' activities, scores and progress:

- **Link to the built in ADL LRS interface:** ADL's LRS provides access to the statements with chronological order. No filtering or searching is provided.
- A **statement viewer** to meet the research needs; it is a module based on the xapi-statement-viewer ('adlnet/xAPIWrapper: Wrapper to simplify communication to an LRS', n.d.; 'xapi-statement-viewer', n.d.) but with major extensions; Figure 9 depicts the configured xAPI Statement Viewer.
- **Visualization tools** (charts, pies and reports) created with xAPI Dashboard ('xAPI Dashboard', n.d.)
- **Link** to the monitoring tools that WordPress provides

A thorough presentation of the Learning Analytics module is provided in (Panagiotakis et al., 2019). In the following figures charts from the visualization tools are illustrated. In particular, Figure 10 shows bar charts with the average score per student, while Figure 11 illustrates a report with the interaction of the users with the YouTube xAPI-enabled content. Figure 12 shows a synopsis of statements created by the H5P content.

EVALUATION

The novel implementation presented here was evaluated using specific content that was prepared specifically to this end. The content is designed for fifth and sixth grade of Primary Schools and is an introductory course on spreadsheets. It comprises of a concise introduction on their usability and of a

Figure 9. The xAPI statement Viewer adapted for the implementation

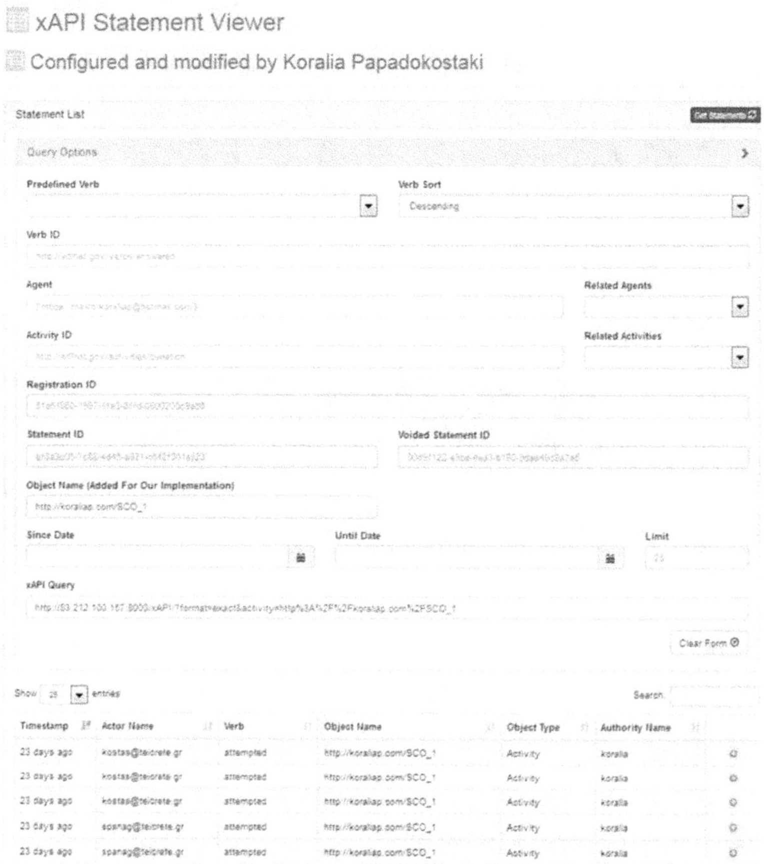

Figure 10. Average score per student

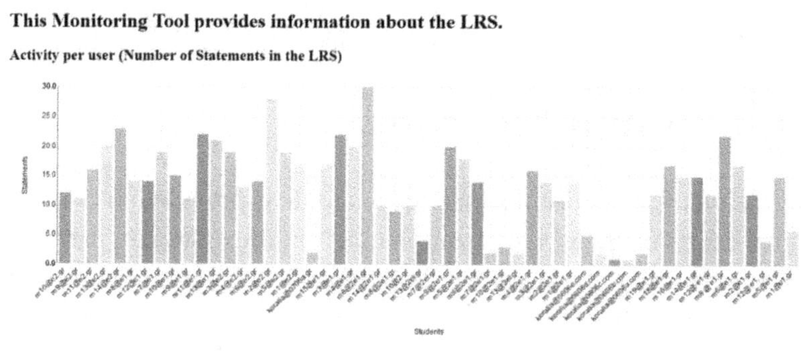

Figure 11. Interaction of users with xAPI-enabled YouTube videos

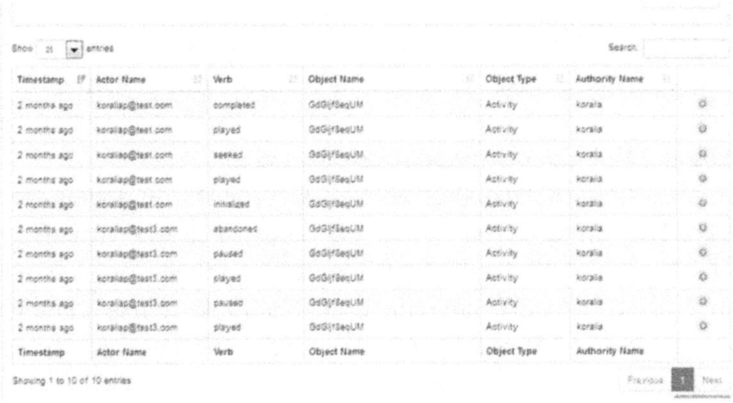

Figure 12. Actions deriving from H5P interactive content

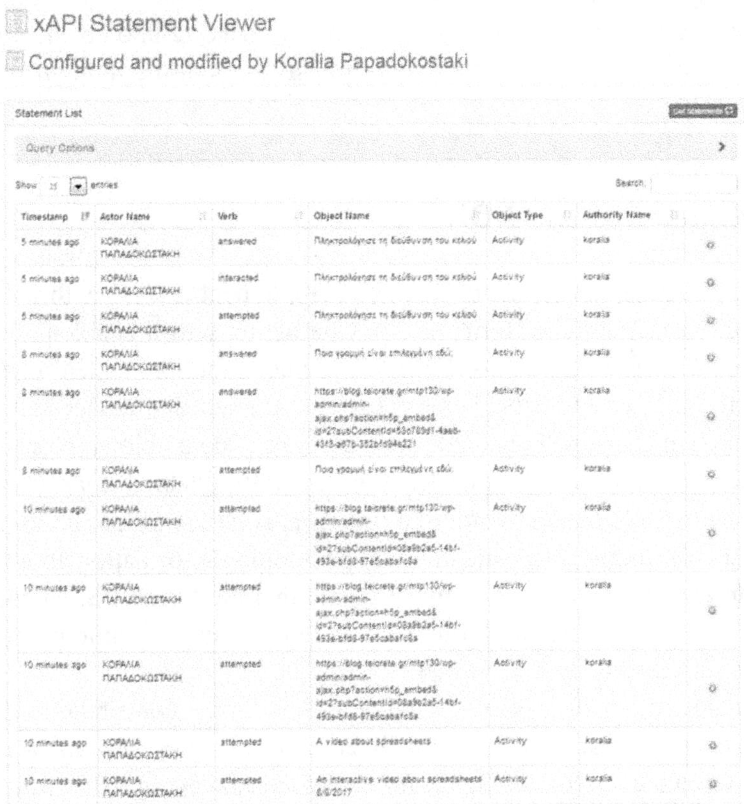

demonstration of their basic features and functionality. Since spreadsheets offer a similar interface to other office suites (basically word processors and presentations software), students find the interface and common functions familiar. With that in mind, the evaluation material consists of two separate web-pages that were different courses. Course 1 meets text formatting functionalities and is generic for word processors, spreadsheets and presentations, whereas Course 2 focuses on spreadsheets. It also includes related YouTube videos and H5P interactive educational content.

The Adaptive Learning System of the implementation was used by 46 students of 5th Grade of Primary School in Crete, Greece. They were given an email address (which was virtual) for their actions to be recorded. One group of 25 was asked to take first Course 1 and then Course 2, whereas the other one was asked to take only Course 2. As expected, the users that took Course 1 automatically skipped the part of Course 2 that was related to formatting (see Figure 5). The students who failed a part (section) had to re-watch the slides regarding this content.

The users showed enthusiasm about the course, as they found the content attractive, user-friendly and highly interactive. Although a large amount of xAPI statements was sent to the LRS in short interval and despite the fact that the communication between the content and the LRS was both ways directed, the implementation responded accurately and stably. The learners did not come across serious delays or inconsistencies during their involvement with the course. Taking into consideration the distributed architecture of the system, one can claim that the responsiveness of the implementation is satisfactory (Panagiotakis et al., 2019).

FUTURE WORK AND DISCUSSION

The described work could be further adapted to fully implement a ULE with IoT devices and adjust its content according to data captured from sensors. Combining sensor enabled ULE with xAPI seems an ideal realization of IoT in education, as it would open new horizons to learning. xAPI can actually be an integral part of the Internet of the Things (IoT), where sensors, devices and wearable's will be interconnected. If they are also xAPI enabled, they might provide precious information upon the user, immediately and seamlessly, without his interference. An xAPI- enabled smart television, for example, might communicate with an LRS to transmit the link of a YouTube video the user watched, while an xAPI enabled GPS-sensor, might automatically send positioning information about the user, that might me related to a conference or seminar. It is also possible that multiple LRSs might cooperate in order to exchange information about a user and offer him more personalized options, e.g. suggested videos, based upon previous activities he may have had.

Generating xAPI statements massively, on the other hand, may contribute to the Big Data revolution that gathers astonishing volume of data from diverse sources with the penetration of the Internet. It complies with the 4-Vs characteristics of Big Data (Papadokostaki et al., 2017) which stand for Volume, Variety, Velocity and Value, as each action of a user may be represented with a statement (Volume), statements may come from a vast spectrum of actions, e.g. learning, training or everyday actions (Variety), statements are transmitted online and with great speed (Velocity) and provide significant information upon a user's actions (Value). The researchers believe xAPI has a great potential if integrated with IoT and may significantly expand the value of Big Data.

Additionally, the authors have proved that xAPI is a technology much more flexible than SCORM and offers autonomy from systems, such as LMSs, realizes the 'lifelong and life wide learning' motto and proves that learning may take place literally everywhere. It captures data in a consistent format about a person or group's activities from many technologies and realizes ubiquitous (Murray et al., 2012) and lifelong (Kevan & Ryan, 2016) learning. With the use of xAPI, learning is freed from the use of LMS and the learning content may be distributed across disparate systems. As Kitto et al (Kitto, Cross, Waters, Lupton, & Street, 2015) support 'Learning beyond the LMS is both desirable and achievable' through xAPI. Additionally, the students' activities like interaction with mobile learning applications, museum exhibits, or any data that can be pushed to the cloud can be gathered to a single store, a concept similar to Personal Learning Environments (PLEs) (Kevan & Ryan, 2016). The authors claim that xAPI can indeed implement the vision of Ubiquitous learning and is a paradigm that will be further adopted in the years to come.

From another perspective, xAPI with its interactivity with social media, serious games and collaborative applications may encourage collaborative learning. The use of xAPI promotes teamwork and may lead the way to the transformation ofpersonal learning to team-learning in the field of e-learning (Kitto et al., 2015) .Therefore it may be used to support the social constructivist learning theory, where social interactions and relations are the key to learning (Kevan & Ryan, 2016; Kitto et al., 2015). Along with the activities theories that they implement they seem to be the future of constructivist-aligned strategies (Kevan & Ryan, 2016), (Kitto et al., 2015).

Combining the arguments presented above in Early Childhood Education, one cannot doubt the countless effects Ubiquitous Learning might have in the learning procedure of very young learners. Toddlers and pre-school aged children are becoming experts in the use of mobile devices and these also can provide charming learning resources for them. Additionally, they often visit zoos, exhibitions and museums, where they interact with artifacts, involving themselves in an informal educational process. Finally, they are mastering collaboration and team-learning through games and group play. Ubiquitous learning promises to compose those learning activities in a vast learning procedure where learning could be tracked and information given to the children could be filtered according to their needs and background.

To conclude with, as adoption of xAPI is extended in learning Industry and other fields, the researchers of the present chapter believe their work will be one more step towards the exploitation and application of xAPI in the provision of adaptive learning for very young learners.

CONCLUSION

In this chapter the authors have contributed towards the use of xAPI in education and more importantly towards its contribution to Ubiquitous and Adaptive Learning. Their work is an Adaptive Learning System that is totally independent of LMSs or other learning platforms and is completely based on the open source xAPI specification. It is a complete learning system that consists of content, infrastructure and teacher's dashboard that can adapt learning content according to the user's history. In essence, the learning content generates xAPI statements when a user makes a learning activity; the infrastructure seamlessly gathers learning activities from diverse (educational) sources and stores them in a single repository. At the same time, the content is dynamicaly adapted to comply with the user's past activities. The bidirectional exchange of data between the LRS and the content is not only seamless but also very responsive. Based on the statements aggregated in the repository, the researchers' implementation may provide the instructor with a Dashboard with powerful statistics and charts. The presented work manifests xAPI, its vast range of uses and applications and the wide possibilities and promising future of the specification. Its potential implementations in Early Childhood learning are also vast and bright, as it can ideally fit all future mobile and wireless technologies that children are proficient in. The authors argue that this paper offers significant contribution to the realization of Ubiquitous Learning with emphasis on the development of synchronous Ubiquitous Learning Environments for all ages and educational needs.

REFERENCES

7. Ways, H. T. W. M. L. M. F. (n.d.). Retrieved 6 June 2019, from http://www.emergingedtech.com/2012/11/7-ways-holographic-technology-will-make-learning-more-fun/

Abdel-Basset, M., Manogaran, G., Mohamed, M., & Rushdy, E. (2019). Internet of things in smart education environment: Supportive framework in the decision-making process. *Concurrency and Computation, 31*(10), 1–12. doi:10.1002/cpe.4515

ADDIE. (n.d.). Retrieved from https://xapi.com/addie/

ADL LRS. (n.d.). Retrieved from https://www.adlnet.gov/adl-lrs

adlnet/xAPIWrapper: Wrapper to simplify communication to an LRS. (n.d.). Retrieved from https://github.com/adlnet/xAPIWrapper

Ally, M., & Prieto-Blázquez, J. (2014). What is the future of mobile learning in education? *International Journal of Educational Technology in Higher Education, 11*(1), 142–151. doi:10.7238/rusc.v11i1.2033

Anatomy of an Experience API Statement. (n.d.). Retrieved from https://xapi.com/ebook/

Balacheff, N., De Jong, T., Lazonder, A., Sharples, M., Milrad, M., & Vavoula, G. (2008). MOBILE LEARNING Small devices, big issues. In *Technology-enhanced learning: Principles and products* (pp. 1–20). Berlin, Germany: Springer-Verlag.

Ballagas, R., Rohs, M., Sheridan, J. G., & Borchers, J. (2004). BYOD: Bring your own device.

BootStrap. (n.d.). Retrieved from http://getbootstrap.com/

Chen, N., Hung, I., & Wei, C. (2010). Developing ubiquitous learning system with robots for children's learning. In *2010 Third IEEE International Conference on Digital Game and Intelligent Toy Enhanced Learning* (pp. 61–68). 10.1109/DIGITEL.2010.17

Christopoulou, E., & Ringas, D. (2013). Learning activities in a sociable smart city. *Interaction Design and Architecture(s) Journal*, 29–42.

Chrysoulas, C. (2017). Building an adaptive e-learning system building an adaptive e-learning system. In *9th International Conference on Computer Supported Education*. 10.5220/0006326103750382

Ciampa, K. (2014). Learning in a mobile age: An investigation of student motivation. *Journal of Computer Assisted Learning, 30*(1), 82–96. doi:10.1111/jcal.12036

CMI-5 and the Experience API - Experience API. (n.d.). Retrieved from https://experienceapi.com/cmi5/

Cope, B., & Kalantzis, M. (2008). Ubiquitous learning : an agenda for educational transformation. In *6th International Conference on Networked Learning* (pp. 576–582).

Corbi, A., & Burgos, D. (2014). Review of current student-monitoring techniques used in elearning-focused recommender systems and learning analytics. The Experience API & LIME model case study. *IJIMAI*, *2*(7), 44–52. doi:10.9781/ijimai.2014.276

Dafoulas, G. A., Maia, C. C., Clarke, J. S., Ali, A., & Augusto, J. (2018). Investigating the role of biometrics in education -- the use of sensor data in collaborative learning. *International Conference on E-Learning*, 115–123. Retrieved from http://search.ebscohost.com/login.aspx?direct=true&db=a9h&AN=131764741&site=ehost-live

Del Blanco, A., Serrano, A., Freire, M., Martinez-Ortiz, I., & Fernandez-Manjon, B. (2013). E-Learning standards and learning analytics. Can data collection be improved by using standard data models? *IEEE Global Engineering Education Conference, EDUCON*, 1255–1261. 10.1109/EduCon.2013.6530268

Edmodo. (n.d.). Retrieved from https://www.edmodo.com/?simplified_landing_page=1&go2url=%2Fuser%3Fuid%3D3742546

Elias, T. (2011). *Learning analytics*. Retrieved from https://pdfs.semanticscholar.org/732e/452659685fe3950b0e515a28ce89d9c5592a.pdf

Experience, A. P. I. (n.d.). Appendix A. Revision history. (n.d.). Retrieved from https://github.com/adlnet/xAPI-Spec/blob/master/xAPI-About.md#Appendix1A

ExperienceA. P. I. (n.d.). Retrieved from https://xapi.com/

Experience, A. P. I. Adopters. (n.d.). Retrieved from https://xapi.com/adopters/

Fleck, B. K. B., Beckman, L. M., Sterns, J. L., & Hussey, H. D. (2014). YouTube in the classroom: helpful tips and student perceptions. *The Journal of Effective Teaching*, *14*(3), 21–37. Retrieved from http://uncw.edu/cte/et/articles/Vol14_3/Fleck.pdf

Glahn, C. (2013). Using the ADL experience API for mobile learning, sensing, informing, encouraging, orchestrating. In *7th International Conference on Next Generation Mobile Applications, Services, and Technologies* (pp. 268–273). 10.1109/NGMAST.2013.55

Godwin-Jones, R. (2011). Emerging technologies mobile apps for language learning. *Language Learning & Technology, 15*(2), 2–11. 1094-3501.

H5P and xAPI. (n.d.). Retrieved from https://h5p.org/documentation/x-api

H5P. (n.d.). Retrieved from https://h5p.org

Horváth, I. (2019). Evolution of teaching roles and tasks in VR / AR-based education. *9th IEEE International Conference on Cognitive Infocommunications, CogInfoCom 2018 - Proceedings*, 355–360. doi:10.1109/CogInfoCom.2018.8639907

Hruska, M., Amburn, C., Long, R., & Poeppelman, T. R. (2014). Experience API and team evaluation : evolving interoperable performance assessment. In The Interservice/Industry Training, Simulation \& Education Conference (I/ITSEC) (pp. 1–11).

Hung, D. (2001). Theories of learning and computer-mediated instructional technologies. *Educational Media International*, *38*(4), 281–287. doi:10.1080/09523980110105114

Hwang, G. J. (2006). Criteria and strategies of ubiquitous learning. *Proceedings - IEEE International Conference on Sensor Networks, Ubiquitous, and Trustworthy Computing, 2006 I*, 72–77. doi:10.1109/SUTC.2006.49

Jones, V., & Jo, J. H. (2004). Ubiquitous learning environment: An adaptive teaching system using ubiquitous technology. *Beyond the comfort zone: Proceedings of the 21st ASCILITE Conference*, (January 2004), 468–474.

Kalogiannakis, M., Ampartzaki, M., Papadakis, S. J., & Skaraki, E. (2018). Teaching natural science concepts to young children with mobile devices and hands-on activities. A case study. *International Journal of Teaching and Case Studies*, *9*(2), 171–183. doi:10.1504/IJTCS.2018.090965

Kevan, J., & Ryan, P. R. (2016). Experience API : Flexible, decentralized and activity-centric data collection. *Technology, Knowledge, and Learning*, *21*(1), 143–149. doi:10.100710758-015-9260-x

Kitto, K., Cross, S., Waters, Z., Lupton, M., & Street, G. (2015). Learning analytics beyond the LMS : the connected learning analytics toolkit. ACM In (Ed.), *Proceedings of the Fifth International Conference on Learning Analytics And Knowledge2* (pp. 11–15). 10.1145/2723576.2723627

Lim, K. C. (2015). Case studies of xAPI applications to e-Learning. In Case Studies of xAPI Applications to E-Learning.

Mcrae, L., Ellis, K., & Kent, M. (2018). Internet of things (IoT): Education and technology. The relationship between education and technology for students with disabilities, 37. Retrieved from http://www.curtin.edu.au/wp-content/uploads/2018/02/IoTEducation_Formatted_Accessible.pdf

Mellon, C. (2003). Best practices guide for content developers 1.

Moisa, V. (2013). Adaptive learning management system. *Journal of Mobile, Embedded, and Distributed Systems*, *5*(2), 70–77.

Monova-Zheleva, M. (2005). Adaptive learning in web-based educational environments. *Cybernetics and Information Technologies*, *5*(1), 44–55.

Murray, K., Berking, P., Haag, J., & Hruska, N. (2012). Mobile learning and ADL's experience API overview of the xAPI. *Connections: The Quarterly Journal*, *12*(1), 45–49. doi:10.11610/Connections.12.1.05

Nine practical uses of the tin can API (xAPI). (n.d.). Retrieved from https://www.youtube.com/watch?v=8LFhDdqQ13A

OverviewS. C. O. R. M. (n.d.). Retrieved from http://www.adlnet.gov/SCORM

Panagiotakis, S., Papadokostaki, K., Vassilakis, K., & Malamos, A. (2019). Towards a novel and LMS-free pervasive learning system exploiting the experience API. *EAI Endorsed Transactions on Creative Technologies*, 1–13.

Panagiotakis, S. (2012). Context-Awareness via Ubiquitous User Profiling : An Implementation Paradigm. Scientific & Academic Publishing. *Software Engineering*, *2*(3), 77–90. doi:10.5923/j.se.20120203.05

Panagiotakis, S., Papadokostaki, K., Vassilakis, K., & Malamos, A. (2019). Towards a novel and LMS-free Pervasive Learning System exploiting the Experience API. EAI Endorsed Transactions on Creative Technologies, 1–13.

Papadakis, S. J., & Kalogiannakis, M. (2017). Mobile educational applications for children. What educators and parents need to know. [Special Issue on Mobile Learning Applications and Strategies]. *International Journal of Mobile Learning and Organisation*, *11*(3), 256–277. doi:10.1504/IJMLO.2017.085338

Papadakis, S. J., Kalogiannakis, M., & Zaranis, N. (2016). Developing fundamental programming concepts and computational thinking with ScratchJr in preschool education : A case study. *International Journal of Mobile Learning and Organisation*, *10*(3), 187–202. doi:10.1504/IJMLO.2016.077867

Papadokostaki, K., Mastorakis, G., Panagiotakis, S., Mavromoustakis, C. X., Dobre, C., & Batalla, J. M. (2017). Handling big data in the era of internet of things (IoT). In C. X. Mavromoustakis, G. Mastorakis, & C. Dobre (Eds.), *Advances in mobile cloud computing and big data in the 5G era* (pp. 3–22). Cham, Switzerland: Springer International Publishing; doi:10.1007/978-3-319-45145-9_1

Papadokostaki, K., Panagiotakis, S., Vassilakis, K., & Malamos, A. (2017). Implementing an adaptive learning system with the use of experience API. In Interactivity, Game Creation, Design, Learning, and Innovation (pp. 393–402). Springer.

Papadokostaki, K., Spyros, P., & Vassilakis, K. (2018). Katagraphi mathisiakon empirion kai exagoyi analitikon me to protipo Experience API [Tracking learning experience and extracting analytics with experience API]. In *4th Conference for the Promotion of Educational Innovation* (pp. 766–776). Larissa, Greece: EEPEK.

Papadokostaki, K. (2018). Diakhiti Mathisi. Apo Orama Pragmatikotita me ti khrisi tou Experience API. [Ubiquitous learning. From vision to reality with experience API]. In IAKE (Ed.), *Conference Proceedings of 4th International Conference 'Democracy, rights and inequalities in the era of economic crisis. Challenges in the field of research and education'* (pp. 244-253 (Volume B')). Heraklion Crete.

Paramythis, A., & Loidl-Reisinger, S. (2003). Adaptive learning environments and e-learning standards. In *Second European Conference on e-Learning, Vol. 1, No. 2003* (pp. 369–379).

Park, Y. (2011). A pedagogical framework for mobile learning : categorizing educational applications of mobile technologies into four types. *The International Review of Research in Open and Distributed Learning*, *12*(2), 78–102. doi:10.19173/irrodl.v12i2.791

Potential and applications of holograms to engage learners. (n.d.). Retrieved from http://edtechreview.in/trends-insights/trends/521-applications-of-holograms-to-engage-learners

Sequencing, S. C. O. R. M. Learn more about SCORM Sequencing. (n.d.). Retrieved from https://scorm.com/scorm-explained/technical-scorm/sequencing/

SequencingS. C. O. R. M. (n.d.). Retrieved from http://www.scormsoft.com/scorm/sn/sequencing

Sharples, M. (2013). Mobile learning : Research, practice and challenges. *Distance Education in China*, *3*(5), 5–11.

Sharples, M., Taylor, J., & Vavoula, G. (2007). A theory of learning for the mobile age. The Sage handbook of e-learning research, 221–247. doi:10.1007/978-3-531-92133-4_6

Shute, V. J., & Towle, B. (2003). Adaptive e-learning. *Educational Psychologist*, *38*(2), 105–114. doi:10.1207/S15326985EP3802_5

Siemens, G., & Long, P. (2011). Penetrating the fog: analytics in learning and education. *EDUCAUSE Review*, *46*(5), 30.

The evolution of technology in the classroom | Purdue University Online. (n.d.). Retrieved from https://online.purdue.edu/ldt/learning-design-technology/resources/evolution-technology-classroom

TinCan Javascript Library. (n.d.). Retrieved from http://rusticisoftware.github.io/TinCanJS/

7 . Ways Holographic Technology Will Make Learning More Fun. (n.d.). Retrieved from http://www.emergingedtech.com/2012/11/7-ways-holographic-technology-will-make-learning-more-fun/

Wong, K. Y. (2010). Cell phones as mobile computing devices. *IT Professional*, *12*(3), 40–45. doi:10.1109/MITP.2010.46

wp-h5p-xapi. (n.d.). Retrieved from https://wordpress.org/plugins/wp-h5p-xapi/

xAPI Dashboard. (n.d.). Retrieved from https://github.com/adlnet/xAPI-Dashboard/blob/master/API_dashboard.md

xAPI-Spec. (n.d.). Retrieved from https://github.com/adlnet/xAPI-Spec/

xapi-statement-viewer. (n.d.). Retrieved from https://github.com/adlnet/xapi-statement-viewer

Yahya, S., Ahmad, E. A., & Jalil, K. A. (2010). The definition and characteristics of ubiquitous learning: A discussion. *International Journal of Education and Development Using Information and Communication Technology*, *6*(1), 117–127.

Yuan, L., & Powell, S. (2014). MOOCs and open education: Implications for higher education. *Cetis*, 19.

Zapata-Rivera, D., & Shute, V. J. (2012). Adaptive educational systems. *Adaptive Technologies for Training and Education*, *7*, 4–27. doi:10.1017/CBO9781139049580.004

Zaranis, N., Kalogiannakis, M., & Papadakis, S. (2013). Using mobile devices for teaching realistic mathematics in kindergarten education. *Creative Education*, *04*(07), 1–10. doi:10.4236/ce.2013.47A1001

Zhang, D., Zhao, J. L., Zhou, L., & Nunamaker, J. F. Jr. (2004). Can e-learning replace classroom learning? *Communications of the ACM*, *47*(5), 75–79. doi:10.1145/986213.986216

Zhao, X., & Okamoto, T. (2011). Adaptive multimedia content delivery for context-aware u-learning. *International Journal of Mobile Learning and Organisation*, *5*(1), 46. doi:10.1504/IJMLO.2011.038691

ADDITIONAL READING

Ally, M., & Prieto-Blázquez, J. (2014). What is the future of mobile learning in education? *International Journal of Educational Technology in Higher Education*, *11*(1), 142–151. doi:10.7238/rusc.v11i1.2033

Amrieh, E. A., Hamtini, T., & Aljarah, I. (2015). Preprocessing and analyzing educational data set using X-API for improving student's performance. *2015 IEEE Jordan Conference on Applied Electrical Engineering and Computing Technologies (AEECT)*, 1–5. 10.1109/AEECT.2015.7360581

Benedek, A. (2013). Learning Design Versus Learning Experience Design: Is the Experience Api Making the Difference? In *Edulearn13: 5th International Conference on Education and New Learning Technologies* (pp. 2609–2621).

Berg, A., Scheffel, M., Drachsler, H., & Specht, M. (2016). Dutch Cooking with xAPI Recipes The Good, the Bad, and the Consistent. In *IEEE 16th International Conference on Advanced Learning Technologies (ICALT)* (pp. 234–236). 10.1109/ICALT.2016.48

Chakravarthy, S. S., & Raman, A. C. (2014). Educational Data Mining on Learning Management Systems Using Experience API. *2014 Fourth International Conference on Communication Systems and Network Technologies*, 424–427. 10.1109/CSNT.2014.90

De Nies, T., Salliau, F., Verborgh, R., Mannens, E., & Van de Walle, R. (2015). TinCan2PROV : Exposing Interoperable Provenance of Learning Processes through Experience API Logs. *WWW '15 Companion Proceedings of the 24th International Conference on World Wide Web*, 689–694. 10.1145/2740908.2741744

Griffiths, D., & Hoel, T. (2016). Comparing xAPI and Caliper. *Learning Analytics Review*, (7), 1–27.

Lim, K. (2018). Using the xAPI to Track Learning (pp. 233–242). doi:10.1007/978-981-10-7995-5_21

Moreno López, G. A., Jiménez-Builes, J., & Villamil, S. (2016). Overview of U-Learning. Concepts, Characteristics, Uses, Application Scenarios and Topics for Research. *IEEE Latin America Transactions*, *14*(12), 4792–4798. doi:10.1109/TLA.2016.7817013

Murphy, J., Hannigan, F., Hruska, M., Medford, A., & Diaz, G. (2016). Leveraging interoperable data to improve training effectiveness using the experience api (XAPI). Lecture Notes in Computer Science (Including Subseries Lecture Notes in Artificial Intelligence and Lecture Notes in Bioinformatics), 9744, 46–54. doi:10.1007/978-3-319-39952-2_5

Shih, S. C., Kuo, B. C., & Liu, Y. L. (2012). Adaptively ubiquitous learning in campus math path. *Journal of Educational Technology & Society*, *15*(2), 298–308.

Sottilare, R. A., Long, R. A., & Goldberg, B. S. (2017). Enhancing the Experience Application Program Interface (xAPI) to Improve Domain Competency Modeling for Adaptive Instruction. *Proceedings of the Fourth (2017) ACM Conference on Learning @ Scale - L@S '17*, 265–268. 10.1145/3051457.3054001

Vandewaetere, M., Desmet, P., & Clarebout, G. (2011). The contribution of learner characteristics in the development of computer-based adaptive learning environments. *Computers in Human Behavior*, *27*(1), 118–130. doi:10.1016/j.chb.2010.07.038

Yahya, S., Ahmad, E. A., Jalil, K. A., & Mara, U. T. (2010). The definition and characteristics of ubiquitous learning : A discussion. *International Journal of Education and Development Using Information and Communication Technology*, *6*(1), 1–12. PMID:20852677

Zapata-rivera, D., & Shute, V. J. (2012). Adaptive educational systems. *Adaptive Technologies for Training and Education*, *7*, 4–27. doi:10.1017/CBO9781139049580.004

KEY TERMS & DEFINITIONS

Adaptive Learning: The philosophy of providing personalized educational material/methods to match the personal needs and characteristics of the learner. That means providing the right content to the right person, at the proper time, in the most appropriate way. Adaptive learning means more efficient learning.

Experience API (xAPI/ formerly Tin Can API): an open source, learning technologies interoperability specification that defines explicitly how tracking of learner activities and experiences between technologies should be delivered. It is licensed under the Apache License, Version 2.0 and is used in various Learning Management Systems (LMSs), Authoring Tools and Training Systems.

Learning Analytics: The extraction of knowledge from educational procedure(s). Tracking and processing statistics about the learners' /class's behavior may lead to powerful conclusions and serious decisions.

Ubiquitous Learning (u-Learning): the form of learning where "all students have access to a variety of digital devices and services, including computers connected to the Internet and mobile computing devices, whenever and wherever they need them". Mobile devices, IoT devices and sensors/beacons are the main means of education, where wireless connections are a catalyst for the provision of the educational material. In u-Learning learning happens literally everywhere.

Chapter 14
Social Hazards or Helpers?
The Role of Mobile Media in Early Childhood Social Development

Courtney K. Blackwell
https://orcid.org/0000-0003-0934-0207
Feinberg School of Medicine, Northwestern University, USA

ABSTRACT

This chapter reviews current theoretical and empirical work on associations between young children's mobile media use in early childhood education settings and their social development, including social relationships and foundational social skills such as communication, collaboration, and positive social interactions and engagement. Touchscreen tablets are highlighted given their increased presence in early childhood education as well as their unique affordances specifically for young children. Particular attention is paid to factors influencing whether, what, and how educators integrate tablets into their classroom environments; facilitators and barriers to integration; how such integration may enable or interfere with social skills and relationships; and implications for practice and policy.

INTRODUCTION

Current discussions on the influence of mobile media on children's health and development reflect a long tradition of potential promises and pitfalls of new technology (Lauricella, Blackwell, & Wartella, 2017; Wartella & Jennings, 2000; Wartella & Robb, 2007). While the promise of computers to transform learning and increase academic achievement was championed by many, others purported this innovation to be antithetical to one of the core aims of early childhood education and development – social relationships (Cordes & Miller, 2000).

So too with mobile media. Affordances such as ubiquitous, on-demand access to media content are viewed as both learning transformers and relational disrupters. In particular, concerns over computers being "emotional and social hazards" (Cordes & Miller, 2000, pg. 39) remain salient for mobile media, with hyperbolic fears that early and continued use will result in a new generation devoid of social skills

DOI: 10.4018/978-1-7998-1486-3.ch014

and positive social relationships. The American Academy of Pediatrics (AAP, 2016) has been outspoken on this topic, continually recommending no or limited screen time for young children as "multiple developmental and health concerns continue to exist for young children using all forms of digital media to excess" (p. 3). In addition to citing concerns over harmful effects of media on children's physical health – such as obesity (e.g., Jordan & Robinson, 2008; Robinson et al., 2017), decreased physical activity (e.g., Lacy et al., 2011), and sleep disruption (e.g., LeBourgeois et al., 2017; Magee, Lee, & Vella, 2014) – and emotional well-being – including anxiety and depression (Hoge, Bickham, & Cantor, 2017; James et al., 2017) – the AAP emphasizes the potential negative influence of digital media on children's social and relational health, prioritizing unplugged, hands-on social play. Further, the AAP cautions parents against using screens to calm their children down, citing "concern that using media as strategy to calm could lead to problems with limit setting or the inability of children to develop their own emotion regulation" (p. 4). Such fears are markedly pertinent in early childhood, as such skills and relationships during this time period have tremendous influences on children's short- and long-term outcomes (e.g., Chetty et al., 2011; Conti, Heckman, & Pinto, 2016; Jones, Greenberg, & Crowley, 2015; National Research Council & Institute of Medicine, 2000).

With young children's near ubiquitous access to mobile devices as well as their increased time spent using such technology (Huber, Highfield, & Kaufman, 2018; Ofcom, 2019; Rideout, 2017), understanding whether, how, and to what extent mobile media influences young children's social health and development is critical. In the U.S., 98% of households with 0- to 8-year-olds have a mobile device, and children spend an average of 48 minutes a day using such devices (Rideout, 2017). Similarly, 19% of 3- to 4-year-olds and 42% of 5- to 7-year-olds in the U.K. have their own mobile device (OfCom, 2019), and in Australia, more than 25% of 2- to 8-year-olds use touchscreens daily (Huber et al., 2018a). Few studies exist, however, on the influence of mobile media on children's general learning, and those that do favor academic outcomes or special student populations (e.g., Autism Spectrum Disorder; Herodotou, 2018). Others have begun to explore social outcomes in the home environment (e.g., Roseberry, Hirsh-Pasek, & Golinkoff, 2014), but little is currently known regarding the ways in which mobile media may promote or prohibit social development in early childhood educational settings.

This chapter reviews current theoretical and empirical work on associations between young children's mobile media use in early childhood education and their social development, including social relationships and foundational social skills (e.g., communication, cooperation, social interactions and engagement; Brown, Winsor, & Blake, 2012; Collaboration for Academic, Social, and Emotional Learning, nd). While mobile media encompasses a wide variety of devices and platforms, this chapter focuses on touchscreen tablets given their prevalence in the lives of young children (Rideout, 2017) and in early childhood education settings (Blackwell, Lauricella, Robb, & Wartella, 2015), as well as the noted affordances of tablets specifically for young children (e.g., Geist, 2014; Habler, Major, & Hennessy, 2015; McManis & Gunnewig, 2012; Neumann & Neumann, 2014; Zaranis, Kalongiannakis, & Papadakis, 2013; Tootell, Plumb, Hadfield, & Dawson, 2013). Particular attention is paid to factors influencing whether, what, and how educators integrate tablets into their classroom environments; facilitators and barriers to integration; how such integration may enable or interfere with social skills and relationships; and implications for practice and policy.

THE PRESENCE AND PROMISE OF MOBILE MEDIA FOR EARLY CHILDHOOD EDUCATION

The importance of early childhood education on children's future academic and work success is undeniable. Two landmark experimental studies – the HighScope Perry Preschool Program (Schweinhart et al., 2011) and the Carolina Abecedarian Project (Campbell, Ramey, Pungello, Sparling, & Miller-Johnson, 2002) – showed the long-term effects of high quality early education; children in these intervention programs scored higher on standardized achievement tests, had higher educational attainment, and were more likely to attend college. Continued longitudinal follow-up of Perry participants 40 years after the intervention showed greater earnings and lower arrest rates for program participants (Schweinhart et al., 2011). Groundbreaking evidence from 50 years later also suggests intergenerational effects of high-quality preschool: the *children* of Perry participants fare better compared to children of the non-treatment group on such outcomes as lower school suspensions and participation in crime, and higher educational attainment and employment rates (Heckman & Karapakula, 2019).

Findings from high-quality educational media (e.g., Fisch & Truglio, 2001; Pasnik, 2019) further emphasize the critical importance of early learning environments on children's academic, behavioral, and social success. In one of the largest reviews to date, Hurwitz's (2019) meta-analysis of Ready to Learn (RTL) – an initiative developed by the U.S. Department of Education and the Corporation for Public Broadcasting to promote school readiness through the dissemination of high-quality digital media programming – showed positive effects on 2- to 8-year-old's literacy outcomes equivalent to 1.5 months of learning above and beyond typical literacy development. Further, Anderson and colleagues (2001) showed viewing early educational television programming was associated with higher grades and less aggression in high school, suggesting potential long-term influences of digital media on learning and development. Importantly, children who viewed educational TV in preschool placed higher value on achievement as high schoolers, suggesting these early digitally mediated experiences instilled a love of learning that set children on more positive life trajectories.

With the advent of mobile technologies, the landscape of children's early education experiences is swiftly shifting. In just two years, from 2012 to 2014, the proportion of early childhood educators with access to tablets nearly doubled in U.S.-based early education settings, from 29% to 55% in (Blackwell et al., 2015). Importantly, access was equitable across low-, middle-, and high-income student populations (Blackwell et al., 2015). Recent estimates also show 80% of K-12 classrooms in the U.S. have 1:1 or shared computing devices in their classrooms (Vega & Robb, 2019), and in the U.K., nearly 70% of primary and secondary schools have tablets (Coughlan, 2014). The national governments of Singapore and Thailand have also implemented 1:1 tablet programs across their schools (Pruet, Ang, & Farzin, 2014).

Underlying these large educational technology investments is the assumption that ubiquitous access to Internet-enabled mobile devices will increase student motivation and engagement via more personalized and individualized educational experiences (Greaves, Hayes, Wilson, Gielniak, & Peterson, 2012; Lemke, Coughlin, & Reifsneider, 2009; Park, 2011; Penuel, 2006; Traxler, 2009). For younger children in particular, mobile media – and especially touchscreen tablets – provide additional affordances that make the learning experience more developmentally-appropriate compared to computers. The mobility of tablets enables learning to occur anywhere at any time, such that children are not tethered to desks or even the four walls of a classroom (Blackwell, 2013; Lauricella, Blackwell, & Wartella, 2016). The flexibility afforded by such mobility complements the unique physical setup of early childhood education environments, where educators often set up small group centers or stations with specific activities and have students move around throughout the school day. As Blackwell (2013) noted, while mobility can pose classroom management challenges, educators also leverage this feature to extend learning time during activity transitions or bus rides to field trips.

Furthermore, tablets are easier for young children to use due to the technology's smaller size and touchscreens, which allow for direct manipulation and make it easier to co-use as students are not restricted to one mouse (Chou, Block, & Jesness, 2012; Habler et al., 2015; Neumann & Neumann, 2014; Tootell et al., 2013). They are also cognitively simpler and more intuitive than computers because there is no mouse, such that young children easily and enthusiastically become tablet users (Geist, 2014; Habler et al., 2015; McManis & Gunnewig, 2012; Zaranis et al., 2013). In Lynch and Redpath's (2014) observations of Australian primary classrooms, the authors noted the ease with which young students acquired tablet competencies, such as turning devices on and off, locating and launching apps, and independently troubleshooting when the technology malfunctioned. Further, the authors observed that children encountered fewer technical issues compared to traditional computers and laptops. Similarly, Mourlam and colleagues (2019) confirmed preschool children are quite adept at using iPads, despite educators tending to underestimate young children's ability to use touchscreen tablets.

Evidence also suggests tablets are highly engaging and motivating for young children, all features that have been shown to be important to young children's learning (e.g., Ciampa, 2014; Couse & Chen, 2010; Falloon, 2013; Kirkorian, Choi, & Pempek, 2013). Indeed, Tootell and colleagues (2013) argue that the gestural interface of tablets encourages children's natural curiosity, allows them to be the protagonist in their learning, and can engage learners in real-world learning based on their cultural background and interests.

Theoretical Perspectives

While innate affordances of tablets suggest they may be particularly suited to enhance young children's learning and development, theories from developmental psychology, education, and human-computer interaction provide conceptual groundings for the potential of tablets to enhance young children's social skills and promote rather than interfere with social relationships.

Sociocultural Theory & Digital Scaffolding

Given the importance of a child's environment in shaping his/her development (Bronfenbrenner, 1979; Bronfenbrenner & Morris, 2006; Vygotsky, 1978), and the proliferation of tablets in young children's lives (Blackwell et al., 2015; Rideout, 2017), it stands to reason that such mobile media can have a critical influence on children's learning and development. While some note that technology is merely substituted for traditional learning tools in the classroom (e.g., a tablet becomes a digital notepad; Collins & Halverson, 2009; Eteokleous, 2008), Vygotsky's (1978) sociocultural theory explains how learning tools, including technology, could be used to "scaffold" (Wood, Bruner, & Ross, 1976) children's learning. Described as social and collaborative in nature, learning occurs when children are given tasks that are slightly advanced for their current development and offered guidance from more knowledgeable others (i.e., the "Zone of Proximal Development"; Vygotsky, 1978). Scaffolding techniques are used to temporarily structure learning as a pathway to more individual and self-directed learning.

Just as language, number systems, art, maps, and diagrams are culturally-specific tools that can act as "more knowledgeable others" to scaffold learning, so too can technology. As Wartella and colleagues

(2016) argue, "Media provide children and adolescents with the affordances necessary to learn information above and beyond their given developmental stages" (p. 15). Yelland and Masters (2007) described three types of scaffolding that can occur during digital interactions. First, *cognitive scaffolding* reflects a traditional notion of scaffolding focused on conceptual and procedural understandings where teachers use modeling and questioning, and encourage collaboration to assist the learning process. Second, *affective scaffolding* relates to the positive feedback and encouragement adults provide during the learning process to help children persist and extend their learning to more higher-level thinking. While cognitive and affective scaffolding are not specific to digital technology, such techniques can be used in conjunction with tablets to enhance learning with and from such technology. For example, in classrooms with shared tablets, educators may use cognitive scaffolding by placing "my turn" and "your turn" flashcards in front of young children to remind and encourage them to share the device; similarly, teachers may use affective scaffolding by encouraging students to persist to complete a challenging level in an app instead of giving up and playing a different app. Alternatively, the third type of scaffolding – *technical scaffolding* – describes how features of the technology can mediate learning experiences, such as built-in response features of some apps that provide immediate, contingent feedback. Together, these three types of scaffolding emphasize not just the technology but the dynamic student-teacher interactions that occur around technology. Such an emphasis reflects the strong body of work from education on the value of high-quality student-teacher relationships for children's academic, behavioral, and social outcomes (Chetty et al., 2011; Mashburn et al., 2008). As such, just because tablets enter the learning space, this does not mean that foundational early years pedagogy simply disappears.

Computer Supported Collaborative Learning

In addition to the unique scaffolding opportunities provided by educators and the technology itself, computer-mediated classroom environments may enable novel opportunities for peer collaboration. Computer Supported Collaborative Learning (CSCL) describes how learning is maximized when learners collaborate to complete a task where technology is critical to the collaboration (Dillenbourg, 1999; Goodyear, Jones, & Thompson, 2014). Prior research on CSCL suggests sharing devices can lead to better learning outcomes compared to individual device use (Johnson & Johnson, 2008; Lou, Abrami, & d'Appolonia, 2001), and Larkin (2011) argued that sharing devices, as opposed to one device per child, was ideal for enhancing student learning and engagement. As Freeman and Somerindyke (2001) noted, working together on computers creates a "shared problem space" for children to accomplish tasks together. Similarly, several studies show that social interactions actually increase in the presence of computers and that children exhibit cooperation, communication, and conflict-resolution skills while engaging with peers around computers (Lim, 2012). Given the ease with which students can collaborate around tablets (Falloon, 2015; Fisher, Lucas, & Galstyan, 2013), this "shared problem space" may hold an even greater affordance than prior technology.

Most work on CSCL has focused on computers and older children or adults (Johnson & Johnson, 2008; Lou et al., 2001) and rarely been studied in the context of tablets, let alone in early childhood education contexts (Gomez et al., 2013; McCarrick & Li, 2007; Perlmutter, Behrend, Kuo, & Muller, 1989). However, the core prominence of peer collaboration echoes traditional paradigms that emphasize the role of early childhood education being the time and place where young children develop core social skills and competencies, including learning to work well with others, play cooperatively (e.g., sharing, turn-taking), and resolve peer conflicts through basic problem-solving skills (Bohart & Procopio 2017; Office of Head Start, 2015).

Positive Technological Development

Extending CSCL to a developmental sciences perspective, Bers (2018) put forth the Positive Technological Development (PTD) framework to emphasize the positive potential of technology, particularly with regard to social development. Drawing on Papert's (1993) constructivist learning theory and research on positive youth development (e.g., Lerner et al., 2005), PTD describes how technology can be designed and used to facilitate positive *intra*personal (i.e., content creation, creativity, choices of conduct) and *inter*personal (i.e., communication, collaboration, and community building) skills and competencies. In turn, such technology-mediated behaviors can promote key developmental assets, such as confidence, empathy, and character. Unlike other theoretical perspectives that emphasize how technology can increase academic outcomes, PTD focuses on social-emotional learning and reflects the bedrock of early childhood education. Furthermore, PTD's emphasis on intentionality in the design and use of educational technology reflects recent policy statements from the American Academy of Pediatrics (2016), the National Association for the Education of Young Children and Fred Rogers Center (NAEYC/FRC; 2012), and Early Childhood Australia (2018) on developmentally-appropriate media content and use in the early years.

From Theory to Practice: The What and How of Tablet Use in the Classroom

The specific affordances of tablets, paired with the theoretical values of technology to enhance young children's learning and development, cannot be understood outside the practical realities of the early childhood education context. While tablets may have innate affordances uniquely suited to aid young children's learning, and while theories may illuminate how tablets can enhance social skills and relationships, whether and how educators integrate such mobile media is paramount to realizing such potential.

Frequency

Tablets can have great potential but if they are not used – or used effectively – there is no reason to believe that such mobile media can impact child outcomes. For example, despite a two-fold increase in U.S. early childhood educators' access to tablets from 2012 to 2014, there was no change in frequency of use over that same time period; only half of early childhood educators reported using tablets at least once a week, and 25% said they never use the technology even though they have access it (Blackwell et al., 2015). Such findings are not unique to tablets or early childhood educators and reflect a broader trend in educational technology, with a third of K-12 teachers reporting that they never use technology provided to them (Vega & Robb, 2019).

Content

Beyond hardware access, Baker and Gowda (2018) found K-12 teachers only used 30% of available software licenses, with the most-used being Google Drive and Canvas – two platforms primarily focused on documentation and classroom management. Such findings reflect early childhood educators' use of tablets as well, with the majority reporting they use tablets to document children's learning (Blackwell et al., 2015). When it comes to specific content areas, most early childhood educators use tablets for traditional academic subjects, especially literacy and math (Wartella, Blackwell, Lauricella, & Robb, 2013).

The focus on using tablets for academic learning may in part reflect the types of apps available as well as the overwhelming volume of apps with which educators are faced. With more than 80% of the top-selling paid apps in the Apple iTunes Education Category targeting preschool and elementary-aged children (Shuler, Levine, & Ree, 2012), Guernsey and colleagues described the current digital environment as the "fast evolving and chaotic Wild West of digital apps" (Geurnsey, Levine, Chiong, & Severns, 2012, p. 15). However, recent reviews question the "educational" value of the actual content, finding most apps focus on basic literacy and math skills and are designed as skill and drill practice activities (Vaala, Ly, & Levine, 2015; Papadakis et al., 2018). Data-mining from actual app use in Australian K-6 classrooms confirms that children primarily used rote learning apps rather than those that promote higher-level thinking (Howard, Yang, Ma, Maton, & Rennie, 2018). Furthermore, while Hirsh-Pasek and colleagues (2015) suggest high-quality educational apps promote social interactivity, Papadakis et al. (2018) found available apps do not provide children with opportunities for synchronous play, therefore limiting the potential for social interactions and collaboration. Such findings are reflected in early childhood educators' actual use of tablets to support social-emotional development, with only 56% reporting using the technology for such purposes (Wartella et al., 2013). This contrasts with 79% of educators who reported using digital cameras to support social-emotional development (Wartella et al., 2013). While tablets have built-in cameras and thus can function as a digital camera in addition to other functional capabilities, early childhood educators appear not to value them as much for aiding children's social skills.

Context

Even if the content and design of educational apps are limited in their ability to promote social interactivity, and even if early childhood educators tend to use tablets primarily for academic subjects, the way in which educators integrate tablets into the learning environment can provide unique opportunities for children to develop key social skills and competencies. Habler and colleagues (2015) outlined four different ways tablets can be used in the classroom: *one-to-one without collaboration*, where each student has his/her own device and works individually to accomplish his/her own individual tasks; *one-to-one with collaboration*, where each student works on his/her own device but collaborates with others to accomplish a shared task; *one-to-many individual use*, where students share tablets but each uses the device individually and then passes it onto another student who engages one-on-one with the tablet; and *one-to-many with collaboration*, where small groups of students share one device while working on a single task.

The prevailing integration strategy remains one device per child (1:1; Vega & Robb, 2019), though evidence suggests teachers of younger children use them in more diverse configurations. Indeed, 67% of preschool teachers reported using tablets for paired learning at least sometimes (Blackwell, Lauricella, & Wartella, 2016). One reason may be due to device access, with fewer devices available in earlier grades (Otterborn, Schonborn, & Hulten, 2018; Vega & Robb, 2019). For example, in a Swedish sample of preschool educators, the majority reported having access to only one or two tablets, with only 6.6% having access to five or more devices for their entire classroom (Otterborn et al., 2018).

As Huber and colleagues (2018b) note, 1:1 tablets are often viewed as the most effective integration strategy without consideration for other ways in which devices can be used, and little evidence exists supporting the 1:1 model as the most appropriate implementation strategy for early childhood education. Indeed, research on collaborative tablet use in early childhood education suggests sharing tablets may be more effective at enhancing children's learning and development (Blackwell, 2015; Papadakis et al., 2018). Beschorner and Hutchison (2013) found preschool children's shared use of iPads promoted communication among peers, and Kucirkova and colleagues (2014) showed children engaged in joint problem-solving and collaborative engagement when sharing tablets. Teachers in Blackwell's (2013) observational study of iPad integration in preschool and kindergarten classrooms similarly reported that sharing devices provided unique opportunities for turn-taking and collaborating with others. Sandvik and colleagues (2012) describe a similar finding, noting the portability of tablets made for "smooth turn-taking" and collaboration among kindergarteners. Additionally, a case study investigating 4-year-old's social behaviors during paired and small group activities with iPads showed more instances of prosocial sharing behaviors than nonsocial or antisocial behaviors (Ralph, 2018), suggesting the technology did not interfere with or prohibit positive social interactions between young children. Danby and colleagues (2018) report similar findings, emphasizing the social context of digital game play where young children interact with peers to solve problems, share strategies, and collaborate towards a shared end goal. The authors suggest children learned from each other through peer-to-peer collaboration and scaffolding. Such findings are reflected in Blackwell's (2015) study examining differences in literacy achievement for kindergarteners in 1:1 versus shared iPad and no iPad classrooms. While the main focus of that study was on academic outcomes, the author hypothesized that children who shared iPads outscored their peers on end-of-year literacy assessments in part because of the increased communication and peer-to-peer scaffolding that occurred when sharing iPads.

Facilitators and Barriers to Using Mobile Media to Promote Social Development

The lack of tablet use in early childhood education reflects a broader phenomenon related to teacher practices with technology, whereby access to devices does not necessarily equate with use. As Ertmer (1999) explained, overcoming access to technology is only part of the integration process. Additional extrinsic barriers related to school support, training, and professional development as well as intrinsic barriers related to attitudes toward and confidence using technology play a large role in whether and how teachers integrate technology into the learning environment. In their meta-analysis of 114 studies evaluating the Technology Acceptance Model (TAM), Scherer and colleagues (2019) confirmed these two levels of barriers across a wide range of pre- and in-service educators from across the globe working in grades K-12, college, and university settings. Findings highlighted the critical role that external facilitating conditions (e.g., organizational support, resources) and technology self-efficacy play in educators' perceived usefulness of technology, as well as the role of attitudes in mediating relationships between usefulness and intended use (Scherer, Siddiq, & Tondeur, 2019).

While much of the empirical work on facilitators and barriers to technology integration has been conducted with teachers of older children, Blackwell and colleagues (2013) provided novel insight into how such factors influence early childhood educators' educational technology use. Applying a modified version of TAM to early learning environments and teachers' actual technology use (versus intended use), the authors found that teacher attitudes toward the value of technology to enhance student learning mediated the relationship between extrinsic barriers, perceived confidence using technology, and technology use. Such findings were replicated and extended for tablet use in particular (Blackwell et al., 2016), whereby teacher attitudes were not only the strongest predictor of time spent using tablets for instructional purposes but also of how educators used the technology in their classrooms. Indeed, positive attitudes were more predictive of whether early childhood educators used tablets for individualized learning, whereas confidence was more predictive of whether they used devices for paired instruction; additionally, having school support was significantly associated with paired instruction practices but not with individualized practices (Blackwell et al., 2016).

Given that educator attitudes are crucial to understanding not only whether but how they integrate technology into their classrooms, it is important to understand how they perceive technology as tools to enhance social skills and competencies. In general, educators do not have positive perspectives. In the U.S., the majority of early childhood educators disagree that technology is beneficial for social interactions, with less than a fifth using technology to support social interactions (Blackwell et al., 2015). Similarly, Australian early childhood educators reported concerns over the impact of technology on young children's social skills and relationships (Hatzigianni & Kalaitzidis, 2018). A survey of Swedish preschool teachers revealed less than half believed tablets enabled meaningful opportunities for collaboration, and only 16% reported using them to support social skills (Otterborn et al., 2018).

At the same time, early childhood educators may not have sufficient support and professional development to implement tablets in developmentally-appropriate ways that encourage and enable positive social interactions. Only 31% of U.S. early childhood educators agreed or strongly agreed that their school/program provides developmentally-appropriate models for technology integration (Blackwell et al., 2015). Findings from other countries reflect such sentiments (Hatzigianni & Kalaitzidis, 2018; Otterborn et al., 2018). In Australia, for example, 65% of early childhood educators were "not very satisfied" with the time or quality of training, and 80% reported never receiving training to use technology with 0- to 3-year-olds (Hatzigianni & Kalaitzidis, 2018). Alternatively, educators who received educational technology professional development tended to be more confident in finding developmentally-appropriate content and more likely to use technology to complement social interactions (Blackwell et al., 2015).

In addition to school support and professional development, negative teacher attitudes toward technology as a tool for social development may be due to a disconnect between research and practice. Academic research may provide evidence for the positive potential of tablets to foster social skills, but often it is inaccessible to the general public, not translated for lay audiences, and lacks practical guidance or solutions that can be readily implemented "in the wild." Further conflating matters is the emphasis on negative outcomes from technology use that permeates the popular press, such that educators may be unaware of current research findings on the positive potential of tablets while at the same time they are inundated with messages on the negative consequences for child outcomes.

SOLUTIONS AND RECOMMENDATIONS

Despite recent initiatives focused on scaling up 1:1 tablet computer initiatives in education contexts across the U.S. and around the globe, the effectiveness of tablets for aiding children's learning remains largely unknown. At the same time, calls for developmentally-appropriate technology use in early childhood education raise important questions regarding not just whether or how much children use tablets, but how they use them, for what purposes, and in what contexts (AAP, 2016; NAEYC/FRC, 2012).

One such context – shared tablet use – appears to set up opportunities for young children to engage in social interactions and foster core social competencies of communication and collaboration (Arnott, 2016; Fleer, 2018; Hatzigianni, Gregoriadis, Karagiorgou, & Chatzigeorgiadou, 2018; McManis & Gunnewig,

2012). Instead of assuming 1:1 tablet computers are more developmentally-appropriate and will enhance student outcomes, early childhood educators and policymakers would benefit from a child first, technology second perspective, where the foundational concepts of early learning and development – including emphasis on social interactions and relationships – are practiced *within the unique context of tablet use.*

In practice, teachers might apply this notion by revising their classroom use of tablets to include more paired and small group activities where children share devices. This context allows for peer-to-peer scaffolding opportunities, which fundamentally requires social and relational skills. Additionally, schools may reconsider their tablet deployment efforts, given the large financial investment needed to provide 1:1 tablets for all students, and policymakers may reconsider the framing of local and national educational technology recommendations and efforts in early childhood contexts. Instead of assuming one device per child is the optimal integration strategy across every grade level, schools and policymakers should evaluate tablets in the context of developmental sensitivity.

Additionally, if educators do not believe tablets can aid children's social development or do not have the confidence to integrate tablets in such a way that enables positive social interactions, it is unreasonable to expect such outcomes to materialize. Though shifting teacher attitudes is undoubtedly challenging, the child first, technology second perspective may alleviate some concern because it aligns more closely with traditional early childhood pedagogies by making technology another tool rather than *the* learning resource. As emerging work suggests positive associations between sharing tablets and young children's social outcomes, pre- and in-service training programs can focus on increasing teachers' confidence as a practical opportunity to shift towards more effective tablet integration practices.

Importantly, early childhood educators cannot be expected to effectively integrate tablets on their own without quality training and support. In addition to lacking general professional development opportunities, early childhood educators do not feel like they have sufficient support for integrating technology in developmentally-appropriate ways. Thus, stronger pre- and in-service training programs are needed, particularly for helping teachers understand best practices for using tablets with young children. One particularly effective strategy may be providing educators with up-front and on-going training in conjunction with a curated curricula of high-quality digital media content (Wartella, Lauricella, & Blackwell, 2016). By combining training with specific and sequenced materials, Pasnik and Llorente (2013) found that preschool teachers were more likely to integrate technology into their classrooms, required less long-term support for integrating new instructional strategies, and perhaps most importantly, exhibited shifts in attitudes towards the positive potential of technology to aid children's learning. Additionally, strengthening school support may be one avenue for enhancing teachers' use of shared tablets, which may be more appropriate for young children (Blackwell et al., 2016).

Furthermore, if early childhood educators are not aware of or have access to the research literature on the potential of tablets to promote positive social development, shifting attitudes and use may be unrealistic. Finding novel ways to "market" the research to those on the ground and provide a holistic perspective on the influence of mobile devices on young children's social skills and relationships may be one way to start this shift. One such method is to develop brief, teacher-facing reports written in non-scientific language and with practical guidance on how to effectively integrate tablets in developmentally-appropriate ways into classroom practice. Additionally, local presentations onsite at schools as well as presentations at educator-facing national and international conferences (e.g., NAEYC annual conference) can help translate the research into practice.

FUTURE RESEARCH DIRECTIONS

Most studies examining associations of tablets and young children's learning treat social skills as a secondary outcome or mechanism through which academic achievement is enhanced. While such work is important and provides insight into the potential of mobile media to promote rather than prohibit social development, more research is needed that specifically examines whether and how touchscreen tablets influence social-emotional learning and relationships. For example, does the use of tablets promote or prohibit empathic behaviors? Do tablets interfere with whether and how children interact with others face-to-face? To what extent does tablet use impact young children's social cognition? Does earlier exposure to tablets influence how children interact online at older ages? Is tablet use associated with better digital literacy? Such questions are particularly critical in early childhood when foundational social-emotional and relational skills emerge.

From a software and hardware perspective, a recent report by Microsoft (2018) suggests three types of technology purported to enhance social-emotional learning: 1) digital collaboration platforms, including software and online programs than enable students to work together across time and place; 2) artificial intelligence (e.g., machine learning, robotics, natural language processing) for such uses as tailoring curricula to individual student needs; and 3) virtual reality, which offers opportunities to observe and practice social-emotional skills. While the report focused on K-12 learning environments, such tools may be applicable to early childhood but additional research is necessary to fully understand the developmental appropriateness for young children.

Additionally, future work on digital games may be one salient line of research (Blumberg et al., 2019; Danby, Evaldsson, Melander, & Aarsand, 2018; McNaughton, Rosedale, Jesson, Hoda, & Teng, 2018). While negative consequences of violent video games have permeated the literature, De Simone (2013) showed that such games can have positive influences as well, such as promoting prosocial behaviors and learning about civics. Indeed, in addition to the fundamental affordance of play, which reflects the core of early learning curricula and is associated with cognitive and social learning, digital games provide young children with the opportunity to engage in joint problem-solving (Danby et al., 2018). McNaughton and colleagues (2018) also suggest that digital games offer unique affordances, such as automated feedback and collaborative play, to support intra- and interpersonal skills.

In addition to examining how specific types and content of apps influence young children's early social development, the context of use must be examined to better understand how best to integrate tablets into early learning settings. This not only includes studies examining how sharing tablets may present natural opportunities for peer engagement but also whether and how tablets might be used to enhance more targeted interventions for promoting social-emotional learning (for review of interventions, see Taylor, Oberle, Durlak, &Weissberg, 2017).

Furthermore, few studies have specifically examined whether 1:1 tablet use interferes with young children's social skills or positive relationships. While emerging work and policy recommendations in the home environment suggest mobile media can both promote (Roseberry et al., 2014; AAP, 2016) and interfere (McDaniel & Radesky, 2018) with child-family relationships, few have investigated this in early learning environments. The one case study that does exist showed 1:1 tablet use did not interfere with children's social interactions (Ralph, 2018), but no quantitative empirical work currently exists. Thus, while fears continue over the negative interference of mobile media on children's social skills and interactions, little evidence actually exists to support such concerns.

CONCLUSION

Deep-seated concerns over the negative potential of traditional technology (e.g., television, computers) to interfere with young children's social development has led to an even greater fear over mobile media. Despite more than half of U.S. early educators reporting access to tablets and near ubiquitous access to mobile devices in the home, there is a dearth of research examining if such concerns are warranted. Indeed, foundational theories and emerging research provide initial evidence that collaborative touch-screen tablet use in early childhood education can enable opportunities for young children to develop social skills and engage in positive, supportive relationships. Importantly, however, it may not be whether young children engage with mobile media but the content and context of such engagement that promotes or prohibits young children's social development.

REFERENCES

American Academy of Pediatrics. (2016). Media and young minds. *Pediatrics*, *138*(5). doi:10.1542/peds.2016-2591 PMID:27940793

Anderson, D. R., Huston, A. C., Schmitt, K. L., Linebarger, D. L., Wright, J. C., & Larson, R. (2001). Early childhood television viewing and adolescent behavior: The recontact study. *Monographs of the Society for Research in Child Development*, 154. PMID:11326591

Arnott, L. (2016). An ecological exploration of young children's digital play: Framing children's social experiences with technologies in early childhood. *Early Years*, *36*(3), 271–288. doi:10.1080/09575146.2016.1181049

Baker, R. S., & Gowda, S. M. (2018). *Towards understanding app effectiveness and cost*. San Francisco, CA: BrightBytes; Retrieved from http://pages.brightbytes.net/LOSnapappInfographic_LOCalculatorLP.html?utm_source=Snapapp&utm_campaign=LOReportInfographicadv

Bers, M. U. (2017). *Coding as a playground: Programming and computational thinking in the early childhood classroom*. Routledge. doi:10.4324/9781315398945

Beschorner, B., & Hutchison, A. C. (2013). iPads as a literacy teaching tool in early childhood. *International Journal of Education in Mathematics, Science, and Technology*, *1*(1), 16.

Blackwell, C. K. (2013). Teacher practices with mobile technology: integrating iPads into the early childhood classroom. *The Journal of Educational Research*, *7*(4), 1–25.

Blackwell, C. K. (2015, May). iPads in kindergarten: The effect of tablet computers on young children's academic achievement. *International Communication Association*, San Juan, Puerto Rico.

Blackwell, C. K., Lauricella, A. R., & Wartella, E. (2016). The influence of TPACK contextual factors on early childhood educators' tablet computer use. *Computers & Education*, *98*, 57–69. doi:10.1016/j.compedu.2016.02.010

Blackwell, C. K., Lauricella, A. R., Wartella, E., Robb, M., & Schomburg, R. (2013). Adoption and use of technology in early education: The interplay of extrinsic barriers and teacher attitudes. *Computers & Education, 69*, 310–319. doi:10.1016/j.compedu.2013.07.024

Blackwell, C. K., Wartella, E., Lauricella, A. R., & Robb, M. (2015). *Technology in the lives of educators and early childhood programs: Trends in access, use, and professional development from 2012 to 2014*. Center on Media and Human Development.

Blumberg, F. C., Deater-Deckard, K., Calvert, S. L., Flynn, R. M., Green, C. S., Arnold, D., & Brooks, P. J. (2019). Digital games as a context for children's cognitive development: research recommendations and policy considerations. *Social Policy Report, 32*(1), 1–33. doi:10.1002op2.3

Bohart, H., & Procopio, H. (2017). *Spotlight on young children: social and emotional development*. Washington, DC: NAEYC.

Bronfenbrenner, U. (1979). *The ecology of human development*. Harvard University Press.

Bronfenbrenner, U., & Morris, P. A. (2006). *The bioecological model of human development*. Handbook of child psychology.

Brown, J., Winsor, D. L., & Blake, S. (2012). Technology and social-emotional development in the early childhood environments. In S. Blake, D. Winsor, & L. Allen (Eds.), *Child development and the use of technology: perspectives, applications, and experiences* (pp. 112–129). Hersey, PA: IGI Global. doi:10.4018/978-1-61350-317-1.ch006

Campbell, F. A., Ramey, C. T., Pungello, E., Sparling, J., & Miller-Johnson, S. (2002). Early childhood education: Young adult outcomes from the Abecedarian Project. *Applied Developmental Science, 6*(1), 42–57. doi:10.1207/S1532480XADS0601_05

Chetty, R., Friedman, J. N., Hilger, N., Saez, E., Schanzenbach, D., & Yagan, D. (2011). How does your kindergarten classroom affect your earnings? Evidence from project STAR. *The Quarterly Journal of Economics, 126*(4), 1593–1660. doi:10.1093/qje/qjr041 PMID:22256342

Chou, C. C., Block, L., & Jesness, R. (2012). A case study of mobile learning pilot project in K12 schools. *Journal of Educational Technology Development and Exchange, 5*(2), 11–26. doi:10.18785/jetde.0502.02

Ciampa, K. (2014). Learning in a mobile age: An investigation of student motivation. *Journal of Computer Assisted Learning, 30*(1), 82–96. doi:10.1111/jcal.12036

Collaboration for academic, social, and emotional learning. (n.d.). Framework for systematic social and emotional learning. Chicago, IL: Author. Retrieved from https://casel.org/what-is-sel/

Collins, A., & Halverson, R. (2009). *Rethinking education in the age of technology: the digital revolution and schooling in America*. New York, NY: Teachers College Press.

Conti, G., Heckman, J. J., & Pinto, R. (2016). The effects of two influential early childhood interventions on health and healthy behaviour. *Economic Journal (London), 126*(596), F28–F65. doi:10.1111/ecoj.12420 PMID:28260805

Cordes, C., & Miller, E. (2000). *Fool's gold: A critical look at computers in childhood.* College Park, MD: Alliance for Childhood.

Coughlan, S. (2014). *Tablet computers in 70% of schools.* Retrieved from http://www.bbc.com/news/education/30216408

Couse, L. J., & Chen, D. W. (2010). A tablet computer for young children? Exploring its viability for early childhood education. *Journal of Research on Technology in Education, 43*(1), 75–96. doi:10.1080/15391523.2010.10782562

Danby, S., Evaldsson, A. C., Melander, H., & Aarsand, P. (2018). Situated collaboration and problem solving in young children's digital gameplay. *British Journal of Educational Technology, 49*(5), 959–972. doi:10.1111/bjet.12636

De Simone, J. J. (2013). What is good can also be bad: The prosocial and antisocial in-game behaviors of young video game players. *Atlantic Journal of Communication, 21*(3), 149–163. doi:10.1080/15456870.2013.801756

Dillenbourg, P. (1999). *Collaborative learning: Cognitive and computational approaches. Advances in learning and instruction series.* New York, NY: Elsevier Science.

Early Childhood Australia (ECA). (2018). *Statement on young children and digital technologies.* Canberra, ACT: ECA. Retrieved from http://www.earlychildhoodaustralia.org.au/wp-content/uploads/2018/10/Digital-policy-statement.pdf

Ertmer, P. A. (1999). Addressing first-and second-order barriers to change: Strategies for technology integration. *Educational Technology Research and Development, 47*(4), 47–61. doi:10.1007/BF02299597

Eteokleous, N. (2008). Evaluating computer technology integration in a centralized school system. *Computers & Education, 51*(2), 669–686. doi:10.1016/j.compedu.2007.07.004

Falloon, G. (2013). Young students using iPads: App design and content influences on their learning pathways. *Computers & Education, 68*, 505–521. doi:10.1016/j.compedu.2013.06.006

Falloon, G. (2015). What's the difference? Learning collaboratively using iPads in conventional classrooms. *Computers & Education, 84*, 62–77. doi:10.1016/j.compedu.2015.01.010

Fisch, S. M., & Truglio, R. T. (Eds.). (2001). *"G" is for growing: Thirty years of research on children and Sesame Street.* Mahwah, NJ: Lawrence Erlbaum Associates.

Fisher, B., Lucas, T., & Galstyan, A. (2013). The role of iPads in constructing collaborative learning spaces. *Technology, Knowledge, and Learning, 18*(3), 165–178. doi:10.100710758-013-9207-z

Fleer, M. (2018). Digital animation: New conditions for children's development in play-based setting. *British Journal of Educational Technology, 49*(5), 943–958. doi:10.1111/bjet.12637

Freeman, N. K., & Somerindyke, J. (2001). Social play at the computer: Preschoolers scaffold and support peers' computer competence. *Information technology in childhood education annual*, 1, 203–213.

Geist, E. (2014). Using tablet computers. *Young Children, 69*(1), 58–63.

Gomez, F., Nussbaum, M., Weitz, J. F., Lopez, X., Mena, J., & Torres, A. (2013). Co-located single display collaborative learning for early childhood education. *Computer-Supported Collaborative Learning, 8*(2), 225–244. doi:10.100711412-013-9168-1

Goodyear, P., Jones, C., & Thompson, K. (2014). Computer-supported collaborative learning: Instructional approaches, group processes and educational designs. In *Handbook of research on educational communications and technology* (pp. 439–451). New York, NY: Springer. doi:10.1007/978-1-4614-3185-5_35

Greaves, T., Hayes, J., Wilson, L., Gielniak, M., & Peterson, R. (2012). *Revolutionizing education through technology: The project RED roadmap for transformation.* Eugene, OR: International Society for Technology in Education.

Guernsey, L., Levine, M., Chiong, C., & Severns, M. (2012). *Pioneering literacy in the digital wild west: Empowering parents and educators.* Washington, DC: Campaign for Grade-Level Reading; Retrieved from http://www.joanganzcooneycenter.org/wp-content/uploads/2012/12/ GLR_TechnologyGuide_final.pdf

Habler, B., Major, L., & Hennessy, S. (2015). Tablet use in schools: a critical review of the evidence for learning outcomes. *Journal of Computer Assisted Learning, 32*(2), 138–156.

Hatzigianni, M., Gregoriadis, A., Karagiorgou, I., & Chatzigeorgiadou, S. (2018). Using tablets in free play: The implementation of the digital play framework in Greece. *British Journal of Educational Technology, 49*(5), 928–942. doi:10.1111/bjet.12620

Hatzigianni, M., & Kalaitzidis, I. (2018). Early childhood educators' attitudes and beliefs around the use of touchscreen technologies by children under three years of age. *British Journal of Educational Technology, 49*(5), 883–895. doi:10.1111/bjet.12649

Heckman, J. J., & Karapakula, G. (2019). *Intergenerational and Intragenerational Externalities of the Perry Preschool Project (No. w25889).* National Bureau of Economic Research. doi:10.3386/w25889

Herodotou, C. (2018). Young children and tablets: A systematic review of effects on learning and development. *Journal of Computer Assisted Learning, 34*(1), 1–9. doi:10.1111/jcal.12220

Hirsh-Pasek, K., Zosh, J. M., Golinkoff, R. M., Gray, J. H., Robb, M. B., & Kaufman, J. (2015). Putting education in "educational" apps: Lessons from the science of learning. *Psychological Science in the Public Interest, 16*(1), 3–34. doi:10.1177/1529100615569721 PMID:25985468

Hoge, E., Bickham, D., & Cantor, J. (2017). Digital media, anxiety, and depression in children. *Pediatrics, 140*(Supplement 2), S76–S80. doi:10.1542/peds.2016-1758G PMID:29093037

Howard, S. K., Yang, J., Ma, J., Maton, K., & Rennie, E. (2018). App clusters: Exploring patterns of multiple app use in primary learning contexts. *Computers & Education, 127*, 154–164. doi:10.1016/j.compedu.2018.08.021

Huber, B., Highfield, K., & Kaufman, J. (2018a). Detailing the digital experience: Parent reports of children's media use in the home learning environment. *British Journal of Educational Technology, 49*(5), 821–833. doi:10.1111/bjet.12667

Huber, B., Yeates, M., Meyer, D., Fleckhammer, L., & Kaufman, J. (2018b). The effects of screen media content on young children's executive functioning. *Journal of Experimental Child Psychology, 170*, 72–85. doi:10.1016/j.jecp.2018.01.006 PMID:29448235

James, C., Davis, K., Charmaraman, L., Konrath, S., Slovak, P., Weinstein, E., & Yarosh, L. (2017). Digital life and youth well-being, social connectedness, empathy, and narcissism. *Pediatrics, 140*(Supplement 2), S71–S75. doi:10.1542/peds.2016-1758F PMID:29093036

Johnson, R. T., & Johnson, D. W. (2008). Active learning: Cooperation in the classroom. *The annual report of educational psychology in Japan, 47*, 29-30.

Jones, D. E., Greenberg, M., & Crowley, M. (2015). Early social-emotional functioning and public health: The relationship between kindergarten social competence and future wellness. *American Journal of Public Health, 105*(11), 2283–2290. doi:10.2105/AJPH.2015.302630 PMID:26180975

Kirkorian, H. L., Choi, K., & Pempek, T. A. (2016). Toddlers' word learning from contingent and noncontingent video on touch screens. *Child Development, 87*(2), 405–413. doi:10.1111/cdev.12508 PMID:27018327

Kucirkova, N., Messer, D., Sheehy, K., & Flewitt, R. (2013). Sharing personalised stories on iPads: A close look at one parent–child interaction. *Literacy, 47*(3), 115–122. doi:10.1111/lit.12003

Larkin, K. (2011). You use! I use! We use! Questioning the orthodoxy of one-to-one computing in primary schools. *Journal of Research on Technology in Education, 44*(2), 101–120. doi:10.1080/1539 1523.2011.10782581

Lauricella, A. R., Blackwell, C. K., & Wartella, E. (2017). The "New" technology environment: The role of content and context on learning and development from mobile media. In R. Barr & D. Linebarger (Eds.), *Media exposure during infancy and early childhood* (pp. 1–23). Cham, Switzerland: Springer International Publishing. doi:10.1007/978-3-319-45102-2_1

LeBourgeois, M. K., Hale, L., Chang, A. M., Akacem, L. D., Montgomery-Downs, H. E., & Buxton, O. M. (2017). Digital media and sleep in childhood and adolescence. *Pediatrics, 140*(Supplement 2), S92–S96. doi:10.1542/peds.2016-1758J PMID:29093040

Lemke, C., Coughlin, E., & Reifsneider, D. (2009). *Technology in schools: What the research says: An update*. Culver City, CA: Commissioned by Cisco.

Lerner, R. M., Lerner, J. V., Almerigi, J. B., Theokas, C., Phelps, E., Gestsdottir, S., ... Smith, L. M. (2005). Positive youth development, participation in community youth development programs, and community contributions of fifth-grade adolescents: Findings from the first wave of the 4-H study of positive youth development. *The Journal of Early Adolescence, 25*(1), 17–71. doi:10.1177/0272431604272461

Lim, E. M. (2012). Patterns of kindergarten children's social interaction with peers in the computer area. *International Journal of Computer-Supported Collaborative Learning, 7*(3), 399–421. doi:10.100711412-012-9152-1

Lou, Y., Abrami, P. C., & d'Apollonia, S. (2001). Small group and individual learning with technology: A meta-analysis. *Review of Educational Research, 71*(3), 449–521. doi:10.3102/00346543071003449

Magee, C. A., Lee, J. K., & Vella, S. A. (2014). Bidirectional relationships between sleep duration and screen time in early childhood. *JAMA Pediatrics*, *168*(5), 465–470. doi:10.1001/jamapediatrics.2013.4183 PMID:24589672

Marsh, J., Plowman, L., Yamada-Rice, D., Bishop, J., Lahmar, J., & Scott, F. (2018). Play and creativity in young children's use of apps. *British Journal of Educational Technology*, *49*(5), 870–882. doi:10.1111/bjet.12622

Mashburn, A. J., Pianta, R. C., Hamre, B. K., Downer, J. T., Barbarin, O. A., Bryant, M., ... Howes, C. (2008). Measures of classroom quality in prekindergarten and children's development of academic, language, and social skills. *Child Development*, *79*(3), 832–749. doi:10.1111/j.1467-8624.2008.01154.x PMID:18489424

McCarrick, K., & Li, X. (2007). Buried treasure: The impact of computer use on young children's social, cognitive, language development and motivation. *AACE Journal*, *15*(1), 73–95.

McDaniel, B. T., & Radesky, J. S. (2018). Technoference: Longitudinal associations between parent technology use, parenting stress, and child behavior problems. *Pediatric Research*, *84*(2), 210–218. doi:10.103841390-018-0052-6 PMID:29895837

McManis, L. D., & Gunnewig, S. B. (2012). Finding the education in educational technology with early learners. *Young Children*, *67*(3), 14–24.

McNaughton, S., Rosedale, N., Jesson, R. N., Hoda, R., & Teng, L. S. (2018). How digital environments in schools might be used to boost social skills: Developing a conditional augmentation hypothesis. *Computers & Education*, *126*, 311–323. doi:10.1016/j.compedu.2018.07.018

Microsoft. (2018). The class of 2030 and life-ready learning: The technology imperative. Redmond, WA: Author. Retrieved from https://education.minecraft.net/wp-content/uploads/13679_EDU_Thought_Leadership_Summary_revisions_5.10.18.pdf

Mourlam, D. J., Strouse, G. A., Newland, L. A., & Lin, H. (2019). Can they do it? A comparison of teacher candidates' beliefs and preschoolers' actual skills with digital technology and media. *Computers & Education*, *129*, 82–91. doi:10.1016/j.compedu.2018.10.016

National Association for the Education of Young Children and the Fred Rogers Center. (2012). *Position statement: Technology and young children*. Washington, DC: Author. Retrieved from http://www.naeyc.org/content/technology-and-young-children

National Research Council and Institute of Medicine. (2000). From neurons to neighborhoods: The science of early childhood development. In J. P. Shonkoff, & D. A. Phillips (Eds.), Committee on integrating the science of early childhood development, board on children, youth, and families, commission on behavioral and social sciences. National Academy Press.

Neumann, M. M., & Neumann, D. L. (2014). Touch screen tablets and emergent literacy. *Early Childhood Education Journal*, *42*(4), 231–239. doi:10.100710643-013-0608-3

Ofcom. (2019). Children and parents: Media Use and Attitudes Report 2018. London, UK: Author.

Ofcom. (n.d.). Children and parents: Media use and attitudes report 2018.

Office of Head Start. (2015). *Head Start early learning outcomes framework – ages birth to five*. Washington, DC: U.S. Department of Health & Human Services, Administration for Children & Families. Retrieved from https://eclkc.ohs.acf.hhs.gov/sites/default/files/pdf/elof-ohs-framework.pdf

Otterborn, A., Schönborn, K., & Hultén, M. (2018). Surveying preschool teachers' use of digital tablets: General and technology education related findings. *International Journal of Technology and Design Education*, 1–21.

Paciga, K. A., & Donohue, C. (2017). *Technology and interactive media for young children: a whole child approach connecting the vision of Fred Rogers with research and practice*. Latrobe, PA: Fred Rogers Center for Early Learning and Children's Media at Saint Vincent College; Retrieved from https://www.fredrogerscenter.org/wp-content/uploads/2017/07/Technology-and-Interactive-Media-for-Young-Children.pdf

Papadakis, S., Kalogiannakis, M., & Zaranis, N. (2018). Educational apps from the Android Google Play for Greek preschoolers: A systematic review. *Computers & Education, 116*, 139–160. doi:10.1016/j.compedu.2017.09.007

Papert, S. (1993). *The children's machine: Rethinking school in the age of the computer*. New York, NY: BasicBooks.

Park, Y. (2011). A pedagogical framework for mobile learning: Categorizing educational applications of mobile technologies into four types. *The International Review of Research in Open and Distributed Learning, 12*(2), 78-102.

Pasnik, S. (Ed.). (2019). *Getting ready to learn: creating effective, educational children's media*. New York, NY: Routledge.

Pasnik, S., & Llorente, C. (2013). *Preschool teachers can use a PBS KIDS transmedia curriculum supplement to support young children's mathematics learning: Results of a randomized controlled trial. Summative Evaluation of the CPB-PBS Ready to Learn Initiative*. Waltham, MA: Education Development Center.

Penuel, W. R. (2006). Implementation and effects of one-to-one computing initiatives: A research synthesis. *Journal of Research on Technology in Education, 38*(3), 329–348. doi:10.1080/15391523.2006.10782463

Perlmutter, M., Behrend, S. D., Kuo, F., & Muller, A. (1989). Social influences on children's problem solving. *Developmental Psychology, 25*(5), 744–754. doi:10.1037/0012-1649.25.5.744

Pruet, P., Ang, C. S., & Farzin, D. (2016). Understanding tablet computer usage among primary school students in underdeveloped areas: Students' technology experience, learning styles and attitudes. *Computers in Human Behavior, 55*, 1131–1144. doi:10.1016/j.chb.2014.09.063

Ralph, R. (2018). Media and technology in preschool classrooms: manifesting prosocial sharing behaviours when using iPads. *Technology, Knowledge, and Learning, 23*(2), 199–221. doi:10.100710758-017-9342-z

Rideout, V. (2017). *The Common Sense census: Media use by kids age zero to eight* (pp. 263–283). San Francisco, CA: Common Sense Media.

Robinson, T. N., Banda, J. A., Hale, L., Lu, A. S., Fleming-Milici, F., Calvert, S. L., & Wartella, E. (2017). Screen media exposure and obesity in children and adolescents. *Pediatrics*, *140*(Supplement 2), S97–S101. doi:10.1542/peds.2016-1758K PMID:29093041

Roseberry, S., Hirsh-Pasek, K., & Golinkoff, R. M. (2014). Skype me! Socially contingent interactions help toddlers learn language. *Child Development*, *85*(3), 956–970. doi:10.1111/cdev.12166 PMID:24112079

Sandvik, M., Smørdal, O., & Østerud, S. (2012). Exploring iPads in practitioners' repertoires for language learning and literacy practices in kindergarten. *Nordic Journal of Digital Literacy*, *7*(03), 204–221.

Scherer, R., Siddiq, F., & Tondeur, J. (2019). The technology acceptance model (TAM): A meta analytic structural equation modeling approach to explaining teachers' adoption of digital technology in education. *Computers & Education*, *128*, 13–35. doi:10.1016/j.compedu.2018.09.009

Schweinhart, L. J., Montie, J., Xiang, Z., Barnett, W. S., Belfield, C. R., & Nores, M. (2011). *Lifetime effects: The High/Scope Perry Preschool study through age 40: Summary, 157 Conclusions, and Frequently Asked Questions*. Ypsilanti, MI: High/Scope Press; Retrieved from http://www.highscope.org/file/Research/PerryProject/specialsummary_rev2011_02_2.pdf

Shuler, C., Levine, Z., & Ree, J. (2012). *iLearn II: An analysis of the education category of Apple's App Store*. New York, NY: Joan Ganz Cooney Center. Retrieved from http://www.joanganzcooneycenter.org/wp-content/uploads/2012/01/ilearnii.pdf

Taylor, R. D., Oberle, E., Durlak, J. A., & Weissberg, R. P. (2017). Promoting positive youth development through school-based social and emotional learning interventions: A meta-analysis of follow-up effects. *Child Development*, *88*(4), 1156–1171. doi:10.1111/cdev.12864 PMID:28685826

Tootell, H., Plumb, M., Hadfield, C., & Dawson, L. (2013, January). Gestural interface technology in early childhood education: a framework for fully engaged communication. In *2013 46th Hawaii International Conference on System Sciences* (pp. 13-20). IEEE. 10.1109/HICSS.2013.241

Traxler, J. (2009). Current state of mobile learning. In M. Ally (Ed.), *Mobile learning: Transforming the delivery of education and training* (pp. 9–24). Edmonton, AB: AU Press.

Vaala, S., Ly, A., & Levine, M. H. (2015). *Getting a read on the app stores: A market scan and analysis of children's literacy apps*. New York, NY: The Joan Ganz Cooney Center at Sesame Workshop.

Vega, V., & Robb, M. B. (2019). *The Common Sense census: Inside the 21st-century classroom*. San Francisco, CA: Common Sense Media.

Vygotsky, L. (1978). *Mind in society*. Cambridge, MA: Harvard University Press.

Wartella, E., Beaudoin-Ryan, L., Blackwell, C. K., Cingel, D. P., Hurwitz, L. B., & Lauricella, A. R. (2016). What kind of adults will our children become? The impact of growing up in a media-saturated world. *Journal of Children and Media*, *10*(1), 13–20. doi:10.1080/17482798.2015.1124796

Wartella, E., Lauricella, A. R., & Blackwell, C. K. (2016). [Policy brief. Report for the Center on Media and Human Development School of Communication, Northwestern University.]. *The Ready To Learn Program*, *2010*, 2015.

Wartella, E., & Robb, M. (2007). Young children, new media. *Journal of Children and Media, 1*(1), 35–44. doi:10.1080/17482790601005207

Wartella, E. A., & Jennings, N. (2000). Children and computers: New technology. Old concerns. *The Future of Children, 10*(2), 31–43. doi:10.2307/1602688 PMID:11255707

Wood, D., Bruner, J. S., & Ross, G. (1976). The role of tutoring in problem solving. *Journal of Child Psychology and Psychiatry, and Allied Disciplines, 17*(2), 89–100. doi:10.1111/j.1469-7610.1976.tb00381.x PMID:932126

Yelland, N., & Masters, J. (2007). Rethinking scaffolding in the information age. *Computers & Education, 48*(3), 362–382. doi:10.1016/j.compedu.2005.01.010

Zaranis, N., Kalogiannakis, M., & Papadakis, S. (2013). Using mobile devices for teaching realistic mathematics in kindergarten education. *Creative Education, 4*(7), 1–10. doi:10.4236/ce.2013.47A1001

ADDITIONAL READING

Donohue, C. (Ed.). (2015). *Technology and Digital Media in the Early Years: Tools for Teaching and Learning. New York, NY: Routledge, and Washington.* DC: National Association for the Education of Young Children.

Ertmer, P. A., Ottenbreit-Leftwich, A. T., Sadik, O., Sendurur, E., & Sendurur, P. (2012). Teachers beliefs and technology integration practices: A critical relationship. *Computers & Education, 59*(2), 423–435. doi:10.1016/j.compedu.2012.02.001

Gülay Ogelman, H., Güngör, H., Körükçü, Ö., & Erten Sarkaya, H. (2018). Examination of the relationship between technology use of 5-6-year-old children and their social skills and social status. *Early Child Development and Care, 188*(2), 168–182. doi:10.1080/03004430.2016.1208190

Henderson, S., & Yeow, J. (2012). iPad in education: A case study of iPad adoption and use in a primary school. In *45th Hawaii International Conference on Systems Science* (HICSS–45 2012) Proceedings (pp. 78–87). Maui, HI: IEEE Computer Society. 10.1109/HICSS.2012.390

Lim, E. M. (2015). The factors influencing young children's social interaction in technology integration. *European Early Childhood Education Research Journal, 23*(4), 545–562. doi:10.1080/1350293X.2013.810484

Mercier, E. M., & Higgins, S. E. (2013). Collaborative learning with multi-touch technology: Developing adaptive expertise. *Learning and Instruction, 25*, 13–23. doi:10.1016/j.learninstruc.2012.10.004

Parette, H. P., Quesenberry, A. C., & Blum, C. (2010). Missing the Boat with Technology Usage in Early Childhood Settings: A 21st Century View of Developmentally Appropriate Practice. *Early Childhood Education Journal, 37*(5), 335–343. doi:10.100710643-009-0352-x

Parnell, W., & Barnett, J. (2012). iDocument: How Smartphones and Tablets Are Changing Documentation in Preschool and Primary Classrooms. *Young Children, 67*(3), 50–57.

Radesky, J. S., Schumacher, J., & Zuckerman, B. (2015). Mobile and interactive media use by young children: The good, the bad, and the unknown. *Pediatrics*, *135*(1), 1–3. doi:10.1542/peds.2014-2251 PMID:25548323

Zevenbergen, R. (2007). Digital natives come to preschool: Implications for early childhood practice. *Contemporary Issues in Early Childhood*, *8*(1), 19–29. doi:10.2304/ciec.2007.8.1.19

KEY TERMS AND DEFINITIONS

Developmentally-Appropriate Practice: early childhood education approach based on child development theories that accounts for a child's age, developmental status, and individual differences to support and promote early learning and development in all domains (cognitive, social, emotional, behavioral, physical).

Early Childhood Education: formal and informal educational programs/settings for children birth to age 8, including school-based (i.e., public and private pre-Kindergarten programs within a formal K-12 school system), center-based (i.e., for- or non-profit organizations not associated with a formal K-12 school system), and home-based childcare (i.e., providers care for 2 or more children in their home).

Extrinsic Barriers: structural limitations external to educators related to resources such as technology access, professional development, training, support, and time to learn/use technology.

Intrinsic Barriers: personal limitations internal to educators related to attitudes toward the perceived value of technology, confidence using technology, and general pedagogical beliefs.

Mobile Media: internet-enabled, interactive handheld technology devices (e.g., Smartphones, tablets) and associated platforms and software (e.g., apps) that can be used anytime, anywhere.

Social Skills: foundational competencies including communication, cooperation, social interactions and engagement with others.

Chapter 15
The Utilization of Van Hiele's Levels on the Instruction of English as a Foreign Language With the Aid of ICT

Elissavet Chlapana
University of Crete, Greece

Nicholas Zaranis
University of Crete, Greece

Evangelia Tzagkourni
University of Crete, Greece

ABSTRACT

This chapter investigates students' comprehension of the alphabet and its accompanying vocabulary, using an ICT-oriented teaching approach that is based upon Van Hiele's levels of geometric thought. The participating sample consisted of Greek students attending the second grade of primary school in Crete. The sample was divided into two groups, which were distributed among an experimental and a control group. The control group was taught following the traditional instruction, according to the pedagogical principles of Curriculum for the second grade of primary school. The experimental group was taught through an ICT-oriented intervention based upon Van Hiele's levels. The results of the research will assist greatly in achieving a more thorough understanding of the instruction of English as a Foreign Language (EFL) and possible implementation of Information and Communication Technologies (ICT), to maximize its effectiveness.

DOI: 10.4018/978-1-7998-1486-3.ch015

INTRODUCTION

The integration of ICT into primary education can play an essential role in achieving the objectives of the first and second-grade curriculum in all sectors and subjects if supported by developmentally appropriate software applications (Haugland, 1999; Lee, 2009). Research focusing on best practices for the integration of technology in Early Childhood Education (ECE) has shown that the use of Information and Communication Technologies (ICT) can lead to improvements in engagement, motivation, insistence, curiosity and attention of children of preschool and primary school age (Fokides & Mastrokoukou, 2018; Feehan, Shamir, & Yoder 2017; Keong, Horani, & Daniel, 2005; Schacter & Jo, 2017). Progress in digital technologies and technological devices dramatically increases the tools available to teachers and students, including primary education (Mango, 2015; Zaranis & Valla, 2017). Also, tablets have three new features (Kucirkova, 2014) with the ability to make a positive difference in initial training: they are portable and lightweight, they eliminate the need for separate input devices, such as the mouse and keyboard, and they host a series of apps, many of which have a child-friendly intuitive design. Regarding the pedagogical use of tablets in primary education, recent studies have concluded that tablets may be able to function as a valuable tool for educational use (Fokides & Atsikpasi, 2017; Fokides & Mastrokoukou, 2018), especially in learning mathematical concepts (Papadakis, Kalogiannakis, & Zaranis, 2018) which are currently correlated with the procedures of learning and acquisition of a second language (Bossé, Bayaga, Ringler, Fountain, & Young, 2018; Edmonds-Wathen, Owens, & Bino, 2019; Prior, Katz, Mahajna, & Rubinsten, 2015; Stacey, 2016). Hence, taking all the above into consideration, the authors propose an English language teaching (ELT) method, based upon an already existent method of geometric thought, reinforced with the principles of blended learning.

THEORETICAL BACKGROUND

ICTs' Educational Use and Blended Learning as Contemporary Practices

ICTs for both students and teachers are a cognitive explanatory tool, which given the usage of quality software, is capable to provide a dynamic learning environment with many opportunities of interaction and direct feedback allowing both individual and group learning (Tseles, Malafantis, & Pamouktsoglou, 2011). Additionally, the learning process is perceived as more manageable when ICTs are involved and they seem to reinforce already existent pedagogical factors, even though their sole existence doesn't seem to lead to cognitive improvement (Genlott & Grönlund, 2016). The ICTs can play an imperative educational role compensating for the major changes occurring, such as the rising numbers of students and the demand towards less rigid learning experiences (Eynon, 2005) with the incorporation of real-life experiences and motivational alternatives to the already existent practices, especially within the context of teaching English as a foreign language (Sharndama & IJemofwu, 2013). In this age of multimodality, which is referred as the combination of multiple literacies that support the new-founded needs of learning and teaching (Meng, 2016), teachers are called upon to question the practices of the past in regards to English language teaching and accept the ICTs as an educational tool exhibiting tremendous opportunities for "socially-situated rhetorical and pedagogical practices" (Grabill & Hicks, 2005, p. 303). Concerning what described above, blended learning constitutes a new instructional practice which can be defined as "the thoughtful integration of classroom face-to-face learning experiences with online learning experiences" (Graham 2013; Jani, Muszali, Nathan, & Abdullah, 2018, p.1032). This can happen simultaneously within the teaching process and it "combines multiple delivery media that are designed to complement each other and promote learning and application-learned behavior" (Singh, 2003, p. 52). Dziuban, Graham, Moskal, Norberg and Sicilia (2018) claim that blended learning is nowadays perceived as the "new traditional model" referring to Ross and Gage (2006) or as the "new normal" referring to Norberg et al. (2011). Multiple studies carried out on national level have exhibit that the use of blended learning leads to students' successful education performance and the development of feelings of satisfaction, as it is evaluated as a convenient means promoting positive interactions and autonomous learning by both students and teachers (Dziuban et. al, 2018; Güzer & Caner, 2014). The charm of blended learning lies in the fact it does not demand any radical turn away the traditional forms of learning, but it rather encourages their preservation enhanced by the ICTs (Nazarenko, 2015).

Mathematics and Foreign Language Teaching with the Use of ICT

Studies have shown that when computers are used in developmental ways in mathematics education, new opportunities for understanding mathematical concepts and processes are introduced (Calder, 2015), as they reinforce the conceptual and procedural knowledge of mathematics, understanding numeracy, counting, shape recognition and composition and sorting (Larkin, 2015). As mathematical activities using ICT combine monotonous aspects of mathematical learning instruction with animation, they attract the interest of young children, giving another dimension to the teaching of mathematics in the primary school classroom (Papadakis, Kalogiannakis, & Zaranis 2016). The benefits of using ICT in mathematics teaching include the application of higher level mathematical and geometric skills development, such as classification and numbering (Clements & Sarama, 2013; Zaranis & Valla, 2019).

More specifically, even though every individual elaborates geometric concepts in a very different way, we all seem to possess the ability to develop and evolve thinking within the context of geometry (Van de Walle, Karp, & Bay-Williams, 2004). Upon this idea, Van Hiele proposed his five leveled model, which focuses predominantly on geometric thought rather than knowledge per se. The model is divided into the following levels of Visualization, Analysis, Informal deduction, Formal deduction, and Rigor. In regards to Visualization the geometric shapes are apprehended holistically, therefore the students can recognize their different morphological characteristics; however, they are unable to pinpoint their more specific features. Moving to the level of Analysis, students can divide the geometric shapes in terms of their features in an empirical manner, based upon their observations and constructions. Moving on to Informal deduction, the students can define each geometric shape in regards to its features and realize their logical relations. Similarly, in Formal deduction, the students utilize induction to reform already given shapes' definitions. Moving on to the last level of Rigor, students are situated in the highest level of geometric thought, upon which they are capable to form a theory, without referring to any tangible geometric features of the shapes, but rather than to their specific relations (Kontogiannis & Ntziaxristos, 1999; Van de Walle et al., 2004). Alan Hoffer later proposed some modifications to reinforce each level, proposing five accompanying skills for each Van Hiele level, which are Inquiry/ Information, Directed Orientation, Explication, Free Orientation, and Integration. Inquiry/ Information focuses on the holistic recognition of shapes based upon observations. Directed Orientation aims to the design of a geometric shape, based upon its features. Explication reinforces previous experiences and students are encouraged to observe and converse about the apparent relations between shapes. Free orientation requests students to fulfill more complex tasks using their logic. And lastly, Integration aims to the review of the newly obtained knowledge about objects and relations (Crowley, 1987; Kontogiannis & Ntziaxristos, 1999; Zaranis, 2000).

Numerous studies are implementing the method above, some of which also utilize the ICTs. In a study conducted in Malaysia, year three students were taught geometry using Geometer Sketchpad in comparison to conventional instruction, both based upon Van Hiele's levels of geometric thinking. The experimental group exhibited significant improvement in regards to Van Hiele's levels of geometric thinking in comparison to the control group (Tieng & Eu, 2014). Similarly, in a study conducted in Greek first graders, Realistic Mathematics Education (RME) and the Van Hiele model in combination with the use of ICTs were implemented. The results of this study showed that the combination referred above led to significant improvement of the experimental group students' geometrical understanding, in comparison to the students of the control group (Zaranis, 2014). Nowadays, in Greece, the instruction of geometry concepts constitutes part of the official kindergarten curriculum about mathematics teaching. Taking this into consideration, the control group in a relevant study was taught geometry concepts only by traditional science instruction and the experimental group by instruction reinforced by the use of ICTs. Results showed a positive effect on geometry education on children of preschool age and highlighted the already known interactive role of ICTs (Zaranis, 2012).

Similarly, with regards to literacy instruction and foreign language teaching, ICTs seem to have claimed a vital role as supplementary means of providing knowledge which reinforces the ever-changing role of the teacher (Aslanabadi & Rasouli, 2013; Rahimi & Yadollahi, 2011). The learning process of a foreign language seems to happen both on a conscious and unconscious level during the use of ICTs due to the playful and pleasant experience they can offer as it leads to feelings of motivation and anticipation towards learning (Aslanabadi & Rasouli, 2013). In their study Dziuban et. al (2018) refer to Floridi (2014, 2008, 2014) who argues that humanity tends to perceive the world in terms of information rather than as "physical entities", and hence it is constantly evolving in terms of it, within the space of "infosphere" amplified by the ICTs . ICTs can play a major or supplementing role in English Language teaching, either when utilizing a software (Aslanabadi & Rasouli, 2013) or when playing the role of an essential educational tool (Coyle & Gómez Gracia, 2014), always targeting to a reinforced learning experience.

In the field of foreign language teaching, helping students develop spelling skills constitutes a challenge for many language educators. More particularly, spelling is a skill of imperative meaning as "it elucidates pupils' cognitive and linguistic strategies involved in literacy acquisition" and it facilitates writing and written expression (Niolaki & Masterson, 2012, p.757). Georgiou, Torppa, Manolitsis, Lyytinen, & Parrila (2012) claim that "letter knowledge is considered the single best predictor of reading acquisition" and more specifically, in regards to English, letter knowledge "provides children with a nameable referent to associate with phonemes" and it "reflects accuracy in the representations and discrimination of individual letters", both necessary to fulfill the process of recognizing a word effectively (p. 322). Referring to Koda (2005, 2008), Niolaki and Masterson (2012) state that there seems to be a transfer of certain cognitive and linguistic abilities that are employed within the literary process from L1 to L2.

Alphabetic languages display a difference in terms of their orthographic depth, meaning the consistency between sounds and letters, and their syllabic complexity (Andreou & Baseki, 2012; Georgiou et al., 2012; Niolaki & Masterson, 2012). In languages like English, a grapheme can have numerous pronunciations, whereas, in other languages, like Greek, there is an almost absolute correspondence between phoneme and grapheme. Greek is considered to be a rather transparent language in terms of reading, exhibiting little opaqueness in terms of spelling. In English case, it is considered an opaque language both in terms of reading and spelling (Niolaki & Masterson, 2012). In his research, Cook (1997) examined a variety of language speakers using English as a foreign language and found a large amount of errors regarding substitution of /e/ and /i/ in terms of spelling, while /y/ was frequently substituted with /i/ and /u/ with /e/ within the context of the word. Additionally, only Greek speakers that participated in the research confused /b/ with /d/, which were accustomed to the mechanics of the formation of the letter, as it constitutes a horizontal mirroring of a letter. Andreou and Baseki (2012) referring to existent literature, state that "spelling studies have determined that Greek children use phonological strategies at the phoneme and syllable level from the earliest stages, gradually augmented with morphological strategies to allow spelling of grammatical morphemes, and are sensitive also to morpheme frequency in applying such strategies" (p. 599).

Hence, taking all the above into consideration, one might come to the realization that geometrical understanding poses similarities to the understanding of a second language. Like geometry, English, which is nowadays considered to be a lingua franca, is surrounding us. However, although individuals have constant stimuli of it on a day to day basis, its official instruction begins within the school context. This equivalence supports the writers' decision to propose an already established method of geometric thought within the context of English language education, and more specifically in order to teach spelling and letters to primary students. This target group was carefully chosen, due to the fact that it is in its initial steps of English language learning, similar to a blank canvas, with no previous misinterpretations and misconceptions about the taught language. Moreover, the alphabet and spelling instruction were similarily chosen due to that. Students had little or no interaction with individual alphabet letters before the intervention. Additionally, the contemporary needs of such instruction are not limited to the instruction itself, but also to the selected means, in order to transmit knowledge. ICTs were implemented within the Van Hiele's level in order to result in a modernized proposed theory, capable of abiding and serving the contemporary students' educational needs in the most effective way.

Therefore, there seems to be a great need for studies utilizing the findings above and examining ways of implementing the ICTs in order to overcome such spelling difficulties. ICTs have proved to be a useful tool in altering false predisposed students' concepts (Fokides & Mastrokoukou, 2018; Tay, Lim, Lim, & Koh, 2012) and thus upon their implementation, issues as referred above, could be greatly improved.

A Correlation between EFL's and Mathematics' Teaching and Learning Processes

Mathematics teaching is currently correlated with processes taking place within the context of Foreign Language teaching (EFL). The use of two different languages seems to greatly influence the flexibility in the use of complex mathematical representations. Prior, Katz, Mahajna, and Rubinsten (2015) suggest a possible correlation between linguistic expression and learning mathematics or vice versa (Stacey, 2016). Both second language and mathematics/ geometry learning and teaching seem to be interrelated since they both contain similar processes of locating, interpreting and analyzing information (Edmonds-Wathen, Owens, & Bino, 2019; Rosa & Orey, 2015). Similarly, the absence of proficiency in a second language, in a context that its use was mandatory, seems to greatly affect the mathematical outcomes of the students (Robertson & Graven, 2018). These two disciplines, despite seemingly different, exhibit profound conceptual similarities, and thus the teaching methods utilized in either of them can benefit both their teaching and learning procedures (Bossé, Bayaga, Ringler, Fountain, & Young, 2018).

Van Hiele's method of geometric thought poses certain characteristics that could be implemented in literacy teaching, and in more specific, in foreign language instruction. To be more specific, referring to the first level of Van Hiele's model, namely Visualization, its description could easily be transformed so as to abide by the needs of literacy instruction, and more specifically in the instruction focusing on the recognition of the English alphabet letters. Hence, like geometric shapes, the individual letters of the English alphabet are apprehended as a whole within the context of a word. As the first Van Hiele level stresses, students of this level can tell the difference between each level, but they cannot yet distinguish independently the different phonological relations and alterations that can occur within the context of the word or when two letters are combined. Therefore, in this first level of Visualization, what matters is the recognition and acquisition of the ability to shape its individual letter, likewise shapes. Through this process of recognition, as it occurs within the lexical and textual context, students begin to realize and categorize the letters, based on their differences or similarities. In regards to Hoffer's skills describing this level, they can likewise be altered in order to describe skills that students can acquire in relation to literacy. To be more specific, in terms of literacy, the skill of Inquiry/ Information can focus on the holistic recognition of letters. Directed Orientation can aim to the phonological representation of each letter, within a lexical context. Explication, based upon the two previous skills, can target in reinforcing and encouraging the designing of letters. Free Orientation, similarly to its initial geometric goal, can request students to fulfill more complex tasks using their logic in order to categorize letters in regards to their phonological representation or use their preexistent knowledge in order to make logical decisions about letters. And lastly, Integration can aim to the review of the newly obtained knowledge about the letters and their relations, within the lexical and textual context.

Van Hiele's method of geometric thought presents all the necessary steps that can reinforce the learning process in all four learning skills, which is speaking, listening, writing and reading, in an effortless way. To be more specific, this way is characterized as effortless as it focuses primarily on the knowledge itself. Students' are encouraged to acquire a fully-fledged knowledge regarding the English alphabet in this first level of Recognition, by their movement from one Hoffer's skill to the other. This claim is almost idyllic, as most students of this proposed age, seem to have a sole holistic understand of the alphabet which later may pose difficulties in reading or spelling, as claimed above (Cook, 1997). With the incorporation of the ICTs, this movement through Hoffer's skills and the consecutive Van Hiele's levels is accomplished in a motivated and entertaining manner. Students' understanding of each letter comes effortlessly, making its spelling in different lexical context a much easier task. Thus, this proposed method, although it seems to be equally useful without the use of ICTs, it is greatly reinforced in numerous ways due to the use of them.

Following the theoretical framework presented above, the authors combined the curriculum of English teaching in the second grade of primary school and the use of ICT at a novice level and the authors designed a three-phase teaching method which is based upon Van Hiele's levels of geometric thought and the further alterations by Hoffer. The majority of previous studies regarding foreign language teaching examined various traditional teaching methods, such as ones that solely focus on grammar or pronunciation and listening skills, or more contemporary teaching methods which utilize online games or songs (e.g. Aslanabadi & Rasouli, 2013; McDonough & Shaw, 2012; Waters, 2012). However, a small number of studies (Basoz & Can, 2016, Coyle & Gómez Gracia, 2014; Kubota, 1999) have dealt with the initial grades of primary school and their focus was turned towards the implemented educational methods, such as the use of educational online games (Aslanabadi & Rasouli, 2013) or the use of songs (Coyle & Gómez Gracia, 2014) in order to teach English vocabulary.

In order to better understand this proposed method, a chronological review of the already existent ELT methods is rendered necessary. In his essay, Dinçay (2010) is briefly accumulating all existing

ELT methods, beginning with the Grammar – Translation Method, which was initially used in Latin and Greek instruction, and that was mainly focused on the study of the grammar of the taught language in the students' mother tongue. Continuing, the Direct Method followed when the previous method was found ineffective in communicative terms. In contrary to the previous, no mother tongue was allowed and translation was prohibited altogether. Grammar was approached inductively and literary texts were taught only for comprehensive reasons. Following the Direct Method, Audio-Lingual Method was solely focusing on the oral use of the target language, utilizing a series of drills and patterns. However, this method was soon found to be ineffective in the endeavor of the target language acquisition. The Silent Way soon followed, based upon the principles of the Cognitive Approach, where students were responsible for their learning and teachers were intervening when absolutely necessary. Desuggestopedia later was proposed, as a method that helped students surpass their fear of failure and hence reinforcing their foreign language learning, facilitated in a homely environment where role-playing was encouraged. Later on, Community Language Learning was presented and it was solely focused on the formation of interpersonal relationships between the teacher and students, which was based on a story reading within the classroom space and further discussion upon it. Total Physical Response followed and it was based on the coordination of speech and action, which means that the target language was given meaning through actions, which activated students' memory. The target language use was mainly orally used and the correction was in an unobtrusive manner. Lastly, Communicative Language Teaching was established, giving great emphasis to all four language skills that are listening, speaking, reading and writing. It mainly focused on the acquisition of meaning rather than the perfect development of grammatical structures and native-like pronunciation, which "means that successfully learning a foreign language is assessed in terms of how well learners have developed their communicative competence, which can loosely be defined as their ability to apply knowledge of both formal and sociolinguistic aspects of a language with adequate proficiency to communicate" (Dinçay, 2010, pp. 55-56). The current dominant English language teaching method, that arises from all the above is called Eclecticism and is an appropriation of all the existent methods in accordance with the context and audience taught. Hence, it is a method blending aspects of the already used practices abiding by their students' educational needs.

The purpose of this present research was based upon this notion of Eclecticism and: (a) the lack of research studies in which Van Hiele's theoretical model is used for foreign language teaching and (b) the restricted research data regarding teaching English as a foreign language to students that attend the first two classes of primary school within the Greek educational system. The intervention proposed in this study is designed specifically for the fulfillment of an educational goal, which is derived from the curriculum provided by the Greek Ministry of Education for second-grade students, and that is the letter recognition of the English alphabet. The writers' decision to utilize this suggested method in regards to teaching the English alphabet, was carefully made due to the vital role that alphabetical knowledge plays in the vocabulary learning process, which is on-going when learning a foreign language. Additionally, based on empirical knowledge, the authors pinpointed a known phenomenon that students of the examined age seem to exhibit. That is an almost mechanical by-heart learning of the English alphabet, which seems to be learned as a whole, rather than as individual letters. The present study aims to examine the effect of an intervention based on the first Van Hiele level of Recognition, with the utilization of the ICTs, on the skills of letter recognition in English language for children attending the second class of primary school, focusing on mainly non-consecutive letters and to letter pairs that students seem to confuse, based upon known literature. Specifically, the question addressed by this study was:

- Can an intervention based upon Van Hiele levels improve the recognition of the English Alphabet by Greek second grade students as it is compared to the traditional taught method?

Taking into consideration existent studies suggesting a link between English language and Mathematics teaching and learning and the proved dynamic role of ICTs within the spectrum of education, the authors can hypothesize that students' taught with a methodological approach which is based upon the Van Hiele's model with the use of ICTs will improve English alphabet knowledge in comparison to students' taught with traditional instruction.

METHODOLOGY

The sample included 53 students consisting of 34 girls and 19 boys aged eight, all of which attended the second grade of primary school. The study was carried out during the 2019 school grade in two public primary schools located in the city of Heraklion on the island of Crete (Greece). The classes that took part in this study belonged to two different schools. The students that participated in this present study were the ones that, according to teachers' estimation or any existent diagnose, did not pose any learning or language disability or difficulty. To investigate the aim of the present study an experimental research was conducted. In more particular it was compared to the teaching process of blended learning accompanied by ICTs with the traditional teaching which is based on the second grade's curriculum.

A pretest-posttest design was used consisted of an experimental group (n=25) and a control group (n=28). The experimental group had continuous exposure to the ICTs throughout the intervention period, as there were tablet computers available for daily use by them as a part of the teaching procedure. The control group, despite the fact that there was a computer available for use, did not use it for the needs of the English Language instruction. For the uniformity of the survey, instructions were given to the teachers who taught in the experimental and control groups.

First Phase

To isolate the effects of the treatment by looking for inherent inequities in the general intelligence of the two groups, in March of 2019 sample students were pretested in the Raven's Progressive Matrice (RPM) (Raven, 1956) which assesses non-verbal cognitive ability. The test is comprised of three progressively more difficult groups of twelve items, with every item presenting a different pattern with a missing component and six possible alternatives given below the pattern. In order to fulfill the test greater abilities of encoding and analyzing information is required. Thus, the test measures the cognitive processes of each student that are linked to reasoning and clear-thinking, without any support of preexistent knowledge and experience (Carpenter, Just, & Shell, 1990). The test was individually administered to each student.

Furthermore, in order to measure the variable of alphabet recognition, a researcher-based test *(Figure 1)* was designed which is based on an adapted edition of The Test of Early Reading Ability-Fourth Edition (TERA-4) was used (Reid et al., 2018). This adapted version incorporated also elements of activities used by the Cambridge English Language Assessment (Weir, Vidaković, & Galaczi, 2013) for the evaluation of letter recognition skills. This adapted test was formed in order to evaluate the skills acquired in the first Van Hiele level and the Hoffer levels accompanying it. Therefore, the students were asked to recognize six targeted letters in terms of skills of inquiry/information, directed orientation, explication,

Figure 1.

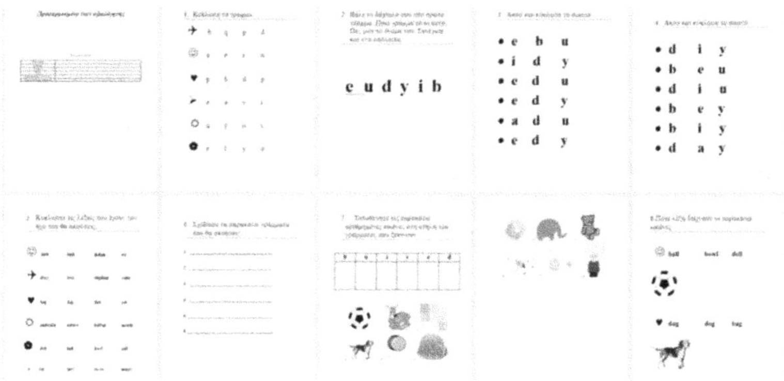

free orientation and integration, all of which were sectioned under the umbrella of the first Van Hiele level of Recognition. The examiner guided the examined students throughout the procedure using oral instructions and the test was administered individually to each student, lasting from 10 to 15 minutes. Similarly, after the intervention, the same test (adapted assessment test) was given to all the students in both the experimental and the control group as a post-test, at the beginning of June 2019. The aim of this post-examination was to measure students' improvement on letter recognition, in accordance with Van Hiele's first level of Recognition and Hoffer's proposed skills.

Second Phase

In the second phase, both groups were taught groups of letters that, according to existent literature, are proved to be perplexed by students. These groups were <b> - <d>, <e> - <i> - <u> - <y>. Group and individual activities were conducted to children of both groups. The teaching of the six selected pairs of letters was divided into six teaching hours for both groups. The control group was taught each letter in a teaching hour with a traditional instruction that follows the guidelines of the curriculum regarding English language teaching of the second class of the primary school. This instruction entailed activities that were focused on the songs provided as materials and their subsequent lyrics concerning the targeted letters, tracing exercises and reading drills.

In the experimental group, the same instructional material was covered according to the suggested teaching method. The instruction of the target letters was divided into six levels, based upon the first Van Hiele level of recognition and the six accompanying Hoffer's skills. In this first level, the letters were taught as a whole. The six levels were distributed within six school hours of four school days. All software activities were designed using Flash CS6 Professional Edition and were presented with tablet computers in the classroom. The intervention program in the experimental group was implemented as follows.

1st Day of Intervention – 45 Minutes

The intervention in the experimental group began with the utilization of a mascot that is familiar to the students and which invited them to "travel to the Letters' city". The letters targeted were placed on the whiteboard of the class on flashcards and they were presented as the little Misters and Misses that the students would get to know. After that, the teacher presented each letter with some certain asserting characteristics, for instance "Mr. <b> eats a lot of burgers, with great stress given to the initial letter of the word, and hence that's why he has a big belly in front of him". After this playful letter presentation, the teacher distributed photocopies of these letters to the students, so that they too had the chance to draw the letter assigned to them with characteristics of their choice. When they had fulfilled this activity, each student was asked to present the letter they had prepared. For example, the letter <d>was presented as a giraffe and it was drawn accordingly. Then, the students were asked to play the game "What's on my head". The students were divided into two groups. In each turn of the game, two different students stood in front of the whiteboard wearing special headbands made out of paper, on which the teacher placed a letter of her choice. The rest of the students had to name the letter on the head of the student belonging to their group and he/ she should write it on the whiteboard. The student that fulfilled the task would add a point to his/her group's rank, whereas otherwise he/she wouldn't. Therefore, the group that had collected the most points was the winner. These activities were both targeting the skills of inquiry/ information as described by Hoffer. Thus, the students were to recognize the targeted letters as a whole in regards to their name and design.

2nd Day of Intervention – 90 Minutes

The second day of intervention began with the distribution of tablets to the students. It is wise to say that the tablets were distributed to groups of two people, and thus every software activity was fulfilled at least twice by the group. This day's activities were focused on the acquisition of skills based upon directed orientation and explication, as described by Hoffer. Therefore, students were asked to fulfill activities that would result in the connection of the name with the grapheme of each letter, both individually but also in the context of a word.

The first activity requested students to choose the letter requested each time through oral instructions *(Figure 2)*. The students had to match the name of the letter requested to its "design". When they succeeded in finding the letter requested, they continue to the next screen, until the activity was fulfilled. The next activity requested students to repeat the procedure of linking the letter's name with its grapheme, but this time within the context of a letter sequence *(Figure 3)*. The oral instructions asked the students to drag three of the provided letters, six in total, from one place of the screen to the other, in a specific order. After that, the teacher asked the students to use a drawing application from their tablets, which gave them the opportunity both to draw but also to take photographs of the classroom around them and after that to draw the six targeted letters, using different colors and placing them however they liked within their photograph.

The last activity of the day requested from the students to design the letter that the teacher assigned them with oral instructions. After that, when the design and the drawing of the letters were done, the teacher asked the students to form combinations of letters, which they would present to the rest of the students. For instance, students with code-names B1, B4 and B7 had designed the letters <b>, <d>, <e> respectively and they presented them by moving to the front of the classroom as "We are <b>, <d>, <e>". The students were encouraged to refer to different characteristics of the letters that were given in the previous day's activity.

Figure 2.

Figure 3.

3rd Day of Intervention – 90 Minutes

This day of intervention aims at the development of skills of free orientation, which entails the ability to categorize words and letters in regards to their name and written representation. Hence, at this point, students were asked to perceive possible differences and similarities between the taught letters and use their logic in order to discern them within the context of a word *(Figure 4)*.

This day began with an activity on a tablet, which requested students to categorize given words in regards to their initial letter. In this case, the letter that is requested was given to them only orally. The students should drag the available words beginning with the requested letter to a certain space on the screen in order to continue to the next screen and to finally fulfill the activity.

After the end of this activity, the teacher distributed randomly three words to each student that began with one of the six letters-targets. Furthermore, she asked the students to find the words beginning with one of the target letters and stick them on the whiteboard. At the end of this activity, six columns of words were formed on the whiteboard. Six different students were asked to design the letter of each column. After the end of this activity, students were distributed with different colored paper on which they were asked to write whichever word they preferred and then draw the representation of the word. For example, in the case of the word elephant, the student wrote the word and drew an elephant.

Figure 4.

4th Day of Intervention – 45 Minutes

On the last day of the intervention, the students were requested to form skills of integration, which means to incorporate all the already acquired skills and apply them in the context of reality. This was succeeded with the use of an assessment activity *(Figure 5 & 6)* that revised the skills acquired in the previous days of the intervention and also added an extra activity of letter recognition in the context of words within the textual space. The texts used were songs known to the children, as they were a part of the curriculum.

After the end of this evaluating activity, students were asked to create the "City of Letters" that they had "visited", on a big cardboard. In this artistic activity, students were left to freely write either the target letters or words containing them within the landscape of the city.

Figure 5.

Figure 6.

Third Phase

During the third and final phase of the study, after the teaching intervention, the same test (adapted assessment test) was given to all students as a post-test in both the experimental and control group at the beginning of June 2019. This post-examination aimed to measure students' improvement on letter recognition in accordance with Van Hiele's first level of Recognition and Hoffer's skills.

Results

The analysis of the data was carried out using the SPSS statistical analysis computer program. Examining the equivalence of the groups regarding gender, a Pearson chi-square test was conducted. Results show showed that the two groups (see Table 1 for descriptives) did not differ significantly in terms of the students' gender, x^2 (1) = 2,889, p > 0,05.

Continually, the equivalence of the groups was examined regarding students' non-verbal cognitive ability and the recognition of target letters in the English language. Therefore, independent samples t-test tests were performed on students' pre-test scores in the Raven test (Raven, 1956) and in the test designed for measuring letter recognition in order to determine if the performance of both groups had roughly the same baseline in regards to these two variables. Results showed that the difference in students' scores in the Raven test (see Table 2 for descriptives) approaches significance, t (43.96) = -2,02, p=0,05. Moreover, results showed that the students' pretest scores in the letters recognition test (see Table 3 for descriptive) did not differ significantly between the two groups, t (51) = -1,601, p>0,05.

Subsequently, in order to examine the impact that the intervention program had on the recognition of the target letters in English language, independent sample t-test test was performed on students' post-test scores in the test designed for measuring letters recognition. Results showed that students in the

Table 1. Gender distribution among the two groups

Group		Gender		Total
		boy	girl	
experimental	f	6	19	25
	%	24	76	
control	f	13	15	28
	%	46.4	53.6	
Total		19	34	53
%		35.8	64.2	100

Table 2. Group statistics of pretest scores in Raven test

group	N	Mean	Std. Deviation	Std. Error Mean
experimental	25	24.44	1.734	0.347
control	28	25.79	3.011	0.569

Table 3. Group statistics of pretest scores in the letters recognition test

group	N	Mean	Std. Deviation	Std. Error Mean
experimental	25	20.44	9.609	1.922
control	28	24.57	9.167	1.732

Table 4. Group statistics of posttest scores in the letters recognition test

group	N	Mean	Std. Deviation	Std. Error Mean
experimental	25	35.84	5.48	1.09
control	28	28,14	9.66	1.83

experimental group outperformed students in the control group (see Table 4 for descriptives) and the difference in children's scores was statistically significant, $t(51) = 3{,}51$, $p<0{,}05$.

The results regarding the approximately significant difference between the two groups on students' performance in the measure of non-verbal cognitive ability lead us to examine further if this difference might have an effect on the impact that the intervention had on the improvement of letters recognition by students. In order to examine this hypothesis, an ANCOVA analysis was processed, setting the Raven test performance as a covariable and posttest scores in the letters recognition test as the dependent variable. Results showed a statistically significant main effect for type of intervention, $F(1, 50) = 15.319$, $p < .001$, Partial Eta Squared = 0.235, and an insignificant main effect for the Raven test performance, $F(1, 50) = 3.042$, $p > 0.005$, Partial Eta Squared = 0.057, on the post-test scores. Conclusively, the results showed that the impact that the intervention had in improving students' recognition of target letters in the English language remained even after controlling any possible differences in children's non-verbal cognitive ability.

DISCUSSION

The general purpose of this study was to investigate the impact of a proposed instructional intervention, which is based upon Van Hiele levels of geometric thought and Hoffer's accompanying skills, in the recognition of English alphabet letters by students attending the second grade of primary school. The results of this research suggest that the implementation of the proposed instructional model and the use of technological means in addition to conventional educational activities can have a significant effect in students' skills of letter recognition.

Our findings agree with the notion that Van Hiele levels' implementation can occur in fields other than geometry teaching and learning (Burger & Shaughnessy, 1986) and with studies that suggest that English Language Teaching can be greatly influenced by the use of ICTs (Aslanabadi & Rasouli, 2013; Basoz & Can, 2016; Coyle & Gómez Gracia, 2014; Kubota, 1999; McDonough & Shaw, 2012; Niolaki & Masterson, 2012; Waters, 2012). As a result, both the research question and hypothesis of this study were answered positively. Therefore, an intervention based upon Van Hiele levels can improve the recognition of the English Alphabet by Greek second grade students when compared to the traditional teaching method. To be more specific, the traditional English language instruction in Greece is based upon the given curriculum. However, very few, if any, instructors choose to follow that curriculum to the letter in the first two grades of primary school, concluding to ineffective instruction and poor performance. Moreover, despite the fact that most of the public schools in Greece include computer labs and ICTs facilities, a small percentage of instructors use them on a day to day basis, as they are considered to be time-consuming (Keong, Horani & Daniel, 2005). Hence, this proposed method comes to prove the apparent beneficial effect of ICTs, when utilized software designed for specific needs and purposes. Moreover, it poses a great contradiction to the already established traditional instruction, as its organized manner seems to greatly affect students' performance.

This can be attributed to the notion on which Van Hiele's method and Hoffer's skills are based upon; that is targeting the acquisition of knowledge. In addition to that, the notion of Eclecticism, which is allowing the utilization of already existent methods of ELT depending on the students' educational needs, seems to be equally efficient when alternative methods of ELT are utilized. More specifically, despite their previous use for purposes of geometric instruction, Hoffer's skills carefully lead the student through a series of necessary skills, which when acquired fully, can help him step from one Van Hiele's level to the other. This meticulous process seems to induce similar effects when used in English language teaching since geometry and mathematics seem to require similar skills, such as decoding, categorizing and recoding. These skills seem to be greatly reinforced when ICTs are utilized in both cases of geometry-mathematics and English language teaching and learning (Clements & Sarama, 2013; Zaranis & Valla, 2019). ICTs' role can be far from just an educational tool, as it highly motivates and excites students. Additionally, when incorporated in a blended learning environment, ICTs greatly strengthen students' positive attitudes towards conventional activities, which they previously find boring or unnecessary. Throughout the intervention, students that were labeled as indifferent prior to the study, by both the teachers and the English language teacher, participated actively. Students' seemed to be enthralled with the idea of using ICTs as a part of their learning process, which encouraged them to be focused and motivated towards it. Our decision to incorporate ICTs in a blended learning environment was an endeavor to deliver multiple chances of learning and applications of it, with means both novel and familiar to the students. Our goal was not to examine solely the educational value of ICTs, but rather to examine its dynamic when incorporated in an environment that combines traditional means of learning, organized upon a method characterized by consecutive levels of learning.

CONCLUSION

This study highlights the need of incorporation of technological means in education, in order to achieve a great impact in learning (Calder, 2015) which, when utilized as a necessary educational tool, leads away from the dated behavioristic teacher-centered classroom (Fokides & Mastrokoukou, 2018; Mango, 2015). The authors approached the ideas above by implementing Van Hiele's model of geometric thought. This model proposes the existence of five levels of thought which are dynamic and constant, exhibiting learning and teaching characteristics that can be utilized in numerous fields (Burger & Shaughnessy, 1986). Utilizing the referred model, the authors designed educational software targeting the improvement of letters' recognition.

The results of the research, as described above, can shed some light on the major issue of letter recognition of the English alphabet by students attending the Greek public primary schools, and more specifically how it can be resolved even from the first two grades of primary school. Nonetheless, there are certain limitations to the present study. Despite the fact that all the necessary precautions were taken, one cannot surely state that the test has accurately recorded the students' knowledge. Moreover, the data were collected from residents of Heraklion, which poses another limitation for this study, as well as the small size of the sample. This study was also influenced by time constraints imposed by the schools, which limited the duration of the intervention. Furthermore, there was no assessment regarding the teachers' skills of implementation the suggested method and neither regarding students' demographic characteristics that might impact their prior performance. Owing to the small size of the sample and the specific context of the study, any application of the findings should be done with caution.

Further studies are needed in which the suggested method will be implemented in different contexts for a larger time period and with a bigger sample. Additionally, this present study could be conducted implementing different Van Hiele levels in different age groups, in order to teach English, so as to investigate whether all the levels of Van Hiele's method could be utilized for the English instruction and whether there are consistent positive results when different levels are utilized for different educational purposes. Moreover, additional data collection tools can be used, such as teachers' and students' interviews, which could shed light on their personal opinions prior to and while using ICTs for educational uses. Additionally, such interviews could reveal possible issues they face while teaching using the traditional instruction which is foreseen by the primary school curriculum and thus understand better their students' educational needs. It would also be of interest to implement the present suggested method without the use of ICTs, in order to determine whether the results of this study can be attributed to the combination of the method and the instrument or to solely one of the two.

Regarding the educational value of this present study, its findings can be taken into account by a variety of stakeholders such as students, teachers, researchers, curriculum designers. More specifically, this proposed method could be set up as a broad range of study in order to examine whether Van Hiele method and Hoffer skills linked to it, could be implemented in the foreign language teaching and learning, in order to resolve possible obstacles and issues that hinder knowledge improvement and development. The results of this research can be extended by developing various similar studies in English language teaching in the first classes of primary education.

Nonetheless, taking all limitations into consideration, the experimental data reinforced our view that the ICTs' education value is great and it can impressively influence learning, both educationally but also in regards to students' and teachers' behavior towards learning and teaching respectively. Students' motivation and interest was apparently maximized, reinforcing the educational goals of the method proposed.

REFERENCES

Andreou, G., & Baseki, J. (2012). Phonological and spelling mistakes among dyslexic and non-dyslexic children learning two different languages: Greek vs English. *Psychology (Irvine, CA.)*, *3*(08), 595–600. doi:10.4236/psych.2012.38089

Aslanabadi, H., & Rasouli, G. (2013). The effect of games on improvement of Iranian EFL vocabulary knowledge in kindergartens. *International Review of Social Sciences and Humanities*, *6*(1), 186–195.

Basoz, T., & Can, D. T. (2016). The effectiveness of computers on vocabulary learning among preschool children: A semiotic approach. *Cypriot Journal of Educational Sciences*, *11*(1), 2–8. doi:10.18844/cjes.v11i1.266

Bossé, M. J., Bayaga, A., Ringler, M. C., Fountain, C., & Young, E. S. (2018). Acquiring math: Connecting math learning and second language acquisition. *International Journal for Mathematics Teaching & Learning*, *19*(2), 223–252.

Burger, W. F., & Shaughnessy, J. M. (1986). Characterizing the Van Hiele levels of development in geometry. *Journal for Research in Mathematics Education*, *17*(1), 31–48. doi:10.2307/749317

Calder, N. (2015). Apps: Appropriate, applicable, and appealing? In T. Lowrie, & R. Jorgensen Zevenbergen (Eds.), *Digital games and mathematics learning. Mathematics Education in the Digital Era* (pp. 233–250). Dordrecht, The Netherlands: Springer. doi:10.1007/978-94-017-9517-3_12

Carpenter, P. A., Just, M. A., & Shell, P. (1990). What one intelligence test measures: A theoretical account of the processing in the Raven Progressive Matrices Test. *Psychological Review*, *97*(3), 404–431. doi:10.1037/0033-295X.97.3.404 PMID:2381998

Clements, D. H., & Sarama, J. (2013). Rethinking early mathematics: What is research based curriculum for young children? In L. D. English, & J. T. Mulligan (Eds.), *Reconceptualizing early mathematics learning* (pp. 121–147). The Netherlands: Springer. doi:10.1007/978-94-007-6440-8_7

Cook, V. J. (1997). L2 users and English spelling. *Journal of Multilingual and Multicultural Development*, *18*(6), 474–488. doi:10.1080/01434639708666335

Coyle, Y., & Gómez Gracia, R. (2014). Using songs to enhance L2 vocabulary acquisition in preschool children. *ELT Journal*, *68*(3), 276–285. doi:10.1093/elt/ccu015

Crowley, M. L. (1987). The Van Hiele model of the development of geometric thought. In M. M. Lindquist (Ed.), *Learning and teaching geometry, K-12* (pp. 1–16). Reston, VA: NCTM.

Dinçay, T. (2010). A quick chronological review of the ELT methods along with their techniques and principles: choosing eclecticism from among language teaching methods. *Dil Dergisi*, (147), 40-62.

Dziuban, C., Graham, C. R., Moskal, P. D., Norberg, A., & Sicilia, N. (2018). Blended learning: The new normal and emerging technologies. *International Journal of Educational Technology in Higher Education*, *15*(1), 3. doi:10.118641239-017-0087-5

Edmonds-Wathen, C., Owens, K., & Bino, V. (2019). Identifying vernacular language to use in mathematics teaching. *Language and Education*, *33*(1), 1–17. doi:10.1080/09500782.2018.1488863

Eynon, R. (2005, April). The use of the internet in higher education: Academics' experiences of using ICTs for teaching and learning. [Emerald Group Publishing Limited.]. *Aslib Proceedings*, *57*(2), 168–180. doi:10.1108/00012530510589137

Feehan, K., Shamir, H., & Yoder, E. (2017). Does CAI improve early math skills? In *Proceedings of the 9th International Conference on Computer Supported Education* (CSEDU 2017), 2, (pp. 285-292), Porto.

Fokides, E., & Atsikpasi, P. (2017). Tablets in education. Results from the initiative ETiE, for teaching plants to primary school students. *Education and Information Technologies*, *22*(5), 2545–2563. doi:10.100710639-016-9560-3

Fokides, E., & Mastrokoukou, A. (2018). Human anatomy, tablets, and primary school students. Results from a pilot study. *Contemporary Educational Technology*, *9*(2), 154–170. doi:10.30935/cet.414808

Genlott, A. A., & Grönlund, Å. (2016). Closing the gaps–Improving literacy and mathematics by ICT-enhanced collaboration. *Computers & Education*, *99*, 68–80. doi:10.1016/j.compedu.2016.04.004

Georgiou, G. K., Torppa, M., Manolitsis, G., Lyytinen, H., & Parrila, R. (2012). Longitudinal predictors of reading and spelling across languages varying in orthographic consistency. *Reading and Writing*, *25*(2), 321–346. doi:10.100711145-010-9271-x

Grabill, J. T., & Hicks, T. (2005). Multiliteracies meet methods: The case for digital writing in English education. *English Education*, *37*(4), 301–311.

Graham, C. R. (2013). Emerging practice and research in blended learning. In M. G. Moore (Ed.), *Handbook of distance education* (3rd ed., pp. 333–350). New York: Routledge. doi:10.4324/9780203803738.ch21

Güzer, B., & Caner, H. (2014). The past, present and future of blended learning: An in depth analysis of literature. *Procedia: Social and Behavioral Sciences*, *116*, 4596–4603. doi:10.1016/j.sbspro.2014.01.992

Haugland, S. (1999). What role should technology play in young children's learning? *Young Children*, *54*(9), 26–30.

Jani, J., Muszali, R., Nathan, S., & Abdullah, M. S. (2018). Blended learning approach using Frog VLE platform towards students' achievement in teaching games for understanding. *Journal of Fundamental and Applied Sciences*, *10*, 1130–1141.

Keong, C. C., Horani, S., & Daniel, J. (2005). A study on the use of ICT in mathematics teaching. *Malaysian Online Journal of Instructional Technology*, *2*(3), 43–51.

Kontogiannis, D., & Ntziaxristos, B. (1999). *Basic geometry notions* (3rd ed.). Athens.

Kubota, R. (1999). Japanese culture constructed by discourses: Implications for applied linguistics research and ELT. *TESOL Quarterly*, *33*(1), 9–35. doi:10.2307/3588189

Kucirkova, N. (2014). iPads in early education: Separating assumptions and evidence. *Frontiers in Psychology*, *5*(715), 1–3. PMID:25071678

Lee, Y. (2009). Pre-K children's interaction with educational software programs: an observation of capabilities and levels of engagement. *Journal of Educational Multimedia and Hypermedia*, *18*(3), 289–309.

Mango, O. (2015). iPad use and student engagement in the classroom. *The Turkish Online Journal of Educational Technology*, *14*, 53–57.

McDonough, J., & Shaw, C. (2012). *Materials and Methods in ELT*. John Wiley & Sons.

McManis, L. D., & Gunnewig, S. B. (2012). Finding the education in educational technology with early learners. *YC Young Children*, *67*, 14–24.

Meng, X. (2016). *A pedagogy of multiliteracies into practice: a case study in one grade one literacy classroom* (Doctoral dissertation, The University of Western Ontario).

Nazarenko, A. (2015). Blended learning vs traditional learning: what works? (a case study research). *Procedia: Social and Behavioral Sciences*, *200*, 77–82. doi:10.1016/j.sbspro.2015.08.018

Niolaki, G. Z., & Masterson, J. (2012). Transfer effects in spelling from transparent Greek to opaque English in seven-to-ten-year-old children. *Bilingualism: language and cognition*, *15*(4), 757–770. doi:10.1017/S1366728911000721

Norris, C., Hossain, A., & Soloway, E. (2012). Under what conditions does computer use positively impact student achievement? In P. Resta (Ed.), *Supplemental vs. essential use. 2021–2028*. Austin, TX: AACE.

Papadakis, S., Kalogiannakis, M., & Zaranis, N. (2016). Improving mathematics teaching in kindergarten with realistic mathematical education. *Early Childhood Education Journal, 45*(3), 369–378. doi:10.100710643-015-0768-4

Papadakis, S., Kalogiannakis, M., & Zaranis, N. (2018). Educational apps from the Android Google Play for Greek preschoolers: A systematic review. *Computers & Education, 116*, 139–160. doi:10.1016/j.compedu.2017.09.007

Prior, A., Katz, M., Mahajna, I., & Rubinsten, O. (2015). Number word structure in first and second language influences arithmetic skills. *Frontiers in Psychology, 6*, 266. doi:10.3389/fpsyg.2015.00266 PMID:25852591

Rahimi, M., & Yadollahi, S. (2011). ICT use in EFL classes: A focus on EFL teachers' characteristics. *World Journal of English Language, 1*(2), 17. doi:10.5430/wjel.v1n2p17

Raven, J. H. (1956). *Guide to using the coloured progressive matrices. Sets A, AB, B*. London, UK: H. K. Lewis.

Reid, D. K., Hresko, W. P., & Hammill, D. D. (2018). *Test of early reading ability* (4th ed.). Austin, TX: PRO-ED.

Robertson, S. A., & Graven, M. (2018). Using a transdisciplinary framework to examine mathematics classroom talk taking place in and through a second language. *ZDM, 50*(6), 1013–1027. doi:10.100711858-018-0952-2

Rosa, M., & Orey, D. C. (2015). A trivium curriculum for mathematics based on literacy, matheracy, and technoracy: An ethnomathematics perspective. *ZDM, 47*(4), 587–598. doi:10.100711858-015-0688-1

Schacter, J., & Jo, B. (2017). Improving preschoolers' mathematics achievement with tablets: A randomized controlled trial. *Mathematics Education Research Journal, 29*(3), 1–15. doi:10.100713394-017-0203-9

Sharndama, E. C., & IJemofwu, I. A. (2013). Application of ICTs in teaching and learning English (ELT) in large classes. *Journal of Arts and Humanities, 2*(6), 34–39.

Singh, H. (2003). Building effective blended learning programs. *Educational technology-Saddle Brook then Englewood Cliffs NJ, 43*(6), 51–54.

Tay, L. Y., Lim, S. K., Lim, C. P., & Koh, J. H. L. (2012). Pedagogical approaches for ICT integration into primary school English and mathematics: A Singapore case study. *Australasian Journal of Educational Technology, 28*(4), 740–754. doi:10.14742/ajet.838

Tieng, P. G., & Eu, L. K. (2014). Improving students' Van Hiele level of geometric thinking using Geometer's Sketchpad. *Malaysian Online Journal of Educational Technology, 2*(3), 20–31.

Tseles, D., Malafantis, K., & Pamouktsoglou, A. (2011). *Education, Science, and Society*. Athens, Greece: Sygxroni Ekdotiki.

Van de Walle, J. A., Karp, K. S., & Bay-Williams, J. M. (2004). *Elementary and middle school mathematics*. Boston, MA: Allyn and Bacon.

Waters, A. (2012). Trends and issues in ELT methods and methodology. *ELT Journal, 66*(4), 440–449.

Weir, C. J., Vidaković, I., & Galaczi, E. D. (2013). *Measured constructs: A history of Cambridge English examinations*. Cambridge, UK: Cambridge University Press.

Zaranis, N. (2000). *The utilization of the Van Hiele method in compulsory education's geometry instruction with the computer's aid* (I axiopiisi tis theorias Van Hiele sti didaskalia tis geometrias stin ipoxreotiki ekpedeusi) (Doctoral dissertation). National and Kapodistrian University of Athens (NKUoA). Athens.

Zaranis, N. (2012). The use of ICT in preschool education for geometry teaching. In *Proceedings of the 10th international conference on computer based learning in science, learning science in the society of computers*, 256-262. Barcelona, Spain: Centre for Research in Science and Mathematics Education (CRECIM).

Zaranis, N. (2014). Geometry teaching through ICT in primary school. In C. Karagiannidis, P. Politis, & I. Karasavvidis (Eds.), *Research on e-learning and ICT in education* (pp. 59–71). New York, NY: Springer.

Zaranis, N., & Valla, V. (2017). Tablet computer assisted counting and calculating activities for kindergarten children. In L. Gómez Chova, A. López Martínez, & I. Candel Torres (Eds.), *EDULEARN17 Proceedings, 9th International Conference on Education and New Learning Technologies* (pp. 9680-9689), July 3-5, 2017 — Barcelona, Spain. 10.21125/edulearn.2017.0828

Zaranis, N., & Valla, V. (2019). Tablets in learning mathematics for kindergarten students. In D. Linda (Ed.), Didactics of smart pedagogy – smart pedagogy for technology enhanced learning (pp. 267–284). Nature Switzerland AG: Springer. doi:10.1007/978-3-030-01551-0_14

ADDITIONAL READING

Chlapana, E. (2016). An intervention programme for enhancing kindergarteners' cognitive engagement and comprehension skills through reading informational texts. *Literacy, 50*(3), 125–132. doi:10.1111/lit.12085

Chlapana, E., & Tafa, E. (2014). Effective practices to enhance immigrant kindergarteners' second language vocabulary learning through storybook reading. *Reading and Writing, 27*(9), 1619–1640. doi:10.100711145-014-9510-7

Dinçay, T. (2010). A quick chronological review of the ELT methods along with their techniques and principles: choosing eclecticism from among language teaching methods. *Dil Dergisi*, (147), 40-62.

Kalogiannakis, M., Ampartzaki, M., Papadakis, St., & Skaraki, E. (2018). Teaching Natural Science Concepts to Young Children with Mobile Devices and Hands-on Activities. A Case Study. *International Journal of Teaching and Case Studies, 9*(2), 171–183. doi:10.1504/IJTCS.2018.090965

Nazarenko, A. (2015). Blended Learning vs Traditional Learning: What Works? (A Case Study Research). *Procedia: Social and Behavioral Sciences, 200*, 77–82. doi:10.1016/j.sbspro.2015.08.018

Papadakis, St., & Kalogiannakis, M. (2017). Combining mobile technologies in environmental education: A Greek case study. International Journal of Mobile Learning and Organisation, (Special Issue on Mobile Learning Applications and Strategies), 11(3), 256-277. DOI: . doi:10.1504/IJMLO.2017.10003925

Papadakis, St., & Kalogiannakis, M. (2017). Mobile educational applications for children. What educators and parents need to know. [Special Issue on Mobile Learning Applications and Strategies]. *International Journal of Mobile Learning and Organisation, 11*(3), 256–277. doi:10.1504/IJMLO.2017.085338

Papadakis, S., Kalogiannakis, M., & Zaranis, N. (2016). Comparing tablets and PCs in teaching mathematics: An attempt to improve mathematics competence in early childhood education. *Preschool and Primary Education, 4*(2), 241–253. doi:10.12681/ppej.8779

Papadakis, S., Kalogiannakis, M., & Zaranis, N. (2018). The effectiveness of computer and tablet assisted intervention in early childhood students' understanding of numbers. An empirical study conducted in Greece. *Education and Information Technologies, 23*(5), 1849–1871. doi:10.100710639-018-9693-7

Papadakis, S., Kalogiannakis, M., & Zaranis, N. (2019). *Parental involvement and attitudes towards young Greek children's mobile usage*. International Journal of Child-Computer Interaction; doi:10.1016/j.ijcci.2019.100144

Zaranis, N. (2000). *The utilization of the Van Hiele method in compulsory education's geometry instruction with the computer's aid* (I axiopiisi tis theorias Van Hiele sti didaskalia tis geometrias stin ipoxreotiki ekpedeusi) (Doctoral dissertation). National and Kapodistrian University of Athens (NKUoA). Athens.

Zaranis, N. (2014). Geometry Teaching Through ICT in Primary School. In C. Karagiannidis, P. Politis, & I. Karasavvidis (Eds.), *Research on e-Learning and ICT in Education* (pp. 59–71). New York, NY: Springer.

Zaranis, N., & Valla, V. (2017). Tablet computer assisted counting and calculating activities for kindergarten children. In L. Gómez Chova, A. López Martínez, & I. Candel Torres (Eds.), *EDULEARN17 Proceedings, 9th International Conference on Education and New Learning Technologies* (pp. 9680-9689), July 3rd-5th, 2017 — Barcelona, Spain. 10.21125/edulearn.2017.0828

KEY TERMS AND DEFINITIONS

Blended Learning: a style of education in which students learn via electronic and online media as well as traditional face-to-face teaching.

ECE: Early childhood education is a branch of education theory that relates to the teaching of children (formally and informally) from birth up to the age of eight.

ELT: ELT is the teaching of English to people whose first language is not English. ELT is an abbreviation for 'English Language Teaching.

ICTs: Stands for "Information and Communication Technologies." ICT refers to technologies that provide access to information through telecommunications. This includes the Internet, wireless networks, cell phones, and other communication mediums.

Primary School: A school for children between the ages of about five and eleven.

Tablets computers: A small portable computer that accepts input directly on to its screen rather than via a keyboard or mouse.

Van Hiele's Levels: The Van Hiele model is a theory that describes how students can learn geometry and it is divided into levels.

Chapter 16
Fostering Computational Thinking and Creativity in Early Childhood Education:
Play-Learn-Construct-Program-Collaborate

Katerina V. Glezou
Arsakeia Schools Psychikou, Greece

ABSTRACT

This chapter focuses on the design and implementation of coding, programming, and robotics activities in real-world classroom conditions in early childhood education that foster computational thinking and creativity. The study extends the author's experience in designing learning environments through experimentation, exploration, expression, construction, meaning negotiation, and collaboration. The author tries to gain further insight into the potential of implementing unplugged activities, robotics systems, and programming environments in the context of multi-disciplinary, hands-on activities. Using a variety of tools, strategies, and methods as well as qualitative approaches employed in case studies of classroom practice during Afternoon Robotics Club courses, the findings highlight the overall process indicating some special features that contribute to or cause difficulty in the formation of an effective learning experience. Special issues of interest, key aspects, solutions, and recommendations, as well as future research directions, are discussed.

INTRODUCTION

Computational Thinking (CT) and STEAM (Science, Technology, Engineering, Art, and Math) education have received growing attention over the last decade within all school levels. However, the introduction and exploitation of appropriate unplugged activities, educational robotics systems and digital learning environments in the classroom still remains an open, composite and multifactoral issue.

DOI: 10.4018/978-1-7998-1486-3.ch016

Learning to think 'computationally' has long been recognized as an important issue (Papert, 1980). In recent years, the CT concept has regained increased significance and has been considered a fundamental skill for all along with reading, writing and arithmetic (Wing, 2006).

Nowadays, both in the international and Greek educational community, increasing efforts are focused on the integration and utilization of educational robotics and the STEAM approach in teaching practice as a vehicle to develop computational thinking and creativity. Robotics is emerging as an important tool in the teaching-learning process, enabling the formation of a collaborative environment and student-centered constructionist teaching practices, and giving emphasis on the active engagement and building of personal meaning by students. The involvement of students and teachers in robotics and STEAM activities contributes to the development of useful skills needed in the 21st century, which is characterized by an increased demand for high-tech applications, as concepts and values, with computational thinking, problem solving, creativity, and team collaboration taking priority over others.

The term "21st century skills" is used to refer to a range of skills, abilities and competencies that are considered to be required for success in 21st century society and workplaces in a constantly evolving technological landscape. In order to succeed in the 21st century, students must learn much more than the "three Rs" (: reading, writing, arithmetic) and basic computer competency. The Partnership for 21st Century Skills (now the Partnership for 21st Century Learning, or P21) organization identified deeper learning competencies and skills that they called the "Four Cs" of 21st century learning: Collaboration, Communication, Critical thinking, Creativity.

Although today's students belong to digital natives, it becomes mandatory for teachers to develop students' digital competencies, with particular care to the development of computational thinking. There is a widespread recognition of the need to train teachers - digital immigrants in the vast majority - regarding coding, programming and robotics activities in classroom settings.

This chapter presents a constructionist approach to develop computational thinking and creativity by engaging in programming and robotics activities in the early childhood classroom. This chapter describes teaching interventions that have been implemented in real conditions to kindergarten and elementary school students during the Afternoon Robotics Club courses at Arsakeia Schools Psychikou in Athens, Greece. It presents authentic examples regarding specially designed, age- and pedagogically appropriate educational robotics systems for Early Childhood Primary Education. More specifically, the paradigms concern the utilization of Bee-Bot for kindergarten children and students in the 1st grade of elementary school (5-6 years old students), as well as of Lego WeDo 1.0 and Lego WeDo 2.0 packages for the 3rd and 4th grade (8-9 years old students). The use of Lego WeDo packages involves beyond building robotic constructions programming in Scratch environment.

Indicative questions to be explored in this chapter are: (i) Which robotics systems and which relevant learning resources are appropriate for each age stage? (ii) What are the advantages of exploiting robotic systems and the respective presuppositions? and (iii) What are the difficulties that arise and how can they be dealt with in order to achieve the best learning outcomes?

Within the framework of these questions, an exploratory case study, concerning computational thinking by students engaging in programming and robotics activities, has been conducted and is described below.

The chapter is structured as follows: a short overview of similar studies is provided; the research framework, key aspects and the findings of the case study are presented. Main features of teaching interventions concerning the utilization of robotics systems, digital learning environments and unplugged activities in the classroom focused on computational thinking and creativity development are described.

Finally, the interpretations of the findings, solutions and recommendations as well as possible future research directions are discussed.

BACKGROUND

Robotics systems and constructionist digital tools offer students the possibility to express and exploit their thoughts, ideas and intuitions and support the process of building knowledge by creating learning environments rich in speculation and opportunities for experimentation. Learning occurs through a process of continuous changes in the individual's cognitive structures and is directly linked to the effects of the sociocultural environment. At the same time, the context in which learning takes place as well as the tools' mediation play a crucial part providing opportunities for active, exploratory and personally significant learning for the individual.

Since the late 1960's, with Seymour Papert, leader of Artificial Intelligence, father of the Logo programming language and the floor turtle, educational robotics continues to grow rapidly, constantly constructing new robotic devices and systems for all educational levels with emphasis on pre-primary and primary education. The learning environment can function as mental scaffolding that allows the structuring of more complex and composite commands, reinforcing, this way, the subtractive thought that progressively develops. When children try to write a program, they actually try to teach the robot or/and the computer how to "think", something that urges them to explore and probably get to the bottom of their own way of thinking, in order to teach it to the device (Papert, 1980; Harel and Papert, 1991).

In Ackermann's words: "A rich learning environment is one that offers the freedom for genuine exploration, reflection, expression and negotiation, while at the same time providing help and support, when needed. Needless to say, it is not easy to decide how much freedom or guidance makes for a nurturing and yet challenging learning experience. Different people need different kinds of feedback, at different times, in different situations! A good clinician, like a good teacher, is someone who masters the art of providing the "right" amount of elbowroom in each singular case" (Ackermann, 2003). Thus, the creation of interesting and demanding environments encouraging the active and constructive participation of students is a great challenge for teachers.

Educational robotics and the STEAM approach (Science - Technology - Engineering - Art - Mathematics) are increasingly integrated into the modern classroom in both the international and Greek educational reality. Robotics is emerging as an important tool in the teaching-learning process, providing the opportunity to apply constructionist teaching practices with positive learning outcomes at the level of knowledge, attitudes and skills, and active learner-building. Figure 1 illustrates forms of floor robots: "Floor Turtle" by Seymour Papert (1960s) and Modern Robots (2019).

Robotics is considered as a fertile way to engage children in creative interdisciplinary explorations, in collaboration and teamwork and develop fine motor skills and hand-eye coordination (Bers, 2008, 2010; Lee, Sullivan, & Bers, 2013). Robotics also is a promising way to foster problem-solving skills and "debugging" (Bers, Flannery, Kazakoff, & Sullivan, 2014). Robotics systems can help children develop an in depth understanding of mathematical concepts such as number, size, and shape (Bers, 2012; Resnick et al., 1998).

Several researchers have focused on experimenting with robotics activities in early childhood classrooms (Bers, 2008; Bers & Horn, 2010; Kazakoff & Bers, 2012). Research shows that children from the age of four years old can successfully participate, build and program simple robotics projects while

Figure 1. Floor robots: "Floor Turtle" by Seymour Papert (1960s) (Adapted from https://upload.wikimedia.org/wikipedia/commons/7/74/Seymour_Papert.jpg
Attribution: Matematicamente.it [CC BY-SA 3.0]) (on the left) and Modern Robots (2019) (on the right)

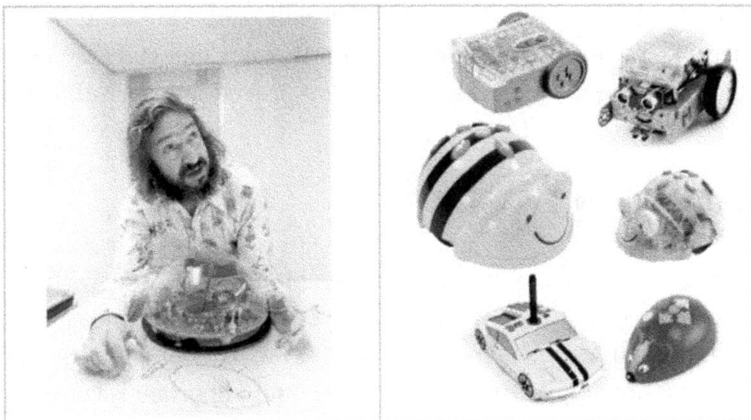

learning a range of engineering and computer science concepts in the process (Bers, Ponte, Juelich, Viera, & Schenker, 2002; Sullivan, Kazakoff, & Bers, 2013).

Recent studies focus on the Bee-Bot robotics system (Glezou, 2017, 2018; Misirli, 2016; Komis & Misirli, 2012; Misirli & Komis, 2012; Kopelke, 2011) and LegoWeDo robotics system (Glezou, 2015, 2018; Elkin, M., Sullivan, A., & Bers, M. U., 2014) and highlight the added value of teaching interventions for its use in kindergarten and elementary school students respectively.

According to Resnick (2010), success in the future will not be based on how much we know, but on our ability to think and act creatively.

COMPUTATIONAL THINKING

Computational thinking (CT, as abbreviation) is a 21st century skill and is considered the backbone of the problem-solving process (Tabesh, 2017). The term computational thinking was first used in 1980 by Seymour Papert in Mindstorms (Papert, 1980) but became widespread and gained a continuously growing interest by Jeannette Wing (Wing, 2006) in 2006.

In recent years, a large body of literature has been produced to conclude to a clear definition of this concept, as well as to provide tools and frameworks to introduce and assess CT in education.

While there is no agreement for a concrete definition of CT concept, a lot of proposed definitions recognize CT as a set of mental processes and transversal skills like creativity, collaboration, tolerance for ambiguity, resilience, and more, and surely not only as a set of technical methods and practices.

As Wing (2006, p. 33) denotes "Computational thinking is using abstraction and decomposition when attacking a large complex task or designing a large complex system. It is separation of concerns. It is choosing an appropriate representation for a problem or modeling the relevant aspects of a problem to make it tractable. It is using invariants to describe a system's behavior succinctly and declaratively. It is having the confidence we can safely use, modify, and influence a large complex system without understanding its every detail."

Computational thinking involves a four-stage problem-solving process as follows (Tabesh, 2017):

Stage One: Decomposition - Analyzing the problem to break it up into smaller parts.
Stage Two: Pattern Recognition - Observing patterns, trends, and regularities in data.
Stage Three: Abstraction - Identifying the underlying principles that generate perceived patterns.
Stage Four: Algorithm Design - Developing step-by-step instructions for solving the problem.

CS - CT PROMOTIONAL INITIATIVES

In recent years, as computation is becoming ever more ubiquitous and pervasive in modern society, a lot of efforts are given to support today's citizens and especially students as future citizens, develop digital, coding and programming competencies. In the following, Computer Science (CS) - Computational Thinking (CT) promotional initiatives are presented and important issues of interest are depicted.

- https://code.org/

Code.org is a nonprofit involved in the promotion of coding and computer science education dedicated to expanding access to computer science and increasing participation of women and underrepresented minorities. Bringing together more than 50 industry, nonprofit, and advocacy organizations, the Code.org Advocacy Coalition works to build and support a movement in the development and implementation of policies to make computer science a fundamental part of the K–12 education system.

- https://codeweek.eu/

EU Code Week is a grass-roots movement which aims to bring coding and digital literacy to everybody in a fun and engaging way. EU Code Week was launched in 2013 by the Young Advisors for the Digital Agenda Europe. The European Commission supports EU Code Week as part of its strategy for a Digital Single Market. In the Digital Education Action Plan the Commission especially encourages schools to join the initiative. The idea is to make programming more visible, show the young, adults and the elderly how you bring ideas to life with code, demystify these skills, and bring motivated people together to learn. EU Code Week is run by volunteers. One, or several, Code Week Ambassadors coordinate the initiative in their countries, but everyone can organize their own activity and add it to the codeweek.eu map. Schools of any level and teachers of all subjects are especially invited to participate in EU Code Week and give the opportunity to their students to explore digital creativity and coding. Learn more about the initiative and how to organize your activity via the webpage dedicated to teachers: CodeWeek.eu/Schools. The goal is to reach 50% of all schools in Europe by 2020.

- http://www.allyouneediscode.eu/about

The European Coding Initiative, or 'all you need is {C<3DE}', brings together a wide array of stakeholders to promote coding and computational thinking at all levels of education, as well as in more informal settings. The project was created in June 2014 under the auspices of the European Commission, and is endorsed by Commission Vice-President Andrus Ansip and by Commissioner Günther Oettinger.

It is supported by partners from the technology industry, including Facebook, Liberty Global, Microsoft, Samsung, and SAP. The initiative also includes not-for-profit organizations involved in the promotion of coding and computer science education, including CoderDojo, Code.org and European Schoolnet.

- https://www.bebras.org/

Bebras is an international initiative aiming to promote Informatics (Computer Science, or Computing) and computational thinking among school students of all ages. Participants are usually supervised by teachers who may integrate the Bebras challenge in their teaching activities. The challenge is performed at schools using computers or mobile devices. The Bebras challenges are made of a set of short problems called Bebras tasks and are delivered online. The tasks are fun, engaging and based on problems that computer scientists often meet and enjoy solving. The tasks can be solved without prior knowledge and instead require logical thinking.

UNPLUGGED ACTIVITIES

Computer Science (CS) Unplugged is an outreach program designed to introduce several key concepts of computational thinking. The original CS Unplugged activities, developed by Tim Bell at University of Canterbury, use a hands-on approach through games, puzzles, magic tricks, and competitions to teach computer science concepts without the use of a computer. The activities are designed to engage students and serve as a vehicle to learn the foundations of computing without actually using the computer. CS Unplugged website (https://csunplugged.org/en/) offers a collection of free teaching material - learning activities that use cards, string, crayons and lots of running around. The activities introduce students to Computational Thinking through concepts such as binary numbers, algorithms and data compression, separated from the distractions and technical details of having to use computers. The activities are divided into six topics: data (representing information), putting computers to work (algorithms), telling computers what to do (representing procedures), really hard problems (intractability), sharing secrets and fighting crime (cryptography), and the human face of computing (interacting with computers). Importantly, no programming is required to engage in these ideas! CS Unplugged is suitable for people of all ages, from elementary school to seniors, and from many countries and backgrounds. Unplugged has been used around the world for over twenty years, in classrooms, science centers, homes, and even for holiday events in a park! The material is available free of charge, and is shared under a Creative Commons BY-NC-SA license, which makes it easy to copy, adapt and share.

More ideas and material for CS Unplugged activities can be found in an increasing number of sites as:

- https://code.org/curriculum/unplugged
- https://www.stem.org.uk/resources/collection/3909/computer-science-unplugged
- http://info.thinkfun.com/stem-education/6-unplugged-coding-activities-for-hour-of-code

Examples of Unplugged Activities

1. Example 1: Feed the turtle.
    a. Fundamental CT Concept: Algorithm

  b. Main Idea: This game teaches children about algorithms. An algorithm is a group of instructions written by the programmer to tell the computer what to do. They will also learn the concept of debugging, which is how to fix a problem in a program. In this game, a path is designed on the floor (or a table) using a deck of cards for the turtle to travel and some flowers are placed for it throughout the path.
  c. Objective: The objective of the game is to move the turtle through the path without missing any flowers on its way.
  d. Materials Needed: A deck of cards, a toy turtle and few toy flowers.
  e. How to play: One child acts as the Computer (who moves the turtle by listening to the programmer's instructions). The other child acts as the Programmer (who gives the commands). The Programmer gives verbal instructions like 'Move Forward' (state how many card spaces), 'Move Backward' (state how many card spaces), and 'Turn Right' or 'Turn Left' to guide the computer/turtle throughout the path, without missing any of the flowers.
  f. Challenge level: The challenge level of the game can be increased by creating a more complex maze and putting obstacles in pathways that the turtle has to escape from or maneuver around.
2. Example 2: If/Then Backyard Coding Game
  a. Fundamental CT Concept: If/Then Statements
  b. Main Idea: This activity introduces children to the conditional statements (If/Then Statements). A conditional statement tells the computer to execute a set of actions depending on a specific event.
  c. Objective: The objective of the game is to follow the programmer's instructions and perform a particular task.
  d. Materials Needed: A backyard or playground, a group of students
  e. How to Play: One child acts as the Programmer (who gives the commands) and everyone else are the Computers. For every round, a different child is the Programmer by rotation. The Programmer stands in front of the Computers and gives them commands: "If I ____ (fill in the blank), Then you _____ (fill in the blank)." For example, the Programmer gave the command "If I turn in a circle, Then you turn in a circle." Or he can give challenging instructions like "If I touch my nose, Then you touch your legs."
  f. Challenge level: The challenge level of the game can be increased by creating a more complex combination of If/Then Statements.

EDUCATIONAL ROBOTICS

Robotics education provides learners with practical experiences for understanding technological and mechanical language and systems; accepting and adapting to constant changes driven by complex environments; and utilizing knowledge in real situations or across time, space, and contexts. In addition, along with the growing attention to STEM education, robotics has been suggested as an innovative solution.

Regardless of the economic and societal needs for new types of innovative and knowledgeable citizens, robotics has the attention of scholars as a means of "empowering learners" and providing "authentic learning". Learners engaging in the process-oriented learning experiences of robotics are transformed from passive knowledge receivers or technology consumers to co-constructors of learning (Jung & Won, 2018).

THE BEE-BOT EDUCATIONAL ROBOTICS SYSTEM

Bee-Bot (or BeeBot), the "smart bee", is a programmable floor robot specially created for the introduction of educational robotics at pre-school and primary school level. Its name comes from the composition of two elements: "Bee" (i.e. the well-known bug) and "Bot" from the word "robot".

With its simple and child-friendly design, Bee-Bot is a good starting point for teaching control, orientation and programming language to young children (Glezou, 2017). Designed for students from kindergarten through second grade, it is a powerful tool for the early stages of introducing and learning programming as it combines learning and play. Basic questions addressing are: a) What is an algorithm? and b) How is a simple program created and executed?

Bee-Bot programming is made by using on-board keys and it can be programmed to move forward, backwards, to turn left and right by 90º, in the same way that the "turtle" of the Logo language moves on the screen of the computer by pressing the corresponding arrow keys on its back. It moves in a 15 cm pitch and turns (left or right) by 90° with precision. By pressing the green GO button Bee-Bot can be watched following the steps. Children are captivated by Bee-Bot and eager to send it on its way. Figure 2 shows Bee-Bot and its command-keys.

Bee-Bot can move on smooth, flat and slightly tilted surfaces of various materials such as: paper, canvas, cement, tile, wood, plastic, carpet.

Through a variety of scenarios and activities within a mission-challenge framework, students learn to collaborate and program the Bee-Bot route to successfully fulfill their mission and solve the problem posed by the teacher or a student/group of students.

Some particular features of Bee-Bot are the following:

- Robust construction and small size.
- Clear command-keys (buttons).
- It confirms the execution of commands with a distinctive sound and by blinking (flashing) its "eyes".
- Easy programming.
- Memory to program a sequence of up to 40 commands-steps.
- It engages students in a wide variety of cross-thematic scenarios and activities, such as:
- Identification of letters, and the alphabet.

Figure 2. Bee-Bot with the available command-keys and their explanation

- Identification of numbers and number sequences.
- Shapes, colors, size and location.
- Virtual travel on the map e.g. of Greece – Europe.

Bee-Bot is considered an appropriate tool for developing computational thinking and skills:

- Introduction to the command sequence and assembly control.
- Development of vocabulary relative to location and orientation.
- Understanding of algorithms.
- Development of logical arguments, assumptions, prediction of simple programs behavior.
- Familiarization with sequence structure, repeat structure.
- Design, creation and debugging of programs.
- Detection and fixing of program bugs.
- Development of fine motor skills.

In order to integrate Bee-Bot into teaching practice, teachers create scenarios as activity frameworks, depending on the subject(s) they want to teach/explore. An activity mat as a framework is simply created by a piece of paper, cardboard or canvas where square cells of 15cm x 15cm are being designed.

Students can draw the surface along with the teacher depending on the activity (e.g. if they are learning math, they can paint numbers or arithmetic operations in the cells, or if they are learning the alphabet, they can draw the letters). Once the activity framework is ready, students can program the robot appropriately according to the challenges posed for example eg getting to know the letters of the English alphabet, the vegetables, the animals, the fruits etc. Figure 3 presents indicative frameworks for Bee-Bot activities, such as: a) recognition of shapes, colors, size and location, b) traveling on the map, c) story-telling of an illustrated fairy tale/story, d) practicing mathematical operations.

THE EDUCATIONAL ROBOTICS SYSTEMS LEGO EDUCATION WEDO 1.0 AND LEGO EDUCATION WEDO 2.0

The educational robotics packages LEGO Education WeDo 1.0 and LEGO Education WeDo 2.0 are educational robotics systems and appropriate tools for activities that serve the STEAM approach, which are designed for elementary school students (age: 7+). They provide an educational suggestion for the introduction and widespread diffusion of robotics in elementary schools. They aim to excite students' curiosity through cross-curricular activities and, at the same time, improve their skills in the fields of natural sciences, technology, engineering, mathematics and programming.

The LEGO Education WeDo 1.0 basic package is delivered in a storage box with a building block tray, and includes the LEGO USB Hub, a medium size motor, a motion sensor, a tilt sensor and 158 LEGO building blocks (Block type: LEGO® System®. Number of units: 158. Manufacturer: LEGO Education). The WeDo 1.0 to Computer interface is achieved through a USB Hub, which is responsible for transmitting/exchanging WeDo 1.0 data with the computer via a cable connected to a USB port.

The LEGO Education WeDo 2.0 basic package is delivered in a storage box with building block trays, accompanying sticker labels, and includes the Smart Hub, a medium size motor, a motion sensor, a tilt sensor and 280 LEGO building blocks (Block type: LEGO® System®, LEGO® Technic. Number of

Figure 3. Indicative frameworks for Bee-Bot activities

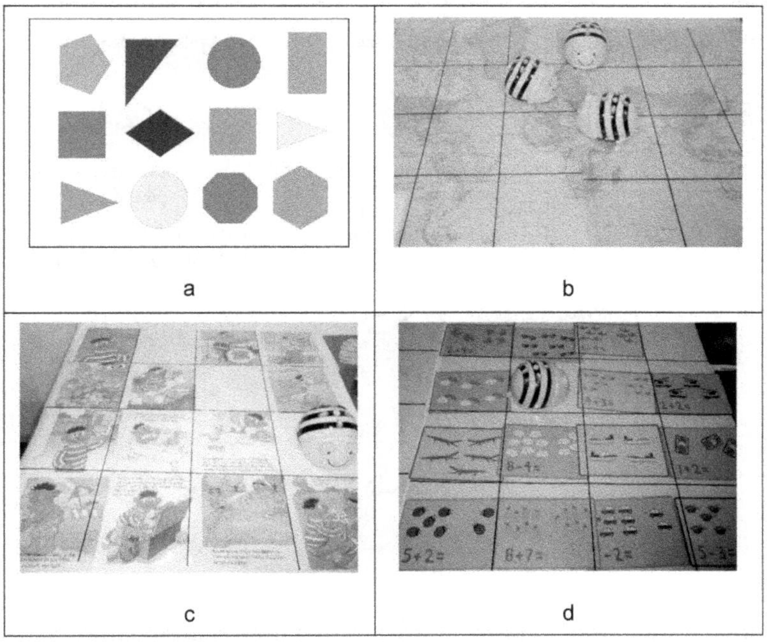

units: 280. Manufacturer: LEGO Education). The WeDo 2.0 to Computer interface is achieved through the Smart Hub, which uses Bluetooth® Low Energy technology (BTLE) technology and is responsible for wirelessly transmitting/exchanging WeDo 2.0 data with a computer or tablet.

Both packages are accompanied by a free software (programming environment), which includes introductory activities that help students familiarize themselves with hardware and software.

The LEGO Education WeDo 2.0 educational robotics system is the descendant, or better said, the modern evolution of LEGO Education WeDo 1.0, always following the philosophy of the STEAM approach but adding new additional features thanks to Smart Hub, which allows wireless communication between robotic construction and computer (Glezou, 2018).

RESEARCH FRAMEWORK

The present study is part of a wider research, which aims to implement a constructionist approach to develop computational thinking and creativity by engaging young students in robotics and programming activities regarding specially designed, age- and pedagogically appropriate educational robotics systems for Early Childhood Primary Education. The aim of this research is to bring forward the basic parameters of an effective constructionist approach, in the level of designing, development and implementation.

In the framework of the present study, the suggested activities concern the utilization of Bee-Bot for kindergarten children and students in the 1st grade of elementary school (5-6 years old students), as well as of Lego WeDo 1.0 and Lego WeDo 2.0 kits for the 3rd and 4th grade (8-9 years old students). The use of Lego WeDo kits involves beyond building robotic constructions programming in Scratch environment.

Table 1. Comparison between WeDo 1.0 and WeDo 2.0 automation elements

Automation elements	Function	WeDo 1.0 elements	WeDo 2.0 elements
Motor	It allows objects to move.	 The motor has medium size and power. It includes connection sockets in the front side.	 The motor has medium size and power. It includes connection sockets in the front side, as well as a 2x2 socket in its top side.
Distance sensor	It interacts when distance from the sensor is increased or		
Tilt sensor	It interacts when the sensor tilt changes.		
Hub	WeDo to Computer Interface.	 WeDo 1.0 to Computer Interface. The two-port USB Hub is responsible for transmitting/exchanging WeDo 1.0 data with a computer connected via a USB port. The USB Hub does not require batteries.	 WeDo 2.0 to Computer Interface. The two-port Smart Hub uses Bluetooth® Low Energy (BTLE) technology and is responsible for transmitting/exchanging WeDo 2.0 data with a computer or tablet. Smart Hub requires 2 AA batteries or a special rechargeable battery.

The suggested activities were implemented in real conditions to kindergarten and elementary school students during the Afternoon Robotics Club courses at Arsakeia Schools Psychikou in Athens, Greece. Arsakeia Schools Psychikou are private schools run by the "Society for the Promotion of Education and Learning" (Philekpaideutiki Etaireia), the oldest educational institution in Greece. Arsakeia Schools Psychikou is a school complex which consists of a Kindergarten, three Primary Schools, two Junior High Schools (Gymnasia) and three Senior High Schools (Lykeia).

In the present study, students from the Kindergarten and the three Primary Schools participated in the Afternoon Robotics Club courses.

The basic research questions of the particular study are: (i) Which robotics system and relevant learning resources are appropriate for each age stage? (ii) What are the advantages of exploiting the specific robotic systems and the respective presuppositions? and (iii) What are the difficulties that arise in the classroom and how can they be dealt with in order to achieve the best learning outcomes?

Within the framework of these questions, an exploratory case study, concerning computational thinking by students engaging in robotics and programming activities, has been conducted and is described below.

It is a case study that uses ethnographic and action research elements since the researcher was also the teacher of the class. In the framework of the study after the setting of the basic teaching objectives, a series of lessons were structured in phases; each phase in stages and each stage in distinct steps. Then, the teacher-researcher developed a) scenarios, b) activity worksheets-lesson plans, c) student worksheets and d) teacher worksheets. Data were collected from the teacher-researcher's notes-diary after each didactic hour, the students' notes-drafts, the filled-in worksheets, the artifacts/projects of the students, as well as semi-structured interviews of students. The data then, underwent a qualitative analysis, whose results led to modifications in the ergonomics, the appearance and the function of the artifacts, as well as the ameliorative reshaping of the lesson plan and the worksheets.

The heterogeneity of the students had to do with the different cognitive level of the students as well as with the basic computer skills and with programming and engineering skills in particular.

ACTIVITIES FOR THE BEE-BOT EDUCATIONAL ROBOTICS SYSTEM

The After-School Robotics Bee-Bot Club began for the first time in the school year 2014-2015 and consisted of mixed groups of fifteen students (5-6 years old) from Kindergarten and 1st grade of the Arsakeia Elementary Schools Psychikou in Athens. During the courses, students had the opportunity to collaborate, experiment and program the Bee-Bot robots, by engaging in a variety of cross-thematic activities.

Initially, the students participated with enthusiasm in the volunteer turtle game, following instructions to proceed on a specific route, utilizing rectangular floor tiles and developing kinesthetic and orientation skills, distance estimation, formulation and execution of clear instructions. Throughout the courses, the students actively participated, experimented, communicated, searched for and managed information, expressed their concerns, made questions, engaged in research process (observation, hypothesis formation, case testing, correction and/or affirmation of hypotheses, drawing conclusions), created, checked and debugged programs, collaborated, and learned through playing (Glezou et al., 2015).

Students worked in teams to create their own mazes or obstacle courses for Bee-Bot to traverse directly on the classroom floor. Each team made a unique obstacle course with its own twists and turns for Bee-Bot to navigate. Careful planning is required to enter the sequence of Bee-Bot commands that send Bee-Bot around the twists and turns of the obstacle course in order to complete it successfully. Teams exchanged their mazes or obstacle courses and challenged one another.

Figure 4 presents typical challenges - problems of graded difficulty within the framework of the Bee-Bot driving activities. On a first level, the teacher uses a framework of activities that utilizes a 2*2 grid with a target object, where additional motion and turn commands are gradually required. The students move into a framework of activities utilizing a 4*4 grid, first with a target object, then with adding more target objects, and enriching with object-barriers that restrict the required algorithm commands. The teacher urges students to look for alternatives to the problem by approaching the optimal algorithm with the lowest number of commands.

Bee-Bot real-time and real-space driving activities can be accompanied/enriched by virtual reality activities on the computer screen by using appropriately designed microworlds, simulators, digital games and other computing applications to further familiarize students with designing and modeling solutions. Moving/switching from tangible objects to virtual objects on the computer screen has a special added value and is usually accepted by the students with enthusiasm.

Figure 4. Indicative challenges-problems in the framework of Bee-Bot activities

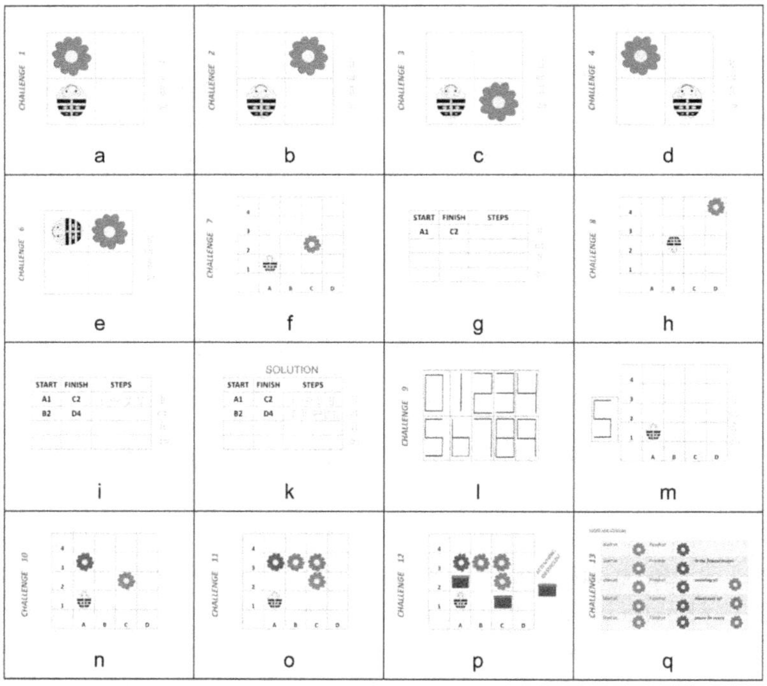

Figure 5 shows screenshots of virtual reality activities on the computer screen: A. Bee-Bot on Scratch website https://scratch.mit.edu/projects/34765070/ (on the left) and B. Bee-Bot Emulator on Terrapin website https://www.terrapinlogo.com/emu/beebot.html (on the right).

A) Bee-Bot / The programmable bee on Scratch

Educational application for the exploration/simulation of the Bee-Bot educational robotics system can be found on Scratch website. Students become familiar with step-by-step programming by giving a move and turn command sequence, in order to successfully complete the Bee-Bot guiding game to the target flower by crossing the suggested path. By completing the solution of a path, students advance on to the next level of graded difficulty, utilizing the newly acquired experience and, via thinking, recognize the evolving of the game and their own progress. There are six gradual difficulty levels. Figure 6 shows

Figure 5. Screenshots for computer screen activities: A. Bee-Bot on Scratch website (on the left) and B. Bee-Bot Emulator on Terrapin website (on the right)

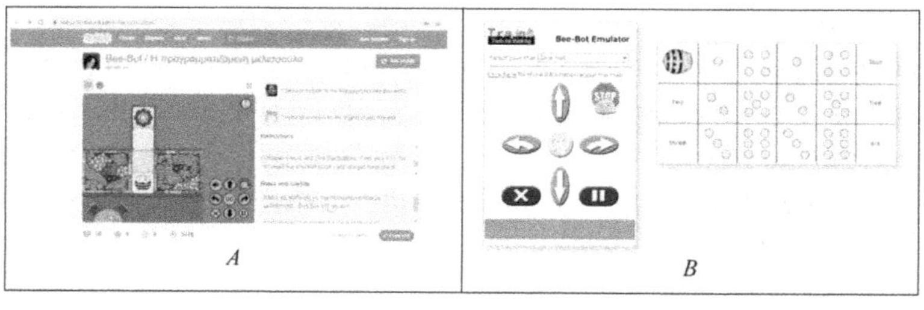

Figure 6. Screenshots for computer screen activities: Bee-Bot on Scratch environment regarding 6 levels of graded difficulty

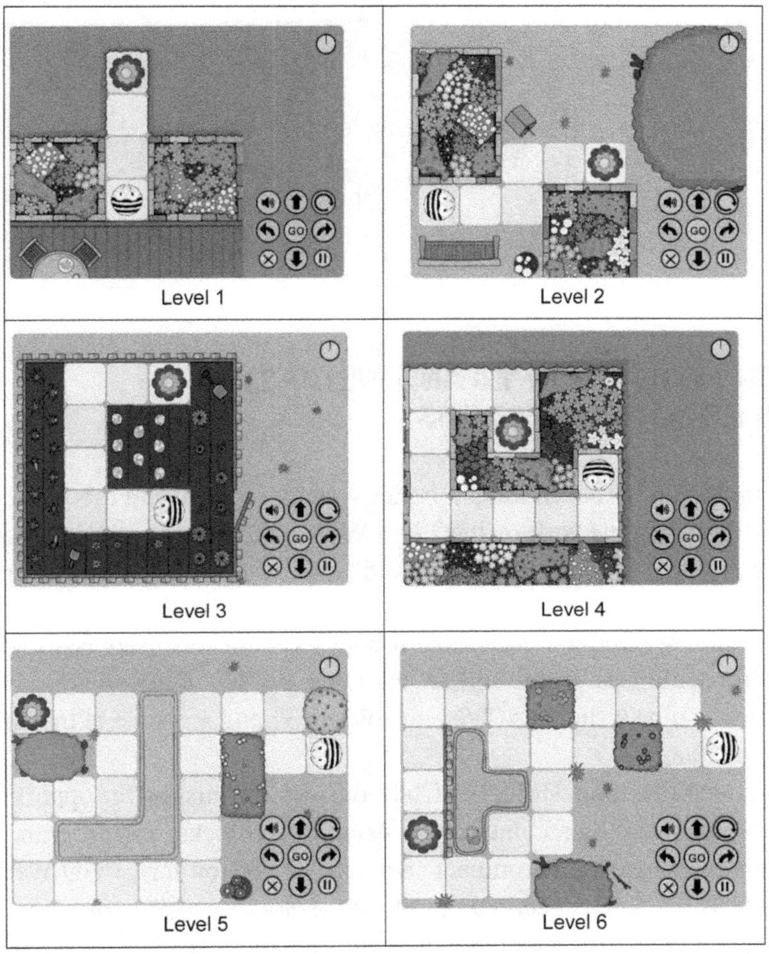

indicative screenshots of virtual reality activities on the computer screen: Bee-Bot on Scratch website - 6 Levels of graded difficulty.

[Project's Instructions: Choose a level and click the buttons. Then click GO. Try to make the shortest script - you will get three stars! X clears the script. Reload returns Bee-Bot into the original position. You can re-enter any level and improve your score (unless you click the green flag).]

Free access is available through the link: https://scratch.mit.edu/projects/34765070/.

B) Bee-Bot Emulator on Terrapin website

Simulation application of the Bee-Bot system can be found on the Terrapin website. Students select the Bee-Bot mat of their choice and familiarize themselves with step-by-step programming by giving a sequence of move and turn commands to successfully complete the bee-robot guiding game in the target, by crossing the suggested path. Free access is available through the link: https://www.terrapinlogo.com/emu/beebot.html. Figure 7 shows indicative screenshots of virtual reality activities on the computer screen with Bee-Bot Emulator selecting different mats as: a. "Go Cart Rally" Mat (on the left) and b. Alphabet Mat (on the right).

Figure 7. Screenshots for computer screen activities with Bee-Bot Emulator - a. "Go Cart Rally" Mat (on the left) and b. Alphabet Mat (on the right)

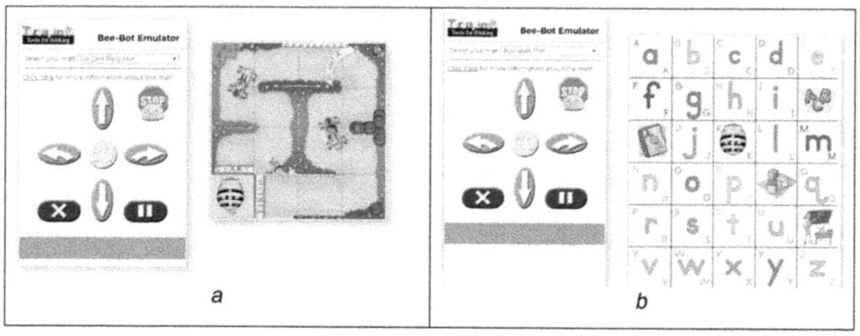

ACTIVITIES FOR LEGO WEDO 1.0 AND WEDO 2.0 EDUCATIONAL ROBOTICS SYSTEMS

The LEGO Education WeDo 1.0 educational robotics system was used as a robotics system in the school year 2014-2015 within the framework of the Lego WeDo 1.0 Afternoon Educational Robotics Club for mixed groups of eighteen students (8-9 years old) in 3rd and 4th grade of the Arsakeia Elementary Schools Psychikou in Athens.

The LEGO Education WeDo 2.0 educational robotics system was used as a robotics system in the school year 2016-2017 within the framework of the Lego WeDo 2.0 Afternoon Educational Robotics Club for mixed groups of nine 9 students (8-9 years old) in 3rd and 4th grade of the Arsakeia Elementary Schools Psychikou in Athens.

During the Afternoon Educational Robotics Club courses, students had the opportunity to collaborate, build projects of personal meaning and interest by using Lego blocks, and program automations in the Scratch programming multimedia environment (Scratch 1.4 version for LEGO WeDo 1.0 and Scratch 2.0 version for LEGO WeDo 2.0 respectively), utilizing motors, as well as distance and tilt sensors.

The students engaged in team collaboration activities that covered the following topics:

- introduction to robotics and programming
- familiarization with the Lego WeDo 1.0 or Lego WeDo 2.0 robotic system respectively
- programming in the Scratch environment (version 1.4 or version 2.0 respectively)
- experiential exploratory learning
- development of algorithmic thinking
- cultivating programming, cognitive and handling skills
- developing communication and collaboration skill
- cultivating creativity and imagination

The students built robotic constructions with particular interest, initially by following step-by-step accompanying instructions (activities from the accompanying LEGO Education software) and then by creating original projects based on their interests.

Groups of students participated in the 1st, 2nd and 3rd Pan-Hellenic Educational Robotics Competition for Elementary School students on themes titled "My City", "Our Galaxy" and "Tomorrow's Vehicles

Figure 8. Indicative photos from the activities with the educational robotics system LEGO Education WeDo 1.0

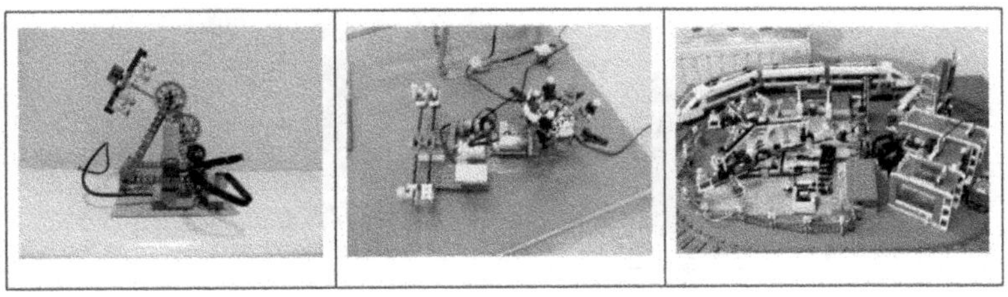

and Transports" in the school years 2014-2015, 2015-2016 and 2016-2017 respectively. After suggestion from their teacher, the students enthusiastically agreed to take part in the Pan-Hellenic Educational Robotics Competition. The students were strongly concerned about the projects they would build in view of the competitions, and suggested various themes. After setting the main goal of the project, with the continuous encouragement and great support of their teachers, students worked together as a team with zeal and enthusiasm for the realization and presentation-demonstration of their constructions-projects. During their preparation, they actively participated, experimented, communicated, searched and managed information, expressed their concerns, posed questions, engaged in research process (observation, hypothesis formation, case testing, correction and/or affirmation of hypotheses, drawing conclusions), built, checked and modified constructions, created, checked and debugged programs, collaborated, and learned through playing.

The students developed communication and collaboration skills by getting actively involved in the creative solving of various problems, as well as in exploratory learning by using STEAM technology. By participating, teachers and parents supported the students and also contributed towards the common effort whose main objective was to promote concepts and values such as cooperation, teamwork, noble competition, progress, innovation, and creativity.

As part of the 3rd Pan-Hellenic Educational Robotics Competition for Elementary School Students on "Vehicles and Transport of Tomorrow" that took place in the school year 2016-2017, the RoboLeaders team competed with a robotic assembly consisting of: (1) a safe driving system; and (2) a traffic lights system on a one-way bridge.

1. Safe driving system. For the safe driving system that works with electronic signals, the students built a robot-car and a (remote) controller with Lego blocks, and programmed its automations in the Scratch 2.0 environment. The robot-car has two smarthubs, two engines and two distance sensors. The controller has a smarthub, a tilt sensor and a distance sensor. The controller is used to drive the vehicle, accelerate forward, move in reverse, turn right and left, start, and stop. The vehicle also has an emergency braking system. More specifically:
    a. The vehicle turns right and left through the tilt sensor on the controller. The sensor emits an electronic signal to the electronic brain (Scratch), which sends appropriate signals (direction and power) to both engines (left-right) so as to turn the vehicle towards the desired direction.
    b. The vehicle starts, moves forward, accelerates, stops, and moves in reverse at a steady speed via a steering controller that employs a distance sensor (to achieve a more accurate measurement of the controller tilt for forward acceleration). The distance sensor was used as a tilt sensor by employing a wall-obstacle with color grading. The wall was built by using Lego

Figure 9. Indicative screenshots programming in Scratch environment for the final project LEGO WeDo 2.0

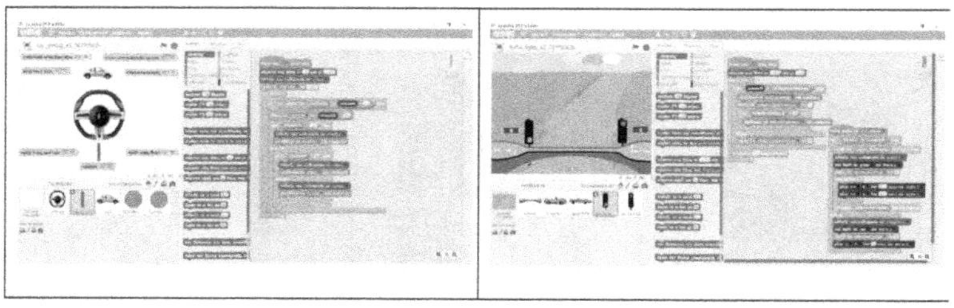

blocks in the appropriate colors. The controller in the vertical position encounters a change from black to red (internally, the optical sensor measures 0 backwards, 10 in the vertical axis, and increases forward - we subtracted 10, so we have -10 in reverse, 0 in the vertical axis and 10+ forward). For backward movement (reverse), we built an extension to the wall so that the sensor perceives the wall-obstacle to be closer, in which case it does not consider one is accelerating. A Lego block was placed on the controller to reduce the range of reverse motion.

c. The vehicle has an automatic emergency braking system. Two autonomous automatic braking subsystems with corresponding safety distance sensors (a front distance sensor and a rear distance sensor) were used. Every subsystem immobilizes the vehicle based on the direction of movement on which it detects an obstacle. When the distance sensor detects an obstacle within the specified safety distance, it sends a signal to the brain. The brain takes control of the vehicle by preventing movement towards the obstacle, but still allowing it to move away from it, while emitting a beeping sound and a warning light.

2. Traffic lights system on a one-way bridge. For the traffic lights system on the one-way bridge, the students built two traffic lights using Lego blocks, two smarthubs and two distance detectors from the Lego WeDo 2.0 robotic kit, and they programmed the automations in Scratch 2.0. Each traffic light has a smarthub and a distance sensor. Students programmed a periodic color shift of the traffic lights (green - red), unless a vehicle is waiting to pass, so it is given priority. If vehicles are waiting on both sides of the bridge, priority is given on one side (left).

When the team completed the construction and programming of the robotic assembly, it went on to construct a scale model with the bridge and the motorway by using foam board, wooden sticks, and various other scale modeling materials. In the scale model, the students chose to build a traditional one-way stone bridge, as they wanted to send the message of exploiting technology while paying respect to traditional architecture and the environment. The scale model is modular, with reusable components.

For more details and information regarding the robotics activities, it is suggested to watch the relevant videos below:

- https://vimeo.com/209129033
- https://vimeo.com/262594363
- https://vimeo.com/262714123

Figure 10. Photos regarding the final project LEGO WeDo 2.0

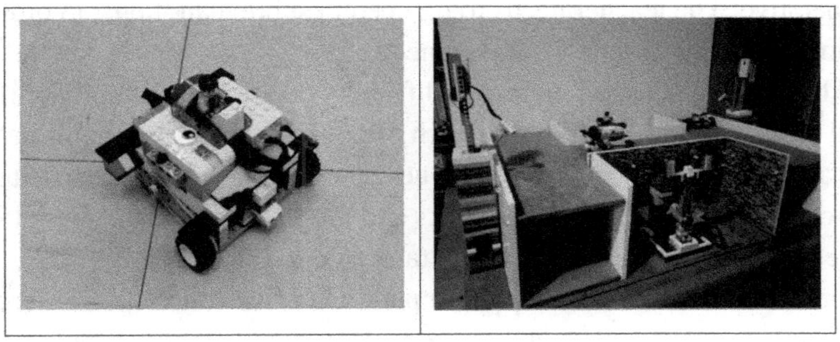

DISCUSSION

The utilization of educational robotics systems in the classroom, such as Bee-Bot for pre-school and elementary school students, LEGO Education WeDo 1.0 and WeDo 2.0 for older elementary school students, as well as a wide variety of other robotics systems (e.g. Edison, Thymio), can substantially contribute to computational thinking development as well as the promotion of the STEAM educational approach and the upgrading of teaching practice with positive learning outcomes at the level of knowledge, attitudes and skills.

Students have the opportunity to learn through playing, experiment, become familiar with robot construction, acquire robotics, engineering and programming skills, and develop computational thinking, algorithmic thinking, communication and teamwork skills.

This pilot implementation of teaching interventions, activities and accompanying material and devices in actual classroom circumstances provided us with important feedback that led to modifications in the ameliorative reshaping of the lesson plans and of the worksheets. The activities functioned as a good starting point, a solid open ground for explorations-modifications-extensions, and a vehicle for collaboration that led to various alternative constructions of personal and social meaningful artifacts. The lesson plans and activities with a gradually increasing degree of complexity encourage the systematization of knowledge and bridge the gap between the simple and the more complex.

The teaching process should exploit the excitement, inherent curiosity and self-interest of students, as well as emphasize on the promotion of social and emotional skills of children. The combination of theory and practice in the direction of forming an open and creative teaching-learning process centered on hands-on, student-engaging learning leads to optimal results, so that students are encouraged and supported to participate actively, to work together nicely, to develop initiatives, to face challenges, to develop the skill of problem solving and decision making, to discover new knowledge, to express themselves freely and create artifacts.

Students should be able to use engineering and programming experiences as a mirror of their thought processes in order to learn how to think in problem solving. Actions such as the planning of the desired interface, the recognition of a command sequence and the definition of a procedure are considered to be indicative of the high level skills and goals' development, such as the recognition of a procedure function and the definition of new procedures. The repetitive shift between the code analysis and its extension, which allows the shift from simple to complex and the gradual familiarization with the programming language within the scaffolding process, has proved to be equally effective.

During the process of developing code and of debugging, the questions posed by the students were often fixed and imperative. The teacher needs to proceed in subtle handlings, in order to accommodate the students with no more support than what they need.

It was observed that students had an increased difficulty due to their lack of familiarity with mechanisms structure and programming, something which often led to their disappointment and their urge to quit. This applies particularly to students who had never wished or had the chance to familiarize themselves with constructing artifacts with LEGO bricks and programming. Curricular tools for CT need to follow a clear process, where their lessons can be transfered to other areas.

The participation in Educational Robotics Competitions often functions as a source of excitement and provides additional motivation for many students. However, sometimes non-qualification leads to a great deal of frustration among students. Cultivating the idea of the value of participation rather than focusing on victory should be a primary concern of the teacher, who should prioritize the students' mental health, as well as their balanced emotional and intellectual development.

SOLUTIONS AND RECOMMENDATIONS

During a lesson the teacher can understand a lot about a child. If he/she possess leadership skills, if he/she can participate in group games, if he/she knows how to lose, win, discipline himself/herself in irritation and anger; determinant elements that will shape his/her adult life.

Preventing disinterest is central to implementing effective CT activities. It can be easy to overstep boundaries and break a child's interest in learning CT. Therefore, it is very important to allow the student to lead the lessons by self-directing their learning and focusing on issues that they find intriguing.

Examples of how to keep learning in a fun, casual, and interesting way are needed. Once students have mastered a skill, such as designing their own game, then they should be able to use the CT competencies they developed during this experience to answer questions in their other classwork.

When students run into a barrier while answering a question or working on a project, it's so easy for them to get frustrated and give up. Students should be introduced to the ideas of frustration and persistence, and needed to realize that a) frustration is the feeling of being annoyed or angry at something and b) persistence is choosing not to give up, and attempting something over and over again. They should recognize that frustration can be an important part of learning in the creative process, rather than a sign of failure, and gradually conceptualize that struggle leads to learning and that persistence can lead to success.

Students need to understand the debugging process of coding. It is crucial to help the students identify when something goes wrong, so if they don't catch it the first time, they'll run it again, and again. Let the students share hypotheses about what is going wrong and how to fix it. Students should feel free to try things that the teacher knows in advance will be incorrect. If students misidentify solutions, they should use the bug finding formula on their configurations repeatedly until a working run is succeeded. Important helpful questions to be asked:

- What happened?
- What was supposed to happen?
- Did it work at the first step?
- Did it work at the second step?

- Where did it go wrong?

Encouragement is a basic key. If things don't work right away, the teacher should praise the class for being so persistent and choosing not to give up. If the students start to get frustrated, the teacher should encourage them to persist a bit longer, promising them that they will get it soon if they just hang in there.

Students need to understand the importance of the activity they just completed by reflecting on it verbally. Students should be encouraged to share their thoughts and questions either with the whole class or with an elbow partner.

The educational emphasis should always be on challenging and helping pupils understand, rather than use, the robotic devices, the programming language and digital tools.

FUTURE RESEARCH DIRECTIONS

Further analysis is needed in order to generalize the conclusions concerning the features of an effective instructional approach, in the level of planning, development and implementation, which responds to the special characteristics of the students of each particular class and of different classes as well. More research is needed for the identification of the unplugged activities, robotics systems and programming environments special design characteristics that promote engagement, "thinking about thinking", construction, meaning negotiation and collaboration. Future research plans focus on exploring the use of different robotics and programming paradigms by implementing alternate instructional strategies which might lead to the most effective combinations to support student engagement in computational thinking development.

Based on the age, cognitive level, personal interest, and prior experience of students, teachers are invited to experiment, get acquainted and become familiar with the various robotic systems, recognizing in this way their particular educational value, in order to develop skills in integrating robotics in the classroom by adopting alternative teaching strategies.

CONCLUSION

The present study aims to bring forward the basic parameters of an effective constructionist approach, in the level of designing, development and implementation of a framework for unplugged, robotics and programming activities in order to foster CT skills and creativity at early childhood and first primary classes.

We argue that a combination of unplugged, robotics and programming developmentally appropriate activities better support student CT development. These activities have varied features, with main characteristic the ability to promote computing concepts and to generate a framework through which students are engaged in CT under suitable learner-centered pedagogical approach.

During the above presented lessons, the students showed interest in their interaction especially with the robotics systems and digital tools remaining active and frequently showing and expressing enthusiasm. They demonstrated their inventiveness; their center of interest is transferred easily, tracing new exploration paths; they built alternative personal and social meaningful artifacts with different levels of detail and variations of methods. Follow-up discussions with students revealed greater willingness to see CT as relevant to everyday life and increased willingness to approach coding and programming challenges and furthermore computational thinking development with a positive attitude.

As computation is becoming ever more ubiquitous and pervasive in modern society, researchers should focus on how to support future citizens towards their transformation from passive consumers to active creators. The research community is gradually articulating the challenges and design considerations relating to the design and implementation of developmentally appropriate tools, activities and content to foster CT skills for all students of different ages from early childhood years.

ACKNOWLEDGMENT

The author is grateful to all the children who participated in the Club for their active engagement and passion during the lessons. It is always a pleasure to work with motivated and enthusiastic students. Also, the author would like to thank George Birbilis for his ideas, advice and assistance with projects' design and implementation.

This study did not receive any specific grant from funding agencies of the public domain, commercial or not-for-profit sectors.

REFERENCES

Ackermann, E. (2003). Hidden drivers of pedagogic transactions: teachers as clinicians and designers. [Porto: Cnotinfor, Lda.]. *Proceedings of Eurologo, 2003*, 29–37.

Bers, M. U. (2008). *Blocks to robots: Learning with technology in the early childhood classroom.* New York, NY: Teachers College Press.

Bers, M. U. (2010). Beyond computer literacy: Supporting youth's positive development through technology. *New Directions for Youth Development, 128*(128), 13–23. doi:10.1002/yd.371 PMID:21240949

Bers, M. U. (2012). Designing digital experiences for positive youth development: From playpen to playground. Cary, NC: Oxford. doi:10.1093/acprof:oso/9780199757022.001.0001

Bers, M. U., Flannery, L. P., Kazakoff, E. R., & Sullivan, A. (2014). Computational thinking and tinkering: Exploration of an early childhood robotics curriculum. *Computers & Education, 72*, 145–157. doi:10.1016/j.compedu.2013.10.020

Bers, M. U., & Horn, M. (2010). Tangible programming in early childhood: Revisiting developmental assumptions through new technologies. In I. R. Berson, & M. J. Berson (Eds.), *High-tech tots: Childhood in a digital world* (pp. 49–70). Greenwich, CT: Information Age Publishing.

Bers, M. U., Ponte, I., Juelich, K., Viera, A., & Schenker, J. (2002). Teachers as designers: Integrating robotics into early childhood education. *Information Technology in Childhood Education, 1*, 123–145.

Brennan, K., Resnick, M., & Monroy-Hernandez, A. (2010). Making projects, making friends: Online community as a catalyst for interactive media creation. *New Directions for Youth Development, 2010*(128), 75–83. doi:10.1002/yd.377 PMID:21240955

Elkin, M., Sullivan, A., & Bers, M. U. (2014). Implementing a robotics curriculum in an early childhood Montessori classroom. *Journal of Information Technology Education: Innovations in Practice, 13*, 153–169. Retrieved from http://www.jite.org/documents/Vol13/JITEv13IIPvp153-169Elkin882.pdf

Fessakis, G., Gouli, E., & Mavroudi, E. (2013). Problem solving by 5–6 years old kindergarten children in a computer programming environment: A case study. *Computers & Education, 63*, 87–97. doi:10.1016/j.compedu.2012.11.016

Glezou, K. (2017). Program and learn - Playing with the BeeBot educational robotics system. In *Proceedings of the 4th Pan-Hellenic eTwinning Conference on Co-operative Programs in Primary and Secondary Education "Exploiting Information and Communication Technologies in Collaborative School Programs"*. Patras, Greece. (In Greek)

Glezou, K. (2018). *Utilizing educational robotics systems in didactic practice.* Paper presented at the Pan-Hellenic Scientix Conference on STEM Education, NTUA of Zografou, Zografou, Athens, Greece. (In Greek)

Glezou, K., Savvidaki, A., & Birbilis, G. (2015). Lego WeDo - Scratch: Constructing and programming. In *Proceedings of the 8th Panhellenic Conference of ICT Teachers "Exploiting Information and Communication Technologies in Teaching Practice"* (pp. 1392-1401). Syros, Greece. (In Greek)

Harel, I., & Papert, S. (1991). *Constructionism: Research reports & essays, 1985-1990 by the epistemology & learning research group. Norwood*, NJ: Ablex Publishing.

Jung, S. E., & Won, E. (2018). Systematic review of research trends in robotics education for young children. *Sustainability, 10*(4), 905. doi:10.3390u10040905

Kafai, Y., & Resnick, M. (Eds.). (1996). *Constructionism in practice: Designing, thinking, and learning in a digital world.* Mahwah, NJ: Lawrence Erlbaum Associates.

Kazakoff, E., & Bers, M. U. (2012). Programming in a robotics context in the kindergarten classroom: The impact on sequencing skills. *Journal of Educational Multimedia and Hypermedia, 21*(4), 371–391.

Kazakoff, E., Sullivan, A., & Bers, M. U. (2013). The effect of a classroom-based intensive robotics and programming workshop on sequencing ability in early childhood. *Early Childhood Education Journal, 41*(4), 245–255. doi:10.100710643-012-0554-5

Komis, V., & Misirli, A. (2011). Robotique pédagogique et concepts préliminaires de la programmation à l'école maternelle: une étude de cas basée sur le jouet programmable Bee-Bot. Georges-Louis Baron, Eric Bruillard, Vassilis Komis. *Sciences et technologies de l'information et de la communication en milieu éducatif: Analyse de pratiques et enjeux didactiques.* Oct. 2011, Patras, Grèce. Athènes: New Technologies Éditions, pp. 271-281, 2011. Retrieved from https://edutice.archives-ouvertes.fr/edutice-00676143/document

Komis, V., & Misirli, A. (2012). L'usage des jouets programmables à l'école maternelle: concevoir et utiliser des scenarios pédagogiques de robotique éducative. *Revue Scholé,* 17, 1, pp. 143-154. Retrieved from https://espe.univ-amu.fr/sites/espe.univ-amu.fr/files/article/skhole_2012_vol_17_.pdf

Kopelke, K. (2011). *Making your classroom buzz with Bee-Bots: Ideas and Activities for the Early Phase, ICT Learning Innovation Centre.* Retrieved from http://elresources.skola.edu.mt/wp-content/uploads/2010/06/doc_669_2468_Bee-BotguideA4v2.pdf

Lee, K., Sullivan, A., & Bers, M. U. (2013). Collaboration by design: Using robotics to foster social interaction in Kindergarten. *Computers in the Schools, 30*(3), 271–281. doi:10.1080/07380569.2013.805676

Papert, S. (1980). *Mindstorms: Children, computers and powerful ideas.* New York, NY: Basic Books.

Papert, S. (1991). *Constructionism.* US: Ablex Publishing.

Resnick, M. (2010). *Rethinking learning in the digital age.* Retrieved from http://www.media.mit.edu/~mres/papers/wef.pdf

Sullivan, A., Kazakoff, E. R., & Bers, M. U. (2013). The wheels on the bot go round and round: Robotics curriculum in pre-kindergarten. *Journal of Information Technology Education: Innovations in Practice, 12*, 203–219. Retrieved from http://www.jite.org/documents/Vol12/JITEv12IIPp203219Sullivan1257.pdf

Tabesh, Y. (2017). Computational thinking: A 21st century skill. Olympiads in Informatics, 11(SI), 65-70. Retrieved from https://ioinformatics.org/journal/v11si_2017_65_70.pdf

Wing, J. M. (2006). *Viewpoint. Computational thinking. Communications of the ACM, 49(3), 33-35.* Retrieved from https://www.cs.cmu.edu/~15110-s13/Wing06-ct.pdf

ADDITIONAL READING

Brennan, K., & Resnick, M. (2012). *New frameworks for studying and assessing the development of computational thinking. Paper presented at Annual American Educational Research Association meeting*. BC, Canada: Vancouver.

Kalogiannakis, M., & Papadakis, S. (2019). Pre-service kindergarten teacher's acceptance of "ScratchJr" as a tool for learning and teaching computational thinking and Science education. [JES]. *The Journal of Emergent Science, 15*, 31–34.

Papadakis, St., & Kalogiannakis, M. (2017). Mobile educational applications for children. What educators and parents need to know. [Special Issue on Mobile Learning Applications and Strategies]. *International Journal of Mobile Learning and Organisation, 11*(3), 256–277. doi:10.1504/IJMLO.2017.085338

Papadakis, St., Kalogiannakis, M., & Zaranis, N. (2016). Developing fundamental programming concepts and computational thinking with ScratchJr in Preschool Education. A case study. *International Journal of Mobile Learning and Organisation, 10*(3), 187–202. doi:10.1504/IJMLO.2016.077867

Papadakis, S., Kalogiannakis, M., & Zaranis, N. (2019). Parental involvement and attitudes towards young Greek children's mobile usage. *International Journal of Child-Computer Interaction* doi: . doi:10.1016/j.ijcci.2019.100144

Resnick, M. (2002). Rethinking learning in the digital age. In G. Kirkman (Ed.), *The global information technology report: Readiness for the networked world* (pp. 32–37). Oxford, England: Oxford University Press.

Resnick, M., Maloney, J., Monroy-Hernandez, A., Rusk, N., Eastmond, E., Brennan, K., ... Kafai, Y. (2009). Scratch: Programming for all. *Communications of the ACM*, *52*(11), 60–67. http://web.media.mit.edu/~mres/papers/Scratch-CACM-final.pdf Retrieved September 27 2017. doi:10.1145/1592761.1592779

Sullivan, A., & Bers, M. U. (2016). Robotics in the early childhood classroom: Learning outcomes from an 8-week robotics curriculum in pre-kindergarten through second grade [Electronic version]. *International Journal of Technology and Design Education*, *26*(1), 3–20. doi:10.100710798-015-9304-5

KEY TERMS AND DEFINITIONS

Algorithm: A sequence of instructions to solve a problem or perform a task.

Computational Thinking: A set of problem-solving skills, methods and practices that involve understanding problems and their solutions in ways that a computer could execute.

Computer Programming: The process of designing and building an executable computer program for accomplishing a specific computing task.

Computer Science: The study of processes, methods and tools that interact with data to manipulate, store and communicate digital information.

Creativity: The ability to use imagination or original ideas to create a tangible or intangible new item by making connections between seemingly unrelated topics and generating solutions to a problem.

Educational Robotics: The use of programmable robot constructs in the teaching and learning process in order to engage groups of students in playful collaborative problem solving activities.

Programming Language: A formal language, which comprises a set of instructions that produce various kinds of output in order to implement algorithms and communicate with a computer.

STEAM: An educational approach to learning that uses Science, Technology, Engineering, the Arts and Mathematics as access points for guiding student inquiry, dialogue, critical thinking, problem solving, experiential learning and collaboration.

Compilation of References

7 . Ways Holographic Technology Will Make Learning More Fun. (n.d.). Retrieved from http://www.emergingedtech.com/2012/11/7-ways-holographic-technology-will-make-learning-more-fun/

Abdel-Basset, M., Manogaran, G., Mohamed, M., & Rushdy, E. (2019). Internet of things in smart education environment: Supportive framework in the decision-making process. *Concurrency and Computation*, *31*(10), 1–12. doi:10.1002/cpe.4515

Academia Mexicana de Pediatría. (2019). *El desarrollo del lenguaje*. Retrieved from https://www.healthychildren.org/spanish/ages-stages/baby/paginas/language-development-4-to-7-months.aspx

Ackermann, E. (2003). Hidden drivers of pedagogic transactions: teachers as clinicians and designers. [Porto: Cnotinfor, Lda.]. *Proceedings of Eurologo*, *2003*, 29–37.

Acquah, E. O., & Katz, H. T. (2020). Digital game-based L2 learning outcomes for primary through high-school students: A systematic literature review. *Computers & Education*, *143*, 103667. doi:10.1016/j.compedu.2019.103667

ADDIE. (n.d.). Retrieved from https://xapi.com/addie/

Adel Ali, R., & Rafie Mohd Arshad, M. (2018). Empirical analysis on factors impacting on intention to use M-learning in basic education in Egypt. *International Review of Research in Open and Distance Learning*, *19*(2), 253–270. doi:10.19173/irrodl.v19i2.3510

ADL LRS. (n.d.). Retrieved from https://www.adlnet.gov/adl-lrs

adlnet/xAPIWrapper: Wrapper to simplify communication to an LRS. (n.d.). Retrieved from https://github.com/adlnet/xAPIWrapper

Ailwood, J. O. (2003). Governing early childhood. *Contemporary issues in early childhood*, *4*(3), 286–299. doi:10.2304/ciec.2003.4.3.5

Akerson, V. L., Flick, L. B., & Lederman, N. G. (2000). The influence of primary children's ideas in science on teaching practice. *Journal of Research in Science Teaching: The Official Journal of the National Association for Research in Science Teaching*, *37*(4), 363-385. doi:10.1002/(sici)1098-2736(200004)37:4<363::aid-tea5>3.0.co;2-#

Aladé, F., Lauricella, A. R., Beaudoin-Ryan, L., & Wartella, E. (2016). Measuring with Murray: Touchscreen technology and preschoolers' STEM learning. *Computers in Human Behavior*, *62*, 433–441. doi:10.1016/j.chb.2016.03.080

Alalwan, A. A., Dwivedi, Y. K., & Rana, N. P. (2017). Factors influencing adoption of mobile banking by Jordanian bank customers: Extending UTAUT2 with trust. *International Journal of Information Management*, *37*(3), 99–110. doi:10.1016/j.ijinfomgt.2017.01.002

Al-emran, M., Elsherif, H. M., & Shaalan, K. (2016). Investigating attitudes towards the use of mobile learning in higher education. *Computers in Human Behavior*, *56*, 93–102. doi:10.1016/j.chb.2015.11.033

Alha, K., Koskinen, E., Paavilainen, J., & Hamari, J. (2019). Why do people play location-based augmented reality games: A study on Pokémon GO. *Computers in Human Behavior*, *93*, 114–122. doi:10.1016/j.chb.2018.12.008

Al-Huneini, H., Walker, S. A., & Badger, R. (2020). Introducing tablet computers to a rural primary school: An Activity Theory case study. *Computers & Education*, *143*, 103648. doi:10.1016/j.compedu.2019.103648

Aliagas, C., & Margallo, A. M. (2017). Children's responses to the interactivity of storybook apps in family shared reading events involving the iPad. *Literacy*, *51*(1), 44–52. doi:10.1111/lit.12089

Aliagas, C., Matsumoto, M., Morgade, M., Correro, C., Galera, N., & Poveda, D. (2017). *Young children (0-8) and digital technology 0-8. What changes in one year?* National report SPAIN. *Papers Infancia*, *20*, 1–65.

Alif, L., & Lara, F. (2018). Influence of gadget usage and parent involvement to children's prosocial behavior. *Journal Imiah Pendidikan Dan Pembelajaran*, *2*(2), 174–179. doi:10.23887/jipp.v2i2.15366

Allen, K. E., & Marotz, L. R. (1989). *Developmental profiles: Birth to six.* Albany, NY: Delmar.

Allen, M. (2015). Preschool children's taxonomic knowledge of animal species. *Journal of Research in Science Teaching*, *52*(1), 107–134. doi:10.1002/tea.21191

Ally, M., & Prieto-Blázquez, J. (2014). What is the future of mobile learning in education? *International Journal of Educational Technology in Higher Education*, *11*(1), 142–151. doi:10.7238/rusc.v11i1.2033

Alper, M. (2013). Developmentally appropriate new media literacies: Supporting cultural competencies and social skills in early childhood education. *Journal of Early Childhood Literacy*, *13*(2), 175–196. doi:10.1177/1468798411430101

Al-Zu'bi, M. A. A., Omar-Fauzee, M. S., & Kaur, A. (2017). The effect of iPad apps on creative thinking among preschoolers in Jordan. *International Journal of Multidisciplinary Education and Research*, 2(4), 29-35.

American Academy of Pediatrics (AAP). (2014). The benefits of limiting TV. Retrieved from http://www.healthychildren.org/English/family-life/Media/Pages/The-Benefits-of-Limiting-TV.aspx

American Academy of Pediatrics (AAP). (2016). American Academy of Pediatrics announces new recommendations for children's media use. Retrieved from https://www.aap.org/en-us/about-the-aap/aap-press-room/Pages/American-Academy-of-Pediatrics-Announces-New-Recommendations-for-Childrens-Media-Use.aspx

American Academy of Pediatrics. (2016). Media and young minds. *Pediatrics*, *138*(5), e20162591. doi:10.1542/peds.2016-2591 PMID:27940793

Anatomy of an Experience API Statement. (n.d.). Retrieved from https://xapi.com/ebook/

Anderson, D. R., Huston, A. C., Schmitt, K. L., Linebarger, D. L., Wright, J. C., & Larson, R. (2001). Early childhood television viewing and adolescent behavior: The recontact study. *Monographs of the Society for Research in Child Development*, 154. PMID:11326591

Andreou, G., & Baseki, J. (2012). Phonological and spelling mistakes among dyslexic and non-dyslexic children learning two different languages: Greek vs English. *Psychology (Irvine, CA.)*, *3*(08), 595–600. doi:10.4236/psych.2012.38089

Angeli, C. (2004). The effects of case-based learning on early childhood pre-service teachers' beliefs about the pedagogical uses of ICT. *Journal of Educational Media*, *29*(2), 139–151. doi:10.1080/1358165042000253302

Angeli, C., & Valanides, N. (2005). Preservice teachers as ICT designers: an instructional design model based on an expanded view of pedagogical content knowledge. *Journal of Computer Assisted Learning*, *21*(4), 292–302. doi:10.1111/j.1365-2729.2005.00135.x

APNI. (2019). Centro de Adiestramiento e Información para Padres. https://www.parentcenterhub.org/

Apple. (2015). Retrieved from http://www.apple.ca

Apple. (2017). App Store shatters records on New Year's Day. Retrieved from http://www.apple.com/newsroom/2017/01/app-store-shatters-records-on-new-years-day.html

Aranda, G., & Ferguson, J. P. (2018). Unplugged programming: The future of teaching computational thinking? *Pedagogika, 68*(3), 279–292. doi:10.14712/23362189.2018.859

Arango-López, J., Valdivieso, C. C. C., Collazos, C. A., Vela, F. L. G., & Moreira, F. (2018). CREANDO: Tool for creating pervasive games to increase the learning motivation in higher education students. *Telematics and Informatics, 38*, 62–73. doi:10.1016/j.tele.2018.08.005

Archer, K. (2017). *Infants, toddlers and mobile technology: examining parental choices and the impact of early technology introduction on cognitive and motor development. (Theses and Dissertations (Comprehensive)*, Wilfrid Laurier University).

Ardies, J., De Maeyer, S., Gijbels, D., & van Keulen, H. (2014). Students' attitudes towards technology. *International Journal of Technology and Design Education, 25*(1), 43–65. doi:10.100710798-014-9268-x

Armstrong, T. (2000). Multiple intelligences in the classroom. In *Intelligence* (2nd ed.). USA: Association for Supervision and Curriculum Development.

Arnott, L. (2016). An ecological exploration of young children's digital play: Framing children's social experiences with technologies in early childhood. *Early Years, 36*(3), 271–288. doi:10.1080/09575146.2016.1181049

Arnott, L., Grogan, D., & Duncan, P. (2016). Lessons from using iPads to understand young children's creativity. *Contemporary Issues in Early Childhood, 17*(2), 157–173. doi:10.1177/1463949116633347

Arnott, L., Palaiologou, I., & Gray, C. (2018, September). Digital devices, internet-enabled toys and digital games: The changing nature of young children's learning ecologies, experiences and pedagogies. *British Journal of Educational Technology, Special Issue, 49*(5), 803–806. doi:10.1111/bjet.12676

Arnott, L., Palaiologou, I., & Gray, C. (2019). Internet of toys across home and early childhood education: Understanding the ecology of the child's social world. *Technology, Pedagogy and Education*, 1–12. doi:10.1080/1475939X.2019.1656667

Aronin, S., & Floyd, K. K. (2013). Using an iPad in inclusive preschool classrooms to introduce STEM concepts. *Teaching Exceptional Children, 45*(4), 34–39. doi:10.1177/004005991304500404

Ashton, P. (2015). Historical overview and theoretical perspectives of research on teachers' beliefs. In H. Fives & M. G. Gills (Eds.), *International handbook of research on teachers' beliefs* (pp. 31–47). London: Routledge.

Aslanabadi, H., & Rasouli, G. (2013). The effect of games on improvement of Iranian EFL vocabulary knowledge in kindergartens. *International Review of Social Sciences and Humanities, 6*(1), 186–195.

Aubrey, C., & Dahl, S. (2008). A review of the evidence on the use of ICT in the Early Years Foundation Stage [Ebook]. Retrieved from https://dera.ioe.ac.uk//1631/2/becta_2008_eyfsreview_report.pdf

Ávila, A. (1999). Inteligencias Múltiples: Una aproximación a la teoría de Howard Gardner. *Horizontes Pedagógicos, 1*, 19–27.

Axelsson, A., Andersson, R., & Gulz, A. (2016). Scaffolding executive function capabilities via play-&-learn software for preschoolers. *Journal of Educational Psychology, 108*(7), 969–981. doi:10.1037/edu0000099

Bajgoric, N., & Moon, Y. B. (2017). Business model for digital economy era: A framework based on the Churchman's theory of design integrity. *International Journal of Business and Systems Research, 11*(3), 284–308. doi:10.1504/IJBSR.2017.085466

Baker, R. S., & Gowda, S. M. (2018). *Towards understanding app effectiveness and cost.* San Francisco, CA: BrightBytes; Retrieved from http://pages.brightbytes.net/LOSnapappInfographic_LOCalculatorLP.html?utm_source=Snapapp&utm_campaign=LOReportInfographicadv

Balacheff, N., De Jong, T., Lazonder, A., Sharples, M., Milrad, M., & Vavoula, G. (2008). MOBILE LEARNING Small devices, big issues. In *Technology-enhanced learning: Principles and products* (pp. 1–20). Berlin, Germany: Springer-Verlag.

Balisane, H. (2015). Investigation of teachers' attitudes towards computer use in primary education in Kurdistan. *Journal of Education and Practice, 6*(14), 37–49.

Ballagas, R., Rohs, M., Sheridan, J. G., & Borchers, J. (2004). BYOD: Bring your own device.

Barak, M. (2018). Are digital natives open to change? Examining flexible thinking and resistance to change. *Computers & Education, 121*, 115–123. doi:10.1016/j.compedu.2018.01.016

Barendregt, W., & Bekker, T. M. (2011). The influence of the level of free - choice learning activities on the use of an educational computer game. *Computers & Education, 56*(1), 80–90. doi:10.1016/j.compedu.2010.08.018

Basoz, T., & Can, D. T. (2016). The effectiveness of computers on vocabulary learning among preschool children: A semiotic approach. *Cypriot Journal of Educational Sciences, 11*(1), 2–8. doi:10.18844/cjes.v11i1.266

Baydas, O., & Yilmaz, R. M. (2018). Pre-service teachers' intention to adopt mobile learning: A motivational model. *British Journal of Educational Technology, 49*(1), 137–152. doi:10.1111/bjet.12521

Bebell, D., & Pedulla, J. (2015). A quantitative investigation into the impacts of 1: 1 iPads on early learner's ELA and math achievement. *Journal of Information Technology Education: Innovations in Practice, 14*(1), 191–215. doi:10.28945/2175

Beck, J. C., & Wade, M. (2004). *Got game: How the gamer generation is reshaping business forever.* Boston, MA: Harvard Business School Press.

Bedford, R., de Urabain, S., Irati, R., Cheung, C. H., Karmiloff-Smith, A., & Smith, T. J. (2016). Toddlers' fine motor milestone achievement is associated with early touchscreen scrolling. *Frontiers in Psychology, 7*, 1108. doi:10.3389/fpsyg.2016.01108 PMID:27531985

Bekebrede, G., Warmelink, H. J. G., & Mayer, I. S. (2011). Reviewing the need for gaming in education to accommodate the net generation. *Computers & Education, 57*(2), 1521–1529. doi:10.1016/j.compedu.2011.02.010

Bers, M. U. (2012). Designing digital experiences for positive youth development: From playpen to playground. Cary, NC: Oxford. doi:10.1093/acprof:oso/9780199757022.001.0001

Bers, M. U. (2018b). Coding and computational thinking in early childhood: the impact of ScratchJr in Europe. *European Journal of STEM Education, 3*(3), 08.

Bers, M. U. (2008). *Blocks to robots: Learning with technology in the early childhood classroom.* New York, NY: Teachers College Press.

Bers, M. U. (2010). Beyond computer literacy: Supporting youth's positive development through technology. *New Directions for Youth Development, 128*(128), 13–23. doi:10.1002/yd.371 PMID:21240949

Bers, M. U. (2017). *Coding as a playground: Programming and computational thinking in the early childhood classroom.* Routledge. doi:10.4324/9781315398945

Bers, M. U. (2018). Coding and computational thinking in early childhood: the impact of ScratchJr in Europe. *European Journal of STEM Education, 3*(3), 1–13. doi:10.20897/ejsteme/3868

Bers, M. U. (2018a). Coding, playgrounds and literacy in early childhood education: The development of KIBO robotics and ScratchJr. In *Global Engineering Education Conference (EDUCON)*, 2018 *IEEE* (pp. 2094-2102). IEEE. 10.1109/EDUCON.2018.8363498

Bers, M. U., Flannery, L., Kazakoff, E. R., & Sullivan, A. (2014). Computational thinking and tinkering: Exploration of an early childhood robotics curriculum. *Computers & Education, 72*, 145–157. doi:10.1016/j.compedu.2013.10.020

Bers, M. U., González-González, C., & Armas–Torres, M. B. (2019). Coding as a playground: Promoting positive learning experiences in childhood classrooms. *Computers & Education, 138*, 130–145. doi:10.1016/j.compedu.2019.04.013

Bers, M. U., & Horn, M. (2010). Tangible programming in early childhood: Revisiting developmental assumptions through new technologies. In I. R. Berson, & M. J. Berson (Eds.), *High-tech tots: Childhood in a digital world* (pp. 49–70). Greenwich, CT: Information Age Publishing.

Bers, M. U., Ponte, I., Juelich, K., Viera, A., & Schenker, J. (2002). Teachers as designers: Integrating robotics into early childhood education. *Information Technology in Childhood Education, 1*, 123–145.

Bers, M. U., & Resnick, M. (2015). *The official ScratchJr book.* San Francisco, CA: No Starch Press.

Bers, M. U., & Sullivan, A. (2018). *ScratchJr coding cards: creative coding activities.* San Francisco, CA: No Starch Press.

Beschorner, B., & Hutchison, A. (2013). iPads as a literacy teaching tool in early childhood. *Education Publications, 26*, 16–24.

Beschorner, B., & Hutchison, A. (2013). iPads as a literacy teaching tool in early childhood. *International Journal of Education in Mathematics, Science, and Technology, 1*(1), 16–24.

Bigelow, E. C. (2013). *iWrite: Digital Message Making Practices of Young Children.* (PhD Thesis, Vanderbilt University). Available at https://etd.library.vanderbilt.edu

Billinghurst, M., Clark, A., & Lee, G. (2015). A survey of augmented reality. *Foundations and trends in human-computer interaction, 8*(2-3), 73–272. doi:10.1561/1100000049

Bishop, J. (2014). Gamification for human factors integration: Social, education, and psychological issues. In *Gamification for human factors integration.* Social, Education, and Psychological Issues; doi:10.4018/978-1-4666-5071-8

Blackwell, C. K. (2015, May). iPads in kindergarten: The effect of tablet computers on young children's academic achievement. *International Communication Association*, San Juan, Puerto Rico.

Blackwell, C. (2013). Teacher practices with mobile technology integrating tablet computers into the early childhood classroom. *The Journal of Educational Research, 7*(4), 231–255.

Blackwell, C. K. (2013). Teacher practices with mobile technology: integrating iPads into the early childhood classroom. *The Journal of Educational Research, 7*(4), 1–25.

Blackwell, C. K., Lauricella, A. R., & Wartella, E. (2016). The influence of TPACK contextual factors on early childhood educators' tablet computer use. *Computers & Education, 98*, 57–69. doi:10.1016/j.compedu.2016.02.010

Blackwell, C. K., Lauricella, A. R., Wartella, E., Robb, M., & Schomburg, R. (2013). Adoption and use of technology in early education: The interplay of extrinsic barriers and teacher attitudes. *Computers & Education*, *69*, 310–319. doi:10.1016/j.compedu.2013.07.024

Blackwell, C. K., Wartella, E., Lauricella, A. R., & Robb, M. (2015). *Technology in the lives of educators and early childhood programs: Trends in access, use, and professional development from 2012 to 2014*. Center on Media and Human Development.

Blumberg, F. C., Deater-Deckard, K., Calvert, S. L., Flynn, R. M., Green, C. S., Arnold, D., & Brooks, P. J. (2019). Digital games as a context for children's cognitive development: research recommendations and policy considerations. *Social Policy Report*, *32*(1), 1–33. doi:10.1002op2.3

Bogdan, R., & Biklen, S. K. (2007). *Qualitative research for education: an introduction to theory and methods* (5th ed.). Boston, MA: Pearson.

Bohart, H., & Procopio, H. (2017). *Spotlight on young children: social and emotional development*. Washington, DC: NAEYC.

BootStrap. (n.d.). Retrieved from http://getbootstrap.com/

Bossé, M. J., Bayaga, A., Ringler, M. C., Fountain, C., & Young, E. S. (2018). Acquiring math: Connecting math learning and second language acquisition. *International Journal for Mathematics Teaching & Learning*, *19*(2), 223–252.

Bourgonjon, J., Valcke, M., Soetaert, R., & Schellens, T. (2010). Students' perceptions about the use of video games in the classroom. *Computers & Education*, *54*(4), 1145–1156. doi:10.1016/j.compedu.2009.10.022

Bowman, M., & Treiman, R. (2004). Stepping stones to reading. *Theory into Practice*, *43*(4), 295–303. doi:10.120715430421tip4304_8

Boyle, E. A., Connolly, T. M., Hainey, T., & Boyle, J. M. (2012). Engagement in digital entertainment games: A systematic review. *Computers in Human Behavior*, *28*(3), 771–780. doi:10.1016/j.chb.2011.11.020

BPS-Statistics. (2017). Telecommunication Statistics in Indonesia-2017 (Directorate of Communications and Information Technology). Jakarta: BPS-Statistics Indonesia.

Bratitsis, T. (2018). An Attempt for Critical Categorization of Android Applications Available for the Greek Kindergarten. In M. Auer & T. Tsiatsos (Eds.), *Interactive Mobile Communication Technologies and Learning. IMCL 2017. Advances in Intelligent Systems and Computing* (Vol. 725, pp. 56–68). Dordrecht: Springer. doi:10.1007/978-3-319-75175-7_7

Braund, M. (1998). *The development of children's ideas on animal classification, form and function; Is school experience becoming increasingly impoverished?* (Report No. 143). East Lansing, MI: National Center for Research on Teacher Learning. (ERIC Document Reproduction Service No. ED436423).

Brennan, K., & Resnick, M. (2012). New frameworks for studying and assessing the development of computational thinking. In *Proceedings of the 2012 annual meeting of the American Educational Research Association*, (Vol. 1, p. 25). Vancouver, Canada.

Brennan, K., Resnick, M., & Monroy-Hernandez, A. (2010). Making projects, making friends: Online community as a catalyst for interactive media creation. *New Directions for Youth Development*, *2010*(128), 75–83. doi:10.1002/yd.377 PMID:21240955

Briz-Ponce, L., Pereira, A., Carvalho, L., Juanes-Méndez, J. A., & García-Peñalvo, F. J. (2017). Learning with mobile technologies – Students' behavior. *Computers in Human Behavior*, *72*, 612–620. doi:10.1016/j.chb.2016.05.027

Brom, C., Levcic, D., Buchtova, M., & Klement, D. (2015). Playing educational micro-games at high schools: Individually or collectively? *Computers in Human Behavior, 48*, 682–694. doi:10.1016/j.chb.2015.02.025

Bronfenbrenner, U. (1979). *The ecology of human development.* Harvard University Press.

Bronfenbrenner, U., & Morris, P. A. (2006). *The bioecological model of human development.* Handbook of child psychology.

Brown, J., Winsor, D. L., & Blake, S. (2012). Technology and social-emotional development in the early childhood environments. In S. Blake, D. Winsor, & L. Allen (Eds.), *Child development and the use of technology: perspectives, applications, and experiences* (pp. 112–129). Hersey, PA: IGI Global. doi:10.4018/978-1-61350-317-1.ch006

Brown, M. B., & Forsythe, A. B. (1974). Robust tests for the equality of variances. *Journal of the American Statistical Association, 69*(346), 364–367. doi:10.1080/01621459.1974.10482955

Bryman, A. (1989). *Research methods and organizational studies.* London: Unwin Hyman.

Buitrago Flórez, F., Casallas, R., Hernández, M., Reyes, A., Restrepo, S., & Danies, G. (2017). Changing a generation's way of thinking: Teaching computational thinking through programming. *Review of Educational Research, 87*(4), 834–860. doi:10.3102/0034654317710096

Burger, W. F., & Shaughnessy, J. M. (1986). Characterizing the Van Hiele levels of development in geometry. *Journal for Research in Mathematics Education, 17*(1), 31–48. doi:10.2307/749317

Bybee, R. W., Taylor, J. A., Gardner, A., Van Scatter, P., Carlson-Powell, J., Westbrook, A., & Landes, N. (2006). *BSCS SE instructional model: Origins and effectiveness. A report prepared for the Office of Science Education, National Institutes of Health.* Colorado Springs, CO: BSCS.

Byers, P., & Hadley, J. (2013). Traditional and novel modes of activity in touch screen math apps. In J. P. Hourcade, N. Sawhney, & E. Reardon (Eds.), *Proceedings of the 12th International Conference on Interaction Design and Children.* New York: ACM.

Cacciamani, S., Villani, D., Bonanomi, A., Carissoli, C., Olivari, M. G., Morganti, L., & Confalonieri, E. (2018). Factors affecting students' acceptance of tablet PCs: A study in Italian high schools. *Journal of Research on Technology in Education, 50*(2), 120–133. doi:10.1080/15391523.2017.1409672

Cadoret, G., Bigras, N., Lemay, L., Lehrer, J., & Lemire, J. (2018). Relationship between screen-time and motor proficiency in children: A longitudinal study. *Early Child Development and Care, 188*(2), 231–239. doi:10.1080/03004430.2016.1211123

Çakıroğlu, Ü., & Öztürk, M. (2017). Pre-service computer teachers' tendencies towards the use of mobile technologies: A technology acceptance. *European Journal of Open Distance and E-Learning, 20*(1), 175–190.

Calder, N. (2015). Apps: Appropriate, Applicable, and Appealing? In T. Lowrie & R. Jorgensen (Zevenbergen) (Eds.), Mathematics Education in the Digital Era 4, Digital games and mathematics learning (pp. 233-250). Dordrecht: Springer.

Calder, N. (2015). Apps: Appropriate, applicable, and appealing? In T. Lowrie, & R. Jorgensen Zevenbergen (Eds.), *Digital games and mathematics learning. Mathematics Education in the Digital Era* (pp. 233–250). Dordrecht, The Netherlands: Springer. doi:10.1007/978-94-017-9517-3_12

Callaghan, M. (2018). *Educational apps designed for kids: Avoiding missed opportunities.* Retrieved from https://bold.expert/educational-apps-designed-for-kids-avoiding-missed-opportunities

Callaghan, M. N., & Reich, S. M. (2018). Are educational preschool apps designed to teach? An analysis of the app market. *Learning, Media and Technology, 43*(3), 280–293. doi:10.1080/17439884.2018.1498355

Camilleri, A. C., & Camilleri, M. A. (2019). The students' perceived use, ease of use and enjoyment of educational games at home and at school. *13th Annual International Technology, Education and Development Conference*. Valencia, Spain (March, 2019). International Academy of Technology, Education, and Development (IATED). Retrieved from https://papers.ssrn.com/sol3/papers.cfm?abstract_id=3339163

Campbell, F. A., Ramey, C. T., Pungello, E., Sparling, J., & Miller-Johnson, S. (2002). Early childhood education: Young adult outcomes from the Abecedarian Project. *Applied Developmental Science*, *6*(1), 42–57. doi:10.1207/S1532480XADS0601_05

Cardak, O. (2009). Science students' misconceptions about birds. *Scientific Research and Essays*, *4*(12), 1518–1522.

Carey, S. (2000). Science education as conceptual change. *Journal of Applied Developmental Psychology*, *21*(1), 13–19. doi:10.1016/S0193-3973(99)00046-5

Carpenter, P. A., Just, M. A., & Shell, P. (1990). What one intelligence test measures: A theoretical account of the processing in the Raven Progressive Matrices Test. *Psychological Review*, *97*(3), 404–431. doi:10.1037/0033-295X.97.3.404 PMID:2381998

Cascales, A., Laguna, I., Pérez-López, D., Perona, P., & Contero, M. (2013). An experience on natural sciences augmented reality contents for preschoolers. *Proceedings of the International Conference on Virtual, Augmented and Mixed Reality*, 103-112. Berlin, Germany: Springer. 10.1007/978-3-642-39420-1_12

Cavus, N., & Ibrahim, D. (2009). m-Learning: An experiment in using SMS to support learning new English language words. *British Journal of Educational Technology*, *40*(1), 78–91. doi:10.1111/j.1467-8535.2007.00801.x

Caws, P. (2015). General systems theory, its past and potential. *Systems Research and Behavioral Science*, *32*(5), 514–521. doi:10.1002res.2353

Cayton-Hodges, G. A., Feng, G., & Pan, X. (2015). Tablet-based math assessment: what can we learn from math apps? *Journal of Educational Technology & Society, 18*(2), 3-20.

Chaka, J. G., & Govender, I. (2017b). Students' perceptions and readiness towards mobile learning in colleges of education. A Nigerian perspective. *South African Journal of Education, 37*(1), 1–12.

Chaka, J. G., & Govender, I. (2017a). Students' perceptions and readiness towards mobile learning in colleges of education: A Nigerian perspective. *South African Journal of Education*, *37*(1), 1–12. doi:10.15700aje.v37n1a1282

Chang, C. C., Yan, C. F., & Tseng, J. S. (2012). Perceived convenience in an extended technology acceptance model. *Mobile Technology and English Learning for College Students, 28*(5), 809–826.

Chao, C. M. (2019). Factors determining the behavioral intention to use mobile learning: an application and extension of the UTAUT model. *Frontiers in Psychology, 10*(July), 1–14. PMID:31379679

ChatterPix Kids App. (2019). Retrieved from http://www.duckduckmoose.com/educational-iphone-itouch-apps-for-kids/chatterpixkids/

Chau, C. L. (2014). *Positive technological development for young children in the context of children's mobile apps* (Doctoral dissertation). Tufts University.

Chaudron, S. (2015). Young children (0-8) and digital technology: a qualitative exploratory study across seven countries. Luxembourg: Publications Office of the European Union.

Checkland, P., & Scholes, J. (1994). *La Metodología de los Sistemas Suaves en Acción*. México: Grupo Noriega.

Chen, G., Shen, J., Barth-Cohen, L., Jiang, S., Huang, X., & Eltoukhy, M. (2017). Assessing elementary students' computational thinking in everyday reasoning and robotics programming. *Computers & Education*, *109*, 162–175. doi:10.1016/j.compedu.2017.03.001

Cheng, G. (2019). Exploring factors influencing the acceptance of visual programming environment among boys and girls in primary schools. *Computers in Human Behavior*, *92*, 361–372. doi:10.1016/j.chb.2018.11.043

Chen, J., & Chang, C. (2006). Using computers in early childhood classrooms: Teachers' attitudes, skills and practices. *Journal of Early Childhood Research*, *4*(2), 169–188. doi:10.1177/1476718X06063535

Chen, N., Hung, I., & Wei, C. (2010). Developing ubiquitous learning system with robots for children's learning. In *2010 Third IEEE International Conference on Digital Game and Intelligent Toy Enhanced Learning* (pp. 61–68). 10.1109/DIGITEL.2010.17

Chen, P. S. D., Lambert, A. D., & Guidry, K. R. (2010). Engaging online learners: The impact of web-based technology on college student engagement. *Computers & Education*, *54*(4), 1222–1232. doi:10.1016/j.compedu.2009.11.008

Chen, R. (2010). Investigating models for preservice teachers' use of technology to support student-centered learning. *Computers & Education*, *55*(1), 32–42. doi:10.1016/j.compedu.2009.11.015

Chen, S. H., & Ku, C. H. (1998). Aboriginal children's alternative conceptions of animals and animal classification. *Proceedings of the National Science Council Republic of China Part D Mathematics Science and Technology Education*, *8*(2), 55–67.

Chen, W., Teo, M. H., & Nguyen, D. (2019). Singapore parents' use of digital devices with young children: Motivations and uses. *The Asia-Pacific Education Researcher*, *28*(2), 1–12. doi:10.100740299-019-00432-w

Cherner, T., Dix, J., & Lee, C. (2014). Cleaning up that mess: A framework for classifying educational apps. *Contemporary Issues in Technology & Teacher Education*, *14*(2), 158–193.

Chetty, R., Friedman, J. N., Hilger, N., Saez, E., Schanzenbach, D., & Yagan, D. (2011). How does your kindergarten classroom affect your earnings? Evidence from project STAR. *The Quarterly Journal of Economics*, *126*(4), 1593–1660. doi:10.1093/qje/qjr041 PMID:22256342

Childwise. (2016). *The monitor pre-school report: Key behaviour patterns among 0-4 year olds*. Retrieved from http://www.childwise.co.uk/

Chi, M. T. H. (2009). Active-constructive-interactive: A conceptual framework for differentiating learning activities. *Topics in Cognitive Science*, *1*(1), 73–105. doi:10.1111/j.1756-8765.2008.01005.x PMID:25164801

Ching, Y. H., Hsu, Y. C., & Baldwin, S. (2018). Developing computational thinking with educational technologies for young learners. *TechTrends*, *62*(6), 563–573. doi:10.100711528-018-0292-7

Chiong, C., Ree, J., Takeuchi, L., & Erickson, I. (2012). *Print Books vs e-Books*. New York, NY: The Joan Ganz Cooney Center.

Chou, C. C., Block, L., & Jesness, R. (2012). A case study of mobile learning pilot project in K12 schools. *Journal of Educational Technology Development and Exchange*, *5*(2), 11–26. doi:10.18785/jetde.0502.02

Christopoulou, E., & Ringas, D. (2013). Learning activities in a sociable smart city. *Interaction Design and Architecture(s) Journal*, 29–42.

Chrysoulas, C. (2017). Building an adaptive e-learning system building an adaptive e-learning system. In *9th International Conference on Computer Supported Education*. 10.5220/0006326103750382

Chyleńska, Z. A., & Rybska, E. (2018). Understanding students ideas about animal classification. *EURASIA Journal of Mathematics. Science and Technology Education, 14*(6), 2145–2155. doi:10.29333/ejmste/86612

Ciampa, K. (2014). Learning in a mobile age: An investigation of student motivation. *Journal of Computer Assisted Learning, 30*(1), 82–96. doi:10.1111/jcal.12036

Cingel, D. P., Sumter, S. R., Stoeten, E., & Mann, S. (2019). Can television help to decrease stigmatization among young children? The role of Theory of Mind and general and explicit inserts. *Media Psychology*, 15213269.2019.1601570. doi:10.1080/15213269.2019.1601570

Cinici, A. (2013). Turkish high school students' ideas about invertebrates: General characteristics and classification. *International Journal of Environmental and Science Education, 8*(4), 645–661.

Clarke, B., & Svanaes, S. (2014). *An updated literature review on the use of tablets in education. Tablets for Schools.* UK: Family Kids & Youth.

Clarke, L., & Abbott, L. (2016). Young pupils', their teachers' and classroom assistants' experiences of iPads in a Northern Ireland school: Four and five years old, who would have thought they could do that? *British Journal of Educational Technology, 47*(6), 1051–1064. doi:10.1111/bjet.12266

Clay, M. M. (1998). *By different paths to common outcomes.* Portland, ME: Stenhouse.

Clements, D. H., & Sarama, J. (2013). Rethinking early mathematics: What is research based curriculum for young children? In L. D. English, & J. T. Mulligan (Eds.), *Reconceptualizing early mathematics learning* (pp. 121–147). The Netherlands: Springer. doi:10.1007/978-94-007-6440-8_7

CMI-5 and the Experience API - Experience API. (n.d.). Retrieved from https://experienceapi.com/cmi5/

Coley, J. D., Arenson, M., Xu, Y., & Tanner, K. D. (2017). Intuitive biological thought: Developmental changes and effects of biology education in late adolescence. *Cognitive Psychology, 92*, 1–21. doi:10.1016/j.cogpsych.2016.11.001 PMID:27865155

Collaboration for academic, social, and emotional learning. (n.d.). Framework for systematic social and emotional learning. Chicago, IL: Author. Retrieved from https://casel.org/what-is-sel/

Collins, A., & Halverson, R. (2009). *Rethinking education in the age of technology: the digital revolution and schooling in America.* New York, NY: Teachers College Press.

Common Sense Media. (2013). *Zero to eight: Children's media use in America 2013.* Retrieved from https://www.commonsensemedia.org/research/zero-to-eight-childrens-media-use-in-america-2013

Common Sense Media. (2017). Zero to eight: Children's media use in America: A common sense research study. Retrieved from https://www.commonsensemedia.org/sites/default/files/uploads/research/csm_zerotoeight_fullreport_release_2.pdf

Connolly, T. M., Boyle, E. A., MacArthur, E., Hainey, T., & Boyle, J. M. (2012). A systematic literature review of the empirical evidence on computer games and serious games. *Computers & Education, 59*(2), 661–686. doi:10.1016/j.compedu.2012.03.004

Conti, G., Heckman, J. J., & Pinto, R. (2016). The effects of two influential early childhood interventions on health and healthy behaviour. *Economic Journal (London), 126*(596), F28–F65. doi:10.1111/ecoj.12420 PMID:28260805

Cook, V. J. (1997). L2 users and English spelling. *Journal of Multilingual and Multicultural Development, 18*(6), 474–488. doi:10.1080/01434639708666335

Cope, B., & Kalantzis, M. (2008). Ubiquitous learning : an agenda for educational transformation. In *6th International Conference on Networked Learning* (pp. 576–582).

Corbi, A., & Burgos, D. (2014). Review of current student-monitoring techniques used in elearning-focused recommender systems and learning analytics. The Experience API & LIME model case study. *IJIMAI*, *2*(7), 44–52. doi:10.9781/ijimai.2014.276

Cordes, C., & Miller, E. (2000). *Fool's gold: A critical look at computers in childhood.* College Park, MD: Alliance for Childhood.

Coughlan, S. (2014). *Tablet computers in 70% of schools.* Retrieved from http://www.bbc.com/news/education/30216408

Courage, M. L. (2019). *Interactive technologies: tablets, smartphones, and learning. Reference module in neuroscience and biobehavioral psychology.* Elsevier; doi:10.1016/B978-0-12-809324-5.23561-X

Couse, L. J., & Chen, D. W. (2010). A tablet computer for young children? Exploring its viability for early childhood education. *Journal of Research on Technology in Education*, *43*(1), 75–96. doi:10.1080/15391523.2010.10782562

Coyle, Y., & Gómez Gracia, R. (2014). Using songs to enhance L2 vocabulary acquisition in preschool children. *ELT Journal*, *68*(3), 276–285. doi:10.1093/elt/ccu015

Crescenzi, L., Jewitt, C., & Price, S. (2014). The role of touch in preschool children's learning using iPad versus paper interaction. *Australian Journal of Language and Literacy*, *37*(2), 86–95.

Crompton, H., & Burke, D. (2018). The use of mobile learning in higher education: A systematic review. *Computers & Education*, *123*, 53–64. doi:10.1016/j.compedu.2018.04.007

Crowley, M. L. (1987). The Van Hiele model of the development of geometric thought. In M. M. Lindquist (Ed.), *Learning and teaching geometry, K-12* (pp. 1–16). Reston, VA: NCTM.

Cubelic, C. J. (2013). *iPad 2 applications and emergent literacy: do they have an impact on the acquisition of early literacy skills?* (PhD Thesis, Youngstown State University). Available at http://rave.ohiolink.edu/etdc/

Curenton, S. M., & Justice, L. M. (2008). Children's preliteracy skills: Influence of mothers' education and beliefs about shared-reading interactions. *Early Education and Development*, *19*(2), 261–283. doi:10.1080/10409280801963939

Dafoulas, G. A., Maia, C. C., Clarke, J. S., Ali, A., & Augusto, J. (2018). Investigating the role of biometrics in education -- the use of sensor data in collaborative learning. *International Conference on E-Learning*, 115–123. Retrieved from http://search.ebscohost.com/login.aspx?direct=true&db=a9h&AN=131764741&site=ehost-live

Danby, S., Evaldsson, A. C., Melander, H., & Aarsand, P. (2018). Situated collaboration and problem solving in young children's digital gameplay. *British Journal of Educational Technology*, *49*(5), 959–972. doi:10.1111/bjet.12636

Dashti, F. A., & Yateem, A. K. (2018). Use of mobile devices: A case study with children from Kuwait and the United States. *International Journal of Early Childhood*, *50*(1), 121–134. doi:10.100713158-018-0208-x

Davis, F. D. (1993). User acceptance of information technology: System characteristics, user perceptions and behavioral impacts. *International Journal of Man-Machine Studies*, *38*(3), 475–487. doi:10.1006/imms.1993.1022

De Bruijn, H., & Herder, P. (2009). System and actor perspectives on sociotechnical systems. *IEEE Transactions on systems, man, and cybernetics. Part a, systems and humans*, *39*(5), 981–992. doi:10.1109/TSMCA.2009.2025452

De Curtis, L. L., & Ferrer, D. (2011). Toddlers and technology: Teaching the techniques. *ASHA Leader*, *16*. Retrieved from http://leader.pubs.asha.org/article.aspx?articleid=2280052

De Simone, J. J. (2013). What is good can also be bad: The prosocial and antisocial in-game behaviors of young video game players. *Atlantic Journal of Communication, 21*(3), 149–163. doi:10.1080/15456870.2013.801756

Del Blanco, A., Serrano, A., Freire, M., Martinez-Ortiz, I., & Fernandez-Manjon, B. (2013). E-Learning standards and learning analytics. Can data collection be improved by using standard data models? *IEEE Global Engineering Education Conference, EDUCON*, 1255–1261. 10.1109/EduCon.2013.6530268

Desai, R. A., Krishnan-Sarin, S., Cavallo, D., & Potenza, M. N. (2010). Video-gaming among high school students: Health correlates, gender differences, and problematic gaming. *Pediatrics, 126*(6), 1414–1424. doi:10.1542/peds.2009-2706 PMID:21078729

Deterding, S., Khaled, R., Nacke, L., & Dixon, D. (2011). Gamification: toward a definition. *Chi 2011*, 12–15.

DfES. (2007). *Department for Education and Skills, Special Educational Needs Code of Practice.* [Ebook]. London, UK. Retrieved from https://assets.publishing.service.gov.uk/government/uploads/system/uploads/attachment_data/file/228675/7092.pdf

Dickinson, D. K., & McCabe, A. (2001). Bringing it all together: The multiple origins, skills, and environmental supports of early literacy. *Learning Disabilities Research & Practice, 16*(4), 186–202. doi:10.1111/0938-8982.00019

Diemer, T. T., Fernandez, E., & Streepey, J. W. (2012). Student perceptions of classroom engagement and learning using iPads. *Journal of Teaching and Learning with Technology, 1*(2), 13–25.

Dillenbourg, P. (1999). *Collaborative learning: Cognitive and computational approaches. Advances in learning and instruction series.* New York, NY: Elsevier Science.

Dinçay, T. (2010). A quick chronological review of the ELT methods along with their techniques and principles: choosing eclecticism from among language teaching methods. *Dil Dergisi*, (147), 40-62.

Dindar, M. (2018). An empirical study on gender, video game play, academic success and complex problem solving skills. *Computers & Education, 125*, 39–52. doi:10.1016/j.compedu.2018.05.018

Domingo, M. G., & Garganté, A. B. (2016). Exploring the use of educational technology in primary education: Teachers' perception of mobile technology learning impacts and applications' use in the classroom. *Computers in Human Behavior, 56*, 21-28.

Donahoo, D. (2012). *The Future of Apps for Young Children: Beyond ABC & 123.* Retrieved from https://www.huffpost.com/entry/the-future-of-apps-for-yo_b_1771395

Donahoo, D. (2013). *Seeking better apps for young children.* Retrieved from https://www.fredrogerscenter.org/2013/03/seeking-better-apps-for-young-children/

Dorouka, P., Papadakis, St., & Kalogiannakis, M. (2019). ((Forthcoming). Tablets & apps for promoting robotics, mathematics, STEM education and literacy in early childhood education. *International Journal of Mobile Learning and Organisation*. Retrieved from https://www.inderscience.com/info/ingeneral/forthcoming.php?jcode=ijmlo

Drigas, A., & Kokkalia, G. (2014). ICTs in kindergarten. *International Journal of Emerging Technologies in Learning, 9*(2), 52–58. doi:10.3991/ijet.v9i2.3278

Driver, R. (1985). *Children's ideas in science.* UK: McGraw-Hill Education.

Driver, R., Squires, A., Rushworth, P., & Wood-Robinson, V. (2014). *Making sense of secondary science: Research into children's ideas*. Routledge.

Dubé, A. K., Alam, S. S., Xu, C., Wen, R., & Kacmaz, G. (2019). Tablets as Elementary Mathematics Education Tools: Are They Effective and Why. In K. Robinson, H. Osana, & D. Kotsopoulos (Eds.), *Mathematical Learning and Cognition in Early Childhood: Integrating Interdisciplinary Research into Practice* (pp. 223–248). Dordrecht: Springer. doi:10.1007/978-3-030-12895-1_13

Dubé, A. K., Kacmaz, G., Wen, R., Xu, C., & Alam, S. S. (2019). Mathematics apps in the App Store: Low transparency impedes educator's ability to find quality apps. *Proceeding of the American Educational Research Association (AERA).*

Dündar, H., & Akçayır, M. (2014). Implementing tablet PCs in schools: Students' attitudes and opinions. *Computers in Human Behavior, 32*, 40–46. doi:10.1016/j.chb.2013.11.020

Dunn, S. E. (2000). Technology: Where is it taking us? A "writing" perspective. *Montessori Life, 12*(1), 34–35.

Dziuban, C., Graham, C. R., Moskal, P. D., Norberg, A., & Sicilia, N. (2018). Blended learning: The new normal and emerging technologies. *International Journal of Educational Technology in Higher Education, 15*(1), 3. doi:10.118641239-017-0087-5

Early Childhood Australia (ECA). (2018). *Statement on young children and digital technologies*. Canberra, ACT: ECA. Retrieved from http://www.earlychildhoodaustralia.org.au/wp-content/uploads/2018/10/Digital-policy-statement.pdf

Early Learning Advisory Group. (2008). *British Columbia early learning framework*. British Columbia, Canada: Ministry of Health and Ministry of Children and Family Development.

Ebbeck, M., Yim, H., Chan, Y., & Goh, M. (2016). Singaporean parents' views of their young children's access and use of technological devices. *Early Childhood Education Journal, 44*(2), 127–134. doi:10.100710643-015-0695-4

Edmodo. (n.d.). Retrieved from https://www.edmodo.com/?simplified_landing_page=1&go2url=%2Fuser%3Fuid%3D3742546

Edmonds-Wathen, C., Owens, K., & Bino, V. (2019). Identifying vernacular language to use in mathematics teaching. *Language and Education, 33*(1), 1–17. doi:10.1080/09500782.2018.1488863

Edwards, S., Nolan, A., Henderson, M., Mantilla, A., Plowman, L., & Skouteris, H. (2018). Young children's everyday concepts of the internet: A platform for cyber-safety education in the early years. *British Journal of Educational Technology, 49*(1), 45–55. doi:10.1111/bjet.12529

Edwards, S., Nolan, A., Henderson, M., Skouteris, H., Mantilla, P., & Bird, J. (2016). Developing a measure to understand young children's internet cognition and cybersafety awareness: A pilot test. *Early Years, 36*(3), 322–335. doi:10.1080/09575146.2016.1193723

Eisen, S., & Lillard, A. S. (2017). Young children's thinking about touchscreens versus other media in the US. *Journal of Children and Media, 11*(2), 167–179. doi:10.1080/17482798.2016.1254095

Ekinci, N. E., Yalcin, I., & Ayhan, C. (2019). Analysis of loneliness levels and digital game addiction of middle school students according to various variables. *World Journal of Education, 9*(1), 20–27. doi:10.5430/wje.v9n1p20

Elias, T. (2011). *Learning analytics*. Retrieved from https://pdfs.semanticscholar.org/732e/452659685fe3950b0e515a28ce89d9c5592a.pdf

Elkin, M., Sullivan, A., & Bers, M. U. (2014). Implementing a robotics curriculum in an early childhood Montessori classroom. *Journal of Information Technology Education: Innovations in Practice, 13*, 153–169. Retrieved from http://www.jite.org/documents/Vol13/JITEv13IIPvp153-169Elkin882.pdf

Ellis, R. (2005). Principles of instructed language learning. *System*, 33.

English, L. D. (2016). STEM education K-12: Perspectives on integration. *International Journal of STEM Education, 3*(3), 1–8.

Ericsson. (2015, June). *Ericsson Mobility Report, On the pulse of the network society*. Retrieved from https://www.ericsson.com/4980fe/assets/local/news/2015/6/ericsson-mobility-report-june-20152.pdf

Ericsson. (2016, June). *Ericsson Mobility Report, On the pulse of the network society*. Retrieved from https://www.abc.es/gestordocumental/uploads/internacional/EMR_June_2016_D5%201.pdf

Ertmer, P. A. (1999). Addressing first-and second-order barriers to change: Strategies for technology integration. *Educational Technology Research and Development, 47*(4), 47–61. doi:10.1007/BF02299597

ESA. (2015). *Essential facts about the computer and video game industry*. Retrieved from http://www.theesa.com/wp-content/uploads/2015/04/ESA-Essential-Facts-2015.pdf

Eshach, H. (2006). *Science literacy in primary schools and pre-schools* (Vol. 1). Springer Science & Business Media; doi:10.1007/1-4020-4674-X

Estapa, A. T., & Tank, K. M. (2017). Supporting integrated STEM in the elementary classroom: a professional development approach centered on an engineering design challenge. *International Journal of STEM education, 4*(1), 6.

Eteokleous, N. (2008). Evaluating computer technology integration in a centralized school system. *Computers & Education, 51*(2), 669–686. doi:10.1016/j.compedu.2007.07.004

Eurostat, Statistical books. (2015). *Being young in Europe today*. Retrieved from http://ec.europa.eu/eurostat/documents/3217494/6776245/KS-05-14-031-EN-N.pdf/18bee6f0-c181-457d-ba82-d77b314456b9

Experience, A. P. I. (n.d.). Appendix A. Revision history. (n.d.). Retrieved from https://github.com/adlnet/xAPI-Spec/blob/master/xAPI-About.md#Appendix1A

Experience, A. P. I. Adopters. (n.d.). Retrieved from https://xapi.com/adopters/

ExperienceA. P. I. (n.d.). Retrieved from https://xapi.com/

Eynon, R. (2005, April). The use of the internet in higher education: Academics' experiences of using ICTs for teaching and learning. [Emerald Group Publishing Limited.]. *Aslib Proceedings, 57*(2), 168–180. doi:10.1108/00012530510589137

Fadhli, M. (2016). Pemikiran Howard Gardner dalam Pendidikan Anak Usia Dini. *Jurnal Indria, 1*(1), 80–87.

Fadhli, M. (2015). *Implementasi beyond center and circle time (BCCT) di TK Aisyiah Dagangan Madiun*. Ponorogo.

Falkner, K. (2015). *Computational thinking as the 'new literacy': professional development opportunities*. Retrieved from https://www.scisdata.com/connections/issue-95/computational-thinking-as-the-new-literacy-professional-development-opportunities/

Falloon, G. (2013). Young students using ipads: App design and content influences on their learning pathways. *Computers & Education, 68*, 505–521. doi:10.1016/j.compedu.2013.06.006

Falloon, G. (2014). What's going on behind the screens? Researching young students' learning pathways using iPads. *Journal of Computer Assisted Learning, 30*(4), 318–336. doi:10.1111/jcal.12044

Falloon, G. (2015). What's the difference? Learning collaboratively using iPads in conventional classrooms. *Computers & Education, 84*, 62–77. doi:10.1016/j.compedu.2015.01.010

Falloon, G., & Khoo, E. (2014). Exploring young students' talk in iPad-supported collaborative learning environments. *Computers & Education, 77*, 13–28. doi:10.1016/j.compedu.2014.04.008

Farhana, L., Borhan, N., & Yatim, M. (2013). A field study of understanding child's knowledge, skills and interaction towards capacitive touch technology (iPad). In *38th International Conference Information Technology in Asia (CITA)*. IEEE. 1-5.

Feehan, K., Shamir, H., & Yoder, E. (2017). Does CAI improve early math skills? In *Proceedings of the 9th International Conference on Computer Supported Education* (CSEDU 2017), 2, (pp. 285-292), Porto.

Feldman, R., Bamberger, E., & Kanat-Maymon, Y. (2013). Parent-specific reciprocity from infancy to adolescence shapes children's social competence and dialogical skills. *Attachment & Human Development, 15*(4), 407–423. doi:10.1080/14616734.2013.782650 PMID:23544455

Fernandez, J., & Lopez, M. (2013). Communication, destination brands and mobile applications. *Communicatio Socialis, 26*(2), 95–112.

Fessakis, G., Gouli, E., & Mavroudi, E. (2013). Problem solving by 5–6 years old kindergarten children in a computer programming environment: A case study. *Computers & Education, 63*, 87–97. doi:10.1016/j.compedu.2012.11.016

Findlay, J. (2016). Retrieved from https://trainingindustry.com/articles/learning-technologies/game-based-learning-vs-gamification-do-you-know-the-difference/

Fink, M. A. (2018). *Effective iPad integration in the kindergarten literacy curriculum through creation-based literacy tasks: an action research study*. (PhD Thesis, University of Pittsburgh). Available at http://d-scholarship.pitt.edu/35369/1/

Fisch, S. M., & Truglio, R. T. (Eds.). (2001). *"G" is for growing: Thirty years of research on children and Sesame Street*. Mahwah, NJ: Lawrence Erlbaum Associates.

Fisher, B., Lucas, T., & Galstyan, A. (2013). The role of iPads in constructing collaborative learning spaces. *Technology, Knowledge, and Learning, 18*(3), 165–178. doi:10.100710758-013-9207-z

Fisher, T., Denning, T., Higgins, C., & Loveless, A. (2012). Teachers' knowing how to use technology: Exploring a conceptual framework for purposeful learning activity. *Curriculum Journal, 23*(3), 307–325. doi:10.1080/09585176.2012.703492

Flannery, L. P., & Bers, M. U. (2013). Let's dance the 'robot hokey-pokey!' Children's programming approaches and achievement throughout early cognitive development. *Journal of Research on Technology in Education, 46*(1), 81–101. doi:10.1080/15391523.2013.10782614

Fleck, B. K. B., Beckman, L. M., Sterns, J. L., & Hussey, H. D. (2014). YouTube in the classroom: helpful tips and student perceptions. *The Journal of Effective Teaching, 14*(3), 21–37. Retrieved from http://uncw.edu/cte/et/articles/Vol14_3/Fleck.pdf

Fleer, M. (2018). Digital animation: New conditions for children's development in play-based setting. *British Journal of Educational Technology, 49*(5), 943–958. doi:10.1111/bjet.12637

Flewitt, R., Kucirkova, N., & Messer, D. (2014). Touching the virtual, touching the real: iPads and enabling literacy for students experiencing disability. *Australian Journal of Language and Literacy, 37*(2), 107–116.

Flewitt, R., Messer, D., & Kucirkova, N. (2014). New directions for early literacy in a digital age: The iPad. *Journal of Early Childhood Literacy, 15*(4), 1–22.

Flewitt, R., Messer, D., & Kucirkova, N. (2015). New directions for early literacy in a digital age: the iPad. *Journal of Early Childhood Literacy, 15*(3), 289–310. doi:10.1177/1468798414533560

Fokides, E., Kaimara, P., Deliyiannis, I., & Atsikpasi, P. (2019). Development of a scale for measuring the learning experience in serious games. Preliminary results. *Proceedings of the Digital Culture & Audiovisual Challenges International Conference*. Corfu, Greece: Ionian University.

Fokides, E. (2018). Tablets in primary schools: Results of a study for teaching the human organ systems. *International Journal of Smart Education and Urban Society*, *9*(3), 1–16. doi:10.4018/IJSEUS.2018070101

Fokides, E., & Atsikpasi, P. (2017). Tablets in education. Results from the initiative ETiE, for teaching plants to primary school students. *Education and Information Technologies*, *22*(5), 2545–2563. doi:10.100710639-016-9560-3

Fokides, E., & Mastrokoukou, A. (2018). Results from a study for teaching human body systems to primary school students using tablets. *Contemporary Educational Technology*, *9*(2), 154–170. doi:10.30935/cet.414808

Fonseca, M. S. (2005). *Comunicación oral. Fundamentos y práctica estratégica*. México: Pearson.

Frahasini, Marhaeni, T., Astuti, P., & Atmaja, H. T. (2018). *The impact of the use of gadgets in school of school age towards children' s social behavior in Semata Village, 7*(2), 161–168.

Fred Rogers Center. (2014). Retrieved from http://www.fredrogerscenter.org/

Freeman, N. K., & Somerindyke, J. (2001). Social play at the computer: Preschoolers scaffold and support peers' computer competence. *Information technology in childhood education annual*, 1, 203–213.

Gardner, H. (1988). *Frames of mind* (2nd ed.). doi:10.2307/3324261

Gardner, H. (2008). *Multiple intelligences: new horizons in theory and practice*. USA: Basic Group.

Gardner, H. (2011). *Frames of mind the theory of multiple intelligences* (2nd ed.). New York: Basic Books.

Geist, E. (2014). Using tablet computers with toddlers and young pre-schoolers. *Young Children*, *69*, 58–63.

Geist, E. (2014). Using tablet computers. *Young Children*, *69*(1), 58–63.

Gelman, S. A., & Meyer, M. (2011). Child categorization. *Wiley Interdisciplinary Reviews: Cognitive Science*, *2*(1), 95–105. doi:10.1002/wcs.96 PMID:23440312

Genlott, A. A., & Grönlund, Å. (2016). Closing the gaps–Improving literacy and mathematics by ICT-enhanced collaboration. *Computers & Education*, *99*, 68–80. doi:10.1016/j.compedu.2016.04.004

Georgiou, G. K., Torppa, M., Manolitsis, G., Lyytinen, H., & Parrila, R. (2012). Longitudinal predictors of reading and spelling across languages varying in orthographic consistency. *Reading and Writing*, *25*(2), 321–346. doi:10.100711145-010-9271-x

Gialamas, V., & Nikolopoulou, K. (2010). In-service and pre-service early childhood teachers' views and intentions about ICT use in early childhood settings: A comparative study. *Computers & Education*, *55*(1), 333–341. doi:10.1016/j.compedu.2010.01.019

Gibbs, G. R. (2007). Thematic coding and categorizing. In Analyzing qualitative data, 38-56. London, UK: Sage. doi:10.4135/9781849208574.n4

Glahn, C. (2013). Using the ADL experience API for mobile learning, sensing, informing, encouraging, orchestrating. In *7th International Conference on Next Generation Mobile Applications, Services, and Technologies* (pp. 268–273). 10.1109/NGMAST.2013.55

Glezou, K. (2018). *Utilizing educational robotics systems in didactic practice*. Paper presented at the Pan-Hellenic Scientix Conference on STEM Education, NTUA of Zografou, Zografou, Athens, Greece. (In Greek)

Glezou, K. (2017). Program and learn - Playing with the BeeBot educational robotics system. In *Proceedings of the 4th Pan-Hellenic eTwinning Conference on Co-operative Programs in Primary and Secondary Education "Exploiting Information and Communication Technologies in Collaborative School Programs"*. Patras, Greece. (In Greek)

Glezou, K., Savvidaki, A., & Birbilis, G. (2015). Lego WeDo - Scratch: Constructing and programming. In *Proceedings of the 8th Panhellenic Conference of ICT Teachers "Exploiting Information and Communication Technologies in Teaching Practice"* (pp. 1392-1401). Syros, Greece. (In Greek)

Godfrey, B. J., & Reed, M. (2013). *App store after five years.* Retrieved from http://actonline.org/2013/07/19/app-store-after-five-years/

Godwin-Jones, R. (2011). Emerging technologies mobile apps for language learning. *Language Learning & Technology, 15*(2), 2–11. 1094-3501.

Gomes, T. C. S., Falcão, T. P., & Tedesco, P. C. D. A. R. (2018). Exploring an approach based on digital games for teaching programming concepts to young children. *International Journal of Child-Computer Interaction, 16*, 77-84.

Gomez, F., Nussbaum, M., Weitz, J. F., Lopez, X., Mena, J., & Torres, A. (2013). Co-located single display collaborative learning for early childhood education. *Computer-Supported Collaborative Learning, 8*(2), 225–244. doi:10.100711412-013-9168-1

Goodwin, K., & Highfield, K. (2012). *iTouch and iLearn: an examination of 'educational' Apps.* Paper presented at the Early Education and Technology for Children conference, Salt Lake City, UT.

Goodyear, P., Jones, C., & Thompson, K. (2014). Computer-supported collaborative learning: Instructional approaches, group processes and educational designs. In *Handbook of research on educational communications and technology* (pp. 439–451). New York, NY: Springer. doi:10.1007/978-1-4614-3185-5_35

Gouws, L. A., Bradshaw, K., & Wentworth, P. (2013). Computational thinking in educational activities. *Proceedings of the 18th ACM Conference on Innovation and Technology in Computer Science Education. ACM, New York, NY.*

Grabill, J. T., & Hicks, T. (2005). Multiliteracies meet methods: The case for digital writing in English education. *English Education, 37*(4), 301–311.

Graham, C. R. (2013). Emerging practice and research in blended learning. In M. G. Moore (Ed.), *Handbook of distance education* (3rd ed., pp. 333–350). New York: Routledge. doi:10.4324/9780203803738.ch21

Gray, C., Dunn, J., Moffett, P., & Mitchell, D. (2017). *Mobile devices in early learning: Evaluating the use of portable devices to support young children's learning.* Report commissioned by Irish Education Authority.

Gray, C., Dunn, J., Moffett, P., & Mitchell, D. (2017). *Mobile Devices in Early Learning: Evaluating the Use of Portable Devices to Support Young Children's Learning.* Belfast, UK: Stranmillis University College.

Greaves, T., Hayes, J., Wilson, L., Gielniak, M., & Peterson, R. (2012). *Revolutionizing education through technology: The project RED roadmap for transformation.* Eugene, OR: International Society for Technology in Education.

Grover, S. (2018). *The 5th 'C' of 21st century skills? Try computational thinking (not coding).* Retrieved from https://www.edsurge.com/news/2018-02-25-the-5th-c-of-21st-century-skills-try-computational-thinking-not-coding

Grover, S., & Pea, R. (2013). Computational thinking in K–12: A review of the state of the field. *Educational Researcher, 42*(1), 38–43. doi:10.3102/0013189X12463051

Grunwald Associates LLC. (2013). *Living and learning with mobile devices: What parents think about mobile devices for early childhood and K-12 learning.* USA: Learning First Alliance.

Guernsey, L., & Levine, M. (2015). *Tap Click Read: Growing Readers in a World of Screens*. San Francisco, CA: Jossey Bass.

Guernsey, L., Levine, M., Chiong, C., & Severns, M. (2012). *Pioneering literacy in the digital wild west: Empowering parents and educators*. Washington, DC: Campaign for Grade-Level Reading.

Gumusgul, O. (2019). Investigation of University students' attitudes to play educational games and games consisting of physical activity. *World Journal of Education*, *9*(2), 31–37. doi:10.5430/wje.v9n2p31

Guss, F. (2005). Reconceptualizing play: Aesthetic self-definitions. *Contemporary Issues in Early Childhood*, *6*(3), 233–243. doi:10.2304/ciec.2005.6.3.4

Gutiérrez, A. D., & Cannon, B. Y. (2014). Impacto de los recursos educativos abiertos (REA) en el aprendizaje del inglés de alumnos de preescolar. In Y. Heredia & B. Y. Cannon (Eds.), Aplicaciones de la tecnología educativa en diversos ambientes de Aprendizaje, 27-52. Monterrey: ITESM.

Güzer, B., & Caner, H. (2014). The past, present and future of blended learning: An in depth analysis of literature. *Procedia: Social and Behavioral Sciences*, *116*, 4596–4603. doi:10.1016/j.sbspro.2014.01.992

H5P and xAPI. (n.d.). Retrieved from https://h5p.org/documentation/x-api

H5P. (n.d.). Retrieved from https://h5p.org

Habler, B., Major, L., & Hennessy, S. (2015). Tablet use in schools: a critical review of the evidence for learning outcomes. *Journal of Computer Assisted Learning*, *32*(2), 138–156.

Hadlington, L., White, H., & Curtis, S. (2019). "I cannot live without my [tablet]": Children's experiences of using tablet technology within the home. *Computers in Human Behavior*, *94*, 19–24. doi:10.1016/j.chb.2018.12.043

Ha, I., Yoon, Y., & Choi, M. (2007). Determinants of adoption of mobile games under mobile broadband wireless access environment. *Information & Management*, *44*(3), 276–286. doi:10.1016/j.im.2007.01.001

Hainey, T., Connolly, T. M., Boyle, E. A., Wilson, A., & Razak, A. (2016). A systematic literature review of games-based learning empirical evidence in primary education. *Computers & Education*, *102*, 202–223. doi:10.1016/j.compedu.2016.09.001

Hainey, T., Westera, W., Connolly, T. M., Boyle, L., Baxter, G., Beeby, R. B., & Soflano, M. (2013). Students' attitudes toward playing games and using games in education: Comparing Scotland and the Netherlands. *Computers & Education*, *69*, 474–484. doi:10.1016/j.compedu.2013.07.023

Hair, J. F. Jr, Anderson, R. E., Tatham, R. L., & Black, W. G. (1998). *Multivariate data analysis* (5th ed.). London, UK: Prentice-Hall.

Hamidi, H., & Jahanshaheefard, M. (2019). Essential factors for the application of education information system using mobile learning: A case study of students of the university of technology. *Telematics and Informatics, 38*(December 2017), 207–224.

Hamidi, H., & Chavoshi, A. (2018). Telematics and informatics analysis of the essential factors for the adoption of mobile learning in higher education: A case study of students of the University of Technology. *Telematics and Informatics*, *35*(4), 1053–1070. doi:10.1016/j.tele.2017.09.016

Handal, B., El-Khoury, J., Campbell, C., & Cavanagh, M. (2013). A framework for categorising mobile applications in mathematics education. *Australian Conference on Science and Mathematics Education*. Retrieved from https://researchonline.nd.edu.au/edu_conference/70/

Han, I., & Sug, W. (2016). The use of a mobile learning management system and academic achievement of online students. *Computers & Education, 102*, 79–89. doi:10.1016/j.compedu.2016.07.003

Harel, I. (2016). *American schools are teaching our kids how to code all wrong*. Retrieved from https://qz.com/691614/american-schools-are-teaching-our-kids-how-to-code-all-wrong/

Harel, I., & Papert, S. (1991). *Constructionism: Research reports & essays, 1985-1990 by the epistemology & learning research group. Norwood*, NJ: Ablex Publishing.

Harlen, W. (2018). *The teaching of science in primary schools*. David Fulton Publishers. doi:10.4324/9781315398907

Hatzigianni, M., Gregoriadis, A., Karagiorgou, I., & Chatzigeorgiadou, S. (2018). Using tablets in free play: The implementation of the digital play framework in Greece. *British Journal of Educational Technology, 49*(5), 928–942. doi:10.1111/bjet.12620

Hatzigianni, M., & Kalaitzidis, I. (2018). Early childhood educators' attitudes and beliefs around the use of touchscreen technologies by children under three years of age. *British Journal of Educational Technology, 49*(5), 883–895. doi:10.1111/bjet.12649

Haugland, S. (1999). What role should technology play in young children's learning? *Young Children, 54*(9), 26–30.

Heckman, J. J., & Karapakula, G. (2019). *Intergenerational and Intragenerational Externalities of the Perry Preschool Project (No. w25889)*. National Bureau of Economic Research. doi:10.3386/w25889

Heflin, H., Shewmaker, J., & Nguyen, J. (2017). Impact of mobile technology on student attitudes, engagement, and learning. *Computers & Education, 107*, 91–99. doi:10.1016/j.compedu.2017.01.006

Heikkilä, M., & Mannila, L. (2018). Debugging in programming as a multimodal practice in early childhood education settings. *Multimodal Technologies and Interaction, 2*(3), 42. doi:10.3390/mti2030042

Hellenic Ministry of Greek National Education & Pedagogical Institute (HMNE-PI). (2003). *A cross thematic curriculum framework for compulsory education (Diathematikon Programma)*. Retrieved from http://www.pischools.gr/programs/depps/

Hendela, T. (2014). App Store sales top $10 billion in 2013. *Apple Press Info*. Retrieved from http://www.apple.com/ca/pr/library/2014/01/07App-Store-Sales-Top-10-Billion-in-2013.html

Henderson, S., & Yeow, J. (2012). iPad in education: A case study of iPad adoption and use in a primary school. In *Proceedings 45th Hawaii International Conference on System Science (HICSS)*. IEEE. 78-87.

Herdzina, J. (2018). *TEC play and lesson plans: T+M In STEM: Coding and computational thinking*. Retrieved from http://teccenter.erikson.edu/early_childhood_educators/tecmentors_coding/

Hernandez, M. W., Estrera, E., Markovitz, C. E., Muyskens, P., Bartley, G., Bollman, K., . . . Silberglitt, B. (2015). Uses of technology to support early childhood practice. OPRE Report 2015-38. Washington, DC: Office of Planning, Research and Evaluation, Administration for Children and Families, U.S. Department of Health and Human Services.

Herodotou, C. (2017). Young children and tablets: A systematic review of effects on learning and development. *Journal of Computer Assisted Learning, 34*(1), 1–9. doi:10.1111/jcal.12220

Herodotou, C. (2018). Mobile games and science learning: a comparative study of 4 and 5 years old playing the game Angry Birds. *British Journal of Educational Technology, 49*(1), 6–16. doi:10.1111/bjet.12546

Hershman, A., Nazare, J., Sysoev, I., Fratamico, L., Buitrago, J., Soltangheis, M., ... Roy, D. (2018). Family learning coach: engaging families in children's early literacy learning with computer-supported tools. In *Proceedings of the 26th International Conference on Computers in Education. Philippines: Asia-Pacific Society for Computers in Education*, 637-646.

Highfield, K., & Goodwin, K. (2013). Apps for mathematics learning: A review of 'educational' apps from the iTunes App Store. In V. Steinele, L. Ball & C. Bardini (Eds.), *Mathematics education: Yesterday, today and tomorrow. Proceedings of the 36th annual conference of the Mathematics Education Research Group of Australasia*, (pp. 378-385). Melbourne, Australia: MERGA.

Highfield, K., & Goodwin, K. (2013). Apps for mathematics learning: A review of "educational" apps from the iTunes app store. *Proceedings of the 36th Annual Conference of the Mathematics Education Research Group of Australia*, 378-385. Melbourne, Australia.

Highfield, K., Paciga, K. A., & Donohue, C. (2018). Supporting whole child development in the digital age. In *Digital childhoods, international perspectives on early childhood education and development 22* (pp. 165–182). Singapore: Springer. doi:10.1007/978-981-10-6484-5_11

Hiniker, A. (2018). *Research Suggests Well-Designed Preschool Apps Can Encourage Family Engagement and Self-Regulation*. Retrieved from http://joanganzcooneycenter.org/2018/05/21/research-suggests-well-designed-preschool-apps-can-encourage-family-engagement-and-self-regulation/

Hiniker, A., Sobel, K., Hong, S., Suh, H., Irish, I., Kim, D., & Kientz, J. A. (2015). Touchscreen prompts for preschoolers: Designing developmentally appropriate techniques for teaching young children to perform gestures. *Proceedings of the 14th International Conference on Interaction Design and Children*, 109-118. ACM. 10.1145/2771839.2771851

Hinostroza, J., Labbé, C., & Matamala, C. (2013). The use of computers in preschools in Chile: Lessons for practitioners and policy designers. *Computers & Education*, *68*, 96–104. doi:10.1016/j.compedu.2013.04.025

Hirsh-Pasek, K., Zosh, J. M., Golinkoff, R. M., Gray, J. H., Robb, M. B., & Kaufman, J. (2015). Putting education in 'educational' apps: Lessons from the science of learning. *Psychological Science in the Public Interest*, *16*(1), 3–34. doi:10.1177/1529100615569721 PMID:25985468

Hisrich, K., & Blanchard, J. (2009). Digital media and emergent literacy. *Computers in the Schools*, *26*(4), 240–255. doi:10.1080/07380560903360160

Hoffman, J. L., & Paciga, K. (2013). Click, Swipe, and Read: Sharing e-Books with Toddlers and Preschoolers. *Early Childhood Education Journal*, *42*(6), 379–388. doi:10.100710643-013-0622-5

Hoge, E., Bickham, D., & Cantor, J. (2017). Digital media, anxiety, and depression in children. *Pediatrics*, *140*(Supplement 2), S76–S80. doi:10.1542/peds.2016-1758G PMID:29093037

Holland, L., & Garfield, J. (2016). Linking research and teaching: an applied soft systems methodology case study. *International Journal of Information Technologies and Systems Approach*, *9*(2), 23–38. doi:10.4018/IJITSA.2016070102

Homer, B. D., Plass, J. L., Raffaele, C., Ober, T. M., & Ali, A. (2018). Improving high school students' executive functions through digital game play. *Computers & Education*, *117*, 50–58. doi:10.1016/j.compedu.2017.09.011

Hong, J. C., Cheng, C. L., Hwang, M. Y., Lee, C. K., & Chang, H. Y. (2009). Assessing the educational values of digital games. *Journal of Computer Assisted Learning*, *25*(5), 423–437. doi:10.1111/j.1365-2729.2009.00319.x

Horváth, I. (2019). Evolution of teaching roles and tasks in VR / AR-based education. *9th IEEE International Conference on Cognitive Infocommunications, CogInfoCom 2018 - Proceedings*, 355–360. doi:10.1109/CogInfoCom.2018.8639907

Howard, S. K., Yang, J., Ma, J., Maton, K., & Rennie, E. (2018). App clusters: Exploring patterns of multiple app use in primary learning contexts. *Computers & Education, 127*, 154–164. doi:10.1016/j.compedu.2018.08.021

Hruska, M., Amburn, C., Long, R., & Poeppelman, T. R. (2014). Experience API and team evaluation : evolving interoperable performance assessment. In The Interservice/Industry Training, Simulation \& Education Conference (I/ITSEC) (pp. 1–11).

Hsiao, C. H., & Tang, K. Y. (2016). On the post-acceptance of mobile movie-themed games. *Electronic Commerce Research and Applications, 18*, 48–57. doi:10.1016/j.elerap.2016.06.003

Huber, B., Highfield, K., & Kaufman, J. (2018a). Detailing the digital experience: Parent reports of children's media use in the home learning environment. *British Journal of Educational Technology, 49*(5), 821–833. doi:10.1111/bjet.12667

Huber, B., Tarasuik, J., Antoniou, M. N., Garrett, C., Bowe, S. J., Kaufman, J., & Team, S. B. (2016). Young children's transfer of learning from a touchscreen device. *Computers in Human Behavior, 56*, 56–64. doi:10.1016/j.chb.2015.11.010

Huber, B., Yeates, M., Meyer, D., Fleckhammer, L., & Kaufman, J. (2018b). The effects of screen media content on young children's executive functioning. *Journal of Experimental Child Psychology, 170*, 72–85. doi:10.1016/j.jecp.2018.01.006 PMID:29448235

Huizenga, J., Admiraal, W., ten Dam, G., & Voogt, J. (2019). Mobile game-based learning in secondary education: Students' immersion, game activities, team performance and learning outcomes. *Computers in Human Behavior, 99*, 137–143. doi:10.1016/j.chb.2019.05.020

Hung, D. (2001). Theories of learning and computer-mediated instructional technologies. *Educational Media International, 38*(4), 281–287. doi:10.1080/09523980110105114

Hung, H. T., Yang, J. C., Hwang, G. J., Chu, H. C., & Wang, C. C. (2018). A scoping review of research on digital game-based language learning. *Computers & Education, 126*, 89–104. doi:10.1016/j.compedu.2018.07.001

Hu, P., Clark, T., & Ma, W. (2003). Examining technology acceptance by school teachers: A longitudinal study. *Information & Management, 41*(2), 227–241. doi:10.1016/S0378-7206(03)00050-8

Hwang, G. J. (2006). Criteria and strategies of ubiquitous learning. *Proceedings - IEEE International Conference on Sensor Networks, Ubiquitous, and Trustworthy Computing, 2006 I*, 72–77. doi:10.1109/SUTC.2006.49

Hwang, M. Y., Hong, J. C., Cheng, H. Y., Peng, Y. C., & Wu, N. C. (2013). Gender differences in cognitive load and competition anxiety affect 6th grade students' attitude toward playing and intention to play at a sequential or synchronous game. *Computers & Education, 60*(1), 254–263. doi:10.1016/j.compedu.2012.06.014

Ihmeideh, F., & Shawareb, A. (2014). The association between internet parenting styles and children's use of the internet at home. *Journal of Research in Childhood Education, 28*(4), 411–425. doi:10.1080/02568543.2014.944723

International Society for Technology in Education (ISTE) and the Computer Science Teachers Association. (CSTA). (2011). *Computational Thinking in K–12 Education teacher resources second edition.* Retrieved from https://csta.acm.org/Curriculum/sub/CurrFiles/472.11CTTeacherResources_2ed-SP-vF.pdf

Iqbal, S., & Qureshi, I. A. (2012). M-learning adoption: A perspective from a developing country. *International Review of Research in Open and Distance Learning, 13*(3), 147–164. doi:10.19173/irrodl.v13i3.1152

Irmayanti, Y. (2018). *The rules of parents in assisting the use of gadget in preschool children.* (Thesis, Universitas Muhammadiyah Surakarta).

Isgett, J., & Mann, L. (2018). *Using iPads to increase students' letters and letter sounds knowledge*. Retrieved from https://sophia.stkate.edu/ maed/253

ISTE. (2007). *Standards for Global Learning in the Digital Age*. Available: http://www.iste.org/standards.aspx

Jackman, G.-A. (2014). Investigating the factors influencing students' acceptance of mobile learning: The Cave Hill Campus experience. *Caribbean Educational Research Journal. The University of the West Indies, 2*(2), 14–32. Retrieved from http://www.cavehill.uwi.edu/fhe/hum/publications/EducationCERJ.htm

Jahnke, I., & Kumar, S. (2014). Digital didactical designs: teachers' integration of iPads for learning-centered processes. *Journal of Digital Learning in Teacher Education, 30*(3), 81–88. doi:10.1080/21532974.2014.891876

Jairak, K., Praneetpolgrang, P., & Mekhabunchakij, K. (2009). An acceptance of mobile learning for higher education students in Thailand. In *The Sixth International Conference on eLearning for Knowledge-Based Society*. Retrieved from http://ijcim.th.org/SpecialEditions/v17nSP3/36_Full_Kallaya Jairak_Online.pdf

James, C., Davis, K., Charmaraman, L., Konrath, S., Slovak, P., Weinstein, E., & Yarosh, L. (2017). Digital life and youth well-being, social connectedness, empathy, and narcissism. *Pediatrics, 140*(Supplement 2), S71–S75. doi:10.1542/peds.2016-1758F PMID:29093036

Jani, J., Muszali, R., Nathan, S., & Abdullah, M. S. (2018). Blended learning approach using Frog VLE platform towards students' achievement in teaching games for understanding. *Journal of Fundamental and Applied Sciences, 10*, 1130–1141.

Jawad, H. H., & Hassan, Z. (2015). Applying Utaut to evaluate the acceptance of mobile learning in higher education in Iraq. *International Journal of Science and Research, 4*(5), 2319–7064. Retrieved from www.ijsr.net

Jimoyiannis, A., & Komis, V. (2007). Examining teachers' beliefs about ICT in education: Implications of a teacher preparation program. *Teacher Development, 11*(2), 149–173. doi:10.1080/13664530701414779

Joe, S. (2018). *Out of touch: An investigation of the disconnect between 'educational' app design and research* (Doctoral dissertation). Auckland University of Technology.

Johnson, R. T., & Johnson, D. W. (2008). Active learning: Cooperation in the classroom. *The annual report of educational psychology in Japan, 47*, 29-30.

Jones, V., & Jo, J. H. (2004). Ubiquitous learning environment: An adaptive teaching system using ubiquitous technology. *Beyond the comfort zone: Proceedings of the 21st ASCILITE Conference*, (January 2004), 468–474.

Jones, D. E., Greenberg, M., & Crowley, M. (2015). Early social-emotional functioning and public health: The relationship between kindergarten social competence and future wellness. *American Journal of Public Health, 105*(11), 2283–2290. doi:10.2105/AJPH.2015.302630 PMID:26180975

Judge, S., Floyd, K., & Jeffs, T. (2015). Using mobile media devices and apps to promote young children's learning. In K. L. Heider & M. Renck Jalongo (Eds.), *Young children and families in the information age* (pp. 117–131). Springer Netherlands. doi:10.1007/978-94-017-9184-7_7

Jung, S. E., & Won, E. (2018). Systematic review of research trends in robotics education for young children. *Sustainability, 10*(4), 905. doi:10.3390u10040905

Kabali, H. K., Irigoyen, M. M., Nunez-Davis, R., Budacki, J. G., Mohanty, S. H., Leister, K. P., & Bonner, R. L. (2015). Exposure and use of mobile media devices by young children. *Pediatrics, 136*(6), 1044–1050. doi:10.1542/peds.2015-2151 PMID:26527548

Kafai, Y., & Resnick, M. (Eds.). (1996). *Constructionism in practice: Designing, thinking, and learning in a digital world.* Mahwah, NJ: Lawrence Erlbaum Associates.

Kalas, I. (2010). Recognizing the potential of ICT in early childhood education: analytical survey. Moscow: UNESCO; Retrieved from HTTPS://UNESDOC.UNESCO.ORG/ARK:/48223/PF0000190433

Kaldor, T. (2017). *Toying with tech: early coding and computational thinking in a museum setting.* Retrieved from http://teccenter.erikson.edu/in-practice/toying-with-tech/

Kalogiannakis, M., & Papadakis, S. (2017). *Combining mobile technologies in environmental education : a Greek case study Michail Kalogiannakis * and Stamatios Papadakis, 11*(2), 108–130.

Kalogiannakis, M., & Papadakis, St. (2017a). A proposal for teaching ScratchJr programming environment in preservice kindergarten teachers. In *Electronic Proceedings of the ESERA 2017 Conference. Research, Practice and Collaboration in Science Education, Part 15/Strand 15* (pp. 2095-2105). Dublin, Ireland: Dublin City University.

Kalogiannakis, M., & Papadakis, St. (2017a). A proposal for teaching ScratchJr programming environment in preservice kindergarten teachers. In *Proceedings of the 12th Conference of the European Science Education Research Association (ESERA), Research, practice and collaboration in science education.* Dublin City University and the University of Limerick.

Kalogiannakis, M. (2010). Training with ICT for ICT from the trainer's perspective. A Greek case study. *Education and Information Technologies, 15*(1), 3–17. doi:10.100710639-008-9079-3

Kalogiannakis, M., Ampartzaki, M., Papadakis, St., & Skaraki, E. (2018). Teaching Natural Science Concepts to Young Children with Mobile Devices and Hands-on Activities. A Case Study. *International Journal of Teaching and Case Studies, 9*(2), 171–183. doi:10.1504/IJTCS.2018.090965

Kalogiannakis, M., Nirgianaki, G.-M., & Papadakis, S. (2018). Teaching magnetism to preschool children: The effectiveness of picture story reading. *Early Childhood Education Journal, 46*(5), 535–546. doi:10.100710643-017-0884-4

Kalogiannakis, M., & Papadakis, S. (2017b). Pre-service kindergarten teachers' acceptance of "ScratchJr" as a tool for learning and teaching computational thinking and Science education. In *Proceedings of the 12th Conference of the European Science Education Research Association (ESERA), 'Research, practice, and collaboration in science education'* (pp. 21-25). Dublin City University and the University of Limerick, Dublin, Ireland.

Kalogiannakis, M., & Papadakis, St. (2017b). Pre-service kindergarten teachers acceptance of 'ScratchJr' as a tool for learning and teaching computational thinking and Science education. In *Proceedings of the 12th Conference of the European Science Education Research Association (ESERA), Research, practice and collaboration in science education.* Dublin City University and the University of Limerick.

Kalogiannakis, M., & Papadakis, St. (2017c). Combining mobile technologies in environmental education: A Greek case study. *International Journal of Mobile Learning and Organisation, 11*(2), 108–130. doi:10.1504/IJMLO.2017.084272

Kalogiannakis, M., & Papadakis, St. (2019). Evaluating pre-service kindergarten teachers' intention to adopt and use tablets into teaching practice for natural sciences. *International Journal of Mobile Learning and Organisation, 13*(1), 113–127. doi:10.1504/IJMLO.2019.096479

Kapp, K. M. (2014). *The gamification of learning and instruction fieldbook.* San Francisco, CA: Wiley.

Karadeniz, S., Samur, Y., & Özden, M. Y. (2014). Playing with algorithms to learn programming: A case study on 5 years old children. In *9th International Conference on Information Technology and Applications (ICITA2014)* (pp. 1-4). Sydney, Australia.

Kara, H. G. E. (2018). A case study on reducing children's screen time: the project of screen free week. *World Journal of Education, 8*(1), 100–110. doi:10.5430/wje.v8n1p100

Kärnä, E., Pihlainen, K., Sutinen, E. & Virnes, M. (2007). Technology in Finnish special education-toward inclusion and harmonized school days. *Informatics in Education, 6*(1), 103-114.

Karsenti, T., & Fievez, A. (2013). *The iPad in education: uses, benefits, and challenges - A survey of 6,057 students and 302 teachers in Quebec, Canada.* Montreal, Canada: CRIFPE.

Kasapakis, V., & Gavalas, D. (2015). Pervasive gaming: Status, trends and design principles. *Journal of Network and Computer Applications, 55*, 213–236. doi:10.1016/j.jnca.2015.05.009

Kattmann, U. (2001). Aquatics, flyers, creepers and terrestrials-students' conceptions of animal classification. *Journal of Biological Education, 35*(3), 141–147. doi:10.1080/00219266.2001.9655763

Kazakoff, E., & Bers, M. U. (2012). Programming in a robotics context in the kindergarten classroom: The impact on sequencing skills. *Journal of Educational Multimedia and Hypermedia, 21*(4), 371–391.

Kazakoff, E., Sullivan, A., & Bers, M. U. (2013). The effect of a classroom-based intensive robotics and programming workshop on sequencing ability in early childhood. *Early Childhood Education Journal, 41*(4), 245–255. doi:10.100710643-012-0554-5

Kelley, T. R., & Knowles, J. G. (2016). A conceptual framework for integrated STEM education. *International Journal of STEM Education, 3*(11). doi:10.118640594-016-0046-z

Keong, C. C., Horani, S., & Daniel, J. (2005). A study on the use of ICT in mathematics teaching. *Malaysian Online Journal of Instructional Technology, 2*(3), 43–51.

Kersey, A. J., Braham, E. J., Csumitta, K. D., Libertus, M. E., & Cantlon, J. F. (2018). No intrinsic gender differences in children's earliest numerical abilities. *NPJ Science of Learning, 3*(1), 12.

Kervin, L., Verenikina, I., & Rivera, C. (2018). Digital play and learning in the home: Families' perspective. In S. J. Danby, M. Fleer, C. Davidson, & M. Hatzigianni (Eds.). Digital childhoods: Technologies and children's everyday lives (pp. 117-130). Springer: International Perspectives on Early Childhood Education and Development, 22.

Kevan, J., & Ryan, P. R. (2016). Experience API : Flexible, decentralized and activity-centric data collection. *Technology, Knowledge, and Learning, 21*(1), 143–149. doi:10.100710758-015-9260-x

Khaddage, F., Lattemann, C., & Bray, E. (2011). Mobile apps integration for teaching and learning, *Proceedings of Society for Information Technology & Teacher Education International Conference*, 2545-2552. Chesapeake, VA.

Kim-Soon, N., Ibrahim, M. A., Ahmad, A. R., & Sirisa, N. M. X. (2015). Factors influencing intention to use mobile technologies for learning among technical universities students. *Innovation and Sustainable Economic Competitive Advantage: From Regional Development to Global Growth*, (2010), 1–12.

Kirby, J. R., Parrila, R. K., & Pfeiffer, S. L. (2003). Naming speed and phonological awareness as predictors of reading development. *Journal of Educational Psychology, 95*(3), 453–464. doi:10.1037/0022-0663.95.3.453

Kirkorian, H. L., Choi, K., & Pempek, T. A. (2016). Toddlers' word learning from contingent and noncontingent video on touch screens. *Child Development, 87*(2), 405–413. doi:10.1111/cdev.12508 PMID:27018327

Kirkorian, H., & Pempek, T. (2013). Toddlers and touch screens: Potential for early learning? *Zero to Three, 33*(4), 32–35.

Kirriemuir, J., & Mcfarlane, A. (2004). *Literature review in games and learning.* A NESTA Futurelab Research report - report 8. Retrieved from http://www.coulthard.com/library/Files/kirriemuir%2Dfuturelabs%5F2004%5Fgamesreview%2Epdf

Kitto, K., Cross, S., Waters, Z., Lupton, M., & Street, G. (2015). Learning analytics beyond the LMS : the connected learning analytics toolkit. ACM In (Ed.), *Proceedings of the Fifth International Conference on Learning Analytics And Knowledge2* (pp. 11–15). 10.1145/2723576.2723627

Koerner, J. (1982). *Ferdinand de Saussure. Génesis y evolución de su pensamiento en el marco de la lingüística occidental.* Madrid: Gredos.

Kokkalia, G., & Drigas, A. (2016). Mobile learning for special preschool education. International *Journal of Interactive Mobile Technologies, 10*(1), 67–74. doi:10.3991/ijim.v10i1.5288

Kokkalia, G., Drigas, A. S., & Economou, A. (2016). Mobile learning for preschool education. *International Journal of Interactive Mobile Technologies, 10*(4), 57–64. doi:10.3991/ijim.v10i4.6021

Komis, V., & Misirli, A. (2011). Robotique pédagogique et concepts préliminaires de la programmation à l'école maternelle: une étude de cas basée sur le jouet programmable Bee-Bot. Georges-Louis Baron, Eric Bruillard, Vassilis Komis. *Sciences et technologies de l'information et de la communication en milieu éducatif: Analyse de pratiques et enjeux didactiques.* Oct. 2011, Patras, Grèce. Athènes: New Technologies Éditions, pp. 271-281, 2011. Retrieved from https://edutice.archives-ouvertes.fr/edutice-00676143/document

Komis, V., & Misirli, A. (2012). L'usage des jouets programmables à l'école maternelle: concevoir et utiliser des scenarios pédagogiques de robotique éducative. *Revue Scholé,* 17, 1, pp. 143-154. Retrieved from https://espe.univ-amu.fr/sites/espe.univ-amu.fr/files/article/skhole_2012_vol_17_.pdf

Kontogiannis, D., & Ntziaxristos, B. (1999). *Basic geometry notions* (3rd ed.). Athens.

Kopelke, K. (2011). *Making your classroom buzz with Bee-Bots: Ideas and Activities for the Early Phase, ICT Learning Innovation Centre.* Retrieved from http://elresources.skola.edu.mt/wp-content/uploads/2010/06/doc_669_2468_Bee-BotguideA4v2.pdf

Korat, O. (2010). Reading electronic books as a support for vocabulary, story comprehension and word reading in kindergarten and first grade. *Computers & Education, 55*(1), 24–31. doi:10.1016/j.compedu.2009.11.014

Koutromanos, G., & Avraamidou, L. (2014). The use of mobile games in formal and informal learning environments: A review of the literature. *Educational Media International Journal, 51*(1), 49–65. doi:10.1080/09523987.2014.889409

Koutromanos, G., Sofos, A., & Avraamidou, L. (2015). The use of augmented reality games in education: A review of the literature. *Educational Media International Journal, 52*(4), 253–271. doi:10.1080/09523987.2015.1125988

Krcmar, M., & Cingel, D. P. (2014). Parent-child joint reading in traditional and electronic formats. *Media Psychology, 17*(3), 262–281. doi:10.1080/15213269.2013.840243

Kron, F. W., Gjerde, C. L., Sen, A., & Fetters, M. D. (2010). Medical student attitudes toward video games and related new media technologies in medical education. *BMC Medical Education, 10*(1), 50. doi:10.1186/1472-6920-10-50 PMID:20576125

Kubiatko, M. (2012). Kindergarten Children's perception of animals focusing on the look and fear of animals. *Educational Sciences: Theory and Practice, 12*(4), 3181–3186.

Kubiatko, M., & Prokop, P. (2007). Pupils' misconceptions about mammals. *Journal of Baltic Science Education, 6*(1), 5–14.

Kubota, R. (1999). Japanese culture constructed by discourses: Implications for applied linguistics research and ELT. *TESOL Quarterly*, *33*(1), 9–35. doi:10.2307/3588189

Kucirkova, N. (2014). iPads in early education: Separating assumptions and evidence. *Frontiers in Psychology*, *5*, 715. doi:10.3389/fpsyg.2014.00715 PMID:25071678

Kucirkova, N. (2016). Trans-and intra-apps: innovating the app market and use. In *Apps, Technology and Younger Learners* (pp. 272–284). New York: Routledge. doi:10.4324/9781315682204

Kucirkova, N. (2018). *How and why to read and create children's digital books*. London, UK: UCL Press. Available at www.ucl.ac.uk/ucl-press

Kucirkova, N., & Falloon, G. (2018). iPad apps and visual methodologies: Empirical and ethical issues in achieving authentic data. *Research in Learning Technology*, *26*(2029), 1–18.

Kucirkova, N., & Flewitt, R. (2018). The future-gazing potential of digital personalization in young children's reading: Views from education professionals and app designers. *Early Child Development and Care*, 1–15. doi:10.1080/03004430.2018.1458718

Kucirkova, N., Messer, D., Sheehy, K., & Flewitt, R. (2013). Sharing personalised stories on iPads: A close look at one parent–child interaction. *Literacy*, *47*(3), 115–122. doi:10.1111/lit.12003

Kucirkova, N., Messer, D., Sheehy, K., & Panadero, C. F. (2014). Children's engagement with educational iPad apps: Insights from a Spanish classroom. *Computers & Education*, *71*, 175–184. doi:10.1016/j.compedu.2013.10.003

Kucirkova, N., & Sakr, M. (2015). Child–father creative text-making at home with crayons, iPad collage & PC. *Thinking Skills and Creativity*, *17*, 59–73. doi:10.1016/j.tsc.2015.05.003

Kumar, P., & Kumar, A. (2003). Effect of a web-based project on preservice and inservice teachers' attitude toward computers and their technology skills. *Journal of Computing in Teacher Education*, *19*(3), 87–92.

Kumi-Yeboah, A., & Campell, K. S. (2015). Emerging use of tablets in K-12 environments: Issues and implications in K-12 schools. In Tablets in K-12 education: Integrated experiences and implications (pp. 46-63). Hershey, PA: IGI Global. doi:10.4018/978-1-4666-6300-8.ch004

Kumi-Yeboah, A., & Campbell, K. S. (2019). Emerging use of tablets in K-12 environments: Issues and implications in K-12 Schools. In *Early childhood development: concepts, methodologies, tools, and applications* (pp. 857–876). Hershey, PA: IGI Global. doi:10.4018/978-1-5225-7507-8.ch042

Kumpulainen, K., & Gillen, J. (2017) Young children's digital literacy practices in the home: A review of the literature. *COST ACTION ISI1410 DigiLitEY*. Retrieved from http://www.digilitey.eu

Kyriakides, A. O., Meletiou-Mavrotheris, M., & Prodromou, T. (2016). Mobile technologies in the service of students' learning of mathematics: The example of game application ALEX in the context of a primary school in Cyprus. *Mathematics Education Research Journal*, *28*(1), 53–78. doi:10.100713394-015-0163-x

Labat, H., Ecalle, J., Baldy, R., & Magnan, A. (2014). How can low-skilled 5-year-old children benefit from multisensory training on the acquisition of the alphabetic principle? *Learning and Individual Differences*, *29*, 106–113. doi:10.1016/j.lindif.2013.09.016

Laidlaw, L., & Wong, S. S. H. (2016). Literacy and complexity: on using technology within emergent learning structures with young learners. *Complicity: An International Journal of Complexity and Education*, *13*(1), 30–42. doi:10.29173/cmplct24008

Laili, U. (2017). Influence of gadget on social personal of early childhood. *Proceeding of Surabaya International Health Conference,* July 13-14, 2017. 273–277.

Lam, J., & Duan, G. (2012). A review of mobile learning environment in higher education sector of Hong Kong: Technological and social perspectives. ICHL (pp. 165-173). Springer.

Landerl, K., & Wimmer, H. (2008). Development of word reading fluency and spelling in a consistent orthography: An 8-year follow-up. *Journal of Educational Psychology, 100*(1), 150–161. doi:10.1037/0022-0663.100.1.150

Landis, J. R., & Koch, G. G. (1977). The measurement of observer agreement for categorical data. *Biometrics, 33*(1), 159–174. doi:10.2307/2529310 PMID:843571

Lantolf, J. (1996). Second language theory building: Letting all the flowers bloom! *Language Learning, 46*(4), 713–749. doi:10.1111/j.1467-1770.1996.tb01357.x

Larkin, K. (2015a). The search for fidelity in geometry apps: an exercise in futility? In M. Marshman, V. Geiger, & A. Bennison (Eds.), *Mathematics education in the margins. Proceedings of the 38th annual conference of the Mathematics Education Research Group of Australasia* (pp. 341–348). Sunshine Coast: MERGA.

Larkin, K. (2015b). An app! An app! My kingdom for an app: an 18 month quest to determine whether apps support mathematical knowledge building. In T. Lowrie & R. Jorgensen (Eds.), Digital games and mathematics learning: potential, promises and pitfalls. New York: Springer. doi:10.1007/978-94-017-9517-3_13

Larkin, K. (2011). You use! I use! We use! Questioning the orthodoxy of one-to-one computing in primary schools. *Journal of Research on Technology in Education, 44*(2), 101–120. doi:10.1080/15391523.2011.10782581

Larkin, K. (2013). Mathematics Education: Is there an app for that? In V. Steinle, L. Ball, & C. Bardini (Eds.), *Proceedings of the 36th annual conference of the Mathematics Education Research Group of Australasia. Mathematics education: Yesterday, today and tomorrow*, (pp. 426–433). Melbourne, Australia: MERGA.

Larkin, K., Kortenkamp, U., Ladel, S., & Etzold, H. (2018). Using the ACAT Framework to Evaluate the Design of Two Geometry Apps: An Exploratory Study. *Digital Experiences in Mathematics Education, 5*(1), 59–92. doi:10.100740751-018-0045-4

Lauricella, A. R., Blackwell, C. K., & Wartella, E. (2017). The "New" technology environment: The role of content and context on learning and development from mobile media. In R. Barr & D. Linebarger (Eds.), *Media exposure during infancy and early childhood* (pp. 1–23). Cham, Switzerland: Springer International Publishing. doi:10.1007/978-3-319-45102-2_1

Learning Exchange Parramatta. (2011). *iPads in Schools: Use Testing.* Parramatta, Australia: Catholic Education Diocese. Retrieved from http://www.learningexchange.nsw.edu.au/_resources/ipads-in-schools--usetesting-2.pdf

LeBourgeois, M. K., Hale, L., Chang, A. M., Akacem, L. D., Montgomery-Downs, H. E., & Buxton, O. M. (2017). Digital media and sleep in childhood and adolescence. *Pediatrics, 140*(Supplement 2), S92–S96. doi:10.1542/peds.2016-1758J PMID:29093040

Lee, J. (2016). *Early learning and educational technology policy brief.* Office of Educational Technology, US Department of Education. Retrieved from https://tech.ed.gov/files/2016/10/Early-Learning-Tech-Policy-Brief.pdf

Lee, J. (2016). *Early Learning and Educational Technology Policy Brief. Office of Educational Technology.* US Department of Education.

Lee, J. (2019). Coding in early childhood. *Contemporary Issues in Early Childhood.* doi:10.1177/1463949119846541

Lee, K., Sullivan, A., & Bers, M. U. (2013). Collaboration by design: Using robotics to foster social interaction in Kindergarten. *Computers in the Schools*, *30*(3), 271–281. doi:10.1080/07380569.2013.805676

Lee, U., Han, K., Cho, H., Chung, K. M., Hong, H., Lee, S. J., ... Carroll, J. M. (2019). Intelligent positive computing with mobile, wearable, and IoT devices: Literature review and research directions. *Ad Hoc Networks*, *83*, 8–24. doi:10.1016/j.adhoc.2018.08.021

Lee, Y. (2009). Pre-K children's interaction with educational software programs: an observation of capabilities and levels of engagement. *Journal of Educational Multimedia and Hypermedia*, *18*(3), 289–309.

Leidl, K. D., Bers, M. U., & Mihm, C. (2017). Programming with ScratchJr: a review of the first year of user analytics. In *Proceedings of International Conference on Computational Thinking Education 2017* (pp. 116-121). Hong Kong: The Education University of Hong Kong.

Leidl, K. D., Bers, M. U., & Mihm, C. (2017). Programming with ScratchJr: a review of the first year of user analytics. In *Conference Proceedings of International Conference on Computational Thinking Education 2017*. Hong Kong: The Education University of Hong Kong.

Lemke, C., Coughlin, E., & Reifsneider, D. (2009). *Technology in schools: What the research says: An update*. Culver City, CA: Commissioned by Cisco.

Lenhart, A. (2015, April). *Teen, social media, and technology overview 2015*. Pew Research Center. Retrieved from http://www.pewinternet.org/2015/04/09/teens-social-media-technology-2015/

Lerner, R. M., Lerner, J. V., Almerigi, J. B., Theokas, C., Phelps, E., Gestsdottir, S., ... Smith, L. M. (2005). Positive youth development, participation in community youth development programs, and community contributions of fifth-grade adolescents: Findings from the first wave of the 4-H study of positive youth development. *The Journal of Early Adolescence*, *25*(1), 17–71. doi:10.1177/0272431604272461

Liebermana, D., Batesa, C., & Soa, J. (2009). Young children's learning with digital media. *Computers in Schools*, *26*(4), 271–283. doi:10.1080/07380560903360194

Lim, K. C. (2015). Case studies of xAPI applications to e-Learning. In Case Studies of xAPI Applications to E-Learning.

Lim, E. M. (2012). Patterns of kindergarten children's social interaction with peers in the computer area. *International Journal of Computer-Supported Collaborative Learning*, *7*(3), 399–421. doi:10.100711412-012-9152-1

Lin, V. J. (2015). *Computational thinking and technology toys*. (Honors Thesis Collection, 307, Wellesley College, MA).

Liua, X., Toki, E. I., & Pange, J. (2014). The use of ICT in preschool education in Greece and China: A comparative study. *Procedia: Social and Behavioral Sciences*, *112*, 1167–1176. doi:10.1016/j.sbspro.2014.01.1281

Liu, Y., & Li, H. (2011). Exploring the impact of use context on mobile hedonic services adoption: An empirical study on mobile gaming in China. *Computers in Human Behavior*, *27*(2), 890–898. doi:10.1016/j.chb.2010.11.014

Livingstone, S. (2016). What are preschoolers doing with tablets and is it good for them? Retrieved from http://eprints.lse.ac.uk/

Livingstone, S., Marsh, J., Plowman, L., ... (2014). *Young children (0-8) and digital technology: a qualitative exploratory study national report*. UK: Joint Research Centre, European Commission, Luxembourg. Retrieved from http://eprints.lse.ac.uk/60799/1/__lse.ac.uk_storage_LIBRARY

Livingstone, S. (2012). Critical reflections on the benefits of ICT in education. *Oxford Review of Education*, *38*(1), 9–24. doi:10.1080/03054985.2011.577938

Looi, C. K., Zhang, B., Chen, W., Seow, P., Chia, G., Norris, C., & Soloway, E. (2011). 1: 1 mobile inquiry learning experience for primary science students: A study of learning effectiveness. *Journal of Computer Assisted Learning, 27*(3), 269–287. doi:10.1111/j.1365-2729.2010.00390.x

Lou, Y., Abrami, P. C., & d'Apollonia, S. (2001). Small group and individual learning with technology: A meta-analysis. *Review of Educational Research, 71*(3), 449–521. doi:10.3102/00346543071003449

Loveless, A., & Dore, B. (Eds.). (2002). *ICT in the primary school.* Buckingham, UK: Open University Press.

Lowrie, T., & Jorgensen, R. (2011). Gender differences in students' mathematics game playing. *Computers & Education, 57*(4), 2244–2248. doi:10.1016/j.compedu.2011.06.010

Lu, Y., Ottenbreit-Leftwich, A. T., Ding, A., & Glazewski, K. (2017). Experienced iPad-using early childhood teachers: Practices in the one-to-one iPad classroom. *Computers in the Schools: Interdisciplinary Journal of Practice, Theory, and Applied Research, 34*. doi:. 2017.1287543 doi:10.1080/07380569

Lu, Y., Zhou, T., & Wang, B. (2009). Exploring Chinese users' acceptance of instant messaging using the theory of planned behavior, the technology acceptance model, and the flow theory. *Computers in Human Behavior, 25*(1), 29–39. doi:10.1016/j.chb.2008.06.002

Lynch, J., & Redpath, T. (2014). 'Smart' technologies in early years literacy education: A meta-narrative of paradigmatic tensions in iPad use in an Australian preparatory classroom. *Journal of Early Childhood Literacy, 14*(2), 147–174. doi:10.1177/1468798412453150

MacCallum, K., & Bell, H. R. (2019). Improving teaching practice in early childhood supported by mobile technology. In *Early childhood development: concepts, methodologies, tools, and applications* (pp. 1066–1082). Hershey, PA: IGI Global. doi:10.4018/978-1-5225-7507-8.ch053

Magee, C. A., Lee, J. K., & Vella, S. A. (2014). Bidirectional relationships between sleep duration and screen time in early childhood. *JAMA Pediatrics, 168*(5), 465–470. doi:10.1001/jamapediatrics.2013.4183 PMID:24589672

Malone, T. W. (1980). *What makes things fun to learn? Heuristics for designing instructional computer games.* In *Proceedings of the 3rd ACM SIGSMALL symposium and the first SIGPC symposium on Small systems*, Palo Alto, CA. 10.1145/800088.802839

Mangen, A. (2010). Point and click: Theoretical and phenomenological reflections on the digitization of early childhood education. *Contemporary Issues in Early Childhood, 11*(4), 415–431. doi:10.2304/ciec.2010.11.4.415

Mango, O. (2015). iPad use and student engagement in the classroom. *The Turkish Online Journal of Educational Technology, 14*, 53–57.

Mann, V. A., & Foy, J. G. (2003). Phonological awareness, speech development, and letter knowledge in preschool children. *Annals of Dyslexia, 53*(1), 149–173. doi:10.100711881-003-0008-2

Manolitsis, G. (2016). Emergent literacy in early childhood education: New issues and educational implications. *Preschool and Primary Education*, *4*(1), 3–34. doi:10.12681/ppej.9970

Mares, M.-L., & Woodard, E. (2005). Positive effects of television on children's social interactions: a meta-analysis. *Media Psychology*, 7(3), 301–322. doi:10.1207/S1532785XMEP0703_4

Margot, K. C., & Kettler, T. (2019). Teachers' perception of STEM integration and education: A systematic literature review. *International Journal of STEM Education*, *6*(2). doi:10.118640594-018-0151-2

Marsh, J., Plowman, L., Yamada-Rice, D., Bishop, J., Lahmar, J., & Scott, F. (2018, September). Play and creativity in young children's use of apps. *British Journal of Educational Technology, Special Issue*, *49*(5), 870–882. doi:10.1111/bjet.12622

Martens, M., Rinnert, G. C., & Andersen, C. (2018). Child-Centered Design: Developing an Inclusive Letter Writing App. *Frontiers in Psychology*, *6*(9), 2277. doi:10.3389/fpsyg.2018.02277 PMID:30574104

Martínez-Losada, C., García-Barros, S., & Garrido, M. (2014). How children characterise living beings and the activities in which they engage. *Journal of Biological Education*, *48*(4), 201–210. doi:10.1080/00219266.2013.849281

Masek, M., Murcia, K., & Morrison, J. (2012). Getting serious with iPads: The intersection of game design and teaching principals. *Australian Educational Computing*, *27*(2), 34–38.

Mashburn, A. J., Pianta, R. C., Hamre, B. K., Downer, J. T., Barbarin, O. A., Bryant, M., ... Howes, C. (2008). Measures of classroom quality in prekindergarten and children's development of academic, language, and social skills. *Child Development*, *79*(3), 832–749. doi:10.1111/j.1467-8624.2008.01154.x PMID:18489424

Ma, W., Anderson, R., & Streith, K. (2005). Examining user acceptance of computer technology: An empirical study of student teachers. *Journal of Computer Assisted Learning*, *21*(6), 387–395. doi:10.1111/j.1365-2729.2005.00145.x

McCarrick, K., & Li, X. (2007). Buried treasure: The impact of computer use on young children's social, cognitive, language development and motivation. *AACE Journal*, *15*(1), 73–95.

McCoog, I. (2007). Integrated instruction: Multiple intelligences and technology. *The Clearing House: A Journal of Educational Strategies, Issues and Ideas*, *81*(November), 25–28. doi:10.3200/TCHS.81.1.25-28

McDaniel, B. T., & Radesky, J. S. (2018). Technoference: Longitudinal associations between parent technology use, parenting stress, and child behavior problems. *Pediatric Research*, *84*(2), 210–218. doi:10.103841390-018-0052-6 PMID:29895837

McDonough, J., & Shaw, C. (2012). *Materials and Methods in ELT*. John Wiley & Sons.

McLennan, D. M. P. (2017). Creating coding stories and games. *Teaching Young Children, 10*, 3. Retrieved from https://www.naeyc.org/resources/pubs/tyc/feb2017/creating-coding-stories-and-games

McManis, L. D., & Gunnewig, S. B. (2012). Finding the education in educational technology with early learners. *YC Young Children*, *67*, 14–24.

McManis, L. D., & Gunnewig, S. B. (2012). Finding the education in educational technology with early learners. *Young Children*, *67*(3), 14–24.

McManis, M. H., & McManis, L. D. (2016). Using a touch-based, computer-assisted learning system to promote literacy and math skills for low-income preschoolers. *Journal of Information Technology Education*, *15*, 409–429. doi:10.28945/3550

McNaughton, S., Rosedale, N., Jesson, R. N., Hoda, R., & Teng, L. S. (2018). How digital environments in schools might be used to boost social skills: Developing a conditional augmentation hypothesis. *Computers & Education*, *126*, 311–323. doi:10.1016/j.compedu.2018.07.018

McPake, J., Plowman, L., & Stephen, C. (2013). Pre-school children creating and communicating with digital technologies in the home. *British Journal of Educational Technology*, *44*(3), 421–431. doi:10.1111/j.1467-8535.2012.01323.x

Mcrae, L., Ellis, K., & Kent, M. (2018). Internet of things (IoT): Education and technology. The relationship between education and technology for students with disabilities, 37. Retrieved from http://www.curtin.edu.au/wp-content/uploads/2018/02/IoTEducation_Formatted_Accessible.pdf

Mellon, C. (2003). Best practices guide for content developers 1.

Meng, X. (2016). *A pedagogy of multiliteracies into practice: a case study in one grade one literacy classroom* (Doctoral dissertation, The University of Western Ontario).

Mertala, P. (2018). *Two worlds collide? Mapping the third space of ICT integration in early childhood education.* (Ph.D. Dissertation, University of Oulu).

Miarso, Y. (2007). Jurnal teknologi pendidikan. *Tekno-Pedagogi.*

Michael Cohen Group LLC. (2011). *Young Children, Apps & iPad.* Michael Cohen Group LLC Report. Retrieved from https://mcgroup.wpengine.com/wp-content/uploads/2015/03/ipad-study-cover-page-report-mcg-info_new-online.pdf

Microsoft. (2018). The class of 2030 and life-ready learning: The technology imperative. Redmond, WA: Author. Retrieved from https://education.minecraft.net/wp-content/uploads/13679_EDU_Thought_Leadership_Summary_revisions_5.10.18.pdf

Minchillo, L. V., Vellozo, A., Borin, E., & Borin, J. F. (2018). Towards better tools and methodologies to teach computational thinking to children. In *26º Workshop sobre Educação em Computação (WEI 2018) (Vol. 26, No. 1/2018). SBC.* Universidade do Estado do Rio Grande do Norte, Brazil.

Ministry of Education & University of Ioannina. (2008). *Program for academic & professional upgrade of primary school teachers.* Retrieved from http://protagoras.edu.uoi.gr/eksomoiosi/ypoerg.htm

Mohamad, A. J., Lakulu, M., & Samsudin, K. (2016). The development of mobile applications for kindergarten early reading: challenges and opportunities. *Journal of Engineering and Applied Sciences (Asian Research Publishing Network), 11*(3), 380–383.

Moisa, V. (2013). Adaptive learning management system. *Journal of Mobile, Embedded, and Distributed Systems, 5*(2), 70–77.

Mojarro, A., Duarte, A., Guzmán, M., & Aguaded, I. (2019). Mobile learning in university contexts based on the unified theory of acceptance and use of technology (UTAUT). *Journal of New Approaches in Educational Research, 8*(1), 7–17. doi:10.7821/naer.2019.1.317

Monova-Zheleva, M. (2005). Adaptive learning in web-based educational environments. *Cybernetics and Information Technologies, 5*(1), 44–55.

Montoya, I. A., & Montoya, L. A. (2013). La formación de estrategias deliberadas y emergentes: Una propuesta a partir de definiciones básicas de una metodología de sistemas suaves. *Investigación y Reflexión, 26*(2), 67–96.

Moomaw, S., & Davis, J. (2010). STEM Comes to Preschool. *Young Children, 65*(5), 12–18.

Moon, K., & Reifel, S. (2008). Play and literacy learning in a diverse language pre-kindergarten classroom. *Contemporary Issues in Early Childhood, 9*(1), 49–65. doi:10.2304/ciec.2008.9.1.49

Moore, T. J., Stohlmann, M. S., Wang, H. H., Tank, K. M., Glancy, A. W., & Roehrig, G. H. (2014). Implementation and integration of engineering in K-12 STEM education. In J. Strobel, S. Purzer, & M. Cardella (Eds.), *Engineering in precollege settings: synthesizing research, policy, and practices* (pp. 35–60). Lafayette: Purdue University Press. doi:10.2307/j.ctt6wq7bh.7

Mora, F. (2009). *Cómo funciona el cerebro.* Madrid: Alianza Editorial.

Morgado, L. (2008). TEL practices in preschool and kindergarten education: Integrating computer use and computer programming in off-computer activities. In M. D. Lytras, D. Gasevic, P. Ordonez de Pablos, & W. Huang (Eds.), *Technology enhanced learning: Best practices.* Hershey, PA: IGI Global. doi:10.4018/978-1-59904-600-6.ch002

Morrison, G. (2005). Educación infantil. México. Pearson Education. 2-14.

Moschella, M., & Basso, D. (2018). *Computational thinking at primary school: a research proposal for post millennial teachers.* Retrieved from https://eera-ecer.de/ecer-programmes/conference/23/contribution/44245/

Mourlam, D. J., Strouse, G. A., Newland, L. A., & Lin, H. (2019). Can they do it? A comparison of teacher candidates' beliefs and preschoolers' actual skills with digital technology and media. *Computers & Education, 129*, 82–91. doi:10.1016/j.compedu.2018.10.016

Moyer-Packenham, P. S., Shumway, J. F., Bullock, E., Tucker, S. I., Anderson-Pence, K. L., Westenskow, A., ... Jordan, K. (2015). Young children's learning performance and efficiency when using virtual manipulative mathematics iPad apps. *Journal of Computers in Mathematics and Science Teaching, 34*(1), 41–69.

Mtebe, J., & Raisamo, R. (2014). Investigating students ' behavioural intention to adopt and use mobile learning in higher education in East Africa. *International Journal of Education and Development Using Information and Communication Technology, 10*(3), 4–20.

Mumtaz, S. (2000). Factors affecting teachers' use of information and communications technology: A review of the literature. *Journal of Information Technology for Teacher Education, 9*(3), 319–341. doi:10.1080/14759390000200096

Muñoz-Leiva, F., Climent-Climent, S., & Liébana-Cabanillas, F. (2017). Determinants of intention to use the mobile banking apps: An extension of the classic TAM model. *Spanish Journal of Marketing - ESIC, 21*(1), 25–38.

Murphy, S., & Headley, M. (2018). The role of digital technology in teen mothers' and their children's literacy. *Journal of Early Childhood Literacy*, Online First. doi:10.1177/1468798418783326

Murray, K., Berking, P., Haag, J., & Hruska, N. (2012). Mobile learning and ADL's experience API overview of the xAPI. *Connections: The Quarterly Journal, 12*(1), 45–49. doi:10.11610/Connections.12.1.05

Mysirlaki, S., & Paraskeva, F. (2007). Digital games: Developing the issues of socio-cognitive learning theory in an attempt to shift an entertainment gadget to an educational tool. In *Proceedings of the first IEEE international workshop on digital game and intelligent toy enhanced learning [DIGITEL2007]* (pp. 147–151). Jhongli, Taiwan.

NAEYC. (2012). *Technology and young children – ages 3 through 8. A position statement of the National Association for the Education of Young Children.* Washington, DC: National Association for the Education of Young Children.

Namodi, M. O., Nyaoga, R., Sirma, J., Obegi, F., & Moriasi, J. (2015). The relationship between teacher characteristics and information and communication technology (ICT) integration in public secondary schools in Nakuru Town Sub-County, Kenya. *European Journal of Business and Management, 7*(30), 200–216.

Namodi, M. O., Nyaoga, O., Sirma, J., Obegi, F., & Moriasi, J. (2015). The relationship between teacher characteristics and Information and Communication Technology (ICT) integration in Nasrin & Islahi, F. (2019). Exploring teacher attitude towards information technology with a gender perspective. *Contemporary Educational Technology, 10*(1), 37–54.

Namukasa, I. K., Gadanidis, G., Sarina, V., Scucuglia, S., & Aryee, K. (2016). Selection of apps for teaching difficult mathematics topics: An instrument to evaluate touch-screen tablet and smartphone mathematics apps. In *International perspectives on teaching and learning mathematics with virtual manipulatives* (pp. 275–300). Dordrecht: Springer. doi:10.1007/978-3-319-32718-1_12

Nassuora, A. (2013). Students acceptance of mobile learning for higher education in Saudi Arabia. *International Journal of Learning Management Systems*, *1*(1), 1–9. doi:10.12785/ijlms/010101

National Association for the Education of Young Children & the Fred Rogers Center for Early Learning and Children's Media at Saint Vincent College. (2012). Technology and interactive media as tools in early childhood programs serving children from birth through. *Age*, *8*. Retrieved from https://www.naeyc.org/files/naeyc/file/positions/PS_technology_WEB2.pdf

National Association for the Education of Young Children & the Fred Rogers Center for Early Learning and Children's Media at Saint Vincent College. (2012). Technology and Interactive Media as Tools in Early Childhood Programs Serving Children from Birth through. *Age*, *8*. Retrieved from https://www.naeyc.org/files/naeyc/file/positions/PS_technology_WEB2.pdf

National Association for the Education of Young Children and the Fred Rogers Center. (2012). *Position statement: Technology and young children*. Washington, DC: Author. Retrieved from http://www.naeyc.org/content/technology-and-young-children

National Research Council and Institute of Medicine. (2000). From neurons to neighborhoods: The science of early childhood development. In J. P. Shonkoff, & D. A. Phillips (Eds.), Committee on integrating the science of early childhood development, board on children, youth, and families, commission on behavioral and social sciences. National Academy Press.

Nazarenko, A. (2015). Blended learning vs traditional learning: what works? (a case study research). *Procedia: Social and Behavioral Sciences*, *200*, 77–82. doi:10.1016/j.sbspro.2015.08.018

Neeli, B. K., Reiners, T., & Wood, L. C. (2015). Gamification in education and business. In Gamification in Education and Business. doi:10.1007/978-3-319-10208-5

Neuman, M., & Neuman, D. (2014). Touch screen tablets and emergent literacy. *Early Childhood Education Journal*, *42*(4), 231–239. doi:10.100710643-013-0608-3

Neumann, M. (2018). Parent scaffolding of young children's use of touch screen tablets. *Early Child Development and Care*, *188*(12), 1654–1664. doi:10.1080/03004430.2016.1278215

Neumann, M. M. (2014). An examination of touch screen tablets and emergent literacy in Australian preschool children. *Australian Journal of Education*, *58*(2), 109–122. doi:10.1177/0004944114523368

Neumann, M. M. (2016). Young children's use of touch screen tablets for writing and reading at home: Relationships with emergent literacy. *Computers & Education*, *97*, 61–68. doi:10.1016/j.compedu.2016.02.013

Neumann, M. M. (2018a). Using tablets and apps to enhance emergent literacy skills in young children. *Early Childhood Research Quarterly*, *42*, 239–246. doi:10.1016/j.ecresq.2017.10.006

Neumann, M. M. (2018b). Maternal scaffolding of preschoolers' writing using tablet and paper-pencil tasks: relations with emergent literacy skills. *Journal of Research in Childhood Education*, *32*(1), 67–80. doi:10.1080/02568543.2017.1386740

Neumann, M. M., Hood, M., & Ford, R. (2012). Mother-child joint writing in an environmental print setting: Relations with emergent literacy. *Early Child Development and Care*, *182*(10), 1349–1369. doi:10.1080/03004430.2011.615928

Neumann, M. M., Hood, M., & Neumann, D. L. (2009). The scaffolding of emergent literacy skills in the home environment: a case study. *Early Childhood Education Journal*, *36*(4), 313–319. doi:10.100710643-008-0291-y

Neumann, M. M., Merchant, G., & Burnett, C. (2018). Young children and tablets: The views of parents and teachers. *Early Child Development and Care*, *1550083*. doi:10.1080/03004430

Neumann, M. M., & Neumann, D. L. (2017). The use of touch screen tablets at home and pre-school to foster emergent literacy. *Journal of Early Childhood Literacy*, *17*(2), 203–220. doi:10.1177/1468798415619773

Niklas, F., & Schneider, W. (2013). Home literacy environment and the beginning of reading and spelling. *Contemporary Educational Psychology*, *38*(1), 40–50. doi:10.1016/j.cedpsych.2012.10.001

Nikolopoulou, K. (2019, June). Mobile technologies and early childhood education. In M. Tsitouridou, J. A. Diniz, & T. Mikropoulos (Eds.), *Technology and innovation in learning, teaching and education. TECH-EDU 2018, CCIS 993* (pp. 444–457). Springer Nature Switzerland AG; doi:10.1007/978-3-030-20954-4_33

Nikolopoulou, K., & Gialamas, V. (2015). ICT and play in preschool: Early childhood teachers' beliefs and confidence. *International Journal of Early Years Education*, *23*(4), 409–425. doi:10.1080/09669760.2015.1078727

Nikolopoulou, K., Gialamas, V., & Batsouta, M. (2010). Young children's access to and use of ICT at home. *Review of Science, Mathematics, and ICT Education*, *4*(1), 25–40.

Nikou, S. A., & Economides, A. A. (2017). Mobile-based assessment: Integrating acceptance and motivational factors into a combined model of self-determination theory and technology acceptance. *Computers in Human Behavior*, *68*, 83–95. doi:10.1016/j.chb.2016.11.020

Nine practical uses of the tin can API (xAPI). (n.d.). Retrieved from https://www.youtube.com/watch?v=8LFhDdqQ13A

Niolaki, G. Z., & Masterson, J. (2012). Transfer effects in spelling from transparent Greek to opaque English in seven-to-ten-year-old children. *Bilingualism: language and cognition*, *15*(4), 757–770. doi:10.1017/S1366728911000721

Norris, C., Hossain, A., & Soloway, E. (2012). Under what conditions does computer use positively impact student achievement? In P. Resta (Ed.), *Supplemental vs. essential use. 2021–2028*. Austin, TX: AACE.

Northrop, L., & Killeen, E. (2013). A framework for using iPads to build early literacy skills. *The Reading Teacher*, *66*(7), 531–537. doi:10.1002/TRTR.1155

Notari, M. P., Hielscher, M., & King, M. (2016). Educational apps ontology. In D. Churchill, J. ... (Eds.), *Mobile learning design, lecture notes* (pp. 83–96). Singapore: Springer. doi:10.1007/978-981-10-0027-0_5

Nunnally, J. C. (1978). *Psychometric theory* (2nd ed.). New York: McGraw-Hill.

O'Toole, K. J. (2015). *Emergent literacy skills in print and electronic contexts: The influence of book type, narration style, and attention* (Doctoral dissertation). Available at http:// ecommons.luc.edu/luc_diss/1652

O'Toole, K. J., & Kannass, K. N. (2018). Emergent literacy skills in print and electronic contexts: The influence of book type, narration source, and attention. *Journal of Experimental Child Psychology*, *173*, 100–115. doi:10.1016/j.jecp.2018.03.013 PMID:29702378

OECD. (2015). Computers and learning: making the connection. (PISA, OECD Publishing, 2015).

Ofcom. (2019). Children and parents: Media Use and Attitudes Report 2018. London, UK: Author.

Ofcom. (2019). *Children and parents: Media use and attitudes report 2018*. Making sense of Media.

OFCOM. (2019). *Children and parents: Media use and attitudes report, 2018*. London, UK: OFCOM. Retrieved from https://www.ofcom.org.uk/__data/assets/pdf_file/0024/134907/Children-and-Parents-Media-Use-and-Attitudes-2018.pdf

Ofcom. (n.d.). Children and parents: Media use and attitudes report 2018.

Office of Head Start. (2015). *Head Start early learning outcomes framework – ages birth to five*. Washington, DC: U.S. Department of Health & Human Services, Administration for Children & Families. Retrieved from https://eclkc.ohs.acf.hhs.gov/sites/default/files/pdf/elof-ohs-framework.pdf

Okanlawon, A. E., Fakokunde, J. B., Yusuf, F. A., Abanikannda, M. O., & Oyelade, A. A. (2017). Attitudes towards instructional games on Peace Education among second year students in junior secondary schools in South-west Nigeria. *International Journal of Education and Development Using Information and Communication Technology*, *13*(3), 98–108. Retrieved from https://files.eric.ed.gov/fulltext/EJ1166610.pdf

Oliemat, E., Ihmeideh, F., & Alkhawaldeh, M. (2018). The use of touch-screen tablets in early childhood: Children's knowledge, skills, and attitudes towards tablet technology. *Children and Youth Services Review*, *88*, 591–597. doi:10.1016/j.childyouth.2018.03.028

Onaolapo, S., & Oyewole, O. (2018). Performance expectancy, effort expectancy, and facilitating conditions as factors influencing smartphones use for mobile learning by postgraduate students of the University of Ibadan, Nigeria. *Interdisciplinary Journal of E-Skills and Lifelong Learning*, *14*(June), 95–115.

Orrin, M. T., & Olcese, N. R. (2011). Teaching and learning with iPads, ready or not? *TechTrends*, *55*(6), 42–48. doi:10.100711528-011-0540-6

Otterborn, A., Schönborn, K., & Hultén, M. (2018). Surveying preschool teachers' use of digital tablets: General and technology education related findings. *International Journal of Technology and Design Education*, 1–21.

Outhwaite, L. A., Faulder, M., Gulliford, A., & Pitchford, N. J. (2019). Raising early achievement in math with interactive apps: A randomized control trial. *Journal of Educational Psychology*, *111*(2), 284–298. doi:10.1037/edu0000286 PMID:30774149

Outhwaite, L. A., Gulliford, A., & Pitchford, N. J. (2017). Closing the gap: Efficacy of a tablet intervention to support the development of early mathematical skills in UK primary school children. *Computers & Education*, *108*, 43–58. doi:10.1016/j.compedu.2017.01.011

OverviewS. C. O. R. M. (n.d.). Retrieved from http://www.adlnet.gov/SCORM

Özdener, N., & Demirci, F. (2019). Determining students' views about an educational game-based mobile application supported with sensors. *Technology Knowledge and Learning*, *24*(1), 143–159. doi:10.100710758-018-9368-x

Özeke, V. (2018). Evaluation of Educational Mobile Apps for Turkish Preschoolers from Google Play Store. *European Journal of Education Studies*, *4*(4), 238–259.

Paciga, K. A., & Donohue, C. (2017). *Technology and interactive media for young children: a whole child approach connecting the vision of Fred Rogers with research and practice*. Latrobe, PA: Fred Rogers Center for Early Learning and Children's Media at Saint Vincent College; Retrieved from https://www.fredrogerscenter.org/wp-content/uploads/2017/07/Technology-and-Interactive-Media-for-Young-Children.pdf

Palaigeorgiou, G., Kamarina, K., Bratitsis, T., & Xefteris, S. (2018). Parental mediation of tablet educational use at home and at school: Facilitators or preventers? *International Conference on Interactive Mobile Communication, Technologies and Learning* (IMCL 2017), 339-350. 10.1007/978-3-319-75175-7_90

Palaiologou, I. (2016). Children under five and digital technologies: Implications for early years pedagogy. *European Early Childhood Education Research Journal*, *24*(1), 5–24. doi:10.1080/1350293X.2014.929876

Pamuk, S. (2011). Understanding preservice teachers' technology use through TPACK framework. *Journal of Computer Assisted Learning*, *28*(5), 425–439. doi:10.1111/j.1365-2729.2011.00447.x

Panagiotakis, S., Papadokostaki, K., Vassilakis, K., & Malamos, A. (2019). Towards a novel and LMS-free pervasive learning system exploiting the experience API. *EAI Endorsed Transactions on Creative Technologies*, 1–13.

Panagiotakis, S., Papadokostaki, K., Vassilakis, K., & Malamos, A. (2019). Towards a novel and LMS-free Pervasive Learning System exploiting the Experience API. EAI Endorsed Transactions on Creative Technologies, 1–13.

Panagiotakis, S. (2012). Context-Awareness via Ubiquitous User Profiling : An Implementation Paradigm. Scientific & Academic Publishing. *Software Engineering*, *2*(3), 77–90. doi:10.5923/j.se.20120203.05

Papadakis, S., & Kalogiannakis, M. (2017a). Combining mobile technologies in environmental education: A Greek case study. *International Journal of Mobile Learning and Organisation*, (Special Issue on Mobile Learning Applications and Strategies), 11(2), 108-128. doi:. doi:10.1504/IJMLO.2017.10003925

Papadakis, St., & Kalogiannakis, M. (2019). Evaluating a Course for Teaching Advanced Programming Concepts with Scratch to Preservice Kindergarten Teachers: A Case Study in Greece. In D. Farland-Smith (Ed.), *Early Childhood Education* (pp. 1-19). London: IntechOpen Limited. Retrieved from www.intechopen.com

Papadakis, S. (2017). Creativity and innovation in European education. 10 years eTwinning. Past, present and the future. *International Journal of Technology Enhanced Learning*, *8*(3-4), 279–296.

Papadakis, S. (2018). The use of computer games in classroom environment. *International Journal of Teaching and Case Studies*, *9*(1), 1–25. doi:10.1504/IJTCS.2018.090191

Papadakis, S. (2018a). Gender stereotypes in Greek computer science school textbooks. *International Journal of Teaching and Case Studies*, *9*(1), 48–71. doi:10.1504/IJTCS.2018.090196

Papadakis, S., & Kalogiannakis, M. (2019). Evaluating a course for teaching advanced programming concepts with scratch to preservice kindergarten teachers: a case study in Greece. In D. Farland-Smith (Ed.), *Early Childhood Education* (pp. 1–19). London, UK: IntechOpen Limited. doi:10.5772/intechopen.81714

Papadakis, S., Kalogiannakis, M., Orfanakis, V., & Zaranis, N. (2017). The appropriateness of Scratch and App Inventor as educational environments for teaching introductory programming in primary and secondary education. [IJWLTT]. *International Journal of Web-Based Learning and Teaching Technologies*, *12*(4), 58–77. doi:10.4018/IJWLTT.2017100106

Papadakis, S., Kalogiannakis, M., & Zaranis, N. (2017). Improving mathematics teaching in kindergarten with realistic mathematical education. *Early Childhood Education Journal*, *45*(3), 369–378. doi:10.100710643-015-0768-4

Papadakis, S., Kalogiannakis, M., & Zaranis, N. (2018). Educational apps from the Android Google Play for Greek preschoolers: A systematic review. *Computers & Education*, *116*, 139–160. doi:10.1016/j.compedu.2017.09.007

Papadakis, S., Kalogiannakis, M., & Zaranis, N. (2018). The effectiveness of computer and tablet assisted intervention in early childhood students' understanding of numbers. An empirical study conducted in Greece. *Education and Information Technologies*, *23*(5), 1849–1871. doi:10.100710639-018-9693-7

Papadakis, S., Kalogiannakis, M., & Zaranis, N. (2019). *Parental involvement and attitudes towards young Greek children's mobile usage.* International Journal of Child-Computer Interaction; doi:10.1016/j.ijcci.2019.100144

Papadakis, S., Kalogiannakis, M., Zaranis, N., & Orfanakis, V. (2016). Using Scratch and App Inventor for teaching introductory programming in secondary education. A case study. *International Journal of Technology Enhanced Learning*, *8*(3/4), 217–233. doi:10.1504/IJTEL.2016.082317

Papadakis, St., & Kalogiannakis, M. (2017a). Mobile educational applications for children. What educators and parents need to know. *International Journal of Mobile Learning and Organisation*, *11*(3), 256–277. doi:10.1504/IJMLO.2017.085338

Papadakis, St., & Kalogiannakis, M. (2017b). Evaluation of Greek Android mobile applications for preschoolers. *Preschool and Primary Education, 5*, 65–100. doi:10.12681/ppej.11208

Papadakis, St., Kalogiannakis, M., & Zaranis, N. (2016a). Comparing tablets and PCs in teaching mathematics: An attempt to improve mathematics competence in early childhood education. *Preschool and Primary Education, 4*(2), 241–253. doi:10.12681/ppej.8779

Papadakis, St., Kalogiannakis, M., & Zaranis, N. (2016b). Developing fundamental programming concepts and computational thinking with ScratchJr in preschool education: A case study. *International Journal of Mobile Learning and Organisation, 10*(3), 187–202. doi:10.1504/IJMLO.2016.077867

Papadakis, St., Kalogiannakis, M., & Zaranis, N. (2017). Designing and creating an educational app rubric for preschool teachers. *Education and Information Technologies, 22*(6), 3147–3165. doi:10.100710639-017-9579-0

Papadokostaki, K. (2018). Diakhiti Mathisi. Apo Orama Pragmatikotita me ti khrisi tou Experience API. [Ubiquitous learning. From vision to reality with experience API]. In IAKE (Ed.), *Conference Proceedings of 4th International Conference 'Democracy, rights and inequalities in the era of economic crisis. Challenges in the field of research and education'* (pp. 244-253 (Volume B')). Heraklion Crete.

Papadokostaki, K., Panagiotakis, S., Vassilakis, K., & Malamos, A. (2017). Implementing an adaptive learning system with the use of experience API. In Interactivity, Game Creation, Design, Learning, and Innovation (pp. 393–402). Springer.

Papadokostaki, K., Spyros, P., & Vassilakis, K. (2018). Katagraphi mathisiakon empirion kai exagoyi analitikon me to protipo Experience API [Tracking learning experience and extracting analytics with experience API]. In *4th Conference for the Promotion of Educational Innovation* (pp. 766–776). Larissa, Greece: EEPEK.

Papadokostaki, K., Mastorakis, G., Panagiotakis, S., Mavromoustakis, C. X., Dobre, C., & Batalla, J. M. (2017). Handling big data in the era of internet of things (IoT). In C. X. Mavromoustakis, G. Mastorakis, & C. Dobre (Eds.), *Advances in mobile cloud computing and big data in the 5G era* (pp. 3–22). Cham, Switzerland: Springer International Publishing; doi:10.1007/978-3-319-45145-9_1

Papadopoulou, P., & Athanasiou, K. (2015). Children ideas for "animal" compared to teachers' conceptions. *International Journal of Innovation and Research in Educational Sciences, 2*(6).

Papavlasopoulou, S., Giannakos, M. N., & Jaccheri, L. (2019). Exploring children's learning experience in constructionism-based coding activities through design-based research. *Computers in Human Behavior, 99*, 415–427. doi:10.1016/j.chb.2019.01.008

Papert, S. (1980). *Mindstorms: Children, computers and powerful ideas*. New York, NY: Basic Books.

Papert, S. (1980). *Mindstorms: Children, computers, and powerful ideas*. Basic Books.

Papert, S. (1991). *Constructionism*. US: Ablex Publishing.

Papert, S. (1993). *The children's machine: Rethinking school in the age of the computer*. New York, NY: BasicBooks.

Paramythis, A., & Loidl-Reisinger, S. (2003). Adaptive learning environments and e-learning standards. In *Second European Conference on e-Learning, Vol. 1, No. 2003* (pp. 369–379).

Park, Y. (2011). A pedagogical framework for mobile learning: Categorizing educational applications of mobile technologies into four types. *The International Review of Research in Open and Distributed Learning, 12*(2), 78-102.

Park, S. Y., Nam, M. W., & Cha, S. B. (2012). University students' behavioral intention to use mobile learning: Evaluating the technology acceptance model. *British Journal of Educational Technology, 43*(4), 592–605. doi:10.1111/j.1467-8535.2011.01229.x

Park, Y. (2011). A pedagogical framework for mobile learning : categorizing educational applications of mobile technologies into four types. *The International Review of Research in Open and Distributed Learning, 12*(2), 78–102. doi:10.19173/irrodl.v12i2.791

Parnell, W. A., & Bartlett, J. (2012). iDocument: How smartphones and tablets are changing documentation in preschool and primary classrooms. *Technology and Young Children, 67*(3), 50–59.

Pasnik, S. (Ed.). (2019). *Getting ready to learn: creating effective, educational children's media.* New York, NY: Routledge.

Pasnik, S., & Llorente, C. (2013). *Preschool teachers can use a PBS KIDS transmedia curriculum supplement to support young children's mathematics learning: Results of a randomized controlled trial. Summative Evaluation of the CPB-PBS Ready to Learn Initiative.* Waltham, MA: Education Development Center.

Patchan, M. M., & Puranic, C. S. (2016). Using tablet computers to teach preschool children to write letters: Exploring the impact of extrinsic and intrinsic feedback. *Computers & Education, 102*, 128–137. doi:10.1016/j.compedu.2016.07.007

Patrick, P., Byrne, J., Tunnicliffe, S. D., Asunta, T., Carvalho, G. S., Havu-Nuutinen, S., ... Tracana, R. B. (2013). Students (ages 6, 10, and 15 years) in six countries' knowledge of animals. *Nordic Studies in Science Education, 9*(1), 18–32. doi:10.5617/nordina.624

Pelgrum, W. (2001). Obstacles to the integration of ICT in education: Results from a worldwide educational assessment. *Computers & Education, 37*(2), 163–178. doi:10.1016/S0360-1315(01)00045-8

Pelton, F. L., & Pelton, T. (2012). Sharing strategies with Teachers: iPods in math class. In *Society for Information Technology & Teacher Education International Conference* (Vol. 2012, No. 1, pp. 4363-4366). Academic Press.

Pentimonti, J. M., & Justice, L. M. (2010). Teachers' use of scaffolding strategies during read alouds in the preschool classroom. *Early Childhood Education Journal, 37*(4), 241–248. doi:10.100710643-009-0348-6

Penuel, W. R. (2006). Implementation and effects of one-to-one computing initiatives: A research synthesis. *Journal of Research on Technology in Education, 38*(3), 329–348. doi:10.1080/15391523.2006.10782463

Pérez, P., & Salmerón, T. (2006). Desarrollo de la comunicación y del lenguaje: Indicadores de preocupación. *Revista Pediatría Atención Primaria, VIII*(32), 111–125.

Perlmutter, M., Behrend, S. D., Kuo, F., & Muller, A. (1989). Social influences on children's problem solving. *Developmental Psychology, 25*(5), 744–754. doi:10.1037/0012-1649.25.5.744

Petkovski, M. (2014). *Parent-child interaction in shared-computer activities: an exploratory study examining parent–child interaction with a mobile device. (Master's Thesis*, Wilfrid Laurier University, Canada). Available at http://scholars.wlu.ca/etd

Pew Research Center. (2015, April). *The smartphone difference.* Retrieved from http://www.pewinternet.org/2015/04/01/us-smartphone-use-in-2015/

Piaget, J. (n.d.). *The psychology of intelligence.* London, UK: Taylor & Francis e-Library.

Pierce, G. L., & Cleary, P. F. (2016). The K-12 educational technology value chain: Apps for kids, tools for teachers and levers for reform. *Education and Information Technologies, 21*(4), 863–880. doi:10.100710639-014-9357-1

Pila, S. (2019). *Young children: What's code got to do with it? TEC Center at Erikson Institute.* Retrieved from http://teccenter.erikson.edu/what-were-reading/pila_coding/

Pila, S., Aladé, F., Sheehan, K. J., Lauricella, A. R., & Wartella, E. A. (2018). Learning to code via tablet applications: An evaluation of Daisy the Dinosaur and Kodable as learning tools for young children. *Computers & Education, 11*, 52–62.

Piotrowski, J. T., & Krcmar, M. (2017). Reading with hotspots: Young children's responses to touchscreen. *Computers in Human Behavior, 70*, 328–334. doi:10.1016/j.chb.2017.01.010

Plowman, L., Stevenson, O., Stephen, C., & McPake, J. (2012). Preschool children's learning with technology at home. *Computers & Education, 59*(1), 30–37. doi:10.1016/j.compedu.2011.11.014

Pope, M., Hare, D., & Howard, E. (2002). ICT integration: Closing the gap between what preservice teachers are taught to do and what they can do. *Journal of Technology and Teacher Education, 10*, 191–203.

Portelance, D. J., Strawhacker, A. L., & Bers, M. U. (2016). Constructing the ScratchJr programming language in the early childhood classroom. *International Journal of Technology and Design Education, 26*(4), 489–504. doi:10.100710798-015-9325-0

Potential and applications of holograms to engage learners. (n.d.). Retrieved from http://edtechreview.in/trends-insights/trends/521-applications-of-holograms-to-engage-learners

Prensky, M. (2001a). Digital natives, digital immigrants part 1. *On the Horizon, 9*(5), 1–6. doi:10.1108/10748120110424816

Prensky, M. (2001b). Digital natives, digital immigrants, part II: Do they really think differently? *On the Horizon, 9*(6), 2–6. doi:10.1108/10748120110424843

Prior, A., Katz, M., Mahajna, I., & Rubinsten, O. (2015). Number word structure in first and second language influences arithmetic skills. *Frontiers in Psychology, 6*, 266. doi:10.3389/fpsyg.2015.00266 PMID:25852591

Protalinski, E. (2018). Apple passes 400 million iPads sold. *Venture Beat.* Retrieved from https://venturebeat.com/2018/10/30/apple-passes-400-million-ipads-sold/

Pruet, P., Ang, C. S., & Farzin, D. (2016). Understanding tablet computer usage among primary school students in underdeveloped areas: Students' technology experience, learning styles and attitudes. *Computers in Human Behavior, 55*, 1131–1144. doi:10.1016/j.chb.2014.09.063

Psycharis, S., & Kotzampasaki, E. (2019). The impact of a STEM Inquiry Game Learning scenario on Computational Thinking and Computer self-confidence. *Eurasia Journal of Mathematics, Science & Technology Education, 15*(4), em1689, 1-18, doi:10.29333/ejmste/103071

Psycharis, S. (2018). STEAM in education: A literature review on the role of computational thinking, engineering epistemology, and computational science. Computational STEAM Pedagogy (CSP). *Scientific Culture, 4*(2), 51–72.

Psycharis, S., & Kalia, M. (2017). The Effects of Computer Programming on high school students' problem solving, reasoning skills and self-efficacy in Mathematics. *Instructional Science, 45*(5), 583–602. doi:10.100711251-017-9421-5

Pullen, D., Swabey, K., Abadooz, M., & Sing, T. K. R. (2015). Pre-service teachers' acceptance and use of mobile learning in Malaysia. *Australian Educational Computing, 30*(1).

Rahimi, M., & Yadollahi, S. (2011). ICT use in EFL classes: A focus on EFL teachers' characteristics. *World Journal of English Language, 1*(2), 17. doi:10.5430/wjel.v1n2p17

Ralph, R. (2018). Media and technology in preschool classrooms: manifesting prosocial sharing behaviours when using iPads. *Technology, Knowledge, and Learning, 23*(2), 199–221. doi:10.100710758-017-9342-z

Raman, A., Malik, B., Dzahir Kasa, M., & Sofian,, O. F. M., & Hussin, F. (2015). Teachers' attitude toward computer use in classroom practice. *Scholars Journal of Arts, Humanities, and Social Sciences, 3*(3A), 646–659.

Ramos, J. R. (2015). Evaluación de un programa de maestría en Educación a través de la MSS. *Educación y Humanismo, 17*(29), 236–255.

Raven, J. H. (1956). *Guide to using the coloured progressive matrices. Sets A, AB, B.* London, UK: H. K. Lewis.

Real Academia Española. (2019). *Lenguaje.* Retrieved from https://dle.rae.es/?id=n7bnifo

Reeves, J. L., Gunter, G. A., & Lacey, C. (2017). Mobile learning in pre-kindergarten: using student feedback to inform practice. *Journal of Educational Technology & Society, 20*(1), 37–44.

Reich, S. M., Yau, J. C., & Warschauer, M. (2016). Tablet-based eBooks for young children: what does the research say? *Journal of Developmental and Behavioral Pediatrics, 37*(7), 585–591. doi:10.1097/DBP.0000000000000335 PMID:27575440

Reid, D. K., Hresko, W. P., & Hammill, D. D. (2018). *Test of early reading ability* (4th ed.). Austin, TX: PRO-ED.

Reinhardt, D., & Heinig, C. (2016). Survey-based exploration of attitudes to participatory sensing tasks in location-based gaming communities. *Pervasive and Mobile Computing, 27*, 27–36. doi:10.1016/j.pmcj.2016.01.001

Report, H. (2017). The NMC/CoSN Horizon Report: 2017 K–12. Retrieved from https://cdn.nmc.org/media/2017-nmc-cosn-horizon-report-k12-EN.pdf

Resnick, M. (2010). *Rethinking learning in the digital age.* Retrieved from http://www.media.mit.edu/~mres/papers/wef.pdf

Richter, A., & Courage, M. L. (2017). Comparing electronic and paper story books for preschoolers: Attention, engagement, and recall (2017). *Journal of Applied Developmental Psychology, 48*, 92–102. doi:10.1016/j.appdev.2017.01.002

Rideout, V. (2013). *Zero to eight: children's media use in America 2013.* Commonsense Media.

Rideout, V. (2017). *The Common Sense census: Media use by kids age zero to eight.* San Francisco, CA: Common Sense Media.

Rideout, V. (2017). *The common sense census: Media use by kids age zero to eight.* San Francisco, CA: Common Sense Media; Available at http://www.commonsensemedia.org

Robertson, S. A., & Graven, M. (2018). Using a transdisciplinary framework to examine mathematics classroom talk taking place in and through a second language. *ZDM, 50*(6), 1013–1027. doi:10.100711858-018-0952-2

Robinson, T. N., Banda, J. A., Hale, L., Lu, A. S., Fleming-Milici, F., Calvert, S. L., & Wartella, E. (2017). Screen media exposure and obesity in children and adolescents. *Pediatrics, 140*(Supplement 2), S97–S101. doi:10.1542/peds.2016-1758K PMID:29093041

Rodríguez, P. (2006). Redefiniendo el trabajo metodológico cualitativo con niños: El uso de la entrevista de grupo aplicada al estudio de la tecnología. *EMPIRIA. Metodología de Ciencias Sociales, 12*(12), 65–88. doi:10.5944/empiria.12.2006.1135

Roldán-Álvarez, D., Márquez-Fernández, A., Martín, E., & Guzmán, C. (2016). Learning experiences using tablets with children and people with autism spectrum disorder. In *European Conference on Technology Enhanced Learning* (pp. 640-643). Dordrecht: Springer. 10.1007/978-3-319-45153-4_81

Román-González, M., Pérez-González, J. C., & Jiménez-Fernández, C. (2017). Which cognitive abilities underlie computational thinking? Criterion validity of the Computational Thinking Test. *Computers in Human Behavior, 72*, 678–691. doi:10.1016/j.chb.2016.08.047

Rosa, M., & Orey, D. C. (2015). A trivium curriculum for mathematics based on literacy, matheracy, and technoracy: An ethnomathematics perspective. *ZDM*, *47*(4), 587–598. doi:10.100711858-015-0688-1

Rose, S., Habgood, J., & Jay, T. (2017). An exploration of the role of visual programming tools in the development of young children's computational thinking. *Electronic Journal of e-Learning, 15*(4), 297-309.

Roseberry, S., Hirsh-Pasek, K., & Golinkoff, R. M. (2014). Skype me! Socially contingent interactions help toddlers learn language. *Child Development*, *85*(3), 956–970. doi:10.1111/cdev.12166 PMID:24112079

Roskos, K., Burstein, K., & You, B.-K. (2012). A typology for observing children's engagement with eBooks at preschool. *Journal of Interactive Online Learning*, *11*(2), 47–66.

Rowsell, J., & Harwood, D. (2015). "Let it go": Exploring the image of the child as a producer, consumer, and inventor. *Theory into Practice*, *54*(2), 136–146. doi:10.1080/00405841.2015.1010847

Saalem, M., Mohamed, A., Gazala, Y., & Rajbunisa, A. (2019). Augmented reality based education for the improvement of sustainable learning ability in OMAN educational system. *International Advanced Trends in Computer Science and Engineering*, *8*(12), 110–115.

Sabanci, A., & Omeroglu, M. (2015). Preschool teachers' views and experiences about ICT use in instruction: a case study. *International Journal of Academic Research in Business and Social Science*, *5*(6), 170–183.

Sáez-López, J. M., Román-González, M., & Vázquez-Cano, E. (2016). Visual programming languages integrated across the curriculum in elementary school: A two year case study using 'Scratch' in five schools. *Computers & Education*, *97*, 129–141. doi:10.1016/j.compedu.2016.03.003

Saldaña, J. (2015). The coding manual for qualitative researchers. Atlanta, GA: *Sage*.

Salmon, L. G. (2013). Factors that affect emergent literacy development when engaging with electronic books. *Early Childhood Education Journal*, *42*(2), 85–92. doi:10.100710643-013-0589-2

Sanchez, A. B., Marcos, J. J. M., Gonzalez, M., & Guanlin, H. (2012). In service teachers' attitudes towards the use of ICT in the classroom. *Procedia: Social and Behavioral Sciences*, *46*, 1358–1364. doi:10.1016/j.sbspro.2012.05.302

Sánchez-Prieto, J. C., Hernández-García, Á., García-Peñalvo, F. J., Chaparro-Peláez, J., & Olmos-Migueláñez, S. (2019). Break the walls! Second-order barriers and the acceptance of mLearning by first-year pre-service teachers. *Computers in Human Behavior, 95*(December 2018), 158–167.

Sánchez-Prieto, J. C., Olmos-Migueláñez, S., & García-Peñalvo, F. J. (2016). MLearning and pre-service teachers: An assessment of the behavioral intention using an expanded TAM model. *Computers in Human Behavior*, *72*, 644–654. doi:10.1016/j.chb.2016.09.061

Sánchez-Prieto, J.-C., Hernández-García, H., García-Peñalvo, F.-J., Chaparro-Peláez, J., & Olmos-Migueláñez, S. (2019). Break the walls! second order barriers and the acceptance of mLearning by FIRST-YEAR PRE-SERVICE TEAChers. *Computers in Human Behavior*, *95*, 158–167. doi:10.1016/j.chb.2019.01.019

Sanders, M. (2009). STEM, STEM education, STEMmania. *Technology Teacher*, *68*(4), 20–26.

Sandvik, M., Smørdal, O., & Østerud, S. (2012). Exploring iPads in practitioners' repertoires for language learning and literacy practices in kindergarten. *Nordic Journal of Digital Literacy*, *7*(03), 204–221.

Santos, M., & Osório, J. A. (2008). Las TIC en la primera infancia: valorización e integración en la educación inicial a través del enlace @rcacomum. *Iberoamericana de Educación, (46), 9.*

Sardar. (1988). The touch of Midas - science, values and environment in Islam and the West. *Social Studies of Science*.

Sari, B., Takacs, Z., & Bus, A. (2017). What are we downloading for our children? Best-selling children's apps in four European countries. *Journal of Early Childhood Literacy*; Advance online publication. doi:10.1177/1468798417744057

Saussure, F. (2016). *Curso de lingüística general*. México: Fontamara.

Savater, F. (2005). *Las preguntas de la vida*. Barcelona: Ariel.

Schacter, J., & Jo, B. (2017). Improving preschoolers' mathematics achievement with tablets: A randomized controlled trial. *Mathematics Education Research Journal, 29*(3), 1–15. doi:10.100713394-017-0203-9

Scherer, R., Siddiq, F., & Teo, T. (2015). Becoming more specific: Measuring and modeling teachers' perceived usefulness of ICT in the context of teaching and learning. *Computers & Education, 88*, 202–214. doi:10.1016/j.compedu.2015.05.005

Scherer, R., Siddiq, F., & Tondeur, J. (2019). The technology acceptance model (TAM): A meta analytic structural equation modeling approach to explaining teachers' adoption of digital technology in education. *Computers & Education, 128*, 13–35. doi:10.1016/j.compedu.2018.09.009

Scherer, R., Siddiq, F., & Tondeur, J. (2019). The technology acceptance model (TAM): A meta-analytic structural equation modeling approach to explaining teachers' adoption of digital technology in education. *Computers & Education, 128*(0317), 13–35.

Scherer, R., Tondeur, J., Siddiq, F., & Baran, E. (2018). The importance of attitudes toward technology for pre-service teachers' technological, pedagogical, and content knowledge: Comparing structural equation modeling approaches. *Computers in Human Behavior, 80*, 67–80. doi:10.1016/j.chb.2017.11.003

Schroeder, E. L., & Kirkorian, H. L. (2016). When seeing is better than doing: Preschoolers' transfer of STEM skills using touchscreen games. *Frontiers in Psychology, 7*, 1377. doi:10.3389/fpsyg.2016.01377 PMID:27679590

Schweinhart, L. J., Montie, J., Xiang, Z., Barnett, W. S., Belfield, C. R., & Nores, M. (2011). *Lifetime effects: The High/Scope Perry Preschool study through age 40: Summary, 157 Conclusions, and Frequently Asked Questions*. Ypsilanti, MI: High/Scope Press; Retrieved from http://www.highscope.org/file/Research/PerryProject/specialsummary_rev2011_02_2.pdf

Searle, J. (1994). *Actos de habla, ensayo de filosofía del lenguaje*. Barcelona: Planeta-Agostini.

See, K. T., Madhubala, B. H., & Koo, A. C. (2019). Motivation of parents towards reading multilingual eBooks to preschool children. *International Journal of Interactive Mobile Technologies, 13*(1), 20–36. doi:10.3991/ijim.v13i01.9060

Segers, E., & Verhoeven, L. (2003). Effects of vocabulary training by computer in kindergarten. *Computer Assisted Learning, 19*(4), 557–566. doi:10.1046/j.0266-4909.2003.00058.x

Seipold, J. (2014). Mobile learning: Structures, concepts and practices of the British and German mobile learning discusssion from a media education perspective. *MedienPädagogik, 24*, 30–52.

Selinger, M. (2001). Learning information and communications technology skills and the subject context of the learning. *Journal of Information Technology for Teacher Education, 10*(1-2), 143–154. doi:10.1080/14759390100200108

Sénéchal, M., Le Fevre, J., Smith-Chant, B. L., & Colton, K. V. (2001). On refining theoretical models of emergent literacy. The role of empirical evidence. *Journal of School Psychology, 39*(5), 439–460. doi:10.1016/S0022-4405(01)00081-4

Sequencing, S. C. O. R. M. Learn more about SCORM Sequencing. (n.d.). Retrieved from https://scorm.com/scorm-explained/technical-scorm/sequencing/

SequencingS. C. O. R. M. (n.d.). Retrieved from http://www.scormsoft.com/scorm/sn/sequencing

Shade, D., & Davis, B. (1997). The role of computer technology in early childhood education. In J. Isenberg & M. Jalongo (Eds.), *Major trends and issues in early childhood education* (pp. 90–103). New York: Teachers College Press.

Sharndama, E. C., & IJemofwu, I. A. (2013). Application of ICTs in teaching and learning English (ELT) in large classes. *Journal of Arts and Humanities*, *2*(6), 34–39.

Sharples, M., Taylor, J., & Vavoula, G. (2007). A theory of learning for the mobile age. The Sage handbook of e-learning research, 221–247. doi:10.1007/978-3-531-92133-4_6

Sharples, M. (2013). Mobile learning : Research, practice and challenges. *Distance Education in China*, *3*(5), 5–11.

Sharples, M., Taylor, J., & Vavoula, G. (2007). A theory of learning for the mobile age. In R. Andrews & C. Haythornthwaite (Eds.), *The Sage handbook of e-learning research* (pp. 221–247). London, UK: Sage. doi:10.4135/9781848607859.n10

Sheehan, K. J., Hightower, B., Lauricella, A. R., & Wartella, E. (2018). STEM Media in the Family Context: The effect of STEM career and media use on preschoolers' science and math skills. *European Journal of STEM Education*, *3*(3), 17. doi:10.20897/ejsteme/3877

Sheridan, M., Howard, J., & Alderson, D. (2010). *Play in early childhood: from birth to six years*. Retrieved from https://books.google.com/books?hl=en&lr=&id=gvWsAgAAQBAJ&pgis=1

Shifflet, R., Toledo, C., & Mattoon, C. (2012). Touch tablet surprises: A preschool teacher's story. *Young Children*, *67*(3), 36.

Shipstead, Z., Hicks, K., & Engle, R. (2012). Cogmed working memory training: Does the evidence support the claims? *Applied Research in Memory and Cognition*, *1*(3), 185–193. doi:10.1016/j.jarmac.2012.06.003

Shuler, C. (2009). *iLearn: A content analysis of the iTunes app store's educational section*. The Joan Ganz Cooney Center at Sesame Workshop.

Shuler, C., Levine, Z., & Ree, J. (2012). *iLearn II: An analysis of the education category of Apple's App Store*. New York, NY: Joan Ganz Cooney Center. Retrieved from http://www.joanganzcooneycenter.org/wp-content/uploads/2012/01/ilearnii.pdf

Shuler, C., Levine, Z., & Ree, J. (2012). *iLearn II: An analysis of the education category of Apple's app store*. New York, NY: The Joan Ganz Cooney Center at Sesame Workshop. Retrieved from http://joanganzcooneycenter.org/upload_kits/ilearnii.pdf

Shuler, C., Levine, Z., & Ree, J. (2012). iLearn II: An analysis of the education category of Apple's app store. New York: The Joan Ganz Cooney Center at Sesame Workshop.

Shuler, C. (2009). *Pockets of potential: Using mobile technologies to promote children's learning*. New York, NY: The Joan Ganz Cooney Center.

Shute, V. J. (2008). Focus on formative feedback. *Review of Educational Research*, *78*(1), 153–189. doi:10.3102/0034654307313795

Shute, V. J., & Towle, B. (2003). Adaptive e-learning. *Educational Psychologist*, *38*(2), 105–114. doi:10.1207/S15326985EP3802_5

Siemens, G., & Long, P. (2011). Penetrating the fog: analytics in learning and education. *EDUCAUSE Review*, *46*(5), 30.

Silawati, E., & Rachmania, S. (2016). Character building of early young learners through gadget as learning media in kindergarten. *Proceedings of the 3rd International Conference on Early Childhood Education (ICECE 2016)*. Atlantis Press. doi:10.2991/icece-16.2017.26

Singh, H. (2003). Building effective blended learning programs. *Educational technology-Saddle Brook then Englewood Cliffs NJ*, *43*(6), 51–54.

Siriram, R. (2012). A soft and hard systems approach to business process management. *Systems Research and Behavioral Science*, *29*(1), 87–100. doi:10.1002res.1095

Slutsky, R., Kragh-Müller, G., Rentzou, K., Tuul, M., Gol Guven, M., Foerch, D., & Paz-Albo, J. (2019). A cross-cultural study on technology use in preschool classrooms: Early childhood teacher's preferences, time-use, impact and association with children's play. *Early Child Development and Care*, 1–13. doi:10.1080/03004430.2019.1645135

Soberanes-Martín, A., Castillo, J. L., & Peña-Martín, A. (2018). GIL-1 aplicación para la ejercitación inicial del lenguaje. In M. Prieto-Méndez, S. J. Pech-Campos, & A. Francesa-Alfaro (Eds.), Tecnologías y aprendizaje: Investigación y práctica. 283-291. Ciudad Real, España: Editorial CIATA.org-UCML.

Soberanes-Martín, A., Peña-Martín, A., Castillo-Mendoza, J. L., & Martínez-Reyes, M. (2017). Aplicación para la ejercitación del lenguaje de niños con discapacidad intelectual grado de leve a moderado. *Sistemas y Gestión Educativa*, *4*(11), 14–23.

Sock Puppets App. (2016). Retrieved from https://apps.apple.com/ca/app/sock-puppets/id394504903

Stephen, C., Stevenson, O., & Adey, C. (2013). Young children engaging with technologies at home: The influence of family context. *Journal of Early Childhood Research*, *11*(2), 149–164. doi:10.1177/1476718X12466215

Storch, S. A., & Whitehurst, G. J. (2002). Oral language and code-related precursors to reading: Evidence from a longitudinal structural model. *Developmental Psychology*, *38*(6), 934–947. doi:10.1037/0012-1649.38.6.934 PMID:12428705

Strawhacker, A., & Bers, M. U. (2019). What they learn when they learn coding: Investigating cognitive development and computer programming in young children. *Educational Technology Research and Development*, *67*(3), 541–575. doi:10.100711423-018-9622-x

Strawhacker, A., Lee, M., & Bers, M. U. (2018). Teaching tools, teachers' rules: Exploring the impact of teaching styles on young children's programming knowledge in ScratchJr. *International Journal of Technology and Design Education*, *28*(2), 347–376. doi:10.100710798-017-9400-9

Strouse, G. A., & Ganea, P. A. (2017). A print book preference: Caregivers report higher child enjoyment and more adult–child interactions when reading print than electronic books. *International Journal of Child-Computer Interaction*, *12*, 8–15. doi:10.1016/j.ijcci.2017.02.001

Suhana, M. (2018). *Influence of gadget usage on children's social-emotional development. 169*(Icece 2017), 224–227. doi:10.2991/icece-17.2018.58

Sullivan, A., & Bers, M. U. (2016). Robotics in the early childhood classroom: Learning outcomes from an 8-week robotics curriculum in pre-kindergarten through second grade. *International Journal of Technology and Design Education*, *26*(1), 3–20. doi:10.100710798-015-9304-5

Sullivan, A., Kazakoff, E. R., & Bers, M. U. (2013). The wheels on the bot go round and round: Robotics curriculum in pre-kindergarten. *Journal of Information Technology Education: Innovations in Practice*, *12*, 203–219. Retrieved from http://www.jite.org/documents/Vol12/JITEv12IIPp203219Sullivan1257.pdf

Swaab, D. (2014). *Somos nuestro cerebro*. Barcelona: Plataforma. The iPad.

Synodi, E. (2010). Play in the kindergarten: The case of Norway, Sweden, New Zealand, and Japan. *International Journal of Early Years Education*, *18*(3), 185–200. doi:10.1080/09669760.2010.521299

Tabesh, Y. (2017). Computational thinking: A 21st century skill. Olympiads in Informatics, 11(SI), 65-70. Retrieved from https://ioinformatics.org/journal/v11si_2017_65_70.pdf

Tao, Y.-H., & Rosa Yeh, C.-C. (2008). Typology of teacher perception toward distance education issues – A study of college information department teachers in Taiwan. *Computers & Education*, *50*(1), 23–36. doi:10.1016/j.compedu.2006.03.005

Tapscott, D. (1998). *Growing up digital: The rise of the Net generation*. New York: McGraw-Hill.

Tapscott, D. (2008). *Grown up digital: How the net generation is changing your world*. New York: McGraw-Hill.

Tay, L. Y., Lim, S. K., Lim, C. P., & Koh, J. H. L. (2012). Pedagogical approaches for ICT integration into primary school English and mathematics: A Singapore case study. *Australasian Journal of Educational Technology*, *28*(4), 740–754. doi:10.14742/ajet.838

Taylor, R. D., Oberle, E., Durlak, J. A., & Weissberg, R. P. (2017). Promoting positive youth development through school-based social and emotional learning interventions: A meta-analysis of follow-up effects. *Child Development*, *88*(4), 1156–1171. doi:10.1111/cdev.12864 PMID:28685826

Teo, T. (2012). Examining the intention to use technology among pre-service teachers: An integration of the technology acceptance model and theory of planned behavior. *Interactive Learning Environments*, *20*(1), 3–18. doi:10.1080/10494821003714632

Terras, M., & Ramsay, J. (2012). The five central psychological challenges facing effective mobile learning. *British Journal of Educational Technology*, *43*(5), 820–832. doi:10.1111/j.1467-8535.2012.01362.x

Tey, T. C. Y., & Moses, P. (2018). UTAUT: Integrating achievement goals and learning styles for undergraduates' behavioural intention to use technology. *EAI Endorsed Transactions on E-Learning*, *5*(17), 155573.

The Asian Parent. (2014). *Mobile device usage among young kids, a Southeast Asia study*. Retrieved from https://s3-ap-southeast-1.amazonaws.com/tap-sg-media/theAsianparent+Insights+Device+Usage+A+Southeast+Asia+Study+November+2014.pdf

The evolution of technology in the classroom | Purdue University Online. (n.d.). Retrieved from https://online.purdue.edu/ldt/learning-design-technology/resources/evolution-technology-classroom

The United Nations. (1989). Convention on the rights of the child. *Treaty Series*, *1577*, 3.

Thomas, T. D., Singh, L., & Gaffar, K. (2013). The utility of the UTAUT model in explaining mobile learning adoption in higher education in Guyana. *Journal of Education and Education and Development Using Information and Communication Technology*, *9*(3), 71–85.

Thompson, P. (2013). The digital natives as learners: Technology use patterns and approaches to learning. *Computers & Education*, *6*, 512–533. doi:10.1016/j.compedu.2012.12.022

Tieng, P. G., & Eu, L. K. (2014). Improving students' Van Hiele level of geometric thinking using Geometer's Sketchpad. *Malaysian Online Journal of Educational Technology*, *2*(3), 20–31.

TinCan Javascript Library. (n.d.). Retrieved from http://rusticisoftware.github.io/TinCanJS/

Tondeur, J., van Braak, J., Siddiq, F., & Scherer, R. (2016). Time for a new approach to prepare future teachers for educational technology use: Its meaning and measurement. *Computers & Education*, *94*, 134–150. doi:10.1016/j.compedu.2015.11.009

Tondeur, J., van Braak, J., & Valcke, M. (2007). Towards a typology of computer use in primary education. *Journal of Computer Assisted Learning*, *23*(3), 197–206. doi:10.1111/j.1365-2729.2006.00205.x

Tootell, H., Plumb, M., Hadfield, C., & Dawson, L. (2013, January). Gestural interface technology in early childhood education: a framework for fully engaged communication. In *2013 46th Hawaii International Conference on System Sciences* (pp. 13-20). IEEE. 10.1109/HICSS.2013.241

Tosco, J. E. (2015). *The effects of technology on engagement and retention among upper elementary Montessori students.* (Unpublished Master's Thesis), St. Catherine University.

Traxler, J. (2009). Current state of mobile learning. In M. Ally (Ed.), *Mobile learning: Transforming the delivery of education and training* (pp. 9–24). Edmonton, AB: AU Press.

Traxler, J., & Vosloo, S. (2014). Introduction: The prospects for mobile learning. *Prospects*, *44*(1), 13–28. doi:10.100711125-014-9296-z

Trundle, K. C. (2010). Teaching science during the early childhood years. Best Practices in Science Education. *Pridobljeno*, *18*(10), 2016.

Tseles, D., Malafantis, K., & Pamouktsoglou, A. (2011). *Education, Science, and Society.* Athens, Greece: Sygxroni Ekdotiki.

Tselfes, V., & Paroussi, A. (2008). Science and theatre education: a cross-disciplinary approach of scientific ideas addressed to student teachers of early childhood education. *Science & Education*, *18*(9), 1115–1134. doi:10.100711191-008-9158-2

Tsitouridou, M., & Vryzas, K. (2003). Early childhood teachers' attitudes towards computer and information technology: the case of Greece. *Information Technology in Childhood Education*, 187-207.

Tsitouridou, M., & Vryzas, K. (2004). The prospect of integrating ICT into the education of young children: The views of Greek early childhood teachers. *European Journal of Teacher Education*, *27*(1), 29–45. doi:10.1080/0261976042000211838

Uğur, N. G., Koç, T., & Koç, M. (2016). An analysis of mobile learning acceptance by college students. *Journal of Educational and Instructional Studies in the World*, *6*(2), 2146–7463.

Umar, I. N., & Hassan, A. S. A. (2015). Malaysian teachers' levels of ICT integration and its perceived impact on teaching and learning. *Procedia: Social and Behavioral Sciences*, *197*, 2015–2021. doi:10.1016/j.sbspro.2015.07.586

UNESCO. (2012). Los *Jóvenes y las competencias: Trabajar con la educación.* Paris: Organización de las Naciones Unidas para la Educación, la Ciencia y la Cultura.

UNESCO. (2013). *Directrices de la UNESCO para las políticas de aprendizaje móvil.* Paris: ONU.

UNICEF. (2018). *Learning through play: Strengthening learning through play in early childhood education programmes.* New York: UNICEF.

Uresti, A. S., & Cannon, B. Y. (2014). Uso de Recursos Educativos Abiertos (REA) y su impacto en la adquisición de la lectura convencional en niños de tercer grado de preescolar. In Y. Heredia & B. Y. Cannon (Eds.), Aplicaciones de la tecnología educativa en diversos ambientes de Aprendizaje 53-90. Monterrey: ITESM.

Urgilés, G. (2016). Aula, lenguaje y educación. *Sophia, Colección de Filosofía de la Educación*, (20), 221-244.

Vaala, S., Ly, A., & Levine, M. H. (2015). Getting a Read on the App Stores: A Market Scan and Analysis of Children's Literacy Apps. Full Report. In *Joan Ganz Cooney Center at Sesame Workshop. Joan Ganz Cooney Center at Sesame Workshop.* Academic Press.

Vaala, S., Ly, A., & Levine, M. H. (2015). *Getting a read on the app stores: A market scan and analysis of children's literacy apps.* New York, NY: The Joan Ganz Cooney Center at Sesame Workshop.

Vakili, L. (2013). *The relationship between linguistic intelligence and L2 learning strategies among EFL learners with intermediate level of proficiency, 1*, 89–93.

Valcke, M., Bonte, S., Wever, B., & Rots, I. (2010). Internet parenting styles and the impact on internet use of primary school children. *Computers & Education, 55*(2), 454–464. doi:10.1016/j.compedu.2010.02.009

Van de Walle, J. A., Karp, K. S., & Bay-Williams, J. M. (2004). *Elementary and middle school mathematics*. Boston, MA: Allyn and Bacon.

van Deurse, A. (2016). Tablet use in primary education: Adoption hurdles and attitude determinants. *Education and Information Technologies, 21*(5), 971–990. doi:10.100710639-014-9363-3

Vatalaro, A., McDonald Culp, A., Hahs-Vaughn, D. L., & Barnes, A. C. (2018). A quasi-experiment examining expressive and receptive vocabulary knowledge of preschool head start children using mobile media apps. *Early Childhood Education Journal, 46*(4), 451–466. doi:10.100710643-017-0877-3

Vatavu, R. D., Cramariuc, G., & Schipor, D. M. (2015). Touch interaction for children aged 3 to 6 years: Experimental findings and relationship to motor skills. *International Journal of Human-Computer Studies, 74*, 54–76. doi:10.1016/j.ijhcs.2014.10.007

Vega, V., & Robb, M. B. (2019). *The Common Sense census: Inside the 21st-century classroom*. San Francisco, CA: Common Sense Media.

Venkatesh, V., Morris, M. G., Davis, G. B., & Davis, F. D. (2003). User acceptance of information technology: toward a unified view. *Management Information Systems Quarterly, 27*(3), 425–478. doi:10.2307/30036540

Venkatesh, V., Thong, & Xu. (2012). Consumer acceptance and use of information technology: extending the unified theory of acceptance and use of technology. *Management Information Systems Quarterly, 36*(1), 157–178. doi:10.2307/41410412

Vernadakis, N., Avgerinos, A., Tsitskari, E., & Zachopoulou, E. (2005). The use of computer assisted instruction in preschool education: making teaching meaningful. *Early Childhood Education Journal, 33*(2), 99–104. doi:10.100710643-005-0026-2

Vygotsky, L. (1978). *Mind in society*. Cambridge, MA: Harvard University Press.

Walsh, C., & Campbell, C. (2018). Introducing coding as a literacy on mobile devices in the early years. In Mobile technologies in children's language and literacy: innovative pedagogy in preschool and primary education (pp. 51-66). Emerald Publishing. doi:10.1108/978-1-78714-879-620181004

Wang, Y. S., Wu, M. C., & Wang, H. Y. (2009). Investigating the determinants and age and gender differences in the acceptance of mobile learning. *British Journal of Educational Technology, 40*(1), 92–118. doi:10.1111/j.1467-8535.2007.00809.x

Warren, S., Sauser, B., & Nowicki, D. (2019). A bibliographic and visual exploration of the historic impact of soft systems methodology on academic research and theory. *Systems, 7*(10), 1–15.

Wartella, E. A., & Jennings, N. (2000). Children and computers: New technology. Old concerns. *The Future of Children, 10*(2), 31–43. doi:10.2307/1602688 PMID:11255707

Wartella, E., Beaudoin-Ryan, L., Blackwell, C. K., Cingel, D. P., Hurwitz, L. B., & Lauricella, A. R. (2016). What kind of adults will our children become? The impact of growing up in a media-saturated world. *Journal of Children and Media, 10*(1), 13–20. doi:10.1080/17482798.2015.1124796

Wartella, E., Lauricella, A. R., & Blackwell, C. K. (2016). [Policy brief. Report for the Center on Media and Human Development School of Communication, Northwestern University.]. *The Ready To Learn Program, 2010*, 2015.

Wartella, E., & Robb, M. (2007). Young children, new media. *Journal of Children and Media, 1*(1), 35–44. doi:10.1080/17482790601005207

Waters, A. (2012). Trends and issues in ELT methods and methodology. *ELT Journal, 66*(4), 440–449.

Watlington, D. (2011). Using iPod touch and ipad educational apps in the classroom. In M. Koehler & P. Mishra (Eds.), *Proceedings of Society for Information Technology & Teacher Education International Conference 2011* (pp. 3112-3114). Chesapeake, VA: AACE.

Wedman, J., & Diggs, L. (2001). Identifying barriers to technology-enhanced learning environments in teacher education. *Computers in Human Behavior, 17*(4), 421–430. doi:10.1016/S0747-5632(01)00012-7

Weir, C. J., Vidaković, I., & Galaczi, E. D. (2013). *Measured constructs: A history of Cambridge English examinations.* Cambridge, UK: Cambridge University Press.

West, M., & Voslooo, S. (2013). *Policy guidelines for mobile learning. Policy guidelines for mobile learning*. 978-92-3-001143-7. France

Wetzel, K., Wilhelm, L., & Williams, M. K. (2004). The introductory technology course: A tool for technology integration. *Contemporary Issues in Technology & Teacher Education, 3*, 453–465.

Wheeler, S. (2001). Information and communication technologies and the changing role of the teacher. *Journal of Educational Media, 26*(1), 7–17. doi:10.1080/1358165001200069292

Whitehurst, G. L., & Lonigan, C. J. (1998). Child development and emergent literacy. *Child Development, 69*(3), 848–872. doi:10.1111/j.1467-8624.1998.tb06247.x PMID:9680688

White, N., Ensor, R., Marks, A., Jacobs, L., & Hughes, C. (2014). "It's mine!" Does sharing with siblings at age 3 predict sharing with siblings, friends, and unfamiliar peers at age 6? *Early Education and Development, 25*(2), 185–201. doi:10.1080/10409289.2013.825189

Wing, J. M. (2006). Computational thinking. *Communications of the ACM, 49-3*, 33-35.

Wing, J. M. (2006). *Viewpoint. Computational thinking. Communications of the ACM, 49(3), 33-35.* Retrieved from https://www.cs.cmu.edu/~15110-s13/Wing06-ct.pdf

Wing, J. M. (2011). Computational thinking: What and why. Retrieved from http://www.cs.cmu.edu/~CompThink/resources/TheLinkWing.pdf

Wing, J. M. (2006). Computational thinking. *Communications of the ACM, 49*(3), 33–35. doi:10.1145/1118178.1118215

Wohlwend, K. E. (2017). Toddlers and touchscreens: Learning 'Concepts Beyond Print' with tablet technologies. In R. J. Meyer & K. F. Whitmore (Eds.), *Reclaiming early childhood literacies: Narratives of hope, power, and vision.* Mahwah, NJ: Lawrence Erlbaum.

Wong, G. K., & Jiang, S. (2018, December). Computational thinking education for children: algorithmic thinking and debugging. In *2018 IEEE International Conference on Teaching, Assessment, and Learning for Engineering (TALE)* (pp. 328-334). IEEE.

Wong, K. Y. (2010). Cell phones as mobile computing devices. *IT Professional, 12*(3), 40–45. doi:10.1109/MITP.2010.46

Wong, S. S. H. (2015). Mobile digital devices and preschoolers' home multiliteracy practices. *Language and Literature, 17*(2), 75–90. doi:10.20360/g2cp49

Wood, D., Bruner, J. S., & Ross, G. (1976). The role of tutoring in problem solving. *Journal of Child Psychology and Psychiatry, and Allied Disciplines*, *17*(2), 89–100. doi:10.1111/j.1469-7610.1976.tb00381.x PMID:932126

Wood, E., Petkovski, M., De Pasquale, D., Gottardo, A., Evans, M. A., & Savage, R. S. (2016). Parent scaffolding of young children when engaged with mobile technology. *Frontiers in Psychology*, *7*, 690. doi:10.3389/fpsyg.2016.00690 PMID:27242603

wp-h5p-xapi. (n.d.). Retrieved from https://wordpress.org/plugins/wp-h5p-xapi/

Wu, W. H., Jim Wu, Y. C., Chen, C. Y., Kao, H. Y., Lin, C. H., & Huang, S. H. (2012). Review of trends from mobile learning studies: A meta-analysis. *Computers & Education*, *59*(2), 817–827. doi:10.1016/j.compedu.2012.03.016

xAPI Dashboard. (n.d.). Retrieved from https://github.com/adlnet/xAPI-Dashboard/blob/master/API_dashboard.md

xAPI-Spec. (n.d.). Retrieved from https://github.com/adlnet/xAPI-Spec/

xapi-statement-viewer. (n.d.). Retrieved from https://github.com/adlnet/xapi-statement-viewer

Xu, Z., Chen, Z., Eutsler, L., Geng, Z., & Kogut, A. (2019). A scoping review of digital game-based technology on English language learning. *Educational Technology Research and Development*, 1–28. doi:10.100711423-019-09702-2

Yahya, S., Ahmad, E. A., & Jalil, K. A. (2010). The definition and characteristics of ubiquitous learning: A discussion. *International Journal of Education and Development Using Information and Communication Technology*, *6*(1), 117–127.

Yang, S. (2013). Understanding undergraduate students' adoption of mobile learning model: a perspective of the extended UTAUT2. *Journal of Convergence Information Technology*, *8*(10), 969–979. doi:10.4156/jcit.vol8.issue10.118

Yelland, N., Drake, P., & Sadler, K. (2017). *Early Learning in STEM: Multimodal learning in the 21st century*. Melbourne: Victoria University.

Yelland, N., & Gilbert, C. (2018). Transformative technologies and play in the early years: Using tablets for new learning. *Global Studies of Childhood*, *8*(2), 152–161. doi:10.1177/2043610617734985

Yelland, N., & Masters, J. (2007). Rethinking scaffolding in the information age. *Computers & Education*, *48*(3), 362–382. doi:10.1016/j.compedu.2005.01.010

Yildirim, S. (2000). Effects of an educational computing course on preservice and inservice teachers: A discussion and analysis of attitudes and use. *Journal of Research on Computing in Education*, *32*(4), 479–496. doi:10.1080/08886504.2000.10782293

Yılmaz Genç, M. M., & Fidan, A. (2017). Children, parents and tablets: Preschool children's tablet use. *Egitim ve Ögretim*, *7*(3), 367–398. doi:10.14527/pegegog.2017.014

Yuan, L., & Powell, S. (2014). MOOCs and open education: Implications for higher education. *Cetis*, 19.

Yu, J., & Roque, R. (2018). A survey of computational kits for young children. In *Proceedings of the 17th ACM Conference on Interaction Design and Children* (pp. 289-299). ACM, New York, NY. 10.1145/3202185.3202738

Yu, J., & Roque, R. (2019). A review of computational toys and kits for young children. *International Journal of Child-Computer Interaction*, *21*, 17–36. doi:10.1016/j.ijcci.2019.04.001

Zainol, Z., Yahaya, N., Yahaya, N. M., & Zain, N. N. B. M. (2017). Factors influencing mobile learning among higher education students in Malaysia. *International Journal of Advanced Scientific Research and Management*, *2*(8), 86–91.

Zapata-Rivera, D., & Shute, V. J. (2012). Adaptive educational systems. *Adaptive Technologies for Training and Education, 7*, 4–27. doi:10.1017/CBO9781139049580.004

Zaranis, N. (2000). *The utilization of the Van Hiele method in compulsory education's geometry instruction with the computer's aid* (I axiopiisi tis theorias Van Hiele sti didaskalia tis geometrias stin ipoxreotiki ekpedeusi) (Doctoral dissertation). National and Kapodistrian University of Athens (NKUoA). Athens.

Zaranis, N. (2012). The use of ICT in preschool education for geometry teaching. In *Proceedings of the 10th international conference on computer based learning in science, learning science in the society of computers*, 256-262. Barcelona, Spain: Centre for Research in Science and Mathematics Education (CRECIM).

Zaranis, N., & Oikonomidis, V. (2009) ICT in preschool education. Athens: Grigoris Publications (Text in Greek).

Zaranis, N., & Valla, V. (2017). Tablet computer assisted counting and calculating activities for kindergarten children. In L. Gómez Chova, A. López Martínez, & I. Candel Torres (Eds.), *EDULEARN17 Proceedings, 9th International Conference on Education and New Learning Technologies* (pp. 9680-9689), July 3-5, 2017 — Barcelona, Spain. 10.21125/edulearn.2017.0828

Zaranis, N., Kalogiannakis, M., & Papadakis, S. (2013). Using mobile devices for teaching realistic mathematics in kindergarten education. *Creative Education (Special Issue in Preschool Education), 4*(7A1), 1-10.

Zaranis, N., Kalogiannakis, M., & Papadakis, S. (2013). Using mobile devices for teaching realistic mathematics in kindergarten education. *Creative Education, 4*(7A1), 1-10. doi:10.4236/ce.2013.47a1001

Zaranis, N. (2014). Geometry teaching through ICT in primary school. In C. Karagiannidis, P. Politis, & I. Karasavvidis (Eds.), *Research on e-learning and ICT in education* (pp. 59–71). New York, NY: Springer.

Zaranis, N. (2018). Comparing the effectiveness of using tablet computers for teaching addition and subtraction. In L. Daniela, & M. Lytras (Eds.), *Learning strategies and constructionism in modern education settings*. Hershey, PA: IGI Global. doi:10.4018/978-1-5225-5430-1.ch009

Zaranis, N., Kalogiannakis, M., & Papadakis, S. (2013). Using mobile devices for teaching realistic mathematics in kindergarten education. *Creative Education, 4*(07), 1–10. doi:10.4236/ce.2013.47A1001

Zaranis, N., & Oikonomidis, V. (2014). The main factors of the attitudes of Greek kindergarten teachers towards information and communication technology. *European Early Childhood Education Research Journal*, 1–18. doi:10.1080/1350293X.2014.970853

Zaranis, N., & Oikonomidis, V. (2015). Profiling the attitudes of Greek kindergarten teachers towards computers. *Education and Information Technologies, 20*(1), 201–215. doi:10.100710639-013-9296-2

Zaranis, N., & Valla, V. (2019). Tablets in learning mathematics for kindergarten students. In *Didactics of smart pedagogy* (pp. 267–284). Cham, Switzerland: Springer; doi:10.1007/978-3-030-01551-0_14

Zare, P., & Othman, M. (2013). Classroom debate as a systematic teaching/learning approach. *World Applied Sciences Journal, 28*(11), 1506–1513.

Zhang, D., Zhao, J. L., Zhou, L., & Nunamaker, J. F. Jr. (2004). Can e-learning replace classroom learning? *Communications of the ACM, 47*(5), 75–79. doi:10.1145/986213.986216

Zhao, X., & Okamoto, T. (2011). Adaptive multimedia content delivery for context-aware u-learning. *International Journal of Mobile Learning and Organisation, 5*(1), 46. doi:10.1504/IJMLO.2011.038691

Zipke, M. (2017). Preschoolers explore interactive storybook apps: The effect on word recognition and story comprehension. *Education and Information Technologies*, *22*(4), 1695–1712. doi:10.100710639-016-9513-x

Zomer, N. R., & Kay, R. H. (2016). Technology use in early childhood education: a review of literature. *Journal of Educational Informatics*, *1*, 1–25.

Zomer, N. R., & Kay, R. H. (2016). Technology use in early childhood education: A review of literature. *Journal of Educational Informatics*, *1*, 1–25.

Zrim, L. (2015). Child's play: Connected Aussie kids spend up to equivalent of three school days online every week. *Nielsen Australia e-Generation Report*. Retrieved from http://www.nielsen.com/au/en/insights/news/2015/childs-play-connected-aussie-kidsspend-up-to-equivalent-of-three-school-days-online.html

Zydney, J. M., & Warner, Z. (2016). Computers & education mobile apps for science learning: Review of research. *Computers & Education*, *94*, 1–17. doi:10.1016/j.compedu.2015.11.001

About the Contributors

Stamatios Papadakis has been a postdoctoral researcher in Educational Technology, with emphasis on mobile learning, at the Department of Preschool Education at the University of Crete, Greece since 2016. He has graduated from the Department of Informatics, Athens University of Economics and Business, Athens, Greece. In 2006 he completed a Master's degree (M.Ed.) in Pre-school Education and Educational Design at the University of Aegean, Rhodes, Greece at the School of Humanities, Department of Pre-school Education. In 2016, he completed a Ph.D. at the University of Crete, Department of Education in Rethymnon, Greece under the supervision by Prof. Zaranis Nikolaos. He has worked as an adjunct Lecturer in Education teaching Informatics (2017-2018) at the Department of Preschool Education, School of Education, University of Crete, Greece and Didactics in Programming (2017-2018) at the Department of Computer Sciences, School of Sciences and Engineering at the University of Crete, Greece. His scientific and research interests include the study of mobile learning, especially on the use of smart mobile devices and their accompanying mobile applications (apps) in the use of Preschool and Primary Education, focusing on the development of Computational Thinking and students' understanding of numbers. Furthermore, his research interests also include ICT in education, novice programming environments and teaching of programming in Primary and Secondary Education (novice programmers).

Michail Kalogiannakis is an Associate Professor in the Department of Preschool Education at the University of Crete and Associate Tutor at School of Humanities at the Hellenic Open University. He graduated from the Physics Department of the University of Crete and continued his postgraduate studies at the University Paris 7-Denis Diderot (D.E.A. in Didactic of Physics), University Paris 5-René Descartes-Sorbonne (D.E.A. in Science Education) and received his PhD degree at the University Paris 5-René Descartes-Sorbonne (PhD in Science Education). His research interests include science education in the early childhood, science teaching and learning, e-learning, the use of ICT in education, distant and adult education. He has published many articles in international conferences and journals and has served on the program committees of numerous international conferences.

* * *

Courtney K. Blackwell is a Research Assistant Professor in the Department of Medical Social Sciences at Northwestern University's Feinberg School of Medicine. As a developmental methodologist, she has expertise in survey development and scale validation, including child and parent proxy Person Reported Outcome measures as well as population health surveys. Her research focuses on young children's learning and positive health development, and the complex social environmental factors that contribute to such outcomes. Of particular interest is the developmental origins of well-being, including conceptualizing, identifying, and measuring early indicators of this construct, as well as ascertaining risk and promotive factors in the home and school environment. With a background in early childhood educational media and technology, she is interested in the role that digital and interactive media plays in young children's engagement in and approaches to learning, both of which are cornerstones for well-being. Dr. Blackwell earned her PhD at Northwestern University and EdM at the Harvard Graduate School of Education.

Elissavet Chlapana received her Ph.D. in Sciences of Education with specialization in Preschool Education Teaching Methodology and her MEd in Preschool Pedagogics from University of Crete. In her dissertation she studied issues related to immigrant kindergarten children's second language vocabulary development through literacy practices. She has worked as a Laboratory Teaching Staff in the Department of Preschool Education of University of Crete. She is currently an Assistant Professor in the same University Department. Her research focuses on Preschool Education Teaching Methodology and on teaching practices for the development of young children's literacy skills, especially in the domains of oral language, vocabulary, text comprehension and story structure knowledge.

Jillianne Code is a learning scientist, whose area of research is at the praxis of educational technology, psychology and measurement. Dr. Code is an Assistant Professor in the Department of Curriculum and Pedagogy in the Faculty of Education at the University of British Columbia (UBC). Before coming to the UBC, Jillianne was Associate Professor of Educational Technology and Psychology in the Faculty of Education at the University of Victoria (UVic), and a Post-doctoral Research Fellow at the Harvard Graduate School of Education in Assessment and Learning Technologies. Dr. Code holds a Ph.D in Educational Psychology from Simon Fraser University, a M.Ed in Educational Psychology with a specialization Instructional Technology and a B.Ed in Secondary Science and Art Education from the University of Alberta.

Muhibuddin Fadhli is a researcher from Universitas Muhammadiyah Ponorogo who concern about Early Chiildhood Development. I have co-authored a number of research papers and have articles published in variuos journals. My speciality is the development of instructional media and educational technology for children.

Emmanuel Fokides is an Assistant Professor in the Department of Primary School Education, University of the Aegean, Greece. His courses focus on the educational uses of Virtual Reality, digital storytelling, Augmented Reality, and Serious Games. Since 1994, he is involved in a number of research projects regarding distance and lifelong learning and the educational uses of Virtual and Augmented Reality. His work is published in several conference proceedings, international volumes, and journals. He is also the co-author of two books.

Katerina Glezou received her BA in physics, MSc, MEd and PhD from University of Athens in 1990, 1994, 2000 and 2010 respectively. Currently, she is a teacher of Computer Science/Robotics/Physics in Arsakeia Schools Psychikou in Athens, Greece and teacher trainer. Her research lies on the area of open interactive learning environments which support exploratory and collaborative computer supported learning in the context of Didactics of Informatics and Didactics of Science. In 2009 she created the Greek Educational Online Social Network "Logo in Education: a community of practice and learning" (http://logogreekworld.ning.com/). She has publications in international journals, contributions in book chapters, papers in proceedings of international and Greek conferences.

Grigorakis Ioannis is a Coordinator of Educational Work in Primary Education at the Regional Directorate of School Education of Crete. He owns a PHD in Education Sciences at the University of Crete.Since 2016 he has been teaching as Adjunct Professor in the Department of Preschool Education, Faculty of Education of University of Crete. He has a long-standing interest in metalinguistic skills such as phonological awareness and morphological awareness especially in relation to children's literacy learning. Current interests focus on the study of different aspects of emergent literacy and the longitudinal effects of emergent literacy skills on early reading and spelling development.

George Koutromanos is an Assistant Professor of Information Communication Technologies (ICT) in Education at the Department of Primary Education of the National and Kapodistrian University of Athens. He holds a Bachelor degree in Education, an MSc degree in Computers in Education (National and Kapodistrian University of Athens) and two PhD degrees in ICT in Education (one from National and Kapodistrian University of Athens and one from Kings' College London, University of London). His current research interests include: The impact of ICT on education, the development and evaluation of computer-assisted learning materials, the flipped learning, the use of augmented reality in education, the use of mobile games in formal and informal learning environments and the use of drones in education. He has published over 80 papers in English and Greek journals, edited volumes and conference proceedings.

Michalis Linardakis graduated from the Department of Statistics of the Athens University of Economics and Business. He took the Master and Ph.D. from the same University. He is currently an Assistant Professor in the Department of Preschool Education at the University of Crete. His research interests include Statistics in Education, Multivariate Statistics, Discrete Choice Models, Item Response Theory (IRT) models, Bayesian Statistics and Psychometrics.

Athanasios G. Malamos received a Bsc in Physics from the Univeristy of Crete (1992) and Msc and Phd from the Technical University of Crete at 1995 and 2000 respectively. From 1997 to 2002 he was research assistant and researcher in th ICCS National Technical University of Athens. Since 2002 is with Technological Educational Institute of Crete, Dept. of Informatics Engineering (former name of department Applied Informatics and Multimedia), as assistant professor (2002-2006) and associate professor (2006 until present). Dr. Malamos is head of the Media Networks and Communications Lab, TEI Crete and he has worked as researcher or project coordinator to many EU and National (Hellenic) funded research projects. He is an author of 53 research papers (15 in international journals, 6 chapters in scientific books and 32 in international conferences. Dr. Malamos is author or co-author in more than 36 scientific papers in books, Journals and Conferences in the area of Multimedia services and network performance evaluation. He is an active member of the WEB3D community and he is leading the Spatial Sound group of interest. His team contributed the spatial sound, the Humanoid and the Physics libraries in X3DOM platform. He has supervised 3 PHDs and 8 Master Thesis. He has served as Guest Editor in Graphical Models Journal (Springer), Special Issues Advances in Web3D and as General Chair in 20th International Conference on 3D Web Technology. Dr. Malamos has served as program and organizational committee in several IEEE and ACM international conferences and workshops. He is regular reviewer of IEEE, Springer as well as other international journals. Dr. Malamos is honored with the ACM Recognition of Service Award and by the WEB3D consortium with recognition of his efforts as a member. Dr. Malamos is a member of ACM SIGGRAPH, IEEE computer society and of the WEB3D consortium. His research interests include multimedia services, multimedia semantics, WEB3D.

Kleopatra Nikolopoulou works as E.DI.P (Laboratory and Teaching Personnel) at the Department of Early Childhood Education, School of Education, University of Athens, Greece. She holds a B.Sc in Physics (Aristotelian University of Thessaloniki, Greece), an M.Ed in 'Educational Technology' (University of Bath, UK) and a Ph.D in 'ICT in Education' (School of Education, King's College, University of London, UK). Her research interests include integration and use of ICT in education (early childhood, secondary, higher), teachers' and students' attitudes towards ICT in teaching and learning process, and mobile technologies in education. She has written a book (2018, new enhanced edition) on "ICT in Early Childhood Education" and has published in scientific international journals and conferences.

Vassilios D. Oikonimidis serves as an Associate Professor of Preschool Education in the School of Preschool Education of the University of Crete. He is a graduate of the Department of Early Childhood Education and the Faculty of Philosophy and Social Studies of the University of Crete, holder of a "Postgraduate Diploma in Psychology and the Development of the Human Behavior". He has taught as a kindergarten teacher in Primary and as a lecturer in Secondary Education. Since December 2003 he has been teaching at the School of Preschool Education of the University of Crete. His scientific interests include vocabulary development of infants, teacher education and training, and issues of kindergarten teaching methodology.

Spyros Panagiotakis is an assistant professor at the Department of Electrical and Computer Engineering of the Hellenic Mediterranean University (HMU) and head of the group for Sensor Networks and Telematics within the Multimedia, Networks and Communications Laboratory (MNCLab). He received his PhD in Communication Systems from the Department of Informatics and Telecommunications of the University of Athens, Greece in 2007. He also received a four-year fellowship for postgraduate studies from National Center for Scientific Research "Demokritos". He has participated in several European and national projects. He is the author of over than 90 publications in international refereed books, journals, and conferences. He also serves as peer reviewer in international conferences and journals, as well as as member of technical programme committees for international conferences and workshops. Finally, he serves as evaluator for various research projects calls. His research interests include communications and networking, Internet of Things, pervasive and physical computing, sensor networks, web engineering, mobile applications, location and context awareness, informatics in education.

Koralia Papadokostaki received her BSc in Informatics in 2003 from Athens University of Economics and Business. She worked as a Jr programmer for a while and since 2005 serves as a Computer & ICT Teacher in Primary and Secondary Schools in Eastern Crete. In 2017 she received her MSc in 'Informatics and Multimedia' from Technological Institute of Crete and has since participated in various Conferences. Her research interests include but are not limited to mobile learning, u-learning and e-learning, IoT and Big Data.

Samuel Olmos Peña has the degree of Doctor of Systems Engineering and Master of Science in Systems Engineering awarded by the National Polytechnic Institute; grade in both exams got Honorable

Mention. Recognition also won the Best Academic Performance generation in master's and doctoral studies. His expertise is in the area of risk analysis / accidents, public safety and natural disasters. He is currently attached to the Autonomous University of the State of Mexico as a full professor of computer engineering career and Administrative Computing. Member of the Academic Body "Complex Systems, sustainability and innovation" UAEM in the university center UAEM Valley of Chalco. Together integrates research group Security Risk Analysis, Accidents and Reliability Systems (SARACS; www.saracs.com.mx). His research include publications in high impact journals (JCR), international conferences and book chapters.

Patrick Pennefather is an award-winning sound designer and composer for live and mediated productions having worked on over 250 productions in Canada, the U.S., Europe and Asia including a number of productions with Arts Club Theatre. Recently appointed at UBC's Theatre & Film program in Design Production, he is passionate about bridging partnerships between various stakeholders in the academic, cultural, service and digital media industries. He has organized numerous workshops and symposiums in the resource, entertainment and educational sectors and at conferences like CVR, SXSW, GDC, SIGGRAPH and IndieCade. He has designed and implemented learning with the Master of Digital Media Program, BCIT, Ryerson University, University of Wuhan, Chinese University of Communication, North Chinese University of Technology, UBC and SFU. Within the digital media industry, he has facilitated small teams internationally, with Riot Games, EA, Microsoft Big Park, Fujitsu, Procon Mining Safety, Crystal CG China, the City of Fukuoka, NGX, Ballet BC, British Columbia Museum Association, Arts Club Theatre, and British Columbia Lottery Corporation. He researches and writes about emerging technologies, sound design, the role of rapid prototyping in the development process, consults and works with others on improving the experience of emerging technologies in public spaces. Most recently he co-produced and directed the Carnival of Mixed Realities, a public event/prototype exposing audiences to Mixed Reality installations involving technologies such as Virtual Reality and Augmented Reality that were integrated with interactive performances.

Stephen Petrina specializes in Media and Technology Studies (MTS), Science and Technology Studies (STS), Science, Technology, Engineering, and Mathematics education (STEM), and of course Curriculum Studies (esp. Metaphysics of Curriculum, and the Sociology of Curriculum and Pedagogy). He is a cultural historian with a current projects on Meducation and Twentieth Century Learning in addition to long-term histories of technotheology and culture, media and technology. Stephen's expertise necessarily includes the philosophy of technology and the philosophy of research, as opposed to methodology.

Rachel Ralph is very passionate about the role of women in tech and how she can explore her own role through research and teaching in digital media. In particular, she focuses on the idea of immersive experiences in entertainment and education (edutainment) and the role of thoughtful design.

Magally Martínez Reyes received a degree in Mathematics by the National Autonomous University of Mexico; also Masters in Sciences by the same University; Doctor in Sciences (Educational Mathematics), by CINVESTAV, IPN, belongs to the National System of Researchers (SNI), has the PROMEP profile, and has also been awarded the "Laudatory Note" and "FAAPA Recognition" by the UAEM. She has been teaching at this University Center since 1998, formerly Chalco Valley Professional Academic Unit, where she was Research Coordinator (2005-2009), Government Advisor (2006-2008)

Coordinator of the Degree in Computer Engineering (2000 -2004); currently in charge of the Office of Management. He has published several articles specialized in national and international magazines of recognized prestige in the area of mathematics and computation. In the Research it has been worth obtaining the Clause 90, which grants the UAEM and winner of competitions on innovative didactic material with the use of technology.

Anabelem Soberanes-Martín graduated in Administrative Computer Systems, Teacher in Education, master's in computer science course, Doctor in Educational Sciences, has the Desirable Profile Recognition (PRODEP), member of the National System of Researchers (SNI-I), Certified as Distance Activity Coordinator (San Diego Global Knowledge University), Certified as Instructor of face to face Courses (CONOCER), works in the Centro Universitario UAEM Valle de Chalco of the Universidad Autónoma del Estado de México. She has held the positions of Licentiate Coordinator, Academic Sub-Director and Academic Faculty Leader. She is currently a full-time professor, teaching in the bachelor's degrees in Computer Engineering, Administrative Informatics, in the master's and Doctorate degrees in Computer Science.

Sukriman Sukirman is one of the English lecturers and researchers at Institut Agama Islam Negeri Palopo (IAIN Palopo). As well as an awardee of MoRA Scholarship (5000 doctors program), he is currently taking his P.hD at Queen University Belfast, United Kingdom. His professional interests include curriculum design and evaluation, material design, academic writing, language testing and assessment, and content-based instruction.

Happy Susanto is an assistant professor and researcher at Universitas Muhammadiyah Ponorogo. He is concern about western philosophy and held P.hD at Universitas Gadjah Mada, Yogyakarta Indonesia.

Aldo Syam is a researcher at Universitas Muhammadiyah Ponorogo who focuses on education management. He had some publication to various journal and has a strong determination about education management. He welcomes all possible connection and collaboration to conduct some academic activities.

Evanngelia Tzagkourni graduated from the Department of English Language and Literature of the National and Kapodistrian University of Athens (NKUoA). She holds a Master's of Science in Information and Communication Technology (ICT) in Education from the University of Crete (UoC).

Saida Ulfa is a lecturer of Educational Technology Department, Faculty of Education State University of Malang, Indonesia. She obtained her master and doctoral degree from Saga University, Japan in the year 2005 and 2009, respectively. In the year 2010, she received Keizo Obuchi Research Fellowship Program to conduct research at Information Engineering Department, Faculty of Engineering, Hiroshima University, Japan. Then, she continue to focus her research in learning engineering area. Currently her research interests are adaptive learning, soft computing in education, and personalization in online learning system.

Kostas Vassilakis (MSc, PhD Computer Science) is currently is Professor at the Department of Electrical and Computer Engineering of the Hellenic Mediterranean University (HMU) Apart from his teaching and administrative activities, he is academic responsible of Network Operation Centre of

TEI of Crete, member of the Internal Evaluation Group of Department of Sciences and member of the Administrative Council of the non-profit civil company "Greek Universities Network" (GUnet - Academic Network). One of his current activities is the promotion of Free and Open Source Software at Hellenic Mediterranean University (HMU) . His research interests are in the areas of tele-services, tele-education (systems & applications) computer networks and decision support systems. He is involved in two research labs: (a) "New Technologies in Education" (responsible) and "Artificial Intel-ligence and Software Engineering" (researcher) in Hellenic Mediterranean University (HMU) . He has participated in various European and National funded projects (research, developmental and educational) and he is author/co-author of more than 40 refereed articles published in peer-reviewed international journals, proceedings of conferences and books. He is member of scientific committees in several international conferences (IADIS, CBLIS, NHIBE etc).

Georgios K. Zacharis is Postdoctoral Researcher at the Laboratory of Education and Research in Learning Technologies, School of Early Childhood Education, Aristotle University of Thessaloniki, Greece. He holds a Diploma and a Master of Science in Physics from School of Science, Department of Physics, and a PhD from the Department of Primary Education, both from the University of Ioannina, Greece. He has participated in European and National projects since 2004. His main research interests are on Teaching of Science and New Technologies in Education, Learning Technologies, Educational Neuroscience and Virtual Worlds in Education.

Dimitra Zachristou holds a Master degree in New Technologies in Education. She serves as a teacher in public Greek kindergarten schools. She is involved in research projects examining how ICT tools can enhance kindergarten students' understanding of various subjects.

Nicholas Zaranis graduated from the Department of Electrical and Computer Engineering at the National Technical University of Athens. He took the Master of Science in Computer Studies from the University of Essex. Also, he received his Master of Science and his Ph.D. from the National and Kapodistrian University of Athens. He has worked for a series of years in the private sector as Electrical Engineer in Hewlett Packard Hellas and in the public sector as a teacher of Information Technology in Secondary Education. He is currently a Professor in the Department of Preschool Education at the University of Crete. His research interests include ICT in Education, Educational Software and Teaching of Mathematics using ICT.

Index

G

H

I

K

L

M

N

P

Q

R